I. Chorkendo. ...u
J.W. Niemantsverdriet

**Concepts of Modern
Catalysis and Kinetics**

I. Chorkendorff and J.W. Niemantsverdriet

Concepts of Modern Catalysis and Kinetics

Third, Completely Revised and Enlarged Edition

Verlag GmbH & Co. KGaA

Authors

Prof. Dr. I. Chorkendorff
Department of Physics
Technical University of Denmark
Fysikvej, Building 312
2800 Kongens Lyngby
Denmark

Prof. Dr. J.W. Niemantsverdriet
SynCat@DIFFER
Syngaschem B.V.
De Zaale 20
5612 AJ Eindhoven
The Netherlands

1st Edition 2003
1st Reprint 2005
2nd Edition 2007
1st Reprint 2010
2nd Reprint 2011
3rd Reprint 2013

All books published by Wiley-VCH are carefully produced. Nevertheless, authors, editors, and publisher do not warrant the information contained in these books, including this book, to be free of errors. Readers are advised to keep in mind that statements, data, illustrations, procedural details or other items may inadvertently be inaccurate.

Library of Congress Card No.:
applied for

British Library Cataloguing-in-Publication Data:
A catalogue record for this book is available from the British Library.

Bibliographic information published by the Deutsche Nationalbibliothek
The Deutsche Nationalbibliothek lists this publication in the Deutsche Nationalbibliografie; detailed bibliographic data are available on the Internet at http://dnb.d-nb.de.

© 2017 WILEY-VCH Verlag GmbH & Co. KGaA, Boschstr. 12, 69469 Weinheim, Germany

All rights reserved (including those of translation into other languages). No part of this book may be reproduced in any form – by photoprinting, microfilm, or any other means – nor transmitted or translated into a machine language without written permission from the publishers. Registered names, trademarks, etc. used in this book, even when not specifically marked as such, are not to be considered unprotected by law.

Cover Design Grafik-Design Schulz, Fußgönheim, Germany
Typesetting le-tex publishing services GmbH, Leipzig, Germany
Printing and Binding CPI books GmbH Leck, Germany

Print ISBN 978-3-527-33268-7
ePDF ISBN 978-3-527-69128-9
ePub ISBN 978-3-527-69130-2
Mobi ISBN 978-3-527-69129-6

Printed on acid-free paper.

Contents

Preface *XIII*

List of Acronyms *XVII*

1	**Introduction to Catalysis** *1*	
1.1	What Is Catalysis? *2*	
1.2	Catalysts Can Be Atoms, Molecules, Enzymes, and Solid Surfaces *4*	
1.2.1	Homogeneous Catalysis *5*	
1.2.2	Biocatalysis *5*	
1.2.3	Heterogeneous Catalysis *6*	
1.3	Why Is Catalysis Important? *9*	
1.3.1	Catalysis and Green Chemistry *9*	
1.3.2	Atom Efficiency, E Factors, and Environmental Friendliness *10*	
1.3.3	The Chemical Industry *11*	
1.4	Catalysis as a Multidisciplinary Science *16*	
1.4.1	The Many Length Scales of a "Catalyst" *16*	
1.4.2	Time Scales in Catalysis *17*	
1.5	The Scope of this Book *18*	
1.6	Appendix: Catalysis in Journals *18*	
	References *22*	
2	**Kinetics** *23*	
2.1	Introduction *23*	
2.2	The Rate Equation and Power Rate Laws *25*	
2.3	Reactions and Thermodynamic Equilibrium *28*	
2.3.1	Example of Chemical Equilibrium: The Ammonia Synthesis *31*	
2.3.2	Chemical Equilibrium for a Nonideal Gas *33*	
2.4	The Temperature Dependence of the Rate *35*	
2.5	Integrated Rate Equations: Time Dependence of Concentrations in Reactions of Different Orders *38*	
2.6	Coupled Reactions in Flow Reactors: The Steady State Approximation *41*	
2.7	Coupled Reactions in Batch Reactors *45*	

2.8	Catalytic Reactions	48
2.8.1	The Mean-Field Approximation	52
2.9	Langmuir Adsorption Isotherms	53
2.9.1	Associative Adsorption	53
2.9.2	Dissociative Adsorption	54
2.9.3	Competitive Adsorption	55
2.10	Reaction Mechanisms	55
2.10.1	Langmuir–Hinshelwood or Eley–Rideal Mechanisms	56
2.10.2	Langmuir–Hinshelwood Kinetics	56
2.10.3	The Complete Solution	57
2.10.4	The Steady State Approximation	58
2.10.5	The Quasi-Equilibrium Approximation	59
2.10.6	Steps with Similar Rates	60
2.10.7	Irreversible Step Approximation	61
2.10.8	The MARI Approximation	61
2.10.9	The Nearly Empty Surface	62
2.10.10	The Reaction Order	63
2.10.11	The Apparent Activation Energy	63
2.11	Entropy, Entropy Production, Auto Catalysis, and Oscillating Reactions	67
2.12	Kinetics of Enzyme-Catalyzed Reactions	73
	References	77
3	**Reaction Rate Theory**	**79**
3.1	Introduction	79
3.2	The Boltzmann Distribution and the Partition Function	80
3.3	Partition Functions of Atoms and Molecules	83
3.3.1	The Boltzmann Distribution	83
3.3.2	Maxwell–Boltzmann Distribution of Velocities	86
3.3.3	Total Partition Function of a System	87
3.4	Molecules in Equilibrium	93
3.5	Collision Theory	100
3.5.1	Reaction Probability	104
3.5.2	Fundamental Objection against Collision Theory	105
3.6	Activation of Reacting Molecules by Collisions: The Lindemann Theory	106
3.7	Transition State Theory	107
3.8	Transition State Theory of Surface Reactions	113
3.8.1	Adsorption of Atoms	113
3.8.2	Adsorption of Molecules	118
3.8.3	Reaction between Adsorbates	121
3.8.4	Desorption of Molecules	123
3.9	Summary	124
	References	127

4	**Catalyst Characterization** *129*	
4.1	Introduction *129*	
4.2	X-ray Diffraction (XRD) *131*	
4.3	X-ray Photoelectron Spectroscopy (XPS) *134*	
4.4	X-ray Absorption Spectroscopy (EXAFS and XANES) *139*	
4.4.1	Extended X-ray Absorption Fine Structure (EXAFS) *139*	
4.4.2	X-ray Absorption Near-Edge Spectroscopy (XANES) *143*	
4.5	Electron Microscopy *144*	
4.6	Mössbauer Spectroscopy *148*	
4.7	Ion Spectroscopy: SIMS, LEIS, RBS *151*	
4.8	Temperature-Programmed Reduction, Oxidation, and Sulfidation *155*	
4.9	Infrared Spectroscopy *158*	
4.10	Surface Science Techniques *160*	
4.10.1	Low Electron Energy Diffraction (LEED) *161*	
4.10.2	Scanning Probe Microscopy *164*	
4.11	Concluding Remarks *169*	
	References *170*	
5	**Solid Catalysts** *173*	
5.1	Requirements of a Successful Catalyst *173*	
5.2	The Structure of Metals, Oxides, and Sulfides and Their Surfaces *175*	
5.2.1	Metal Structures *175*	
5.2.2	Surface Crystallography of Metals *176*	
5.2.3	Oxides and Sulfides *182*	
5.2.4	Surface Free Energy *185*	
5.3	Characteristics of Small Particles and Porous Material *187*	
5.3.1	The Wulff Construction *187*	
5.3.2	The Pore System *190*	
5.3.3	The Surface Area *191*	
5.4	Catalyst Supports *197*	
5.4.1	Silica *197*	
5.4.2	Alumina *199*	
5.4.3	Carbon *201*	
5.4.4	Shaping of Catalyst Supports *201*	
5.5	Preparation of Supported Catalysts *203*	
5.5.1	Coprecipitation *203*	
5.5.2	Impregnation, Adsorption, and Ion Exchange *203*	
5.5.3	Deposition Precipitation *205*	
5.6	Unsupported Catalysts *206*	
5.7	Zeolites *206*	
5.7.1	Structure of a Zeolite *207*	
5.7.2	Compensating Cations and Acidity *208*	
5.7.3	Applications of Zeolites *209*	
5.8	Catalyst Testing *210*	

5.8.1	Ten Commandments for Testing Catalysts	*211*
5.8.2	Activity Measurements	*213*
	References	*223*

6	**Surface Reactivity**	*225*
6.1	Introduction	*225*
6.2	Physisorption	*226*
6.2.1	The Van der Waals Interaction	*226*
6.2.2	Including the Repulsive Part	*227*
6.3	Chemical Bonding	*228*
6.3.1	Bonding in Molecules	*229*
6.3.2	The Solid Surface	*233*
6.4	Chemisorption	*246*
6.4.1	The Newns–Anderson Model	*246*
6.4.2	Summary of the Newns–Anderson Approximation in Qualitative Terms	*252*
6.4.3	Electrostatic Effects in Atomic Adsorbates on Jellium	*254*
6.5	Important Trends in Surface Reactivity	*256*
6.5.1	Trend in Atomic Chemisorption Energies	*257*
6.5.2	Trends in Molecular Chemisorption	*261*
6.5.3	Trends in Surface Reactivity	*265*
6.5.4	Universality in Heterogeneous Catalysis	*274*
6.5.5	Scaling Relations	*276*
6.5.6	Appendix: Density Functional Theory (DFT)	*278*
	References	*280*

7	**Kinetics of Reactions on Surfaces**	*283*
7.1	Elementary Surface Reactions	*283*
7.1.1	Adsorption and Sticking	*283*
7.1.2	Desorption	*289*
7.1.3	Lateral Interactions in Surface Reactions	*295*
7.1.4	Dissociation Reactions on Surfaces	*297*
7.1.5	Intermediates in Surface Reactions	*301*
7.1.6	Association Reactions	*301*
7.2	Kinetic Parameters from Fitting Langmuir–Hinshelwood Models	*304*
7.3	Microkinetic Modeling	*306*
7.3.1	Reaction Scheme and Rate Expressions	*307*
7.3.2	Activation Energy and Reaction Orders	*310*
7.3.3	Ammonia Synthesis Catalyst under Working Conditions	*313*
	References	*315*

8	**Catalysis in Practice: Synthesis Gas and Hydrogen**	*319*
8.1	Introduction	*319*
8.2	Synthesis Gas and Hydrogen	*319*
8.2.1	Steam Reforming: Basic Concepts of the Process	*321*

8.2.2	Mechanistic Detail of Steam Reforming 323
8.2.3	Challenges in the Steam Reforming Process 326
8.2.4	The SPARG Process: Selective Poisoning by Sulfur 328
8.2.5	Gold–Nickel Alloy Catalyst for Steam Reforming 329
8.2.6	Direct Uses of Methane 330
8.3	Reaction of Synthesis Gas 332
8.3.1	Methanol Synthesis 332
8.3.2	Fischer–Tropsch Process 343
8.4	Water–Gas Shift Reaction 351
8.5	Synthesis of Ammonia 353
8.5.1	History of Ammonia Synthesis 353
8.5.2	Ammonia Synthesis Plant 355
8.5.3	Operating the Reactor 356
8.5.4	Scientific Rationale for Improving Catalysts 359
8.6	Promoters and Inhibitors 361
8.7	The "Hydrogen Society" 364
8.7.1	The Need for Sustainable Energy 364
8.7.2	Sustainable Energy Sources 366
8.7.3	Energy Storage 368
8.7.4	Hydrogen Fuel Cells 377
	References 385

9	**Oil Refining and Petrochemistry** 391
9.1	Crude Oil 391
9.2	Hydrotreating 394
9.2.1	Heteroatoms and Undesired Compounds 395
9.2.2	Hydrotreating Catalysts 397
9.2.3	Hydrodesulfurization Reaction Mechanisms 399
9.3	Gasoline Production 402
9.3.1	Fluidized Catalytic Cracking 404
9.3.2	Reforming and Bifunctional Catalysis 406
9.3.3	Alkylation 410
9.4	Petrochemistry: Reactions of Small Olefins 412
9.4.1	Ethylene Epoxidation 412
9.4.2	Partial Oxidation and Ammoxidation of Propylene 413
9.4.3	Polymerization Catalysis 415
	References 418

10	**Environmental Catalysis** 421
10.1	Introduction 421
10.2	Air Pollution by Automotive Exhaust 422
10.2.1	The Three-Way Catalyst 423
10.2.2	Catalytic Reactions in the Three-Way Catalyst: Mechanism and Kinetics 429
10.2.3	Concluding Remarks on Automotive Catalysts 436

10.3	Air Pollution by Large Stationary Sources *437*
10.3.1	Selective Catalytic Reduction: The SCR Process *437*
10.3.2	The SCR Process for Mobile Units *443*
	References *444*

Appendix *447*

Questions and Exercises *449*

Index *497*

To Ina and Camilla

*To Marianne, Hanneke & Marten-Jan,
Annemieke & Ronald, Karin & Camiel,
and Peter*

Preface

Catalysis: Conceptually Understood but Still Far Away from Maturity

Catalysis as a phenomenon is reasonably well understood on a conceptual level. Recognized as a phenomenon around 1835, catalysis obtained an extensive empirical basis by the systematic experiments of Mittasch in the early twentieth century. Studies of catalytic mechanisms became feasible when Langmuir–Hinshelwood kinetics became available in the mid 1920s. Since then, for many decades fundamental catalysis became more or less synonymous with kinetic analysis. The advent of spectroscopy, starting with infrared spectroscopy in the late 1950s, followed by a range of other techniques for catalyst characterization and investigation of surface species opened the opportunity to relate catalytic properties with composition and structure of materials. Surface science enabled one to resolve adsorption geometries and reactivity patterns in well-defined structures, culminating in scanning tunneling spectroscopy as the ultimate tool to resolve surface structure and adsorbed species with atomic precision by the end of the twentieth century. Ever increasing computational power enabled the calculation of adsorbate geometries, bond strengths, and even reaction rates. Anno 2016, catalysis is

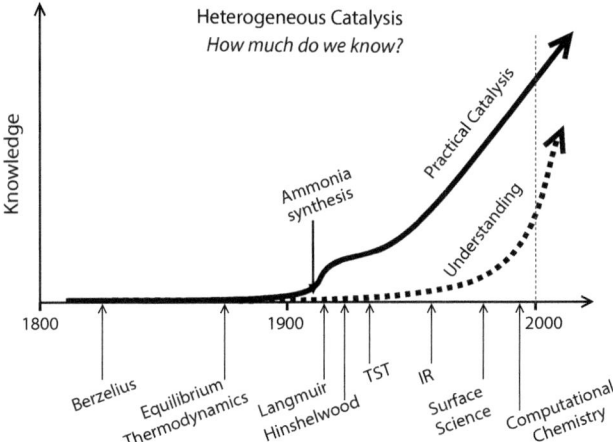

a scientific discipline with a firm conceptual basis. The relation between catalytic activity for a certain reaction and the composition, structure of a surface is in general qualitatively well understood, provided the surfaces are not too complicated in composition.

The prospect of designing a catalyst from first principles is coming closer. The density functional theory method of determining potential energy landscapes of reaction pathways and of calculating activation energy barriers is now possible with reasonable accuracy and is beginning to give predictive power for choices between different systems. Thus, 'in silico' screening is becoming viable in helping experimentalists pick the best choices for further investigation. Close interaction between experimentation and theory remains vital: although we may be able to describe a catalytic reaction on a well-defined single crystal of a metal, or a small crystallite under well-defined and simplified conditions, this becomes tremendously more complicated when the same reaction runs over dynamic catalyst particles on a support in a realistic reactor environment. The world of the ideal surface science laboratory and that of industrial practice are not only separated by the often cited pressure gap, but also by materials and reactor environment gaps. The complexity of small, supported crystallites that dynamically respond to every change in the reaction environment is only beginning to be explored.

Second, the way we describe the kinetics of catalytic reactions is, albeit greatly refined, still based on the adsorption isotherm of Langmuir (1915) and the kinetic formalism of Hinshelwood (1927), based on ideal surfaces with equivalent adsorption sites and adsorbate species that are randomly mixed and do not interact. This represents another gross oversimplification which has recently become recognized and is beginning to become addressed.

This book intends to be an introduction to the fundamentals of heterogeneous catalysis, aiming to explain the phenomenon of catalysis on a conceptual level. Kinetics, being the tool to investigate and describe catalytic reactivity as it expresses itself in a reactor, plays an important role in this book. Also reaction rate theory, providing the formalism to relate reaction rates to molecular structure of reacting species, is described extensively. Next we describe catalytic surfaces, as well as the tools to study them. With this knowledge, we treat surface reactivity in greatly simplified molecular orbital theory, again with the aim to give a conceptual explanation of how a catalyst works. Finally, we show how molecular modeling can be used to make predictions of new catalysts for those reactions of which we have a detailed mechanistic understanding.

The final chapters serve to illustrate catalysis in practice, to give the reader an impression of how catalysis is applied. We also illustrate the importance of catalysis in sustainable energy solutions for the future. We emphasize that the book is a textbook, written as an introduction for students in chemistry, physics and chemical engineering who are interested in understanding the fundamental concepts that underlie catalysis. Therefore, many important details that are worth knowing but are beyond the scope of this book will have to be found in specialist literature. We provide references at the end of each chapter.

The book is based on courses which the authors have taught at Lyngby and Eindhoven for many years. At the end of the book we have added a list of questions for every chapter which students may use to test their knowledge. The exercises are mainly meant to enable students to acquire skills in kinetic modeling. Some of these exercises have been used in written examinations. Solutions and also some figures and presentations are available at the website www.cinf.dtu.dk. Some video lectures plus handouts can be found on www.catalysiscourse.com.

The authors are indebted to many friends and colleagues in the field of catalysis, and in particular their many PhD students and postdocs. We are grateful to Prof. Roel Prins for indicating several errors in the previous version of this book. We also want to thank the numerous students who followed our courses in Lyngby, Eindhoven, and various places over the world, notably Capetown University and the c*change network in South Africa, Tianjin University and SynCat@Beijing in China, and SUNCAT at Stanford University. These audiences have taught us more than they perhaps realize. It is very gratifying that the previous editions of the book have been adopted as textbooks at several universities throughout the world. We hope that this new and updated version will be received as well as its predecessor. We will be grateful for any comments the readers may give us to improve the text further, or to correct the errors that the book undoubtedly still contains.

Finally, we thank our families who patiently endured that we spent a considerable amount of time on this book. We devote it to them.

Lyngby, Denmark *Ib Chorkendorff and*
Nuenen, The Netherlands *Hans Niemantsverdriet*
August 2016

List of Acronyms

AES	Auger electron spectroscopy
AFM	Atomic force microscopy
bcc	Body-centered cubic
BET	Brunauer–Emmet–Teller
CSTR	Continuously stirred tank reactor
DFT	Density functional theory
DFT-GGA	Density functional theory generalized gradient approximation
DOS	Density of states
EDX	Energy dispersive X-ray
ER	Eley–Rideal
ESR	Electron spin resonance
EXAFS	Extended X-ray absorption fine structure
FCC	Fluidized catalytic cracking
fcc	Face-centered cubic
hcp	Hexagonally close packed
HDA	Hydrodearomatization
HDM	Hydrodemetallization
HDO	Hydrodeoxygenation
HDN	Hydrodenitrogenation
HDS	Hydrodesulfurization
HOMO	Highest occupied molecular orbital
IR	Infrared
IRAS	Infrared absorption spectroscopy
ISS	Ion scattering spectroscopy
LCAO	Linear combination of atomic orbitals
LEED	Low energy electron diffraction
LEIS	Low energy ion scattering
LH	Langmuir–Hinshelwood
LPG	Liquefied petroleum gas
LUMO	Lowest occupied molecular orbital
MARI	Most abundant reaction intermediate
MTBE	Methyl tertiary butyl ether

NMR	Nuclear magnetic resonance
PEMFC	Proton exchange membrane fuel cell
PFR	Plug flow reactor
RBS	Rutherford backscattering
RLS	Rate-limiting step
RON	Research octane number
RWGS	Reverse water–gas shift
SATP	Standard ambient temperature and pressure
SCR	Selective catalytic reduction
SEM	Scanning electron microscopy
SIMS	Secondary ion mass spectroscopy
SMSI	Strong metal support interaction
SOFC	Solid oxide fuel cell
SPA-LEED	Spot profile analysis low energy electron diffraction
SPARG	Sulfur passivated reforming process
STM	Scanning tunneling microscopy
STP	Standard temperature and pressure
TEM	Transmission electron microscopy
TPD	Temperature programmed desorption
TPH	Temperature programmed hydrogenation
TPO	Temperature programmed oxidation
TPR	Temperature programmed reduction
TPS	Temperature programmed sulfidation
TS	Transition state
TWC	Three-way catalyst
UHV	Ultrahigh vacuum
UPS	Ultraviolet photoelectron spectroscopy
UV-Vis	Ultraviolet and visible absorption
WGS	Water–gas shift
XANES	X-ray absorption near-edge spectroscopy
XPS	X-ray photoelectron spectroscopy
XRD	X-ray diffraction
YSZ	Yttrium-stabilized zirconia

Chapter 1
Introduction to Catalysis

Ask the average person in the street what a catalyst is, and they will very likely tell you that it is part of a car. Indeed, the automotive exhaust converter represents a very successful application of catalysis; it effectively removes most of the pollutants that leave the engines of cars through exhausts. However, catalysis has a much wider scope of application than abating pollution alone [1–11].

Catalysis in Industry

Catalysts are the workhorses of chemical transformations in industry. Approximately 85–90% of the products of chemical industry are made in catalytic processes. Catalysts are indispensable in

- Production of transportation fuels in approximately 440 oil refineries all over the world.
- Production of bulk and fine chemicals in all branches of chemical industry.
- Prevention of pollution by avoiding formation of waste (unwanted byproducts).
- Abatement of pollution in end-of-pipe solutions (automotive and industrial exhaust).

A catalyst offers an alternative, energetically favorable mechanism for the noncatalytic reaction, thus, enabling processes to be carried out at industrially feasible conditions of pressure and temperature.

The economical importance is immediately evident from the estimate that industrial catalysis contributes about one quarter to the gross domestic product of developed countries. Around 2000, the catalyst world market was estimated to be about 10 billion US$, about equally distributed over refining, polymerization, chemicals, and environmental applications. The products of these processes, however, were valued at 200–300 times that of the catalyst [12].

Outside science and technology, the word catalyst is often used in the sense of 'enabler' for all kinds of processes in society.

Catalysis also plays a key role in nature [13]. Living matter relies on enzymes, which are the most abundant catalysts to be found. Photosynthesis generates sugars and oxygen from carbon dioxide and water by using chlorophyll as the catalyst and is probably the largest catalytic process in nature. Finally, the chemical indus-

Concepts of Modern Catalysis and Kinetics, Third Edition. I. Chorkendorff and J.W. Niemantsverdriet.
© 2017 WILEY-VCH Verlag GmbH & Co. KGaA. Published 2017 by WILEY-VCH Verlag GmbH & Co. KGaA.

try relies heavily on catalysis, which is an indispensable tool in the production of bulk and fine chemicals, as well as fuels. In fact, energy technology also depends largely on catalysts, not only in the processing of conventional fuels, but also in emerging forms of renewable energy, for instance, where hydrogen is produced via photocatalytic or electrocatalytic routes [14, 15].

For scientists and engineers, catalysis is a tremendously challenging and highly multidisciplinary field. To better understand it, let us first see what catalysis is before we proceed to its importance and functionality for mankind.

1.1
What Is Catalysis?

In simple terms, a catalyst accelerates a chemical reaction. It does so by forming bonds with reacting molecules, allowing these to react to form a product, which then detaches from the catalyst (leaving it unaltered such that it is available for the next reaction). In fact, we can describe the catalytic reaction as a cyclic event in which the catalyst participates and is recovered in its original form at the end of the cycle.

Let us consider the catalytic reaction between two molecules A and B to a product P, see Figure 1.1. The cycle starts with the bonding of molecules A and B to the catalyst. The next step is that A and B react, by ways of the catalyst, to create product P. In the final step, product P separates from the catalyst, thus, leaving the reaction cycle in its original state.

In order to see how the catalyst accelerates the reaction, we need to look at the potential energy diagram in Figure 1.2. To fully appreciate such diagrams, it is important to realize that lower energies imply more stable situations. It compares the noncatalytic and the catalytic reaction. For the former, the figure is simply the familiar way to visualize the Arrhenius equation: the reaction proceeds when A and B collide with sufficient energy to overcome the activation barrier ΔE in

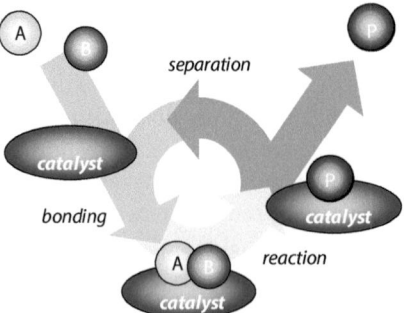

Figure 1.1 Every catalytic reaction is a sequence of elementary steps, in which reactant molecules bind to the catalyst, where they react, after which the product detaches from the catalyst, liberating the latter for the next cycle.

Figure 1.2. The change in Gibbs free energy between the reactants, A + B, and the product P is ΔG, and the change in enthalpy is ΔH.

The catalytic reaction starts by the bonding of reactants A and B to the catalyst, in a spontaneous reaction. Thus, the formation of this complex is exothermic, which in turn means that energy is lowered. Next comes the reaction between A and B while they are bound to the catalyst. This step is associated with an activation energy, ΔE_{cat}. However, it is significantly lower than for the uncatalyzed reaction. Eventually, product P separates from the catalyst in an endothermic step.

The diagram of Figure 1.2 illustrates several important points, in terms of the enthalpy changes in the course of the reaction:

- The catalyst offers an alternative path for the reaction, which is obviously more complex, but energetically much more favorable.
- The activation energy of the catalytic reaction is significantly smaller than that of the uncatalyzed reaction; hence, the rate of the catalytic reaction is much larger (we explain this in greater detail in Chapter 2).
- The overall changes in enthalpy and also in free energy for the catalytic reaction equals that of the uncatalyzed reaction. Hence, the catalyst does not affect the equilibrium constant for the overall reaction of A + B to P. Thus, if a reaction is thermodynamically unfavorable, a catalyst cannot change this situation. A catalyst changes the kinetics but *not* the thermodynamics.
- The catalyst accelerates both the forward and the reverse reaction to the same extent. In other words, if a catalyst accelerates the formation of the product P from A and B, it will do the same for the decomposition of P into A and B.

In addition to the enthalpy, H, one should also consider the entropy, S. Together they determine the free energy, $G = H - TS$. Hence, we can also draw Figure 1.2 in

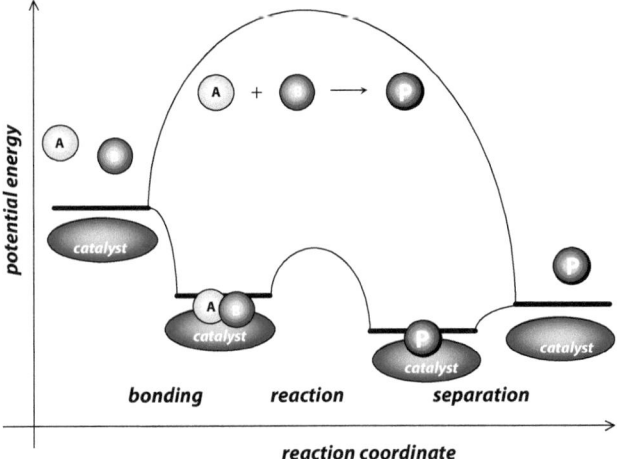

Figure 1.2 Potential energy diagram of a heterogeneous catalytic reaction, with gaseous reactants and products and a solid catalyst. Note that the uncatalyzed reaction has to overcome a substantial energy barrier, whereas the barriers in the catalytic route are much lower.

terms of changes in free energy. While a reaction releasing heat is called *exothermic* ($\Delta H < 0$), one that produces free energy is *exergonic* ($\Delta G < 0$). Spontaneous processes are exergonic, and in catalysis, adsorption is an example. Conversely, a reaction that consumes heat is *endothermic* ($\Delta H > 0$), while a reaction for which the free energy increases is *endergonic* ($\Delta G > 0$). The final state of a process – for example of the reaction $N_2 + 3H_2 \rightleftharpoons 2NH_3$ – is dictated by the equilibrium, and the process can be manipulated to become more exergonic by choosing optimal conditions, in case of the ammonia synthesis low temperature and high pressure. A catalyst only affects the way in which the reaction reaches its final state, and conversions can never lead to product concentrations exceeding those dictated by equilibrium.

Hence, in order to predict whether a reaction will proceed, one needs to consider the free energy, rather than the enthalpy. Doing this on the level of the elementary steps of a catalytic reaction requires that one knows both ΔH and ΔS, which has only recently become possible with the advent of computational chemistry. We discuss this in detail in Chapters 3 and 7.

With our current knowledge of catalytic reactions from Figure 1.2, we can already understand that there will be cases in which the combination of catalyst with reactants or products is not successful:

- If the bonding between reactants and catalyst is too weak, there will hardly be any conversion of A and B into products.
- If the bond between the catalyst and one of the reactants, say A, is too strong, the catalyst will mostly be occupied with species A, thus, not allowing B to form any product. Also, if both A and B form strong bonds with the catalyst, the intermediate situation with A or B on the catalyst may be so stable that reaction becomes unlikely. In terms of Figure 1.2, the second level lies so deep that the activation energy to form P on the catalyst becomes too high. The catalyst is said to be poisoned by (one of) the reactants.
- In the same way, the product P may be too strongly bound to the catalyst for separation to occur. In this case, the product poisons the catalyst.

Hence, we intuitively feel that the successful combination of catalyst and reaction is that in which the interaction between catalytic surface and reacting species is not too weak, but also not too strong. This is a loosely formulated version of Sabatier's Principle, which we encounter in a more precise form in Chapter 2.

The catalyst has thus far been an unspecified, abstract body, so let us now look at the different forms of catalysts that exist.

1.2
Catalysts Can Be Atoms, Molecules, Enzymes, and Solid Surfaces

Catalysts come in a multitude of forms, varying from atoms and molecules to large structures as zeolites or enzymes. In addition, they may be employed in a variety of surroundings, in liquids, gases, or at the surface of solids. Preparing a catalyst in the optimum form and studying its precise composition and shape are an important specialism, which we describe in later chapters.

It is customary to distinguish the following three subdisciplines in catalysis: homogeneous, heterogeneous, and bio catalysis. We illustrate each with an example.

1.2.1
Homogeneous Catalysis

In homogeneous catalysis [16], the catalyst and the reactants are in the same phase, that is, all are molecules in the gas phase or, more commonly, in the liquid phase. One of the simplest examples is found in atmospheric chemistry. Ozone in the atmosphere decomposes, among other routes, via a reaction with chlorine atoms:

$$Cl + O_3 \longrightarrow ClO_3$$
$$ClO_3 \longrightarrow ClO + O_2$$
$$ClO + O \longrightarrow Cl + O_2$$

or overall

$$O_3 + O \longrightarrow 2O_2$$

Ozone can decompose spontaneously, and also under the influence of light, but a Cl atom accelerates the reaction tremendously. As it leaves the reaction cycle unaltered, the Cl atom is a catalyst. Because the reactant and catalyst are both in the same phase (namely the gas phase), the reaction cycle is an example of homogeneous catalysis. This reaction was historically important in the prediction of the ozone hole.

Industry uses a multitude of homogeneous catalysts in all kinds of reactions to produce chemicals. The catalytic carbonylation of methanol to acetic acid

$$CH_3OH + CO \longrightarrow CH_3COOH$$

by $[Rh(CO)_2I_2]^-$ complexes in solution is one of many examples. In homogeneous catalysis, often aimed at the production of delicate pharmaceuticals, organometallic complexes are synthesized in procedures employing molecular control, such that the judicious choice of ligands directs the reacting molecules to the desired products.

1.2.2
Biocatalysis

Enzymes are nature's catalysts [13, 17]. For the moment, it is sufficient to consider an enzyme as a large protein holding an active site within a very specific shape formed by the protein structure (see Figure 1.3). Having shapes that are optimally suited to guide reactant molecules (usually referred to as substrates)

1 Introduction to Catalysis

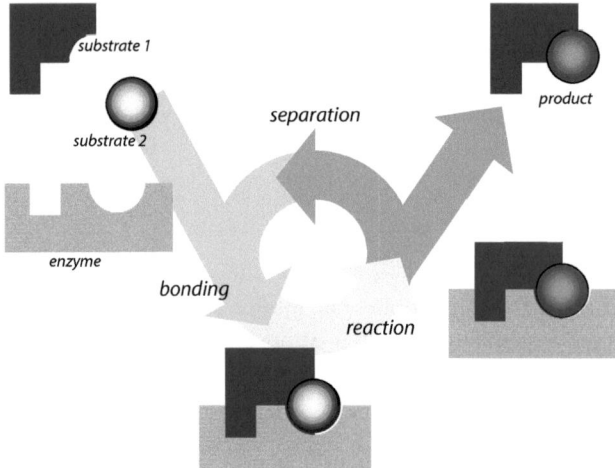

Figure 1.3 Schematic representation of an enzyme-catalyzed reaction. Enzymes often match the shape of the substrates they bind to, or the transition state (TS) of the reaction they catalyze. Enzymes are highly efficient catalysts and represent a great source of inspiration for designing technical catalysts.

in the optimum configuration for reaction, enzymes are highly specific and efficient catalysts. For example, the enzyme catalase, catalyzes the decomposition of hydrogen peroxide into water and oxygen

$$2H_2O_2 \xrightarrow{\text{catalase}} 2H_2O + O_2$$

at an incredibly high rate of up to 10^7 hydrogen peroxide molecules per second!

The enzyme catalase occurs in many living organisms. It consists of four polypeptide chains, each over 500 amino acids long, which all contain iron in a porphyrin heme group that is responsible for its catalytic activity. By decomposing hydrogen peroxide, it protects the cell against damage from reactive oxygen species.

Enzymes let biological reactions occur at rates which are necessary for the maintenance of life. For instance, the build up of proteins and DNA, or the breakdown of molecules, like H_2O_2 or much larger ones, and the storage of energy in sugars. An example that especially appeals to students is the breakdown of alcohol to acetaldehyde inside the body by the enzyme alcohol dehydrogenase. The acetaldehyde in turn is converted to acetate by aldehyde hydrogenase. Some people cannot tolerate alcohol (as revealed by facial flushing) because they lack the form of the enzyme that breaks down the acetaldehyde.

1.2.3
Heterogeneous Catalysis

In heterogeneous catalysis, solids catalyze reactions of molecules in gas or solution. As solids – unless they are porous – are commonly impenetrable, catalytic

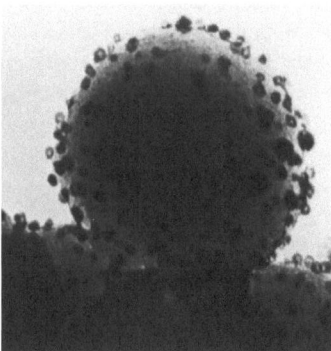

Figure 1.4 Catalysts are nanomaterials and catalysis is, thus, nanotechnology. By defining nanotechnology as the branch of materials science aiming at control over material properties on the nanometer scale, then catalysis represents a field where nanomaterials have been applied commercially for about one century. Many synthetic techniques are available to produce small particles for heterogeneous catalysts and to keep them sufficiently stable that they can withstand the often hostile conditions under which they have to work inside an industrial reactor. Modern catalysis is pre-eminently nanotechnology (figure reproduced from Datye and Long [18] with permission of Elsevier).

reactions occur at the surface. In order to use the often expensive materials (e.g., platinum) in an economical way, catalysts are usually nanometer-sized particles, supported on an inert, porous structure (Figure 1.4). Heterogeneous catalysts can be considered the workhorses of the chemical and petrochemical industry and we will discuss many applications of heterogeneous catalysis throughout this book.

As an introductory example, we take one of the key reactions in cleaning automotive exhaust, the catalytic oxidation of CO on the surface of noble metals such as platinum, palladium, and rhodium. In order to describe the process, we will assume that the metal surface consists of active sites, denoted as "∗" (we define them properly later on). The catalytic reaction cycle begins with the adsorption of CO and O_2 on the surface of platinum, where the O_2 molecule dissociates into two separate O atoms (the subscript "ads" indicates that the atom or molecule is adsorbed on the surface, i.e., bound to the site ∗):

$$O_2 + 2* \Leftrightarrow 2O_{ads}$$
$$CO + * \Leftrightarrow CO_{ads}$$

Next, the adsorbed O atom and the adsorbed CO molecule react on the surface to form CO_2, which being a very stable and relatively unreactive molecule, interacts only weakly with the platinum surface and desorbs almost instantaneously:

$$CO_{ads} + O_{ads} \Leftrightarrow CO_2 + 2*$$

Note that in the latter step, the adsorption sites on the catalyst are liberated, so that these become available for further reaction cycles. Figure 1.5 shows the reaction cycle along with a potential energy diagram.

1 Introduction to Catalysis

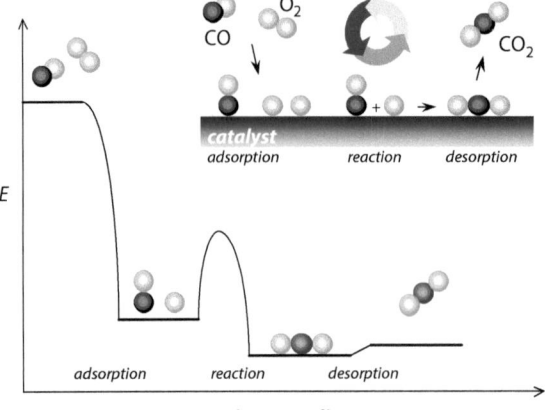

Figure 1.5 Reaction cycle and potential energy diagram for the catalytic oxidation of CO by O_2.

Where in this cycle is the essential influence of the catalyst? Suppose we carry out the reaction in the gas phase without a catalyst. The reaction will proceed if we raise the temperature sufficiently for the O_2 molecule to dissociate into two O atoms (radicals). Once these radicals are available, the reaction with CO to CO_2 follows instantaneously. The activation energy of the gas phase reaction will be roughly equal to the energy required to split the strong O–O bond in O_2 of about 500 kJ/mol. In the catalytic reaction, however, the O_2 molecule dissociates easily – in fact without activation energy – on the surface of the catalyst. The activation energy is associated with the reaction between adsorbed CO and O atoms, which is on the order of 50–100 kJ/mol. Desorption of the product molecule CO_2 costs only about 15–30 kJ/mol (depending on the metal and its surface structure). Hence, if we compare the catalytic and the uncatalyzed reaction, we see that the most difficult step of the homogeneous gas phase reaction, namely the breaking of the O–O bond, is easily performed by the catalyst. As a consequence, the ease at which the CO_2 molecule forms determines the rate at which the overall reaction from CO and O_2 to CO_2 proceeds. This is a very general situation in catalyzed reactions, hence the saying: "a catalyst breaks bonds, and lets other bonds form." The beneficial action of the catalyst is in the dissociation of a strong bond; the following steps might actually proceed faster without the catalyst (which is a hypothetical situation of course). In Chapter 6, we analyze in detail how a surface induces the breaking of intramolecular bonds.

1.3
Why Is Catalysis Important?

It is impossible to imagine how the chemical industry of the twentieth century could have developed to its present form without catalysis (i.e., on the basis of noncatalytic, stoichiometric reactions alone). Reactions in general can be controlled on the basis of temperature, concentration, pressure, and contact time. Raising temperature and pressure will enable stoichiometric reactions to proceed at a reasonable rate of production, but the reactors in which such conditions can be safely maintained become progressively more expensive and difficult to make. In addition, there are thermodynamic limitations to the conditions under which products can be formed, for example, conversion of N_2 and H_2 to ammonia is practically impossible above 600 °C. Nevertheless, higher temperatures are needed to be able to break the very strong N–N bond in N_2. Without catalysts, many reactions that are now common in the chemical industry would not be possible, and many other processes would not be economical.

Catalysts accelerate reactions by orders of magnitude, thus, enabling reactions to be carried out in the thermodynamically most favorable regime, and at much milder conditions of temperature and pressure. In this way, efficient catalysts – in combination with optimized reactor and total plant design – are the key factor in reducing both the investment and operation costs of chemical processes. But that is not all.

1.3.1
Catalysis and Green Chemistry

Technology is termed "green" if it uses raw materials efficiently, such that the use of toxic and hazardous reagents and solvents can be avoided, while formation of waste or undesirable byproducts is minimized. Catalytic routes often satisfy these criteria.

A good example is provided by the selective oxidation of ethylene to ethylene epoxide, an important intermediate towards ethylene glycol (antifreeze) and various polyethers and polyurethanes (Figure 1.6). The old, noncatalytic route (called the epichlorohydrine process) follows a three-step synthesis:

1. $Cl_2 + NaOH \longrightarrow HOCl + NaCl$
2. $C_2H_4 + HOCl \longrightarrow CH_2Cl-CH_2OH$ (epichlorohydrine)
3. $CH_2Cl-CH_2OH + 1/2 Ca(OH)_2 \longrightarrow 1/2 CaCl_2 + C_2H_4O + H_2O$

or in total:

$$Cl_2 + NaOH + 1/2 Ca(OH)_2 + C_2H_4 \longrightarrow C_2H_4O + 1/2 CaCl_2 + NaCl + H_2O$$

Hence, for every molecule of ethylene oxide, 1 1/2 molecule of salt is formed, creating a waste problem that was traditionally solved by dumping it in a river. Such practice is of course now totally unacceptable.

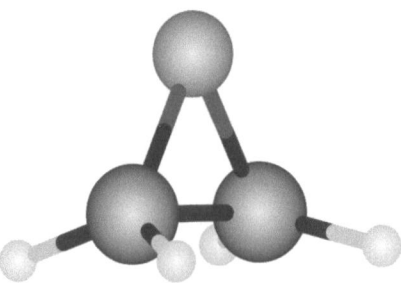

Figure 1.6 Ethylene epoxide is an important intermediate in the chemical industry.

The catalytic route, on the other hand, is simple and clean (although it does produce a small amount of CO_2). By using silver promoted by small amounts of chlorine as the catalyst, ethylene oxide is directly formed from C_2H_4 and O_2 at a selectivity of around 90% (with about 10% of the ethylene ending up as CO_2). Nowadays all production facilities for ethylene oxide use catalysts.

1.3.2
Atom Efficiency, E Factors, and Environmental Friendliness

Organic chemistry is full of synthesis routes that are based on either stoichiometric oxidations of hydrocarbons with sodium dichromate and potassium permanganate, or on hydrogenations with alkali metals, borohydrides, or metallic zinc. In addition, there are reactions such as aromatic nitrations with H_2SO_4 and HNO_3, or acylations with $AlCl_3$ that generate significant amounts of inorganic salts as byproducts.

Fine chemicals are predominantly (although not exclusively) the domain of homogeneous catalysis, where solvents present another issue of environmental concern. According to Sheldon [19]: "The best solvent is no solvent, but if a solvent is unavoidable, then water is a good candidate."

Sheldon has introduced several indicators to measure the efficiency and environmental impact of a reaction [20]. The atom efficiency is the molecular weight of the desired product divided by the total molecular weight of all products. For example, the conventional oxidation of the secondary alcohol

$$3C_6H_5-CHOH-CH_3 + 2CrO_3 + 3H_2SO_4 \longrightarrow$$
$$3C_6H_5-CO-CH_3 + Cr_2(SO_4)_3 + 6H_2O$$

has an atom efficiency of 360/860 = 42%. By contrast, the catalytic route

$$C_6H_5-CHOH-CH_3 + 1/2O_2 \longrightarrow C_6H_5-CO-CH_3 + H_2O$$

offers an atom efficiency of 120/138 = 87%, with water as the only byproduct. The reverse step, a catalytic hydrogenation, proceeds with 100% atom efficiency:

$$C_6H_5-CO-CH_3 + H_2 \longrightarrow C_6H_5-CHOH-CH_3$$

as does the catalytic carbonylation of this molecule:

$$C_6H_5-CHOH-CH_3 + CO \longrightarrow C_6H_5-CH(CH_3)COOH$$

Table 1.1 Environmental acceptability of products in different segments of the chemical industry [20].

Industry segment	Product tonnage	E factor kg waste/kg product
Oil refining	10^6–10^8	≪ 0.1
Bulk chemicals	10^4–10^6	< 1–5
Fine chemicals	10^2–10^4	5–50
Pharmaceuticals	10–10^3	25 up to > 100

Another useful indicator of environmental acceptability is the E factor, which is the weight of waste or undesirable byproduct divided by the weight of the desired product. As Table 1.1 shows, production of fine chemicals and pharmaceuticals generate the highest amounts of waste per unit weight of product. Atom efficiencies and E factors can be calculated alongside each other; however, in practice, E factors tend to be higher due to the yields being less than optimum and reagents that are used in excess. Also, loss of solvents should be included, and perhaps even energy consumption with the associated generation of waste CO_2.

To stress the environmental impact rather than the amount of waste, Sheldon introduced an environmental quotient EQ. This is the E factor multiplied by an unfriendliness quotient, Q, which assigns a value indicating how undesirable a byproduct is. For example, $Q = 0$ for clean water, 1 for a benign salt as NaCl, and 100–1000 for toxic compounds. Evidently, catalytic routes that avoid waste formation are highly desirable, and placing more weight on the economic value of environmental acceptability will spark greater motivation towards catalytic alternatives. After all – and economics permitting – waste prevention is in principle much to be preferred over waste remediation.

1.3.3
The Chemical Industry

Catalysts accelerate reactions and, thus, enable industrially important reactions to be carried out efficiently under practically attainable conditions. Very often catalytic routes can be designed to ensure that raw materials are used efficiently, thus, minimizing waste production. As a consequence, the chemical industry heavily relies on catalysis, roughly 85–90% of all products are made in catalytic processes [12].

The chemical industry produces a large range of base, middle, and end products. The main subsectors are

- Base chemicals, including organic and inorganic chemicals, polymers and plastics, dyes and pigments. Table 1.2 lists the most important ones, but there are many more in each category. Polymers have become enormously important in

Table 1.2 Some of the most important base chemicals and polymers.

Inorganic chemicals	Organic chemicals	Polymers and plastics
Sulfuric acid	Ethylene	Polyethylene
Ammonia	Propylene	– high density
Chlorine	Ethylene chloride	– linear low density
Phosphoric acid	Urea	– low density
Sodium hydroxide	Ethylbenzene	Polypropylene
Nitric acid	Styrene	Polyvinyl chloride
Ammonium nitrate	Ethylene oxide	Polystyrene
Hydrochloric acid	Cumene	Styrene polymers (e.g., ABS)
Ammonium sulfate	1,3-Butadiene	Polyamines, nylons
Titanium oxide	Acrylonitrile	
Aluminum sulfate	Benzene and aromatics	
N_2, O_2, H_2 gas	Aniline	
Sodium chlorate and sulfate		

replacing traditional materials in the automotive, construction, and packaging markets.
- Specialty chemicals, such as paints and coatings, adhesives, plastic additives, and also catalysts. These are performance-oriented products that are usually sold including technical services for customers.
- Agricultural chemicals, for farming and food processing.
- Pharmaceuticals, including diagnostics, drugs, vaccines, vitamins, etc., for humans or animals.
- Consumer products, such as soaps, detergents, cleaners, toiletries, and cosmetics.

Although catalysis is vitally important in all subsectors, in this book we limit ourselves to catalysts in the production of fuels and base chemicals. Figure 1.7 shows where chemicals are produced in the world. Where Western Europe and the USA largely dominated the market in the twentieth century, nowadays the Asia–Pacific region, and notably China, is the largest producer of chemicals in the world.

Some of the largest catalytic processes in this sector are listed in Table 1.3; Tables 1.4 and 1.5 list the top 50 of the chemical industry and the largest companies in the petroleum sector, respectively.

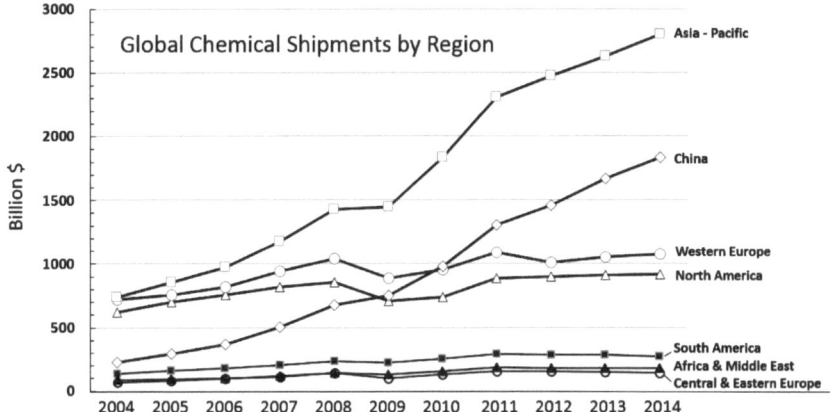

Figure 1.7 China and the Asia–Pacific region (including large producers in Japan, Korea, and India) dominate the chemicals market these days (data from the American Chemistry Council, www.americanchemistry.com).

Table 1.3 Selection of the largest processes based on heterogeneous catalysis.

Reaction	Catalyst
Catalytic cracking of crude oil	Zeolites
Hydrotreating of crude oil	Co-Mo, Ni-Mo, Ni-W (sulfidic form)
Reforming of naphtha (to gasoline)	Pt, Pt-Re, Pt-Ir
Alkylation	H_2SO_4, HF, solid acids
Polymerization of ethylene, propylene, and others	Cr, $TiCl_x/MgCl_2$
Ethylene epoxidation to ethylene oxide	Ag
Vinylchloride (ethylene + Cl_2)	Cu (as chloride)
Steam reforming of methane to $CO + H_2$	Ni
Water–gas shift reaction, and reverse	Fe (oxide), Cu-ZnO
Methanation	Ni
Ammonia synthesis	Fe
Ammonia oxidation to NO and HNO_3	Pt-Rh
Methanol synthesis	$Cu/ZnO/Al_2O_3$
Fischer–Tropsch synthesis	Fe, Co
Acrylonitrile from propylene and ammonia	Bi-Mo, Fe-Sb (oxides)
Hydrogenation of vegetable oils	Ni
Sulfuric acid	V (oxide)
Oxidation of CO and hydrocarbons (car exhaust)	Pt, Pd
Reduction of NO_x (in exhaust)	Rh, Pd, vanadium oxide

Table 1.4 World top 50 of chemical producers [21].

Rank 2014	2010	2005	2001	Company	Country	Chemical sales (million $)	Chemicals (%)
1	1	2	3	BASF	Germany	78 698	80
2	2	1	1	Dow Chemical	USA	58 167	100
3	3	7	26	Sinopec	China	57 953	13
4	7	10	18	SABIC	Saudi Arabia	43 341	86
5	4	4	6	ExxonMobil	USA	38 178	10
6	6	11	30	Formosa Plastics	Taiwan	37 059	60
7	9	12	50	LyondellBasell Industries	Netherlands	34 839	76
8	8	6	2	DuPont	USA	29 945	86
9	11	17	42	Ineos	Switzerland	29 652	100
10	13	8	4	Bayer	Germany	28 120	50
11	10	13	14	Mitsubishi Chemical	Japan	26 342	76
12	5	3	7	Royal Dutch Shell	Netherlands	24 607	6
13	18	50	–	LG Chem	South Korea	21 456	100
14	26	–	–	Braskem	Brazil	19 578	100
15	19	20	22	Air Liquide	France	19 210	94
16	14	18	11	AkzoNobel	Netherlands	19 011	100
17	22	–	–	Linde	Germany	18 539	82
18	16	19	20	Sumitomo Chemical	Japan	17 833	79
19	15	15	13	Mitsui Chemicals	Japan	17 201	100
20	17	14	8	Evonik Industries (< 2007: Degussa)	Germany	17 177	100
21	21	21	31	Toray Industries	Japan	17 006	89
22	20	39	38	Reliance Industries	India	15 870	26
23	29	36	–	Yara	Norway	15 141	100
24	23	30	24	PPG Industries	USA	14 250	93
25	33	32	41	Solvay	Belgium	14 134	100
26	45	–	–	Lotte Chemical	South Korea	14 121	100
27	28	22	23	Chevron Phillips	USA	13 416	100
28	25	26	19	DSM	Netherlands	12 344	100
29	30	35	36	Praxair	USA	12 273	100
30	–	–	–	SK Innovation	South Korea	12 011	19
31	24	25	37	Shin-Etsu Chemical	Japan	11 874	100
32	32	16	12	Huntsman Corp.	USA	11 578	100
33	34	44	32	Syngenta	Switzerland	11 286	75
34	37	46	48	Borealis	Austria	11 076	100
35	31	28	–	Lanxess	Germany	10 646	100
36	27	31	35	Asahi Kasei	Japan	10 628	55
37	43	42	–	Sasol	South Africa	10 299	55
38	36	34	29	Air Products & Chemicals	USA	9 989	96

Table 1.4 (Continued).

Rank 2014	2010	2005	2001	Company	Country	Chemical sales (million $)	Chemicals (%)
39	48	38	33	Eastman Chemical	USA	9 527	100
40	–	–	–	PTT Global Chemical	Thailand	9 522	54
41	44	–	–	Mosaic	USA	9 056	100
42	35	27	15	DIC (Dainippon Ink & Chemicals)	Japan	8 218	100
43	38	–	–	Arkema	France	7 915	100
44	40	–	–	Tosoh	Japan	7 657	100
45	–	–	–	Hanwha Chemical	Korea	7 655	100
46	–	–	–	Siam Cement	Thailand	7 617	51
47	–	–	–	Indorama	Thailand	7 514	100
48	–	9	9	BP	UK	7 284	2
49	–	–	–	Ecolab	USA	7 215	51
50	–	23	10	Johnson Matthey	UK	7 203	43
> 50	12	5	5	Total	France	7 570	3
> 50	39	33	46	ENI	Italy	7 397	5

Table 1.5 The world's largest petroleum companies.

	Company	Product sales (barrels/day)		Company	Distillation capacity (barrels/day)
1	Royal Dutch Shell	6 235 000	1	ExxonMobil	5 375 000
2	ExxonMobil	6 174 000	2	Sinopec	5 239 000
3	BP	5 657 000	3	CNPC	4 421 000
4	Sinopec	3 548 000	4	Royal Dutch Shell	3 360 000
5	Total	3 403 000	5	PDV	2 822 000

BP: British Petroleum; CNPC: China National Petroleum Company; PDV: Petroleos de Venezuela. Data from Energy Intelligence, http://www2.energyintel.com/.

Every chemistry student who is interested in a career in the chemical industry, or in chemical design in general, should know and understand what catalysis is and does. This is the reason that the authors and many of their colleagues teach catalysis in early stages of the chemistry curriculum of their respective universities.

1.4
Catalysis as a Multidisciplinary Science

1.4.1
The Many Length Scales of a "Catalyst"

Catalysis as a field of study is very broad and is closely interwoven with numerous other scientific disciplines. This becomes immediately evident if we realize that catalysis is a phenomenon encompassing many length scales. Figure 1.8 illustrates this for the case of heterogeneous catalysis [22].

When we introduced the cycle of catalysis in Figures 1.1, 1.2, and 1.5, we dealt with catalysis on the molecular level. For heterogeneous, homogeneous, or enzymatic catalysis, this is the level on which the chemistry takes place. Understanding reactions at the elementary level of the rupture of bonds in reactants and the formation of bonds on the way to products is at the heart of the matter and requires the most advanced experimental techniques and theoretical descriptions available. This is the domain of spectroscopy, computational chemistry, and kinetics and mechanism on the level of elementary reaction steps. The length scales of interest are in the subnanometer region. Publications and research at the molecular scale are predominantly found in academic journals specialized in chemistry, physical chemistry, and physics.

The next level is that of small catalytically active particles, with typical dimensions between 1 and 10 nm. These particles exist within the pores of support par-

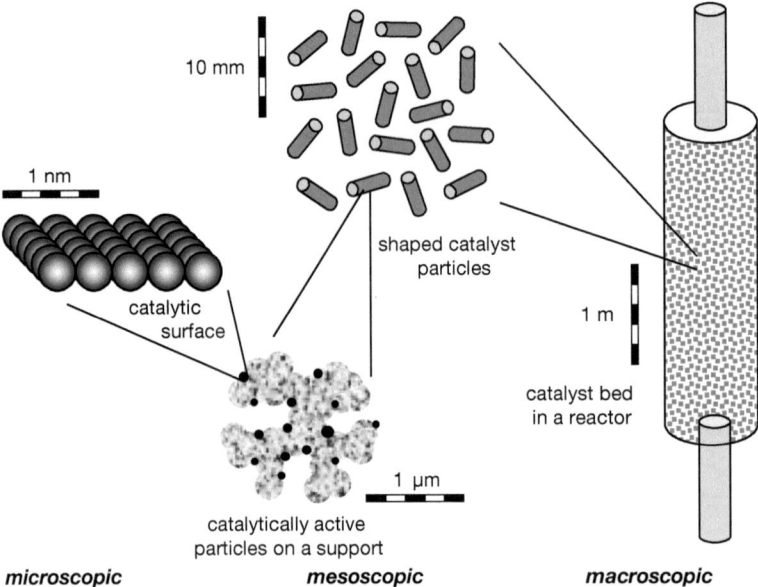

Figure 1.8 The relevant length scales in catalysis range from the subnanometer domain of the atomic and molecular level to the macroscopic domain of an industrial reactor.

ticles in the micrometer range. The key interests at this level are the size, shape, structure, and composition of the active particles, and in particular their surfaces, and how these properties relate to catalytic reactivity. Although we will deal with heterogeneous catalysis, the anchoring of catalytic molecules or even enzymes to supporting structures is also of great interest in homogeneous and biocatalysis. This is the domain of catalyst preparation, characterization, testing on the laboratory scale, and mechanistic investigations. Transport phenomena such as the diffusion of molecules inside pores may affect the rate at which products form and become an important consideration on this level. Much academic research as well as exploratory work in industry occurs on this scale. Journals dedicated to catalysis are largely dealing with phenomena on this length scale.

The mesoscopic level is that of shaped catalysts in the form of extrudates, spheres, or monoliths on length scales varying from millimeters to centimeters, and occasionally even larger. This field is to a large extent materials science. Typical points of interest are porosity, strength, and attrition resistance which enable catalysts to survive the conditions inside industrial reactors. This part of catalysis is mainly (though not exclusively) covered by industry, in particular by catalyst manufacturers. Consequently, much of the knowledge is covered by patents.

The macroscopic level is that of reactors, be it a 10 cm test reactor in a laboratory or a 50-m high reactor vessel in an industrial plant. The catalyst forms the heart of the reactor. Nevertheless, catalysis as a discipline is only one of many other aspects of reaction engineering, together with, for example, the design of efficient reactors that are capable of handling high pressure, offer precise control of temperature, enable optimized contact between reactants and catalyst, and removal of products during operation. In describing the kinetics of catalytic reactions on the scale of reactors, extrinsic factors dealing with the mass and heat transport properties of reactants and products through the catalyst bed are as important as the intrinsic reactivity of the molecules reacting at the catalytic site. For the catalyst, matters as mechanical stability, sensitivity to trace impurities in the reactant feed, and degradation of the particles (e.g., due to exposure to high temperatures) are essential in addition to intrinsic properties as activity and selectivity. Literature on these aspects of catalysis is largely found in chemical engineering journals and patents.

1.4.2
Time Scales in Catalysis

The characteristic times on which catalytic events occur vary more or less in parallel with the different length scales that we discussed above. The activation and breaking of a chemical bond inside a molecule occurs in the picosecond regime; completion of an entire reaction cycle from complexation between catalyst and reactants through separation from the product may take anywhere between microseconds for the fastest enzymatic reactions to minutes for complicated reactions on surfaces. On the mesoscopic level, diffusion in and outside the pores, and through shaped catalyst particles may vary between seconds and minutes. Furthermore, residence times of molecules inside entire reactors may vary from seconds to infinity. This happens if reactants end up in unwanted byproducts such as coke, which stay on the catalyst.

1.5
The Scope of this Book

The emphasis of this book is on the fundamental level of catalytic reactions on the molecular and the mesoscopic scale. In simple terms, catalysis is a kinetic phenomenon which deals with the acceleration of reactions. Hence, we start with a chapter on kinetics to describe how the rate of the catalytic reaction cycle depends on the main external process variables by which reactions can be influenced, namely concentration, pressure, and temperature. Next is a chapter on the theory of reaction rates, which is meant to bridge the connection between the properties of reactant molecules and their reactivity. Owing to the spectacular advances in theoretical chemistry and computational facilities of the last decades, it has become possible to predict reaction rates from first principles (be it for idealized cases only). We intend to provide the reader with the necessary background to help understand the current possibilities of catalysis anno 2016. Consequently, we will describe phenomena such as adsorption and reaction on surfaces, while providing the tools needed to investigate and conduct research. In order to ensure a practical knowledge of catalysts, we will describe what catalysts look like on the mesoscopic level, how they are made, and how they are characterized. Finally, we will describe a number of industrially important catalytic processes, from both applied and fundamental points of view. The emphasis is on concepts and general trends rather than on specific details, while the aim is to provide students with the necessary background to appreciate the more specialized literature on fundamental catalysis. The literature section gives references to a number of general books in catalysis and related disciplines.

1.6
Appendix: Catalysis in Journals

The results of research in catalysis are published in a wide range of general and more specialized journals. This reflects the highly multidisciplinary nature of the field. Referring to the different length scales in Figure 1.8, research in the microscopic domain (which deals with the fundamental level in which adsorbed molecules and elementary reaction steps take place) is often reported in general journals, such as the *Journal of Chemical Physics*, the *Journal of Physical Chemistry*, *Physical Chemistry Chemical Physics*, *Surface Science*, *Langmuir*, and *Physical Review*, and occasionally in the *Journal of the American Chemical Society*, *Angewandte Chemie*, *Energy & Environmental Science*, *Nature*, and its daughters *Nature Chemistry* and *Nature Materials*, and *Science*. Also the more specialized

Journal of Catalysis, ACS Catalysis, ChemCatChem, and *Catalysis Letters* publish articles in this area.

The mesoscopic domain of real catalysts is mostly covered by the typical catalysis periodicals, such as *Applied Catalysis, ACS Catalysis, Catalysis Science and Technology, Catalysis Structure and Reactivity*, the *Journal of Catalysis, Catalysis Letters, Topics in Catalysis, Catalysis Today, Microporous Materials*, and *Zeolites*. However, occasionally articles appear in *Angewandte Chemie, Journal of Physical Chemistry*, and *Physical Chemistry Chemical Physics*, and many others.

The macroscopic domain of reactors and transport processes is largely covered by journals from the chemical engineering field such as *Chemical Engineering Science, Industrial & Engineering Chemistry Research*, and the *Journal of the American Institute of Chemical Engineers (AIChE Journal)* which are some of the best-known periodicals in this field.

Exciting new results, which are not fully understood but are nevertheless important, are published in the form of Letters, Notes, and Rapid Communications. Specialized "Letter Journals" are *Chemical Communications, Catalysis Letters, Catalysis Communications, Chemical Physics Letters*, and *Physical Review Letters*, while several regular journals have sections for letters, such as the Communications and Research Notes in the *Journal of Catalysis*.

In addition, there are the highly appreciated review journals. These publish overviews of the status of certain topics of interest in the field. *Advances in Catalysis, Catalysis Reviews: Science & Engineering*, and *Catalysis, Specialist Periodical Reports* publish the most exhaustive reviews.

As in all scientific fields, conferences are frequently followed up by a book of proceedings, containing short accounts of the presentations. However, these books of proceedings are becoming less popular among scientists. As refereeing procedures tend to be less strict and the amount of information that can be included is usually limited, the quality of these proceedings is not always satisfactory. Midway between conference proceedings, reviews, and regular research papers, there are also topical issues published by *Catalysis Today* and *Topics in Catalysis*. Rather than publishing the proceedings of an entire conference, the editor makes a selection of particularly noteworthy contributions. Presenters are than invited to write a short article about their findings. Such topical issues often give a valuable overview of the status of a certain field, while the quality of the publications is better guaranteed by adequate review procedures.

In order to get an impression which journals are best read and cited, the *Journal Citation Reports*® provides interesting indicators: the impact factor for a given journal is the number of citations in that year to articles published by the journal in the two previous years. Thus, if each article published in either 2013 or 2014 is cited exactly once in 2015, the journal will have an impact factor of 1 in that year. Publishers use impact factors to monitor the success of their journals, while authors use them as a guide as to where to submit their work. For specialized journals, for example, those focusing on catalysis, impact factors of 5 and higher are considered high. Review journals usually have higher impact factors, whereas letters journals usually score lower.

A much less appreciated but in fact highly informative parameter and indicator of quality is the citation half-life of a journal. This is the period in years (going back from the current year) during which cited papers were published, from which it received half of all the citations in the current year. A long citation half-life indicates that the journal has published quality articles that have long kept their value to the scientific community. Care should be exercised with these numbers, as relatively young journals need years before they can reach an appreciable citation half-life to prove that they indeed published papers of long-lasting value.

Table 1.6 lists the impact factor and citation half-life for several journals in the field. Almost all journals have seen a significant increase in citations over the period 2001–2014, reflecting the growing research activity and increased use of scientific journals in the world. The number of publications in the field of catalysis has worldwide gone up substantially, with the emerging countries contributing most to the growth. Another factor of importance is that nowadays articles are available on line, and accessing them has become very easy. Also, several publishers have changed from subscriptions to individual titles to arrangements where groups of institutions subscribe to access to all journals in certain domains, which has facilitated the access to journals greatly. Also the requirement of some universities for PhD students to have journal publications before they can graduate has contributed substantially to the increased volume of publications, and higher impact factors.

The down side is that some scientists (or their universities) regard publishing as a competition, where the reputation of a journal is almost seen as more important than the quality of the article.

One should realize that the impact of the journal is only of immediate interest in the first one or two years after publication, where the title of the journal could indicate an expected impact. But in the end, it is the impact of the published results that counts. Hence, there are many examples of highly cited papers in low impact journals, as well as little cited articles in high impact titles.

Similarly one should be aware that among different fields of science there are large variations in citation practices. Medical journals, for example, in general receive more citations than those in catalysis, physical chemistry, or surface science.

In choosing where to publish, one should realize that each journal has its own characteristic scope, and prospective authors should therefore always consult the guide for authors of a journal, and check if their subject fits.

Table 1.6 Impact factors and citation half-lives for selected journals (source: Journal Citation Reports®, Thomson Reuters, New York, USA).

	Impact factor				$t_{1/2}$
	2015	2010	2005	2001	2015 (years)
Catalysis journals					
ACS Catalysis	9.307	–	–	–	2.4[a]
Applied Catalysis B – Environmental	8.328	4.749	3.809	3.643	5.2
Catalysis Reviews – Science and Engineering	7.526	7.000	5.312	8.471	> 10.0
Journal of Catalysis	7.354	5.415	4.780	3.293	> 10.0
Catalysis Science & Technology	5.287	–	–	–	2.6[a]
ChemCatChem	4.724	3.345	–	–	2.9[a]
Advances in Catalysis	4.500	8.167	2.750	6.846	> 10.0
Catalysis Today	4.312	2.993	2.365	2.333	8.1
Applied Catalysis A – General	4.012	3.384	2.728	2.258	8.4
Journal of Molecular Catalysis A	3.958	2.872	2.348	1.520	8.7
Catalysis Communications	3.389	2.827	2.098	–	5.6[a]
Catalysts	2.964	–	–	–	2.5[a]
Chinese Journal of Catalysis	2.628	0.752	0.665	–	3.3
Topics in Catalysis	2.355	2.359	2.547	2.136	6.9
Catalysis Letters	2.294	1.907	2.088	1.852	9.1
Catalysis Surveys of Asia	2.038	2.432	1.236	–	7.7
Engineering journals					
Chemical Engineering Journal	5.310	3.074	2.034	0.847	3.9
AIChE Journal	2.980	2.030	2.036	1.793	> 10.0
Chemical Engineering Science	2.750	2.379	1.735	1.547	10.0
Industrial & Engineering Chemistry Research	2.567	2.072	1.504	1.351	6.8
General journals					
Nature Materials	38.891	29.920	15.941	–	6.3
Nature	38.138	36.104	29.273	27.955	> 10.0
Science	34.661	31.377	30.927	23.329	> 10.0
Nature Chemistry	27.893	17.927	–	–	3.8[a]
Energy & Environmental Science	25.427	9.488	–	–	3.3[a]
Journal of the American Chemical Society	13.038	9.023	7.419	6.079	8.0
Angewandte Chemie – International Edition	11.709	12.730	9.596	8.255	5.8
Journal of Physical Chemistry Letters	8.539	–	–	–	3.0[a]
Physical Review Letters	7.645	7.622	7.489	6.668	9.4
Chemical Communications	6.567	5.787	4.426	3.902	4.6
Chemistry – A European Journal	5.771	5.476	4.907	4.614	4.6
Journal of Physical Chemistry C	4.509	4.524	–	–	4.7[a]
Physical Chemistry Chemical Physics	4.449	3.454	2.519	1.787	4.1
Langmuir	3.993	4.269	3.705	2.963	7.6
Journal of Physical Chemistry B	3.187	3.603	4.033	3.379	9.8
ChemPhysChem	3.138	3.340	3.607	4.271	5.5

a) Note that a number journals have not existed long enough to establish a meaningful citation half-life; such values are indicated with a footnote.

References

1. Ertl, G., Knoezinger, H., Schueth, F., and Weitkamp, J. (2008) *Handbook of Heterogeneous Catalysis*, Wiley-VCH Verlag GmbH, Weinheim.
2. Bartholomew, C.H. and Farrauto R.J. (2006) *Fundamentals of Industrial Catalytic Processes*, John Wiley & Sons, Inc., Hoboken.
3. Bowker, M. (1998) *The Basis and Applications of Heterogeneous Catalysis*, University Press, Oxford.
4. Gates, B.C. (1992) *Catalytic Chemistry*, John Wiley & Sons, Inc., New York.
5. Thomas, J.M. and Thomas, W.J. (2014) *Principles and Practice of Heterogeneous Catalysis*, Wiley-VCH Verlag GmbH, Weinheim.
6. Rothenberg, G. (2008) *Catalysis: Concepts and Green Applications*, Wiley-VCH Verlag GmbH, Weinheim.
7. Somorjai, G.A. and Li, Y. (2010) *Introduction to Surface Chemistry and Catalysis*, John Wiley & Sons, Inc., Hoboken.
8. Nørskov, J.K., Studt, F., Abild-Pedersen, F., and Bligaard, T. (2014) *Fundamental Concepts in Heterogeneous Catalysis*, John Wiley & Sons, Inc., Hoboken.
9. Murzin, D.Y. (2013) *Engineering Catalysis*, De Gruyter, Berlin.
10. Hagen, J. (2006) *Industrial Catalysis: A Practical Approach*, Wiley-VCH Verlag GmbH, Weinheim.
11. Cornils, B., Herrmann, W.A., Zanthoff, H.-W., and Wong, C.-H. (2013) *Catalysis from A to Z: A Concise Encyclopedia*, Wiley-VCH Verlag GmbH, Weinheim.
12. Schmidt, F. (2004) in *Basic Principles in Applied Catalysis*, (ed. M. Baerns), Springer, Berlin.
13. Bommarius, A.S. and Riebel, B.R. (2004) *Biocatalysis*, Wiley-VCH Verlag GmbH, Weinheim.
14. Lewis, N.S. and Nocera, D.G. (2006) Powering the planet: Chemical challenges in solar energy utilization. *Proceedings of the National Academy of Sciences of the United States of America*, **103**, 15729–15735.
15. Schloegl, R.E. (2012) *Chemical Energy Storage*, De Gruyter, Berlin.
16. Van Leeuwen, P.W.N.M. (2008) *Homogeneous Catalysis; Understanding the Art*, Springer, New York.
17. Buchholz, K., Kasche, V., and Bornscheuer, U.T. (2012) *Biocatalysts and Enzyme Technology*, Wiley-Blackwell.
18. Datye, A.K. and Long, N.J. (1988) The use of nonporous oxide particles for imaging the shape and structure of small metal crystallites in heterogeneous catalysts. *Ultramicroscopy*, **25**, 203–208.
19. Sheldon, R.A. (2005) Green solvents for sustainable organic synthesis: State of the art. *Green Chemistry*, **7**, 267–278.
20. Sheldon, R.A. (2008) E factors, green chemistry and catalysis: An odyssey. *Chemical Communications*, 3352–3365.
21. Tullo, A.H. (2015) Global top 50 chemical companies. *Chemical & Engineering News*, American Chemical Society, pp. 14–26.
22. Ertl, G. (2008) Reactions at surfaces: From atoms to complexity (Nobel lecture). *Angewandte Chemie – International Edition*, **47**, 3524–3535.

Chapter 2
Kinetics

2.1
Introduction

Kinetics provides the framework to describe the rate at which a chemical reaction occurs and enables us to relate the rate to a reaction mechanism that describes how the molecules react via intermediates to the eventual product. It also allows us to relate the rate to macroscopic process parameters such as concentration, pressures, and temperatures. Hence, kinetics provides us with the tools to link the microscopic world of reacting molecules to the macroscopic world of industrial reaction engineering. Obviously, kinetics is a key discipline for catalysis [1–3].

Historically, catalysis has been closely interwoven with kinetics [4]. Table 2.1 lists some of the highlights. Catalysis was first recognized as a phenomenon in the first decades of the nineteenth century, while the first descriptions of the rate of reaction stem from around 1850. The breakthrough in catalysis occurred with the development of the ammonia synthesis and the associated development of high pressure flow reactors. This was possible because equilibrium thermodynamics of van't Hoff towards the end of the nineteenth century enabled Fritz Haber and Walter Nernst to predict the conditions under which ammonia synthesis would be feasible (high pressure and low temperature). Methods for fixation of nitrogen from the air were heavily desired as technology to produce fertilizers needed to be able to feed the ever growing population. Unfortunately, the ammonia synthesis also enabled the production of explosives – a factor of importance in the First World War.

Catalytic reactions (as well as the related class of chain reactions which we describe below) are coupled reactions, and their kinetic description requires methods to solve the associated set of differential equations describing the constituent steps. This stimulated Chapman in 1913 to formulate the steady state approximation which, as we will see, plays a central role in solving kinetic schemes.

Langmuir's research of how oxygen gas deteriorated the tungsten filaments of light bulbs led to a theory of adsorption that relates the surface concentration of a gas to its pressure above the surface (1915). This, together with Taylor's concept of active sites on the surface of a catalyst, enabled Hinshelwood in about 1927 to formulate the Langmuir–Hinshelwood kinetics that we still use today to de-

Concepts of Modern Catalysis and Kinetics, Third Edition. I. Chorkendorff and J.W. Niemantsverdriet.
© 2017 WILEY-VCH Verlag GmbH & Co. KGaA. Published 2017 by WILEY-VCH Verlag GmbH & Co. KGaA.

Table 2.1 Historical development of kinetics in relation to catalysis [3].

1813	Thénard	Ammonia decomposition on several metals
1814	Kirchhoff	Hydrolysis of starch catalyzed by acids
1817	Humphrey Davy	Mixture of coal gas and air makes platinum wire white hot
1818	Thénard	Measurements on rate of H_2O_2 decomposition
1823	Döbereiner	Selective oxidation of ethanol to acetic acid on platinum
1834	Faraday	Comprehensive description of $H_2 + O_2$ reaction on platinum
1835	Berzelius	Definition of catalysis, catalyst, catalytic force
1850	Wilhelmy	First quantitative analysis of reaction rates
1862	Guldberg & Waage	Law of mass action
1865	Harcourt & Esson	Systematic study on concentration dependence of reaction rate
1884	van't Hoff	First concise monograph on chemical kinetics
1887	Ostwald	Definition order of reaction
1889	Arrhenius	Arrhenius equation: $k = \nu \exp(-E_a/k_B T)$
1905	Nernst	Third law of thermodynamics
1908	Haber	Prediction of conditions for ammonia synthesis
1913	Chapman	Steady state approximation
1915	Langmuir	Quantitative theory of adsorption of gases on surfaces
1921	Lindemann	Mechanism of unimolecular reactions – activation by collisions
1925	Taylor	Catalytically active sites on surfaces
1927	Hinshelwood	Kinetic mechanism of reactions in heterogeneous catalysis
1931	Onsager	Nonequilibrium thermodynamics
1935	Eyring; Polanyi and Evans	Transition-state theory

scribe catalytic reactions. Indeed, research in catalysis was synonymous with kinetic analysis until in the second half of the twentieth century, when the spectrum of tools was extended with disciplines such as spectroscopy and computational chemistry.

Hence, one could say that kinetics in the twentieth century widened its scope from a purely empirical description of reaction rates to a discipline which encompasses the description of reactions on all scales of relevance: from the interaction between molecules at the level of electrons and atoms in chemical bonds, to reactions of large quantities of matter in industrial reactors.

Kinetics on the level of individual molecules is often referred to as *reaction dynamics*. Subtle details are taken into account, such as the effect on the orientation of the molecules in a collision that may result in a reaction, and the distribution of energy over a molecule's various degrees of freedom. This is the fundamental level of study needed if we want to link reactivity to quantum mechanics, which is really what rules the game at this fundamental level. This is the domain of molecular beam experiments, laser spectroscopy, *ab initio* theoretical chemistry, and transition state theory. It is at this level that we can learn what determines whether a chemical reaction is feasible.

In practical situations, we are not interested in the single event but in what happens in a large ensemble of molecules. We therefore have to average over all possibilities, that is, over all energies and coordinates of the reacting molecules. This is what *kinetics* is about. Temperature is the parameter that suitably describes the average energy content of the molecules participating in the reaction. Thermodynamics determines the changes in free energy that accompany a reaction and sets limits to the fraction of molecules that can be converted. Thus, kinetics describes the behavior of large ensembles of molecules during a reaction.

2.2
The Rate Equation and Power Rate Laws

Suppose a reaction between molecules A and B gives product molecules C and D according to the reaction equation

$$\nu_a A + \nu_b B \underset{k^-}{\overset{k^+}{\rightleftharpoons}} \nu_c C + \nu_d D \tag{2.1}$$

with forward and backward rate constants, k^+ and k^-, and stoichiometric coefficients ν_a, ν_b, ν_c, and ν_d. The rate of reaction is defined as the rate of disappearance of the reactants, or the rate of formation of products:

$$r \equiv -\frac{1}{\nu_a}\frac{d[A]}{dt} = -\frac{1}{\nu_b}\frac{d[B]}{dt} = \frac{1}{\nu_c}\frac{d[C]}{dt} = \frac{1}{\nu_d}\frac{d[D]}{dt} \tag{2.2}$$

where [X] stands for the concentration of component X. If the reaction occurs in the gas phase, we may replace the concentrations by partial pressures: $[X] \propto p_X/p°$, where $p°$ is the reference pressure ($p° = 1$ bar).

$$r \propto -\frac{1}{\nu_a p°}\frac{dp_A}{dt} = -\frac{1}{\nu_b p°}\frac{dp_B}{dt} = \frac{1}{\nu_c p°}\frac{dp_C}{dt} = \frac{1}{\nu_d p°}\frac{dp_D}{dt} \tag{2.3}$$

Very often, we will not include the reference pressure, $p°$, but implicitly assume that p represents a relative quantity.

A reaction is *elementary* if it occurs in a single step that cannot be divided into any further substeps and proceeds exactly as expressed by the reaction equation (Eq. (2.1)). For an elementary step, the rate equals

$$r = k^+[A]^{\nu_a}[B]^{\nu_b} - k^-[C]^{\nu_c}[D]^{\nu_d} = r^+ - r^- \tag{2.4}$$

in which k^+ and k^- are reaction rate constants; the stoichiometric constants equal the orders of the reaction in the respective molecules. A reaction mechanism is a sequence of elementary steps.

Example

An example of a first-order elementary reaction is the isomerization of cyclopropane to propene, for which the reverse reaction hardly proceeds, and we may write the rate as

$$r = -\frac{dp_{\text{cyclopropane}}}{dt} = \frac{dp_{\text{propene}}}{dt} = k^+ p_{\text{cyclopropane}} \tag{2.5}$$

Another example is the equilibrium

$$N_2O_4 \underset{k^-}{\overset{k^+}{\rightleftharpoons}} 2NO_2 \tag{2.6}$$

which represents a first-order elementary reaction in the forward direction and a second-order elementary reaction in the reverse direction. Adsorption and desorption of molecules on or from a surface may also be elementary reactions, although even such simple reactions may consist of two steps when precursor states are involved.

The rate constants of the forward and reverse reactions are linked through the equilibrium constant K. At equilibrium, the net rate r equals zero:

$$r = k^+[A]^{\nu_a}[B]^{\nu_b} - k^-[C]^{\nu_c}[D]^{\nu_d} \quad \Rightarrow \quad K \equiv \frac{k^+}{k^-} = \frac{[C]_{\text{eq}}^{\nu_c}[D]_{\text{eq}}^{\nu_d}}{[A]_{\text{eq}}^{\nu_a}[B]_{\text{eq}}^{\nu_b}} \tag{2.7}$$

where $[X]_{\text{eq}}$ refers to the concentration of species X at equilibrium. By introducing the equilibrium constant K into the rate expression, we can rewrite r to give

$$r = k^+[A]^{\nu_a}[B]^{\nu_b}\left(1 - \frac{1}{K}\frac{[C]^{\nu_c}[D]^{\nu_d}}{[A]^{\nu_a}[B]^{\nu_b}}\right) = r^+ - r^- \tag{2.8}$$

where the first term expresses the rate far from equilibrium (in the limit of zero conversion), while the part in parentheses represents the affinity of the reaction towards equilibrium. It should be noted that when the reaction is close to equilibrium, the rate decreases exponentially until equilibrium is obtained.

The reactions dealt with in this book are generally overall reactions that consist of a series of elementary steps involving a large number of intermediates. This is true even for such apparently simple reactions as $2O_3 \rightarrow 3O_2$ or $H_2 + Cl_2 \rightarrow 2HCl$. In particular, reactions in heterogeneous catalysis are always a series of steps, including adsorption on the surface, reaction, and desorption back into the gas phase. In the course of this chapter, we will see how the rate equations of overall reactions can be constructed from those of the elementary steps.

Although very often, we will not know *a priori* how a complex reaction proceeds in detail, for the purpose of parameterization it may be advantageous to write the rate as a function of concentrations or partial pressures, in the form of a power rate law:

$$r = k[A]^{n_A}[B]^{n_B}[C]^{n_C}[D]^{n_D} \quad \text{or}$$
$$r = k p_A^{n_A} p_B^{n_B} p_C^{n_C} p_D^{n_D} \tag{2.9}$$

where the orders n_X may assume virtually all values, integer, fractional, positive, negative, or zero. The overall orders of the forward and reverse reactions equal $n_A + n_B$ and $n_C + n_D$, respectively. Note that the order in a particular component is the slope of the line in a log–log plot of the rate versus the concentration of that component. Hence, the formal way to derive the order from a given rate expression is to apply the following derivative:

$$n_X \equiv \frac{\partial \ln r}{\partial \ln [X]} = [X] \frac{\partial \ln r}{\partial [X]} \qquad (2.10)$$

We will often encounter the logarithmic derivative in this book. It provides a very handy way to derive various parameters of interest from rate expressions or thermodynamic equations. The logarithmic derivative of a function f with respect to a variable x is given by

$$\frac{\partial \ln f}{\partial x} = \frac{1}{f} \frac{\partial f}{\partial x}$$

If a detailed reaction mechanism is available, we can describe the overall behavior of the rate as a function of temperature and concentration. In general, it is only of interest to study kinetics far from thermodynamic equilibrium (in the zero conversion limit) and the reaction order is therefore defined as:

$$n_X \equiv \frac{\partial \ln r_+}{\partial \ln [X]} \qquad (2.11)$$

Elementary reactions have integral orders. However, for overall reactions the rate often cannot be written as a simple power law. In this case, orders will generally assume nonintegral values that are only valid within a narrow range of conditions. This is often satisfactory for the description of an industrial process in terms of a power-rate law. The chemical engineer in industry uses it to predict how the reactor behaves within a limited range of temperatures and pressures.

Example

As an example, we quote the formation of hydrocarbons from synthesis gas on Group 8 metal catalysts

$$xCO + yH_2 \xrightarrow{Fe,Co} C_xH_{2x}(+2) + xH_2O \qquad (2.12)$$

for which the order in CO is usually negative, between 0 and −1, and that in H_2 close to 1. We will later discuss how this is caused by the way gases interact with the surface. CO adsorbs in general strongly, causing it to be present in excess of H atoms. The negative order in CO indicates that increasing its pressure would further decrease the chance for H_2 to be on the surface and, hence, the rate would diminish. Lowering the pressure of CO, on the other hand, gives more room for H atoms and, hence, the rate of reaction goes up.

Another interesting case – which immediately illustrates how opportunistic the concept of reaction orders for catalytic reactions may be – is that of the CO oxidation, an important subreaction in automotive exhaust catalysis:

$$CO + 1/2 O_2 \xrightarrow{Pt,Pd,Rh} CO_2 \qquad (2.13)$$

At low temperatures the orders in CO and O_2 are about -1 and $+1/2$, while at high temperatures they become $+1$ and $+1/2$, respectively. Thus, the orders of overall reactions should certainly not be treated as universal constants but rather as a convenient parameterization that is valid for a specific set of reaction conditions. We shall later see how these numbers become meaningful when we construct a detailed model for the overall process in terms of a number of elementary steps. The model should naturally be cable of describing what has been measured.

As mentioned in the first example above, we will often encounter the case that a component interacts strongly with the catalytic surface, causing it to appear in the denominator of the rate expression. If the forward rate is given by the expression

$$r^+ = \frac{kKp}{1+Kp} \qquad (2.14)$$

the reaction order in p is then given by

$$n \equiv p \frac{\mathrm{d}\ln r_+}{\mathrm{d}p} = 1 - \frac{Kp}{1+Kp} \qquad (2.15)$$

Hence, the reaction order is seen to depend both on the equilibrium constant K, which depends on temperature, and the actual partial pressure p. We shall see later that the latter term for a catalyst is related to the extent that the surface is covered by a reactant.

2.3
Reactions and Thermodynamic Equilibrium

For a reaction that has reached equilibrium, the rates of the forward and reverse reaction are equal and the Gibbs free energy is at its minimum value. If we assume the pressure and temperature to be constant, the derivative of G with respect to the reactants and products will be equal to zero for the reaction in Eq. (2.1), that is,

$$dG = +\mu_A \, dn_A + \mu_B \, dn_B + \mu_C \, dn_C + \mu_D \, dn_D = \sum_i \mu_i \, dn_i = 0 \qquad (2.16)$$

where μ_i is the chemical potential of the species i. We introduce the extent of reaction, ξ, such that the changes in the number of moles of the different reactants and products can be written as

$$dn_A = -v_A \, d\xi, \quad dn_B = -v_B \, d\xi, \quad dn_C = v_C \, d\xi, \quad dn_D = v_D \, d\xi \qquad (2.17)$$

2.3 Reactions and Thermodynamic Equilibrium

This leads to

$$\frac{dG}{d\xi} = -v_A\mu_A - v_B\mu_B + v_C\mu_C + v_D\mu_D = \sum_i v_i\mu_i = 0 \tag{2.18}$$

stating the important fact that at equilibrium the stochiometrically weighted chemical potentials add up to zero. Here, we have used the convention that the stoichiometric factors for reactants are negative. The chemical potentials of the involved species are given by

$$\mu_i = \mu_i^\circ + RT\ln(a_i) = \mu_i^\circ + RT\ln\left(\frac{p_i}{p^\circ}\right) \tag{2.19}$$

where a_i refers to the activity of species i, which can be transformed into either a partial pressure p_i/p° or concentration [i], all at the equilibrium conditions (p° is the pressure at standard conditions, $p^\circ = 1$ bar). By utilizing this expression, we obtain

$$\sum_i v_i\mu_i = \sum_i v_i\mu_i^\circ + RT\ln\left(\prod_i a_i^{v_i}\right) = 0 \tag{2.20}$$

where R is the gas constant and T is the temperature. Since the standard Gibbs free energy change ΔG° can be related to the standard chemical potentials μ° we obtain

$$\Delta G^\circ = \sum_i v_i\mu_i^\circ = -v_A\mu_A^\circ - v_B\mu_B^\circ + v_C\mu_C^\circ + v_D\mu_D^\circ \tag{2.21}$$

The equilibrium constant $K(T)$ can then be written as

$$K(T) = e^{-\frac{\Delta G^\circ}{RT}} = \prod_i a_i^{v_i} = \frac{a_C^{v_c} a_D^{v_d}}{a_A^{v_a} a_B^{v_b}} \tag{2.22}$$

The activities are usually approximated by more convenient quantities, for example, pressures if we are dealing with an ideal gas mixture:

$$K(T) = \frac{k_+}{k_-} = \frac{p_C^{v_c} p_D^{v_d}}{p_A^{v_a} p_B^{v_b}} (p^\circ)^{v_a+v_b-v_c-v_d} \tag{2.23}$$

or concentration if we are dealing with an ideal solution

$$K(T) = \frac{k_+}{k_-} = \frac{[C]_{eq}^{v_c}[D]_{eq}^{v_d}}{[A]_{eq}^{v_a}[B]_{eq}^{v_b}} \tag{2.24}$$

In Chapter 3, we will see how fundamental quantities such as μ can be estimated from first principles (via basic knowledge of the molecule like molecular weight, rotational constants, etc.) and how the equilibrium constant is derived by requiring the chemical potentials of the interacting species to add up to zero as

Table 2.2 Thermodynamic data for important catalytic reactions.

Reaction ($T = 298$ K)	$\Delta H°$ (kJ/mol)	$\Delta G°$ (kJ/mol)
$NH_3 \rightarrow 1/2 N_2 + 3/2 H_2$	+45.9	+16.4
$1/2 N_2 + 3/2 H_2 \rightarrow NH_3$	−45.9	−16.4
$N_2 + 3H_2 \rightarrow 2NH_3$	−91.9[a]	−32.8[a]
$2NO \rightarrow N_2 + O_2$	−182.6[a]	−175.2[a]
$CH_4 + H_2O \rightarrow CO + 3H_2$	+205.9	+142.0
$CH_4 + 1/2 O_2 \rightarrow CO + 2H_2$	−35.9	−86.8
$CH_4 + 2O_2 \rightarrow CO_2 + 2H_2O$	−802.6	−801.0
$CH_4 + 1/2 O_2 \rightarrow CH_3OH$	−275.6	−111.77
$CO_2 + H_2 \rightarrow CH_3OH + H_2O$	−49.3	+3.5
$CO + 2H_2 \rightarrow CH_3OH$	−90.5	−25.1
$CO + H_2O \rightarrow CO_2 + H_2$	−41.2	−28.6

a) Per two moles of NH_3 or NO. Data taken from [5, 6].

in Eq. (2.20). The above equations relate kinetics to thermodynamics and enable one to predict the rate constant for a reaction in the forward direction if the rate constant for the reverse reaction as well as thermodynamic data is known.

To avoid writing long equations with similar terms, we introduce a convenient shorthand notation for most of the expressions considered so far. X_i will denote the molecules A, B, etc. and v_i the stoichiometric coefficients, which are negative for reactants and positive for products. This yields the following set of equations:

$$\sum_i v_i X_i = 0 \tag{2.25}$$

$$K(T) = \prod_i (c_{i,\text{eq}})^{v_i} \tag{2.26}$$

$$\Delta G° = \sum_i v_i \mu_i° \tag{2.27}$$

To calculate the equilibrium composition of a mixture at a given temperature, we first need to calculate the equilibrium constant from thermodynamic data valid under the standard conditions of 298 K and 1 bar, as in Table 2.2. Differentiating Eq. (2.22) and using $\Delta G° = \Delta H° - T\Delta S°$, we obtain the Van't Hoff equation:

$$\frac{d \ln K}{dT} = \frac{d}{dT}\left(-\frac{\Delta G°}{RT}\right) = \frac{\Delta H°}{RT^2} \tag{2.28}$$

Integrating from the standard 298 K to the desired temperature T, we obtain

$$\ln K(T) = \ln K(298) + \int_{298}^{T} \frac{\Delta H^\circ}{RT^2} \, dT \tag{2.29}$$

or

$$\ln K(T) = \ln K(298) - \frac{\Delta H^\circ}{R} \left\{ \frac{1}{T} - \frac{1}{298} \right\} \tag{2.30}$$

in which we have ignored the small effect of the heat capacity, that is, we have assumed that ΔH° does not depend on temperature. Hence, logarithmic plots of the equilibrium constant versus the reciprocal temperature yield a straight line. For endothermic reactions, the equilibrium constant increases with temperature, whereas for exothermic processes it decreases. This method can be used if we only need a rough estimate of the equilibrium constant. More accurate estimates require detailed knowledge of the temperature dependence of both ΔH° and ΔS°, which can be found from thermodynamic tables for a wide range of different materials. We shall see in Chapter 3 that this normally rather weak temperature dependence is due to the detailed manner in which both energy and entropy can be distributed over the internal degrees of freedom in a molecule at a specific temperature. Table 2.2 contains values of ΔH° and ΔG° under standard conditions for several important reactions.

2.3.1
Example of Chemical Equilibrium: The Ammonia Synthesis

As an example of how these expressions are used in practical situations, we will calculate the extent to which the ammonia synthesis reaction

$$1/2\, N_2 + 3/2\, H_2 \rightleftharpoons NH_3 \tag{2.31}$$

proceeds at a given temperature from thermodynamic data. The standard Gibbs free energy change of this reaction at 298 K is about -16.4 kJ/mol, and the standard enthalpy of formation -45.9 kJ/mol, both per mole of NH_3.

In our calculation, we assume that the gas mixture approaches equilibrium under conditions where the pressure is constant. This situation corresponds, for instance, to a volume of gas moving through a plug flow reactor with a negligible pressure drop. Note that if the ammonia synthesis were carried out in a closed system, the pressure would decrease with increasing conversion.

By introducing the mole fraction $Y_X = p_X/p_{tot}$ where p_X is the partial pressure of gas X and p_{tot} is the total pressure, Eq. (2.22) can be formulated as

$$K(T) = \frac{Y_C^{v_c} Y_D^{v_d}}{Y_A^{v_a} Y_B^{v_b}} \left(\frac{p_{tot}}{p^\circ} \right)_{v_c + v_d - v_a - v_b} = e^{-\Delta G/RT} \tag{2.32}$$
$$= e^{-\Delta H/RT} e^{+\Delta S/R}$$

This is a useful expression for calculating equilibrium concentrations. One can easily see that for an exothermic process (ΔH negative) the equilibrium concentration of products decreases with temperature, while it will also increase with pressure if the process consumes gas ($\nu_C + \nu_D - \nu_A - \nu_B < 1$). Using Eq. (2.32) for the ammonia synthesis and assuming ideal gas behavior, we immediately obtain

$$K(T) = e^{-\Delta G/RT} = \frac{a_{NH_3}}{a_{N_2}^{1/2} a_{H_2}^{3/2}} \cong \frac{p_{NH_3}}{p_{N_2}^{1/2} p_{H_2}^{3/2}} p° = \frac{Y_{NH_3}}{Y_{N_2}^{1/2} Y_{H_2}^{3/2}} \frac{p°}{p_{tot}} \qquad (2.33)$$

To achieve higher product yields, the industrial synthesis of ammonia is carried out under high pressures (p_{tot} = 100–200 bar) and relatively low temperatures ($T \approx 400\,°C$) in industry. The equilibrium condition is described in terms of mole fractions because this conveniently expresses the changes during the reaction. In reality the initial gas mixture does not consist only of N_2 and H_2, since N_2 is extracted from air that also contains some argon. Moreover, the hydrogen usually contains traces of methane. We also want to include the possibility that the initial gas mixture contains some ammonia. Thus, in general terms the reactor contains a gas mixture; component X is present at n_X moles, and to indicate the initial number of moles at the start of the reaction we use the superscript i. By introducing the extent of reaction, ξ, defined as

$$\xi \equiv \frac{\nu_X - \nu_X^i}{\nu_X} \qquad (2.34)$$

the final number of moles can then easily be calculated knowing the stoichiometric number as shown in Table 2.3.

Since the mole fraction Y_X equals n_X/n_{tot}, the mole fraction for the equilibrium situation has to be normalized with n_f leading to Eq. (2.35)

$$K(T) = \frac{Y_{NH_3}}{Y_{N_2}^{1/2} Y_{H_2}^{3/2}} \frac{p°}{p_{tot}} = \frac{\frac{n_{NH_3}+\xi}{n_i-\xi}}{\left(\frac{n_{N_2}-\frac{1}{2}\xi}{n_i-\xi}\right)^{1/2} \left(\frac{n_{H_2}-\frac{3}{2}\xi}{n_i-\xi}\right)^{3/2}} \frac{p°}{p_{tot}} \qquad (2.35)$$

Table 2.3 Parameters needed to estimate the conversion of hydrogen into ammonia at constant pressure.

Species	ν_X	Initial number of moles	Final number of moles
N_2	$-\frac{1}{2}$	n_{N_2}	$n_{N_2} - \frac{1}{2}\xi$
H_2	$-\frac{3}{2}$	n_{H_2}	$n_{H_2} - \frac{3}{2}\xi$
NH_3	$+1$	n_{NH_3}	$n_{NH_3} + \xi$
Ar	0	n_{Ar}	n_{Ar}
Sum	-1	$n_i = n_{N_2} + n_{H_2} + n_{NH_3} + n_{Ar}$	$n_f = n_i - \xi$

If we now simplify the problem by starting with a gas consisting of only N_2 and H_2 in a stoichiometric ratio of 1 : 3, the expression reduces to

$$K(T) = \frac{16\xi(n_i - \xi)}{\sqrt{27}(n_i - 2\xi)^2} \frac{p^\circ}{p_{tot}} \tag{2.36}$$

Solving for ξ leaves us with, in principle, two solutions

$$\xi = \frac{n_i}{2}\left(1 \pm \frac{1}{\sqrt{\frac{\sqrt{27}K(T)p_{tot}}{4p^\circ} + 1}}\right) \tag{2.37}$$

of which the solution with the + sign is artificial and has no physical meaning.

The partial pressure of ammonia follows as

$$p_{NH_3} = \frac{\xi}{n_i - \xi} p_{tot} = \frac{\sqrt{K(T)p_{tot}\sqrt{27} + 4} - 2}{\sqrt{K(T)p_{tot}\sqrt{27} + 4} + 2} p_{tot} \tag{2.38}$$

Table 2.4 shows the mole fractions of ammonia for this situation (i.e., initially a 1 : 3 mixture of N_2 and H_2) at a number of different temperatures and pressures. The data in Table 2.4 suggest that to obtain high yields of ammonia one should run the reaction at low temperature. However, a very strong bond of about 940 kJ/mol holds the two N atoms in the nitrogen molecule together, and high temperature would be needed to dissociate this bond. Unfortunately, high temperatures limit the conversion of the nitrogen/hydrogen mixture to ammonia to very small fractions, as Table 2.4 shows. Hence, although thermodynamics favors ammonia formation at low temperature, a high kinetic barrier prevents its formation. The solution is to use a catalyst that will dissociate the N_2 molecule at lower temperatures. Enzymes do this effectively at room temperatures where conversion to NH_3 is not at all limited by equilibrium [7]. In industry, iron catalysts enable ammonia synthesis at pressures of around 200 bar and temperatures of 625–675 K, where the equilibrium conversion is somewhere between 60% and 80% per pass. The last column in Table 2.4 shows the effect of taking the nonideality (fugacity) of the gases into consideration.

2.3.2
Chemical Equilibrium for a Nonideal Gas

In the above example, we assumed that the gases behave ideally, which is not always a good approximation. When the pressure is high or the temperature is low the interaction between the molecules is strong and leads to deviation from ideality. In such cases, we can no longer equate activities with concentrations or partial pressures, but have to introduce a correction. This is done by introducing the fugacity [8], which is related to the partial pressure through the relation

$$f_X = \varphi_X p_X \tag{2.39}$$

where φ_X is the fugacity coefficient. The latter is dimensionless; thus, f_X has the dimension of a pressure. Fugacity coefficients can be found from the virial coefficients of the gas or from tables giving the coefficients in terms of reduced pressure and temperature with respect to the critical pressure and temperature. The fugacity coefficient can assume values larger or smaller than 1, depending on the actual pressure and temperature. Ideal gas behavior corresponds to $\varphi_X = 1$. In the following, we describe briefly how to determine equilibrium conversions when fugacity has to be taken into account.

As mentioned above, activities are correctly represented by fugacities and not by partial pressures. Hence, the correct form of Eq. (2.22) is

$$K(T) = \frac{a_C^{v_c} a_D^{v_d}}{a_A^{v_a} a_B^{v_b}} = \frac{f_C^{v_c} f_D^{v_d}}{f_A^{v_a} f_B^{v_b}} (p^\circ)^{a+b-c-d} \tag{2.40}$$

When dealing with an ideal mixture of gases (that do not behave ideally themselves), the fugacity f_X of gas X is given by

$$f_X = \varphi_X p_X = \varphi_X^{p_{tot}} Y_X p_{tot} \tag{2.41}$$

leading to

$$K(T) = \frac{f_C^{v_c} f_D^{v_d}}{f_A^{v_a} f_B^{v_b}} = \frac{\varphi_C^{p_{tot}} \varphi_D^{p_{tot}}}{\varphi_A^{p_{tot}} \varphi_B^{p_{tot}}} \frac{Y_C^{v_c} Y_D^{v_d}}{Y_A^{v_a} Y_B^{v_b}} \left(\frac{p_{tot}}{p^\circ}\right)^{v_c + v_d - v_a - v_b}$$
$$K(T) = K_\varphi^{p_{tot}} \frac{Y_C^{v_c} Y_D^{v_d}}{Y_A^{v_a} Y_B^{v_b}} \left(\frac{p_{tot}}{p^\circ}\right)^{v_c + v_d - v_a - v_b} \tag{2.42}$$

where $\varphi_X^{p_{tot}}$ is the fugacity coefficient of the pure gas X at pressure p_{tot} and at the given temperature. Note that fugacity coefficients depend both on pressure and temperature. Introducing this in the equilibrium constant for the case of ammonia synthesis, we find

$$K(T) = \frac{a_{NH_3}}{a_{N_2}^{1/2} a_{H_2}^{3/2}} = \frac{f_{NH_3}}{f_{N_2}^{1/2} f_{H_2}^{3/2}} p^\circ \tag{2.43}$$

Table 2.4 Equilibrium conversion of N_2 and H_2 to ammonia in stoichiometric mixtures at a constant pressure p_{tot}.

		Mole fraction of NH_3			
T (K)	K_{eq}	$p_{tot} =$ 1 bar	$p_{tot} =$ 10 bar	$p_{tot} =$ 100 bar	$p_{tot} =$ 100 bar with fugacity
298	6.68×10^2	0.934	0.979	0.993	0.997
400	6.04×10^0	0.497	0.798	0.933	0.945
500	3.18×10^{-1}	0.0862	0.387	0.733	0.759
600	4.18×10^{-2}	0.0132	0.108	0.434	0.466
700	9.40×10^{-3}	0.00303	0.0288	0.197	0.222
800	3.00×10^{-3}	0.00097	0.000955	0.0821	0.097

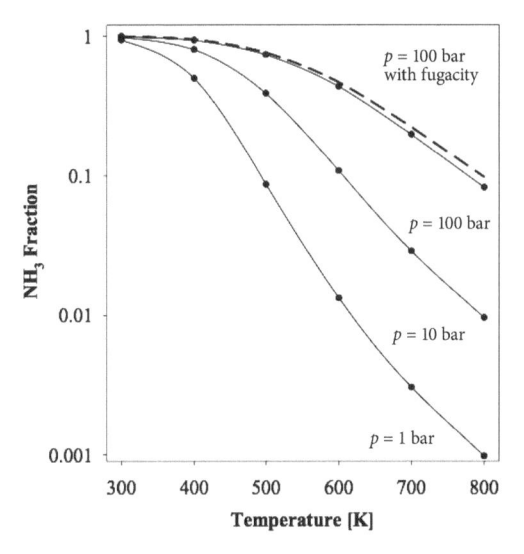

Figure 2.1 Equilibrium concentration of ammonia in a mixture of initially 1 : 3 (N$_2$: H$_2$) as a function of temperature for several total pressures. Note the slight deviation due to nonideality of the gases.

If we make the same assumptions as above, that is, constant total pressure and a stoichiometric gas mixture (N$_2$ + 3 H$_2$ at the start), we obtain

$$\xi = \frac{n_i}{2}\left(1 - \frac{1}{\sqrt{\frac{\sqrt{27}K(T)p_{\text{tot}}}{4K_\varphi^{p_{\text{tot}}} p^\circ} + 1}}\right) \quad (2.44)$$

Please note that the correction corresponds to introducing a corrected equilibrium constant $K(T)/K_\varphi^{p_{\text{tot}}}$. The correction has hardly any influence on the mole fraction of ammonia in the mixture at low pressures, but for $p_{\text{tot}} = 100$ bar and higher the correction becomes significant. The results are presented in the last column of Table 2.4 and in Figure 2.1. It should be noted that correction procedures exist for cases where the *mixture* does not behave ideally, but this goes beyond the scope of the present treatment.

2.4
The Temperature Dependence of the Rate

Rates of reaction usually go up when the temperature increases, although in catalysis this is only partly true as we will see later in this chapter. As a (crude!) rule of thumb, the rate of reaction doubles for every 10 K increase in temperature.

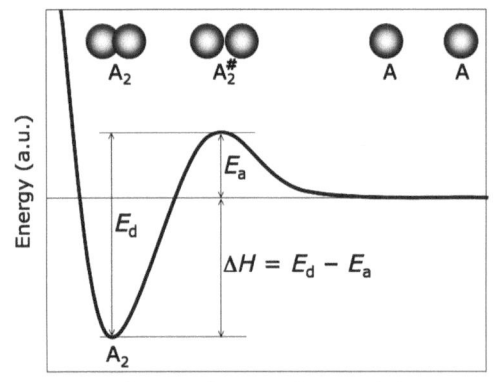

Figure 2.2 Schematic illustration of the potential energy diagram for either the recombination of two atoms or the dissociation of a molecule.

For an elementary reaction, the temperature dependence of the rate constant is given by the Arrhenius equation

$$k(T) = \nu e^{-E_a/RT} \tag{2.45}$$

where ν is called the pre-exponential factor (sometimes called prefactor) and E_a the activation energy (in kJ/mol). One can also express the activation energy in joules per single molecule (a very small number), provided one replaces the gas constant, R, by Boltzmann's constant, k_B.

Arrhenius proposed his equation in 1889 on empirical grounds, justifying it with the hydrolysis of sucrose to fructose and glucose. Note that the temperature dependence is in the exponential term and that the pre-exponential factor is a constant. Reaction rate theories (see Chapter 3) show that the Arrhenius equation is to a very good approximation correct; however, the assumption of a prefactor that does not depend on temperature cannot strictly be maintained as transition state theory shows that it may be proportional to T^x. Nevertheless, this dependence is usually much weaker than the exponential term and is therefore often neglected.

The physical interpretation of the Arrhenius equation is that of an event in which a potential energy barrier has to be surmounted in order to achieve a completed reaction. Figure 2.2 illustrates this for an elementary reaction between two atoms to form a diatomic molecule. When the atoms approach from the right-hand side of the diagram, they first repel each other, implying that the potential energy increases, until they are within reaction distance, where, for example, electrons can rearrange to establish the bond between the two atoms. For the moment, we interpret the height of the barrier as the activation energy (later we will see that this is not quite correct). Also indicated in Figure 2.2 is the reaction enthalpy.

We can determine the activation energy from a series of measurements by plotting the logarithm of the rate constant against the reciprocal temperature, as rearrangement of Eq. (2.45) shows

$$\ln k(T) = \ln \nu - \frac{E_a}{R}\frac{1}{T} \tag{2.46}$$

The slope of such a plot yields the activation energy and the intercept the prefactor. This procedure is represented by the following elegant mathematical expression, which we will often use in the course of this book:

$$E_a = -R\frac{\partial \ln k}{\partial (1/T)} = RT^2\frac{\partial \ln k}{\partial T} \tag{2.47}$$

The reader may now readily calculate the activation energy for which the rule of thumb 'the rate of reaction doubles for every 10 K increase in temperature' is correct.

As explained before, a chemical reaction can seldom be described by a single elementary step, and hence we need to adapt our definition of activation for an overall reaction. Since we are not particularly interested in the effects of thermodynamics, we define the *apparent* activation energy as

$$E_{app} \equiv RT^2\frac{\partial \ln r_+}{\partial T} \tag{2.48}$$

In principle derived from an Arrhenius plot of $\ln r_+$ versus $1/T$, but such a plot may deviate from a straight line. Hence, the apparent activation energy may only be valid for a limited temperature range. Just as for the orders of reaction, one should be very careful with interpreting the activation energy since it depends on the experimental conditions. Below is an example where the forward rate depends both on an activated process and equilibrium steps, representing a situation that occurs frequently in catalytic reactions.

Example

Assume we have an overall reaction consisting of several elementary steps for which the rate expression predicts that the forward rate proceeds as

$$r^+ = \frac{kK_1K_2p_2}{1 + K_2p_2} \tag{2.49}$$

We assume furthermore that k and K_x can be written as

$$k(T) = \nu e^{-E_a/RT} \tag{2.50}$$

and

$$K_x(T) = e^{-\Delta G_x/RT} = e^{-\Delta H_x/RT}e^{\Delta S_x/R} = \nu_x e^{-\Delta H_x/RT} \tag{2.51}$$

Straightforward application of Eq. (2.48) leads to

$$E_{app} = E_a + \Delta H_1 + \Delta H_2 - \frac{K_2p_2}{1 + K_2p_2}\Delta H_2 \tag{2.52}$$

Note that the apparent activation energy is the activation energy of the activated process modified by the equilibrium enthalpies. Thus, the apparent activation energy depends on both the pressure and temperature in this case. Note also that we have neglected any nonexponential temperature dependence. As we shall see in Chapter 3, ν, ΔH, and ΔS are to some degree functions of temperature.

2.5
Integrated Rate Equations: Time Dependence of Concentrations in Reactions of Different Orders

In general the rate of a reaction is determined by monitoring its progress over time. Hence, we need expressions that relate the concentrations of reacting molecules to time, as opposed to the differential equations in the preceding section, which relate the rate of reaction to the concentrations of the participating molecules.

Let us take the following first-order reaction (obviously an isomerization, although a dissociation of R in two products, P_1 and P_2 would yield similar expressions)

$$R \xrightarrow{k} P \tag{2.53}$$

with a corresponding rate equation written in differential form

$$r = -\frac{d[R]}{dt} = k[R] \tag{2.54}$$

To obtain the time-dependent concentrations of R and P, we need to integrate this rate equation, which is simply done by separation of variables:

$$\int_{[R]_0}^{[R]} \frac{d[R]}{[R]} = \int_0^t -k\, dt \quad \Rightarrow \quad \ln\left(\frac{[R]}{[R]_0}\right) = -kt \tag{2.55}$$

The integration limits follow from the consideration that $[R] = [R]_0$ at $t = 0$. Hence, we obtain the following expression for the concentrations of reactant and product:

$$[R] = [R]_0 e^{-kt}; \quad [P] = [R]_0(1 - e^{-kt}) \tag{2.56}$$

Thus, plotting log[R] versus time provides a convenient way to verify that the reaction follows first-order kinetics. Such a plot should yield a straight line and the rate constant follows from the slope of the line.

If it is certain that the reaction is indeed an irreversible first-order reaction, one can also determine how long it takes before 50% of the reactant has converted into products, as for any exponential decay the half-life, $t_{1/2}$, is related to the rate constant k as

$$k = \frac{\ln 2}{t_{1/2}} \tag{2.57}$$

2.5 Integrated Rate Equations: Time Dependence of Concentrations in Reactions of Different Orders

Deriving the integrated rate equation of a second-order reaction is a little more complicated. Let us assume the second-order reaction

$$A + B \xrightarrow{k} AB \tag{2.58}$$

for which the rate equation is

$$r = \frac{d[AB]}{dt} = -\frac{d[A]}{dt} = -\frac{d[B]}{dt} = k[A][B] \tag{2.59}$$

After some time, an amount x has been converted, implying that the concentrations of A and B equal $[A]_0 - x$ and $[B]_0 - x$, respectively. As a consequence, after separating the variables, we are left with the equation

$$\frac{dx}{([A]_0 - x)([B]_0 - x)} = k\,dt \tag{2.60}$$

The standard way to integrate such a differential equation is to split the product on the left into a sum. Hence, we are looking for constants α_1 and α_2 for which

$$\frac{\alpha_1}{[A]_0 - x} + \frac{\alpha_2}{[B]_0 - x} = \frac{1}{([A]_0 - x)([B]_0 - x)} \tag{2.61}$$

With some algebra, we find the solution

$$\int_0^x \frac{1}{[B]_0 - [A]_0} \left\{ \frac{1}{[A]_0 - x} - \frac{1}{[B]_0 - x} \right\} dx = \int_0^t k\,dt \tag{2.62}$$

which is easily integrated, using the boundary conditions that $x = 0$ at $t = 0$, to

$$\frac{1}{[B]_0 - [A]_0} \ln \left\{ \frac{[A]_0\,[B]_0 - x}{[B]_0\,[A]_0 - x} \right\} = kt \tag{2.63}$$

Two special cases are of interest. In the first, A and B are identical species, as, for example, in the recombination of two O radicals to O_2. Obviously, we cannot use Eq. (2.63), as $[A_0] = [B_0]$ leads to division by zero. Thus, we need to derive a special integral rate equation, starting from Eq. (2.60):

$$\frac{dx}{([A]_0 - x)^2} = k\,dt \tag{2.64}$$

which is straightforwardly integrated to

$$\frac{1}{[A]_0 - x} - \frac{1}{[A]_0} = kt \tag{2.65}$$

Thus, plotting $1/([A]_0 - x)$ versus time should produce a straight line with a slope equal to the second-order rate constant.

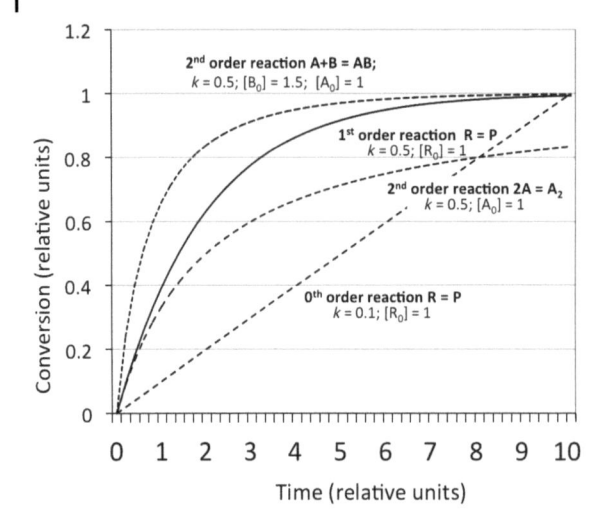

Figure 2.3 Conversion into product as a function of time for different reaction schemes in a batch reactor according to expressions (2.56), (2.63), (2.65) and (2.67).

A second interesting situation occurs if one of the two reactants is present in large excess, for instance, [B] ≫ [A]. In this case, [B] = [B]$_0$ − x ≅ [B]$_0$ and consequently Eq. (2.63) reduces to the much simpler form of a pseudo-first order equation

$$\ln \frac{[A]_0}{[A]_0 - x} = [B]_0 kt \quad \Rightarrow \quad [A] = [A]_0 e^{-k[B]_0 t} \tag{2.66}$$

To determine the rate constant, one uses the same methods as mentioned under the first-order reaction, that is, a plot of ln[A] versus time. The product $k[B]_0$ assumes the role of a pseudo-first order rate constant, from which the true second-order rate constant is easily obtained.

Finally, although rare, we mention the occurrence of zero-order reactions. The special case of a pseudo-zero order reaction arises if a reactant is present in large excess while the reaction does not noticeably change the concentration of the reactant. The differential and integral rate equations for a zero-order reaction R → P are

$$\frac{d[P]}{dt} = -\frac{d[R]}{dt} = k \quad \Rightarrow \quad [R] = [R]_0 - kt \: ; \quad [P] = kt \tag{2.67}$$

Figure 2.3 compares the rate of conversion of reactant to product for different orders of reaction. Table 2.5 summarizes the differential and integral forms of the rate expressions for reactions with various orders.

Table 2.5 Reaction rate expressions in differential and integral form.

Reaction order	Differential form $\dfrac{d[A]}{dt}=$	Integral form $kt=$	Dimension of k
0	$-k$	$[A]_0-[A]$	$\text{mol L}^{-1}\text{s}^{-1}$
1	$-k[A]$	$\ln\dfrac{[A]_0}{[A]}$	s^{-1}
2	$-k[A]^2$	$\dfrac{[1]}{[A]}-\dfrac{[1]}{[A]_0}$	$\text{mol}^{-1}\text{L s}^{-1}$
2	$-k[A][B]$	$\dfrac{[1]}{[B]_0-[A]_0}\ln\dfrac{[A]_0[B]}{[B]_0[A]}$	$\text{mol}^{-1}\text{L s}^{-1}$

2.6 Coupled Reactions in Flow Reactors: The Steady State Approximation

As mentioned above, almost all reactions of practical interest consist of more than one elementary step. The question is then how the kinetics of the elementary steps add up to those of the overall process. Let us consider a two-step reaction, written in generalized form

$$\sum_i v_i^R R_i \underset{k_1^-}{\overset{k_1^+}{\rightleftharpoons}} \sum_i v_i^I I_i \overset{k_2^+}{\rightarrow} \sum_i v_i^P P_i \qquad (2.68)$$

The question of how to deal with the kinetics of such coupled processes depends on the type of reactor employed [9–11]. We will distinguish two cases, namely the flow reactor and the batch reactor (Figure 2.4).

The flow reactor is typically the one used in large-scale industrial processes. Reactants are continuously fed into the reactor at a constant rate, and products appear at the outlet, also at a constant rate. Such reactors are said to operate under steady state conditions, implying that rates of reaction and concentrations be-

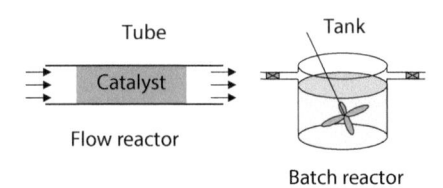

Figure 2.4 Schematic drawings of a cylindrical flow reactor and a batch reactor. Ideally, the flow reactor operates as a plug-flow reactor in which the gas moves as a piston down through the tube. The ideal batch reactor is a well-mixed tank reactor with no concentration gradients. If the valves indicated are open and reactants are continuously fed to the tank, while products are removed, we would have a continuously stirred tank reactor (CSTR).

come independent of time (unless the rate of reaction oscillates around its steady state value).

The batch reactor is generally used in the production of fine chemicals. At the start of the process, the reactor is filled with reactants, which gradually convert into products. As a consequence, the rate of reaction and the concentrations of all participants in the reaction vary with time.

We will first discuss the kinetics of coupled reactions in the steady state regime.

Without limiting the validity of the approach, we may simplify the situation by replacing reactants, intermediates, and products by a single molecule each

$$R \underset{k_1^-}{\overset{k_1^+}{\rightleftharpoons}} I \overset{k_2^+}{\rightarrow} P \tag{2.69}$$

The rate equations for each participant in this process are

$$\frac{d[R]}{dt} = -k_1^+[R] + k_1^-[I] \tag{2.70}$$

$$\frac{d[I]}{dt} = k_1^+[R] - (k_1^- + k_2^+)[I] \tag{2.71}$$

$$\frac{d[P]}{dt} = k_2^+[I] \tag{2.72}$$

Assuming that this process runs under steady state conditions, as for an industrial flow reactor with a constant inflow of reactants and a constant outflow of products, the concentration of the intermediate will be constant, as expressed in the steady state assumption:

$$\frac{d[I]}{dt} = 0 \tag{2.73}$$

This condition can be used to eliminate the concentration of I, which in general will not be easily accessible to measurement, from the rate expressions:

$$[I] = \frac{k_1^+[R]}{k_1^- + k_2^+} \tag{2.74}$$

and the rate of product formation becomes

$$\frac{d[P]}{dt} = \frac{k_1^+ k_2^+[R]}{k_1^- + k_2^+} \tag{2.75}$$

This expression has two limiting cases. If the rate constant of the second step is small, we can ignore it in the denominator of Eq. (2.75), and obtain

$$\frac{[P]}{dt} = \frac{k_1^+ k_2^+[R]}{k_1^-} = k_2^+ K_1[R] \quad (k_2^+ \ll k_1^-) \tag{2.76}$$

The rate of the overall reaction is limited by the conversion of intermediate into product, which is called the rate-determining step. The first step is virtually at equilibrium, and therefore the concentration of the intermediate is determined by the equilibrium constant K_1.

The apparent activation energy of the reaction follows by taking the logarithmic derivative as expressed by Eq. (2.48):

$$E_{app} = RT^2 \frac{\partial \ln(k_2^+ K_1)}{\partial T} = E_{a,2} + \Delta H_1 \tag{2.77}$$

The overall activation energy, usually referred to as *apparent* activation energy, is now a composite property of the two reaction steps. If the enthalpy of the first reaction is sufficiently negative (i.e., the reaction is exothermic), the apparent activation energy may even become negative, expressing that the intermediate concentration decreases more rapidly with increasing temperature then the rate of the second step increases.

The second case arises if the second step is fast and the first step becomes rate determining, in which case the rate equation reduces to

$$\frac{d[P]}{dt} = k_1^+[R] \quad (k_2^+ \gg k_1^-) \tag{2.78}$$

The activation energy of the overall reaction equals that of the first step, $E_{a,1}$. Note that fast elementary steps following the one that limits the rate become kinetically insignificant, whereas fast steps before the rate-determining step do enter the rate equation, as they directly affect the concentration of the intermediate that is converted in the rate-determining step.

Although the two cases represent entirely different reaction mechanisms, the overall rate of reaction maintains the same form with respect to its dependence on reactant concentration. Measurements of the kinetics would in both cases reveal the reaction to be first order in [R]. In general, it is not possible to prove that a mechanism is correct on the basis of kinetic measurements, as one can almost always find a modified mechanism leading to the same behavior of the rate equation. It is often possible, however, to exclude certain mechanisms on the basis of kinetic measurements.

Example

Let us consider the reaction

$$2O_3 \longrightarrow 3O_2 \tag{2.79}$$

which proceeds through the following elementary steps (in which the reverse reactions are ignored)

$$O_3 \xrightarrow{k_1} O_2 + O \tag{2.80}$$

$$O + O_3 \xrightarrow{k_2} 2O_2 \tag{2.81}$$

The rate equations for each participant in this process are

$$\frac{d[O_3]}{dt} = -k_1[O_3] - k_2[O][O_3] \tag{2.82}$$

$$\frac{d[O]}{dt} = k_1[O_3] - k_2[O][O_3] \tag{2.83}$$

$$\frac{d[O_2]}{dt} = k_1[O_3] + 2k_2[O][O_3] \tag{2.84}$$

Application of the steady state assumption yields $[O] = k_1/k_2$ and consequently the overall rate becomes

$$r = \frac{d[O_2]}{dt} = 3k_1[O_3] \tag{2.85}$$

The reaction is thus first order in the ozone concentration. Had we treated the overall reaction as an elementary step, the rate would be second order with respect to ozone. In this case, measurements would distinguish between the two mechanisms.

Historically, the steady state approximation has played an important role in unraveling mechanisms of apparently simple reactions as $H_2 + Cl_2 = 2HCl$, which involve radicals and chain mechanisms [4].

Chain reactions were discovered around 1913, when Bodenstein and Dux found that the reaction between H_2 and Cl_2 could be initiated by irradiating the reaction mixture with photons. They were surprised to find that the number of HCl molecules per absorbed photon, called quantum yield, is around 10^6! Nernst gave the explanation of the phenomenon in 1918: the photon facilitated the dissociation of the Cl_2 into Cl radicals (the initiation step), which then started the following chain process:

$$Cl + H_2 \longrightarrow HCl + H$$
$$H + Cl_2 \longrightarrow HCl + Cl$$

which forms a closed cycle (propagation steps) that continues until two chlorine radicals recombine in what is called the termination step. As the concentration of radicals is low, the chance that termination occurs is only around 1 in 10^6.

We discuss here the formation of NO from N_2 and O_2, responsible for NO formation in the engines of cars. In Chapter 10, we will describe how NO is removed catalytically from automotive exhausts.

The reaction is initiated by the dissociation of O_2 on hot parts of the engine:

$$O_2 \xrightarrow{k_1} 2O \tag{2.86}$$

The reactive O atoms become engaged in a cycle of propagation steps, producing NO:

$$O + N_2 \xrightarrow{k_2} NO + N$$
$$N + O_2 \xrightarrow{k_3} NO + O \qquad (2.87)$$

Finally, the cyclic chain terminates by

$$2O \xrightarrow{k_4} O_2 \quad \text{or} \quad 2N \xrightarrow{k_5} N_2 \qquad (2.88)$$

We shall assume that the latter reaction is negligible and, as the reaction occurs entirely in the gas phase, we will write partial pressures instead of concentrations. The rate of NO formation is

$$\frac{dp_{NO}}{dt} = k_2 p_O p_{N_2} + k_3 p_N p_{O_2} \qquad (2.89)$$

Applying the steady state approximation to the partial pressures of the O and N atoms is valid if the average number of propagation cycles prior to termination is large. Assuming this to be the case, we find

$$\frac{dp_N}{dt} = k_2 p_O p_{N_2} - k_3 p_N p_{O_2} = 0 \qquad (2.90)$$

$$\frac{dp_O}{dt} = 2k_1 p_{O_2} - k_2 p_O p_{N_2} + k_3 p_N p_{O_2} - 2k_4 p_O^2 = 0 \qquad (2.91)$$

It follows directly that

$$p_O = \left(\frac{k_1}{k_4}\right)^{1/2} p_{O_2}^{1/2}, \quad p_N = \frac{k_2}{k_3}\left(\frac{k_1}{k_4}\right)^{1/2} p_{O_2}^{-1/2} p_{N_2} \qquad (2.92)$$

Substituting these results in the rate equation, we obtain

$$\frac{dp_{NO}}{dt} = 2k_2 \left(\frac{k_1}{k_4}\right)^{1/2} p_{O_2}^{1/2} p_{N_2} \qquad (2.93)$$

for the rate of NO production. The order of 1/2 in oxygen derives essentially from the fact that the oxygen molecule has to dissociate to initiate the reaction. We will encounter this situation more often when we discuss the kinetics of catalytic reactions.

2.7
Coupled Reactions in Batch Reactors

The second important environment in which coupled reactions can occur is that of a batch reactor [9–11]. We will assume that our batch reactor behaves as a well-stirred tank reactor, such that all participants are well mixed and concentration gradients do not occur. We will furthermore assume that our coupled reactions proceed only in the forward direction, such that our sequence of elementary steps is reduced to

$$R \xrightarrow{k_1} I \xrightarrow{k_2} P \tag{2.94}$$

For convenience, we have dropped the forward superscript from the rate constants. The rate equations for each participant in this process are

$$-\frac{d[R]}{dt} = k_1[R] \tag{2.95}$$

$$\frac{d[I]}{dt} = k_1[R] - k_2[I] \tag{2.96}$$

$$\frac{d[P]}{dt} = k_2[I] \tag{2.97}$$

Now all concentrations and, hence, the rates depend on time and we want to know how. The rate of conversion of R is easily integrated to (see Section 2.5)

$$[R] = [R]_0 e^{-k_1 t} \tag{2.98}$$

Substituting the result in Eq. (2.96) and separating terms with [I] and [R], we obtain

$$\frac{d[I]}{dt} + k_2[I] = k_1[R]_0 e^{-k_1 t} \tag{2.99}$$

If the left-hand side of this differential equation were the differential of a product, for example, $[I] * f(t)$, integration would be straightforward. Hence, we should multiply both sides with a function, $f(t)$, chosen such that

$$\frac{df(t)}{dt} = k_2 f(t) \quad \Rightarrow \quad f(t) = e^{k_2 t} \tag{2.100}$$

Doing so we find

$$\frac{d\{[I]e^{k_2 t}\}}{dt} = k_1[R]_0 e^{(k_2 - k_1)t} \tag{2.101}$$

and recalling that the concentration of the intermediate is zero at the start of the reaction, we find the solution

$$[I] = [R]_0 \frac{k_1}{k_2 - k_1} (e^{-k_1 t} - e^{-k_2 t}) \tag{2.102}$$

Finally, by inserting the expression for [I] in the rate equation for the product [P] and applying the same boundary that [P] = 0 when $t = 0$, we find the concentration of the product as a function of time:

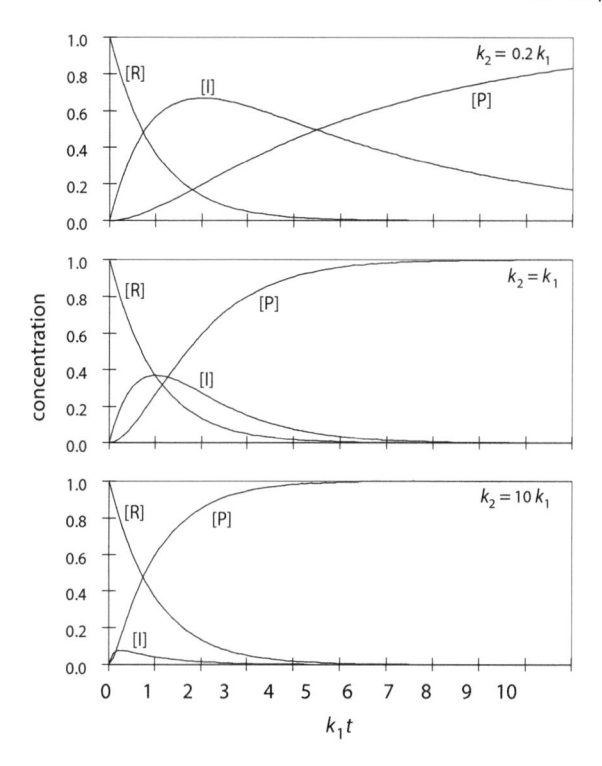

Figure 2.5 Concentrations of reactant, intermediate, and product for a consecutive reaction mechanism for different values of the rate constant.

$$[P] = [R]_0 \left(1 - \frac{k_2}{k_2 - k_1}e^{-k_1 t} + \frac{k_1}{k_2 - k_1}e^{-k_2 t}\right) \quad (2.103)$$

Figure 2.5 shows how the concentrations vary for different rate constants. Note that the intermediate is formed as soon as the reaction starts, as indicated by the positive derivative at $t = 0$, whereas the rate of product formation is zero at the start. The product concentration curve shows an inflection point that corresponds with the maximum intermediate concentration.

Suppose we perform an organic synthesis in a batch reactor where the desired molecule is the intermediate and not the end product. It is then very important that we know how long we should let the reaction run to obtain the highest yield of the intermediate. Setting the differential $d[I]/dt$ in Eq. (2.99) equal to zero and substituting Eq. (2.102) into Eq. (2.99), we find the time, $t_{\max}$, at which the maximum is reached, and by inserting $t_{\max}$ in Eq. (2.102), the corresponding optimal concentration of the intermediate:

$$t_{\max} = \frac{1}{k_1 - k_2} \ln \frac{k_1}{k_2}; \quad [I]_{\max} = \left(\frac{k_1}{k_2}\right)^{\frac{k_2}{k_2 - k_1}} [R]_0 \quad (2.104)$$

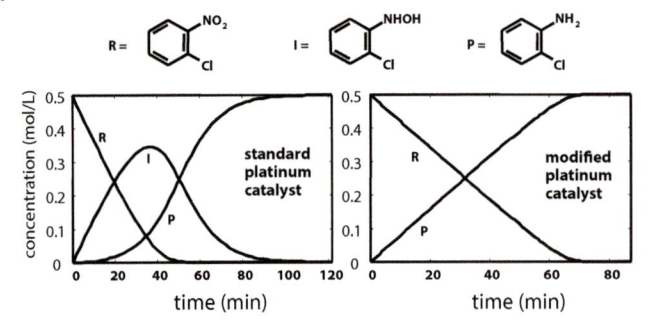

Figure 2.6 The hydrogenation of substituted nitro arenes to amines in a batch reactor involves two consecutive reactions when a standard platinum catalyst is used. Unfortunately, the hydroxylamine intermediate is both explosive and carcinogenic. Application of a modified platinum catalyst, which contains a vanadium promoter, accelerates the second reaction step so much that the undesired hydroxylamine intermediate becomes undetectable (courtesy of Dr. H.-U. Blaser, Novartis, Switzerland).

Example

We now illustrate the opposite case where the intermediate is in fact a highly undesirable substance, as it presents a health or even an explosion hazard. The hydrogenation of aromatic nitro compounds, such as the one shown in Figure 2.6, is industrially important for the production of dyes, whiteners, agrochemicals, and pharmaceuticals. The reaction occurs in the presence of a platinum catalyst and proceeds via intermediates, among which the hydroxylamine (–NHOH) species is particularly hazardous, as it is both carcinogenic and explosive. Unfortunately, standard platinum catalysts give rise to high levels of this undesired intermediate.

By modifying the catalyst with a so-called promoter (in this case vanadium oxide) is it possible to largely eliminate the intermediate. As Figure 2.6 shows, the rate constant of the reaction from the hydroxylamine to the amine is much larger when the promoted catalyst is used, and thus the intermediate reacts instantaneously, resulting in a safer and environmentally friendlier process.

2.8
Catalytic Reactions

A catalyst accelerates the rate of a reaction without itself being consumed in the process.

As explained in Chapter 1, catalytic reactions occur when the reacting species are associated with the catalyst. In heterogeneous catalysis, this happens at a surface, in homogeneous catalysis in a complex formed with the catalyst molecule. In terms of kinetics, the catalyst must be included as a participating species that leaves the reaction unaltered, as schematically indicated in Figure 2.7 (which shows the simplest conceivable catalytic cycle). We will investigate the kinetics

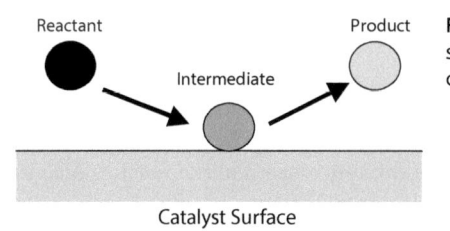

Figure 2.7 Schematic representation of the simplest possible reaction in heterogeneous catalysis.

of the simple two-step mechanism and compare them with the kinetics of the corresponding mechanism of the uncatalyzed reaction discussed in Section 2.6.

We can think of a heterogeneous catalyst as a collection of active sites (denoted by ∗) located at a surface. The total number of these sites is constant and equal to N (if there is any chance of confusion with N atoms, we will use the symbol N_*). The adsorption of the reactant is formally a reaction with an empty site to give an intermediate I∗ (or more conveniently R∗ if we explicitly want to express that it is the reactant R sitting on an adsorption site). All sites are equivalent and each can be occupied by a single species only. We will use the symbol θ_R to indicate the fraction of occupied sites occupied by species R, making $N\theta_R$ the number of occupied sites. Hence, the fraction of unoccupied sites available for reaction will be $1 - \theta_R$. The following equations represent the catalytic cycle of Figure 2.7:

$$R + * \underset{k_1^-}{\overset{k_1^+}{\rightleftharpoons}} R* \quad (2.105)$$

$$R* \overset{k_2^+}{\rightarrow} P + * \quad (2.106)$$

Units and Dimensions

Before deriving the rate equations, we first need to think about the dimensions of the rates. As heterogeneous catalysis involves reactants and products in the three-dimensional space of gases or liquids, but with intermediates on a *two-dimensional* surface, we cannot simply use concentrations as in the case of uncatalyzed reactions. Our choice throughout this book will be to express the macroscopic rate of a catalytic reaction in moles per unit of time. In addition, we will use the microscopic concept of turnover frequency, defined as the number of molecules converted per active site and per unit of time. The macroscopic rate can be seen as a characteristic activity per weight or per volume unit of catalyst in all its complexity with regard to shape, composition, etc., whereas the turnover frequency is a measure for the intrinsic activity of a catalytic site.

With these definitions, the equations for the macroscopic rate of the simple catalytic reaction of Figure 2.7 in (2.105) and (2.106) become

$$-V\frac{d[R]}{dt} = Nk_1^+(1-\theta_R)[R] - Nk_1^-\theta_R \quad (2.107)$$

$$\frac{d\theta_R}{dt} = k_1^+(1-\theta_R)[R] - (k_1^- + k_2^+)\theta_R \quad (2.108)$$

$$V\frac{d[P]}{dt} = Nk_2^+ \theta_R \tag{2.109}$$

Note that the rates of product formation and reactant conversion indeed have the dimensions of mol per unit of time, and that these rates are proportional to the number of sites, or in fact, the amount of catalyst present in the reactor. Also in the case of a second order reaction, for example, between adsorbed species A* and B*, we write the rate in the form $r = Nk\theta_A\theta_B$ by applying the mean-field approximation. Here, the rate is proportional to the total number of sites on the surface and the probability of finding a species A adjacent to a species B on the surface, the latter being proportional to the coverages of A and B. In the mean-field approximation, A and B are distributed randomly over the N available sites; this only tends to be valid when the adsorbents repel each other. Thus, the rate is not $r = k(N\theta_A)(N\theta_B)$ since the reactants need to be on adjacent sites. Another important consideration is that we want the rate to be linearly proportional to the amount of catalyst in the reactor, in accordance with $r = Nk\theta_A\theta_B$ for a second-order surface reaction.

Assuming that the catalytic reaction takes place in a flow reactor under stationary conditions, we may use the steady state approximation to eliminate the fraction of the adsorbed intermediate from the rate expressions to yield:

$$\theta_R = \frac{\frac{k_1^+[R]}{k_1^- + k_2^+}}{1 + \frac{k_1^+[R]}{k_1^- + k_2^+}} \tag{2.110}$$

This leaves us with the following expression for the rate of the simplest catalytic reaction from R to P:

$$V\frac{d[P]}{dt} = N\frac{\frac{k_1^+ k_2^+[R]}{k_1^- + k_2^+}}{1 + \frac{k_1^+[R]}{k_1^- + k_2^+}} = \frac{Nk_2^+[R]}{K_M + [R]} \tag{2.111}$$

The first expression is used in heterogeneous catalysis, the second part is more common in homogeneous and bio-enzymatic catalysis, with the useful constant

$$K_M \equiv \frac{k_1^- + k_2^+}{k_1^+} \tag{2.112}$$

being called the Michaelis constant (see Section 2.12).

Comparing with the corresponding expressions for the noncatalytic reaction (2.75), we see that the equations differ by the denominator in Eq. (2.111), which is a direct consequence of the participation of a catalyst with a constant number of sites. This is easily seen by introducing the coverage of free sites θ_*:

$$\theta_* = 1 - \theta_R = \frac{1}{1 + \frac{k_1^+[R]}{k_1^- + k_2^+}} \tag{2.113}$$

2.8 Catalytic Reactions

Another fundamental difference is that the rate of the uncatalyzed reaction from R to P is always first order in the reactant, whereas the order in R of the catalytic reaction depends on the values of the rate constants in Eq. (2.111), which on their turn depend on the temperature of the reaction. All we can say is that the order will be a fractional number between 0 and 1, depending on the conditions. We have earlier defined the reaction order n_R as

$$n_R \equiv [R]\frac{\partial \ln r_+}{\partial [R]} \tag{2.114}$$

Thus, by taking the logarithmic derivative of Eq. (2.111), we find that the order is

$$n_R \equiv [R]\frac{\partial \ln(V\,d[P]/dt)}{\partial [R]} = \frac{1}{1+\frac{k_1^+[R]}{k_1^-+k_2^+}} = 1 - \theta_R \tag{2.115}$$

implying a rate in the form of a power rate law as

$$V\frac{d[P]}{dt} = Nk'[R]^{(1-\theta_R)} \tag{2.116}$$

in which k' is just an effective rate constant. One should realize how complex the order $(1 - \theta_R)$ is, as it varies with both the concentration of [R] and temperature. It is meaningless to speak about *the order of a catalytic reaction* without stating the exact conditions under which it has been determined.

Let us look at the limiting cases, starting from the complete rate expression in Eq. (2.111). If the conversions from adsorbed intermediate to either product or reactant are fast, then the denominator approaches 1 and the catalyst surface will be mostly empty, $\theta_R \cong 0$. In this case, the rate expression becomes equal to that of the uncatalyzed reaction, and the order in the reactant becomes +1. Under such conditions, it is beneficial to increase the reactant concentration since the surface is mainly unoccupied, and the catalyst inefficiently used.

In the other limiting case, when the denominator of Eq. (2.111) becomes large, the surface is heavily occupied because either the reaction from intermediate to product or the reverse reaction to reactant is slow. In this case, the rate equation is approximately

$$V\frac{d[P]}{dt} \cong Nk_2^+ \tag{2.117}$$

and the rate has become zero order in [R]. Note how the rate is still first order in the number of active sites N. In the following, we will consider the activity per site and simply set $N = 1$. It is, however, important to keep in mind that the efficiency of a catalyst is determined by both the reactivity of the active sites and their number.

The apparent activation energy for the overall reaction, Eqs. (2.105) and (2.106), can be readily derived from Eq. (2.111); activation energies in general are discussed below.

2.8.1
The Mean-Field Approximation

We have already made use of the so-called mean-field approximation by assuming that (1) all adsorbed species are distributed randomly over the surface and (2) there is no interaction between the adsorbed species. This is an approximation that is seldom fulfilled. Usually there will be either an attractive or repulsive interaction between the adsorbed species (Figure 2.8). When the interactions between adsorbates are repulsive and the coverage is low, the mean-field approximation usually works well. For attractive interactions, however, the mean-field approximation may already break down at low coverages, particularly when one of species agglomerates into islands (Figure 2.9), whence reaction occurs only at the edges of these islands. Diffusion of reactants to these sites comes into play and the coupling of transport phenomena to the reaction scheme makes the situation quite complex. High temperatures, however, tend to randomize the adsorbed species across the surfaces, depending on the magnitude of the attractive interaction.

At high coverage, adsorbate interactions will always be present, implying that pre-exponential factors and activation energies are dependent on coverage. In the

(a) $r \propto \theta_A \theta_B$ (b) $r < \theta_A \theta_B$

Figure 2.8 Checkerboard model of a surface with two adsorbates; (a) randomly distributed; (b) attractive interactions give rise to islands of B. In the latter case reaction between A and B is only possible at the perimeter of the island, hence the rate is no longer proportional to the coverage of B.

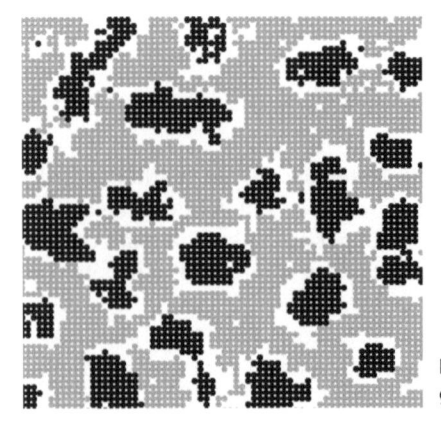

Figure 2.9 Segregated species (black and grey dots) on a checkerboard surface.

following, we shall assume that the mean-field approximation is valid, but one should be aware that it may be a source of error. The alternative to this approximation is to perform Monte Carlo simulations (see Chapter 7).

2.9
Langmuir Adsorption Isotherms

Adsorption of reactants on the catalyst's surface is the first step in every reaction of heterogeneous catalysis. Here, we focus on gases reacting on solid catalysts. Although we will deal with the adsorption of gases in a separate chapter, we need to discuss the relation between the coverage of a particular gas and its partial pressure above the surface. Such relations are called isotherms, and they form the basis of the kinetics of catalytic reactions.

We owe the first quantitative theory of gas adsorption on surfaces to Irving Langmuir (1891–1957), who studied the deterioration of tungsten filaments in electric light bulbs at General Electric. Extending his work to the adsorption of and reaction of gases on metals, he made an essential contribution to the kinetic description of catalytic reactions, for which he received the Nobel Prize in Chemistry in 1932. We will now derive the Langmuir adsorption isotherms for molecular, dissociative, and competitive adsorption.

2.9.1
Associative Adsorption

The Langmuir adsorption isotherm is easy to derive. Again we assume that the catalyst contains equivalent adsorption sites and that the adsorbed molecules do not interact. If the adsorbed molecules are in equilibrium with the gas phase, we may write the reaction equation as

$$A + * \underset{k_A^-}{\overset{k_A^+}{\rightleftharpoons}} A* \tag{2.118}$$

and the rate equation as

$$\frac{d\theta_A}{dt} = p_A k_A^+ (1 - \theta_A) - k_A^- \theta_A \tag{2.119}$$

where the first term is due to adsorption and the latter to desorption. Because the reaction is in equilibrium, we may write

$$\theta_A = K_A\, p_A (1 - \theta_A)\, ; \quad K_A = \frac{k_A^+}{k_A^-} \tag{2.120}$$

which is readily rearranged to the Langmuir adsorption isotherm for associative adsorption of a single gas (e.g., CO, NH_3, NO, provided these do not decompose upon adsorption):

$$\theta_A = \frac{K_A\, p_A}{1 + K_A\, p_A} \tag{2.121}$$

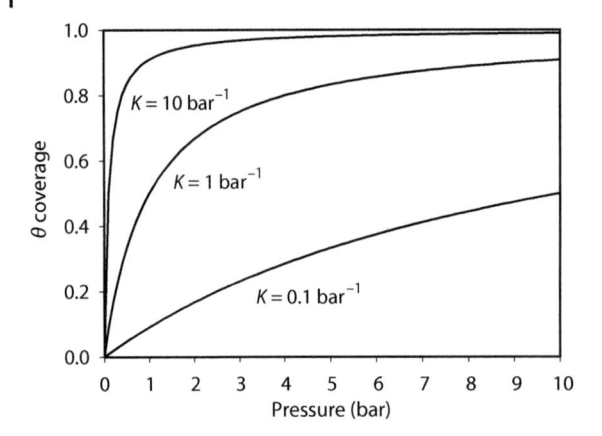

Figure 2.10 Langmuir adsorption isotherms for associative (molecular) adsorption represented as a plot of coverage versus pressure for three values of the equilibrium constant, K.

Note that the fraction of empty sites is given by

$$\theta_* = (1 - \theta_A) = \frac{1}{1 + K_A \, p_A} \tag{2.122}$$

Consequently, one can also write the Langmuir isotherm as $\theta_A = K_A p_A \theta_*$ (a convenient form to use in solving a kinetic scheme if the fraction of unoccupied sites is not yet known).

Figure 2.10 shows a plot of θ_A versus the partial pressure of A, p_A. At low pressure, the coverage is small, and increases linearly with pressure; the derivative of the plot equals the equilibrium constant, K_A. At high pressure, the surface becomes saturated, and the coverage approaches asymptotically its saturation value of 100%.

A series of measurements of coverage against partial pressure can easily be tested for consistency with the Langmuir isotherm, by plotting $1/\theta$ against $1/p$, which should yield a straight line of slope $1/K$.

2.9.2
Dissociative Adsorption

Molecules such as H_2 and O_2 almost always dissociate directly upon adsorption, and in general one may assume equilibrium between the adsorbed atoms and the molecules in the gas phase. In this case, we have

$$A_2 + 2* \underset{k_A^-}{\overset{k_A^+}{\rightleftharpoons}} 2A* \tag{2.123}$$

with the corresponding rate expression

$$\frac{d\theta_A}{dt} = p_A k_A^+ (1 - \theta_A)^2 - k_A^- \theta_A^2 \tag{2.124}$$

which, at equilibrium, becomes

$$\theta_A^2 = K_{A_2} \, p_{A_2} (1 - \theta_A)^2 \qquad (2.125)$$

and

$$\theta_A = \frac{\sqrt{K_{A_2} \, p_{A_2}}}{1 + \sqrt{K_{A_2} \, p_{A_2}}}, \quad \theta_* = \frac{1}{1 + \sqrt{K_{A_2} \, p_{A_2}}} \qquad (2.126)$$

as the Langmuir adsorption isotherm for dissociative adsorption.

2.9.3
Competitive Adsorption

A very important case arises if two species, say A and B, compete for the same sites:

$$\begin{array}{c} A + * \xrightleftharpoons{K_A} A* \\ B + * \xrightleftharpoons{K_B} B* \end{array} \qquad (2.127)$$

for which the equilibrium equations are

$$\theta_A = K_A \, p_A \theta_*, \quad \theta_B = K_B \, p_B \theta_* \qquad (2.128)$$

Conservation of sites requires that

$$\theta_A + \theta_B + \theta_* = 1 \qquad (2.129)$$

and the respective coverages become

$$\theta_A = \frac{K_A \, p_A}{1 + K_A \, p_A + K_B \, p_B}$$

$$\theta_B = \frac{K_B \, p_B}{1 + K_A \, p_A + K_B \, p_B} \qquad (2.130)$$

$$\theta_* = \frac{1}{1 + K_A \, p_A + K_B \, p_B}$$

Competitive adsorption occurs in many catalytic reactions and has important consequences for the conditions under which reactions are carried out. If component A binds more strongly than component B (i.e., $K_A > K_B$), one may increase the concentration of B at the surface by choosing $p_B > p_A$ to compensate for the difference in bond strength.

2.10
Reaction Mechanisms

Here we analyze the kinetics of a general catalytic reaction:

$$A + B \xrightarrow{\text{catalyst}} AB \qquad (2.131)$$

The catalytic process is a sequence of elementary steps that form a cycle from which the catalyst emerges unaltered. Identifying which steps and intermediates have to be taken into account may be difficult, requiring spectroscopic tools and computational approaches, as we described elsewhere (see Chapter 7). Here, we will assume that the elementary steps are known, and we will describe in detail how one derives the rate equation for such processes.

2.10.1
Langmuir–Hinshelwood or Eley–Rideal Mechanisms

In Langmuir–Hinshelwood kinetics, it is assumed that all species are adsorbed and accommodated (in thermal equilibrium) with the surface before they take part in any reactions. Hence, species react in the chemisorbed state on the surface. This is the prevailing situation in heterogeneous catalysis.

Another possibility is the so-called Eley–Rideal mechanism, in which one of the reactants reacts directly out of the gas phase, without being accommodated at the surface. For instance, in the reaction A + B, a gas-phase molecule of B might approach the surface and react with chemisorbed A∗ molecule without being adsorbed itself. An example of an Eley–Rideal process is the reaction of gas-phase atomic hydrogen with a surface saturated with atomic hydrogen. The strongly activated hydrogen atom (which is a free radical) reacts readily with one of the adsorbed hydrogen atoms and, since molecular hydrogen does not usually adsorb strongly to a surface, the molecule desorbs instantaneously.

Whether a catalytic reaction proceeds via a Langmuir–Hinshelwood or Eley–Rideal mechanism has significant implications for the kinetic description, as in the latter case one of the reactants does not require free sites to react. However, Eley–Rideal mechanisms are extremely rare, and we will assume Langmuir–Hinshelwood behavior throughout the remainder of this book.

2.10.2
Langmuir–Hinshelwood Kinetics

Writing out the catalytic reaction between A and B in elementary steps according to the Langmuir–Hinshelwood mechanism, we obtain

$$1)\quad A + * \underset{k_1^-}{\overset{k_1^+}{\rightleftharpoons}} A* \tag{2.132}$$

$$2)\quad B + * \underset{k_2^-}{\overset{k_2^+}{\rightleftharpoons}} B* \tag{2.133}$$

$$3)\quad A* + B* \underset{k_3^-}{\overset{k_3^+}{\rightleftharpoons}} AB* + * \tag{2.134}$$

$$4)\quad AB* \underset{k_4^-}{\overset{k_4^+}{\rightleftharpoons}} AB + * \tag{2.135}$$

Note that in the final desorption step the equilibrium constant for adsorption of AB equals $1/K_4$, whereas for the other adsorption steps it is defined as

$$K_x = \frac{k_x^+}{k_x^-} \tag{2.136}$$

where k^+ describes the adsorption. In step (4), k_4^+ represents the desorption.

In drafting a catalytic cycle as Eqs. (2.132)–(2.135), we naturally have to ensure that the reaction steps are thermodynamically and stochiometrically consistent. For instance, the number of sites consumed in adsorption and dissociation steps must be equal to the number of sites liberated in formation and desorption steps, to fulfill the criterion that a catalyst is unaltered by the catalytic cycle.

For each step, there is a corresponding rate (for convenience we drop the total number of sites from the expressions, that is, r becomes a rate per site, or a turnover frequency):

$$r_1 = k_1^+ p_A \theta_* - k_1^- \theta_A \tag{2.137}$$

$$r_2 = k_2^+ p_B \theta_* - k_2^- \theta_B \tag{2.138}$$

$$r_3 = k_3^+ \theta_A \theta_B - k_3^- \theta_{AB} \theta_* \tag{2.139}$$

$$r_4 = k_4^+ \theta_{AB} - k_4^- p_{AB} \theta_* \tag{2.140}$$

Note that the number of sites on a catalyst is constant and hence all coverages should always add up to unity, as expressed by the following balance of sites:

$$\theta_* + \theta_A + \theta_B + \theta_{AB} = 1 \tag{2.141}$$

We will work out these equations for a number of cases.

2.10.3
The Complete Solution

To solve the kinetics for the most general case, in which, for example, we allow partial pressures to vary with time, we need the full set of differential equations describing the coverage of all species participating in the reaction:

$$\frac{d\theta_A}{dt} = r_1 - r_3 = k_1^+ p_A \theta_* - k_1^- \theta_A - k_3^+ \theta_A \theta_B + k_3^- \theta_{AB} \theta_* \tag{2.142}$$

$$\frac{d\theta_B}{dt} = r_2 - r_3 = k_2^+ p_B \theta_* - k_2^- \theta_B - k_3^+ \theta_A \theta_B + k_3^- \theta_{AB} \theta_* \tag{2.143}$$

$$\frac{d\theta_{AB}}{dt} = r_3 - r_4 = k_3^+ \theta_A \theta_B - k_3^- \theta_{AB} \theta_* - k_4^+ \theta_{AB} + k_4^- p_{AB} \theta_* \tag{2.144}$$

$$\frac{d\theta_*}{t} = -r_1 - r_2 + r_3 + r_4$$
$$= -k_1^+ p_A \theta_* + k_1^- \theta_A - k_2^+ p_B \theta_* + k_2^- \theta_B + k_3^+ \theta_A \theta_B - k_3^- \theta_{AB} \theta_*$$
$$+ k_4^+ \theta_{AB} - k_4^- p_{AB} \theta_* \qquad (2.145)$$

These equations can be solved numerically with a computer, without making any approximations. Naturally all the involved kinetic parameters need to be either known or estimated to give a complete solution capable of describing the transient (time dependent) kinetic behavior of the reaction. However, as with any numerical solution we should anticipate that stability problems may arise and, if we are only interested in steady state situations (i.e., time independent), the complete solution is not the route to pursue.

2.10.4
The Steady State Approximation

In industry, as well as in a test reactor in the laboratory we are most often interested in the situation where a constant flow of reactants enters the reactor, leading to a constant output of products. In this case, all transient behavior due to start up phenomena have died out and coverages and rates have reached a constant value. Hence, we can apply the steady state approximation, and set all differentials in Eqs. (2.142)–(2.145) equal to zero:

$$\frac{d\theta_A}{dt} = r_1 - r_3 = 0 \quad \Rightarrow \quad k_1^+ p_A \theta_* - k_1^- \theta_A - k_3^+ \theta_A \theta_B + k_3^- \theta_{AB} \theta_* = 0 \quad (2.146)$$

$$\frac{d\theta_B}{dt} = r_2 - r_3 = 0 \quad \Rightarrow \quad k_2^+ p_B \theta_* - k_2^- \theta_B - k_3^+ \theta_A \theta_B + k_3^- \theta_{AB} \theta_* = 0 \quad (2.147)$$

$$\frac{d\theta_{AB}}{dt} = r_3 - r_4 = 0$$
$$\Rightarrow \quad k_3^+ \theta_A \theta_B - k_3^- \theta_{AB} \theta_* - k_4^+ \theta_{AB} + k_4^- p_{AB} \theta_* = 0 \qquad (2.148)$$

$$\frac{d\theta_*}{dt} = -r_1 - r_2 + r_3 + r_4 = 0$$
$$\Rightarrow \quad -k_1^+ p_A \theta_* + k_1^- \theta_A - k_2^+ p_B \theta_* + k_2^- \theta_B + k_3^+ \theta_A \theta_B - k_3^- \theta_{AB} \theta_*$$
$$+ k_4^+ \theta_{AB} - k_4^- p_{AB} \theta_* = 0 \qquad (2.149)$$

The last equation is not independent of the others due to the site balance of Eq. (2.141); thus, in general, we have $n - 1$ equations for a reaction containing n elementary steps. Note that 'steady state' does not imply that surface concentrations are low. They just do not change with time. Hence, in the steady state approximation we cannot describe time-dependent phenomena, but the approximation is sufficient to describe many important catalytic processes.

2.10.5
The Quasi-Equilibrium Approximation

In this approximation, we assume that one elementary step determines the rate, while all other steps are sufficiently fast that they can be considered as being in quasi-equilibrium. If we take the surface reaction to AB∗ (step 3, Eq. (2.134)) as the rate-determining step (RDS), we may write the rate equations for steps (1), (2), and (4) as

$$r_1 \cong 0 \quad \Rightarrow \quad k_1^+ p_A \theta_* = k_1^- \theta_A \quad \Rightarrow \quad \theta_A = K_1 p_A \theta_* \tag{2.150}$$

$$r_2 \cong 0 \quad \Rightarrow \quad k_2^+ p_B \theta_* = k_2^- \theta_B \quad \Rightarrow \quad \theta_B = K_2 p_B \theta_* \tag{2.151}$$

$$r_4 \cong 0 \quad \Rightarrow \quad k_4^+ \theta_{AB} = k_4^- p_{AB} \theta_* \quad \Rightarrow \quad \theta_{AB} = K_4^{-1} p_{AB} \theta_* \tag{2.152}$$

In essence, we have used the Langmuir isotherms for the adsorbing and desorbing species. By substituting the coverages into the rate expression for the rate-determining step, we obtain

$$r = r_3 = k_3^+ \theta_A \theta_B - k_3^- \theta_{AB} \theta_*$$
$$= k_3^+ K_1 K_2 p_A p_B \left(1 - \frac{p_{AB}}{p_A p_B K_1 K_2 K_3 K_4}\right) \theta_*^2 \tag{2.153}$$

If we introduce the equilibrium constant for the overall reaction in the gas phase,

$$K_G = K_1 K_2 K_3 K_4 \tag{2.154}$$

and assume that the overall rate equals that of the rate-determining step, we obtain

$$r = r^+ - r^- = k_3^+ K_1 K_2 p_A p_B \left(1 - \frac{p_{AB}}{p_A p_B K_G}\right) \theta_*^2 \tag{2.155}$$

The term in parenthesis, as we saw in Eq. (2.8), expresses the affinity of the reaction towards equilibrium. The striking contrast with Eq. (2.8), however, is the term θ_*^2, which describes the fraction of free sites available for reaction. Thus, even if the rate constants and the affinity towards equilibrium are high, the rate of the process may still be low if there are insufficient free sites, if the surface is blocked such that $\theta_* \to 0$.

The fraction of free sites in Eq. (2.155) is found from the principle of conservation of sites:

$$\sum_{i=1}^{n} \theta_i = 1 \tag{2.156}$$

where n is the number of different species on the surface, including the empty sites. Since we have expressions for each surface species expressed in terms of free sites (Eqs. (2.153)–(2.155)), we can solve Eq. (2.156) for the fraction of free sites:

$$\theta_* = \frac{1}{1 + K_1 p_A + K_2 p_B + K_4^{-1} p_{AB}} \tag{2.157}$$

By utilizing this, we can now express the coverages of all the relevant intermediates and the overall rate in terms of equilibrium constants of the steps in quasi-equilibrium, the pressures of the reactants and the products, and the rate constant of the rate-determining step.

$$\theta_A = \frac{K_1 p_A}{1 + K_1 p_A + K_2 p_B + K_4^{-1} p_{AB}} \tag{2.158}$$

$$\theta_B = \frac{K_2 p_B}{1 + K_1 p_A + K_2 p_B + K_4^{-1} p_{AB}} \tag{2.159}$$

$$\theta_{AB} = \frac{K_4^{-1} p_{AB}}{1 + K_1 p_A + K_2 p_B + K_4^{-1} p_{AB}} \tag{2.160}$$

$$r = r^+ - r^-$$
$$= k_3^+ K_1 K_2 p_A p_B \left(1 - \frac{p_{AB}}{K_G p_A p_B}\right) \left(\frac{1}{1 + K_1 p_A + K_2 p_B + K_4^{-1} p_{AB}}\right)^2 \tag{2.161}$$

It is important to realize that the assumption of a rate-determining step limits the scope of our description. As with the steady state approximation, it is not possible to describe transients in the quasi-equilibrium model. In addition, the rate-determining step does not have to be valid over a wider range of conditions. For example, the rate-determining step in the mechanism might shift to a different step if the reaction conditions change, for example, if the partial pressure of a gas changes markedly. For a surface science study of the reaction A + B in an ultrahigh vacuum chamber with a single crystal as the catalyst, the partial pressures of A and B may be so small that the rates of adsorption become smaller than the rate of the surface reaction.

2.10.6
Steps with Similar Rates

In cases where more than one step has a small rate, we will have to consider the rate for both of these steps. Suppose, for example, that steps (1) and (3) in the scheme of Eqs. (2.132)–(2.135) possess slow rates, whereas (2) and (4) may be considered at quasi-equilibrium, we would have the following set of equations:

$$r_1 = k_1^+ p_A \theta_* - k_1^- \theta_A \tag{2.162}$$

$$r_2 \cong 0 \quad \Rightarrow \quad \frac{d\theta_B}{dt} = 0 \quad \Rightarrow \quad k_2^+ p_B \theta_* = k_2^- \theta_B$$

$$\Rightarrow \quad \theta_B = K_2 p_B \theta_* \tag{2.163}$$

$$r_3 = k_3^+ \theta_A \theta_B - k_3^- \theta_{AB} \theta_* \tag{2.164}$$

$$r_4 \cong 0 \quad \Rightarrow \quad \frac{d\theta_{AB}}{dt} = 0 \quad \Rightarrow \quad k_4^+ \theta_{AB} = k_4^- p_{AB} \theta_*$$

$$\Rightarrow \quad \theta_{AB} = K_4^{-1} p_{AB} \theta_* \tag{2.165}$$

The solution of the resultant set of differential equations is more complex than the situation involving one rate-determining step, but still simpler than the full solution.

2.10.7
Irreversible Step Approximation

This is a further simplification of the quasi-equilibrium approximation, in which we simply neglect the reverse reaction of one or several steps. For instance, we may envisage a situation where the product concentration AB is kept so low that the reverse reaction in step (4) may be neglected. This greatly simplifies Eq. (2.161) since $p_{AB} = 0$:

$$r = r_3 = r^+ = \frac{k_3^+ K_1 K_2 p_A p_B}{(1 + K_1 p_A + K_2 p_B)^2} \tag{2.166}$$

It is important to keep in mind that, in general, the model cannot describe the approach towards equilibrium, since this would violate our assumption that the product concentration is negligible. We note that Eq. (2.166) would also describe the case in which the adsorption–desorption equilibrium lies on the desorption side, that is, if the temperature is such that the molecule AB adsorbs hardly on the surface.

2.10.8
The MARI Approximation

The Most Abundant Reaction Intermediate (MARI) approximation is a further development of the quasi-equilibrium approximation. Often one of the intermediates adsorbs so strongly in comparison to the other participants that it completely dominates the surface. This intermediate is called the MARI [1]. In this case Eq. (2.156) reduces to

$$\sum_{i=1}^{n} \theta_i \approx \theta_* + \theta_{MARI} = 1 \tag{2.167}$$

If molecule A in our reaction scheme (Eqs. (2.132)–(2.135)) binds more strongly to the surface than B or AB, and the pressures of B and AB are comparable or lower than that of A, then A will be the MARI and the expressions for the coverages reduce to:

$$\theta_A \cong \frac{K_1 p_A}{1 + K_1 p_A}, \quad \theta_B \cong 0, \quad \theta_{AB} \cong 0, \quad \theta_* \cong 1 - \theta_A \quad (2.168)$$

and the rate becomes

$$r \cong r^+ - r^- \cong k_3^+ K_1 K_2 p_A p_B \left(1 - \frac{p_{AB}}{K_G p_A p_B}\right)\left(\frac{1}{1 + K_1 p_A}\right)^2 \quad (2.169)$$

2.10.9
The Nearly Empty Surface

In situations where the intermediates are bound very weakly or the temperature is high enough for the equilibrium to be shifted sufficiently towards the gas phase, the surface is mostly empty, and we may use the approximation:

$$\theta_* \cong 1 \quad (2.170)$$

The rate expression, for example, Eq. (2.169), simplifies further since we can neglect the last term. If we also assume that the rate-limiting step (RLS) is irreversible or that the product concentration is low, we only have to consider the forward reaction and the rate reduces to

$$r = r^+ \cong k_3^+ K_1 K_2 p_A p_B \quad (2.171)$$

To determine the composition of the reaction mixture that corresponds to the optimum rate, it is convenient to define a relative concentration, χ_A, as

$$\chi_A = \frac{p_A}{p_A + p_B} = \frac{p_A}{p_{tot}} \quad (2.172)$$

Then

$$p_A = \chi_A p_{tot} \quad \text{and} \quad p_B = (1 - \chi_A) p_{tot} \quad (2.173)$$

The extreme of the rate is readily found by differentiation

$$\frac{dr^+}{d\chi_A} = 0 \quad \Rightarrow \quad k_3^+ K_1 K_2 (1 - 2\chi_A) = 0 \quad \Rightarrow \quad \chi_A = \frac{1}{2} \quad (2.174)$$

Thus, for an almost empty surface, the rate assumes its maximum with equal amounts of reactants, at the limit of zero conversion. Again, we need to assess the validity of the approximations under the conditions employed. Nevertheless, the above procedure for determining the reaction rate as a function of mole fraction can be quite useful in the exploration of reaction mechanisms.

2.10.10
The Reaction Order

The orders of reaction, n_i, with respect to A, B, and AB are obtained from the rate expression by differentiation as in Eq. (2.11). In the rare case that we have a complete numerical solution of the kinetics, as explained in Section 2.10.3, we can find the reaction orders numerically. Here, we assume that the quasi-equilibrium approximation is valid, which enables us to derive an analytical expression for the rate as in Eq. (2.161) and to calculate the reaction orders as

$$n_A \equiv p_A \frac{\partial \ln(r^+)}{\partial p_A}$$

$$= p_A \frac{\partial}{\partial p_A} \left(\ln\left(k_3^+ K_1 K_2 p_B\right) + \ln(p_A) \right.$$

$$\left. - 2\ln\left(1 + K_1 p_A + K_2 p_B + K_4^{-1} p_{AB}\right) \right)$$

$$= 0 + 1 - \frac{2 K_1 p_A}{1 + K_1 p_A + K_2 p_B + K_4^{-1} p_{AB}}$$

$$= 1 - 2\,\theta_A \tag{2.175}$$

$$n_B = 1 - 2\,\theta_B \tag{2.176}$$

$$n_{AB} = -2\,\theta_{AB} \tag{2.177}$$

We will use Eq. (2.161) to illustrate the dependence of the rate on the partial pressure of the reactants and on the temperature. In addition, one further simplifying assumption is made, namely that the product AB desorbs very rapidly at the reaction temperature and that its coverage will thus be negligible with respect to that of the reactants. Hence, we can ignore the term $K_{AB} p_{AB}$ in the denominator of Eq. (2.161). The order of the reaction with respect to AB is zero under these conditions. The result is shown in Figure 2.11 where the coverages of A and B as well as the reaction order n_A and n_B have been plotted as a function of mole fraction of reactant A. Note that the reaction orders are not constant but vary strongly with partial pressure. An example of the corresponding plot of the dependence of the rate on temperature will be made in a later example.

2.10.11
The Apparent Activation Energy

By determining the formula for the apparent activation energy, and recalling the thermodynamic relations (Eq. (2.48)) for the equilibrium constant, we obtain a relation between the apparent activation energy and the coverages:

$$E_a^{app} \equiv RT^2 \frac{\partial \ln(r^+)}{\partial T} \tag{2.178}$$

2 Kinetics

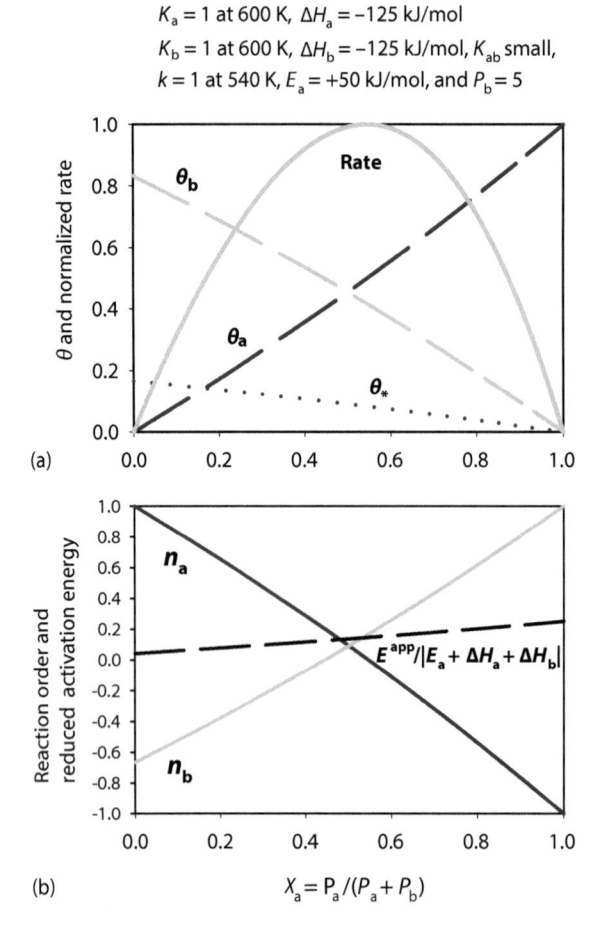

$K_a = 1$ at 600 K, $\Delta H_a = -125$ kJ/mol
$K_b = 1$ at 600 K, $\Delta H_b = -125$ kJ/mol, K_{ab} small,
$k = 1$ at 540 K, $E_a = +50$ kJ/mol, and $P_b = 5$

Figure 2.11 Coverages of A and B and the rate as function of mole fraction of A (a). The orders of reaction are shown together with the apparent activation energy (b).

$$= RT^2 \frac{\partial}{\partial T} \Big(\ln(p_A p_B) + \ln\left(k_3^+\right) + \ln(K_1) + \ln(K_2) \\ - 2\ln(1 + K_1 p_A + K_2 p_B + K_4^{-1} p_{AB}) \Big) \tag{2.179}$$

$$= RT^2 \left(0 + \frac{\partial\left(\frac{-E_a}{RT}\right)}{\partial T} + \frac{\partial\left(\frac{-\Delta H_A}{RT}\right)}{\partial T} + \frac{\partial\left(\frac{-\Delta H_B}{RT}\right)}{\partial T} \right. \\ \left. - \frac{2\left(\frac{\partial(1+K_1 p_A + K_2 p_B + K_4^{-1} p_{AB})}{\partial T}\right)}{1 + K_1 p_A + K_2 p_B + K_4^{-1} p_{AB}} \right) \tag{2.180}$$

$$= E_{\mathrm{a}} + \Delta H_{\mathrm{A}} + \Delta H_{\mathrm{B}} - 2(\theta_{\mathrm{A}}\Delta H_{\mathrm{A}} + \theta_{\mathrm{B}}\Delta H_{\mathrm{B}} - \theta_{\mathrm{AB}}\Delta H_{\mathrm{AB}})$$
$$= E_{\mathrm{a}} + (1 - 2\theta_{\mathrm{A}})\Delta H_{\mathrm{A}} + (1 - 2\theta_{\mathrm{B}})\Delta H_{\mathrm{B}} + 2\theta_{\mathrm{AB}}\Delta H_{\mathrm{AB}} \quad (2.181)$$

where we have used the Arrhenius expression form for the rate constant:

$$k_3^+ = \nu_3^+ \exp\left(\frac{-E_{\mathrm{a}}}{RT}\right) \quad (2.182)$$

along with the well-known expressions for the equilibrium constants:

$$K_x = \exp\left(\frac{-\Delta G_x}{RT}\right) = \exp\left(\frac{\Delta S_x}{R}\right)\exp\left(\frac{-\Delta H_x}{RT}\right) \quad (2.183)$$

We again assume that the pre-exponential factor and the entropy contributions do not depend on temperature. This assumption is not strictly correct but, as we shall see in Chapter 3, the latter dependence is much weaker than that of the energy in the exponential terms. The normalized activation energy is also shown in Figure 2.11 as a function of mole fraction. Notice that the activation energy is not just that of the rate-limiting step. It also depends on the adsorption enthalpies of the steps prior to the rate-limiting step and the coverages.

Hence, the apparent activation energy found from an Arrhenius plot depends strongly on the conditions under which it was determined (pressure and temperature). The same is true for the order of reaction in the participating species.

Example CO Oxidation

CO oxidation, an important step in automotive exhaust catalysis, is a relatively simple reaction, and it has been the subject of numerous fundamental studies [12, 13]. The reaction is catalyzed by noble metals such as platinum, palladium, rhodium, iridium, and even by gold, provided the gold particles are very small. We will assume that the oxidation of CO on such catalysts proceeds through a mechanism in which adsorbed CO, O, and CO_2 are equilibrated with the gas phase, that is, we can use the quasi-equilibrium approximation.

The reaction mechanism is then

1) $CO + * \underset{}{\overset{K_1}{\rightleftharpoons}} CO*$ (2.184)

2) $O_2 + 2* \underset{}{\overset{K_2}{\rightleftharpoons}} 2O*$ (2.185)

3) $CO* + O* \underset{k_3^-}{\overset{k_3^+}{\rightleftharpoons}} CO_2* + *$ (RDS) (2.186)

4) $CO_2* \underset{}{\overset{K_4^{-1}}{\rightleftharpoons}} CO_2 + *$ (2.187)

Step (3), the recombination of adsorbed oxygen and adsorbed CO to give adsorbed CO_2, is assumed to be the rate-determining step (RDS). We could describe step (2) in more detail, as the adsorption of O_2 proceeds via a molecular precursor, O_2*, which subsequently dissociates. However, for simplicity, we neglect this elementary step, leaving it to the reader to determine the consequences of taking it into account.

For each step in quasi-equilibrium, we can either start from the differential equations as before or immediately use the Langmuir isotherm:

$$\theta_{CO} = K_1 p_{CO} \theta_* \tag{2.188}$$

$$\theta_O = \sqrt{K_2 p_{O_2}} \theta_* \tag{2.189}$$

$$\theta_{CO_2} = K_4^{-1} p_{CO_2} \theta_* \tag{2.190}$$

It is now straightforward to find an expression for the fraction of free sites from the site balance:

$$\theta_{CO} + \theta_O + \theta_{CO_2} + \theta_* = 1 \Rightarrow$$

$$\theta_* = \frac{1}{1 + K_1 p_{CO} + \sqrt{K_2 p_{O_2}} + K_4^{-1} p_{CO_2}} \tag{2.191}$$

The rate is that of the rate-determining step:

$$r = k_3^+ \theta_{CO} \theta_O - k_3^- \theta_{CO_2} \theta_*$$

$$= k_3^+ K_1 \sqrt{K_2} p_{CO} \sqrt{p_{O_2}} \left(1 - \frac{p_{CO_2}}{p_{CO} \sqrt{p_{O_2}} K_G}\right) \theta_*^2 \tag{2.192}$$

where

$$K_G = K_1 \sqrt{K_2} K_3 K_4 \tag{2.193}$$

is the equilibrium constant for the overall reaction:

$$CO + \frac{1}{2} O_2 \overset{K_G}{\rightleftharpoons} CO_2 \tag{2.194}$$

Let us consider the limiting cases for this reaction. Usually, CO_2 interacts so weakly with the surface that its presence can be neglected, that is, the desorption of CO_2 is fast and step (4) of (2.187) can be considered irreversible. Hence, terms containing p_{CO_2} are zero.

At low temperatures, the surface will be dominated by adsorbed CO, such that CO is the MARI, implying that the rate can be written as

$$r = \frac{k_3^+ \sqrt{K_2 p_{O_2}}}{K_1 p_{CO}} \tag{2.195}$$

We see immediately that the reaction orders are $n_{O_2} = 0.5$ and $n_{CO} = -1$ in the low temperature limit. The negative order in CO demonstrates that the surface is completely covered by CO. Any further increase in CO pressure will reduce the rate because free sites are blocked, and consequently oxygen cannot adsorb and react.

At high temperatures desorption prevails, implying that the coverages of all species are small and that the surface is nearly empty. This does not mean that the reaction can not take place, but the residence time of any species on the surface before it desorbs or reacts is short. Since the surface is nearly empty, we can set $\theta_* \cong 1$ and obtain

$$r = k_3^+ K_1 \sqrt{K_2} p_{CO} \sqrt{p_{O_2}} \tag{2.196}$$

Note that the reaction order remains 0.5 in oxygen, but becomes +1 in CO. Because the surface is predominantly empty, increasing the partial pressures of both reactants leads to an increase in rate. Thus, the reaction order is strongly dependent not only of the pressure. as discussed in the previous section, but also the temperature.

Figure 2.12 shows the rate, the coverages, the reaction orders, and the normalized apparent activation energy, all as a function of temperature. Note the strong variations of all these parameters with temperature, in particular that of the rate, which initially increases, then maximizes and decreases again at high temperatures. This characteristic behavior is expected for all catalytic reactions, but is in practice difficult to observe with supported catalysts because diffusion phenomena come into play.

The apparent activation energy is positive for low temperatures; hence, the rate increases with increasing temperature. However, at high temperature the apparent activation energy becomes negative, as predicted by Eq. (2.181); thus, the rate decreases with temperature. The reason for this is clearly illustrated by Figure 2.12: there is a lack of adsorbed species at high temperature. The rate is seen to have a maximum when the temperature is high enough that step (3) can proceed at a reasonable rate, while at the same time there are sufficient adsorbed reactants with a favorable distribution among CO and O to react. Hence, one could say that catalytic engineering is all about having the right amount of reacting species on the catalyst at the right temperature.

2.11
Entropy, Entropy Production, Auto Catalysis, and Oscillating Reactions

As we all know from thermodynamics, closed systems in equilibrium have minimum free energy and maximum entropy. If such a system were brought out of equilibrium, that is, in a state with lower entropy and higher free energy, it would automatically decay to the state of equilibrium, and it would lose all information about its previous states. A system's tendency to return to equilibrium is given by its free energy. An example forms the batch reaction that is run to completion.

2 Kinetics

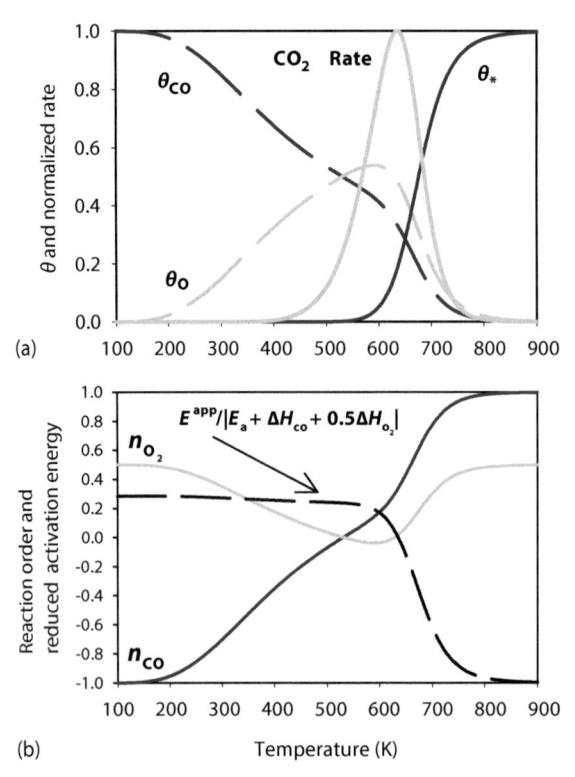

CO oxidation, $K_{CO} = 1$ at 650 K, $\Delta H_{CO} = -135$ kJ/mol
$K_{O_2} = 1$ at 630 K, $\Delta H_{O_2} = -250$ kJ/mol, K_{CO_2} small,
$k = 1$ at 540 K, $E_a = 50$ kJ/mol, $P_{O_2} = 10\, P_{CO}$, and $P_{CO} = 1$

Figure 2.12 Coverage of CO and O and the rate for CO oxidation as a function of temperature (a). Both the reaction order and the overall activation energy both as function of temperature are shown (b). Any influence of CO_2 has been ignored.

A reaction at steady state is not in equilibrium. Nor is it a closed system, as it is continuously fed by fresh reactants, which keep the entropy lower than it would be for the system in equilibrium. In this case the deviation from equilibrium is described by the rate of entropy increase, dS/dt, also referred to as entropy production. It can be shown that a reaction at steady state possesses a minimum rate of entropy production and, when perturbed, it will return to this state, which is dictated by the rate at which reactants are fed to the system [3]. Hence, steady states settle for the smallest deviation from equilibrium possible under the given conditions. Steady state reactions in industry satisfy these conditions and are operated in a regime where linear nonequilibrium thermodynamics holds. Nonlinear nonequilibrium thermodynamics, however, represents a regime where explosions and uncontrolled oscillations may arise. Obviously, industry wants to avoid such situations at all times!

Steady states may also arise under conditions that are far outside equilibrium [14]. If the deviation becomes larger than a critical value, and the system is fed by a steady inflow that keeps the free energy high (and the entropy low), it may become unstable and start to oscillate, or switch chaotically and unpredictably between steady state levels.

Explosions can also be seen as the result of an autocatalytic process:

Explosive + heat $\longrightarrow$ *Product + more heat*

The heat accelerates the exothermic reaction, which produces even more heat. The process goes on until all the explosive has been consumed. Explosions can also occur on surfaces.

Intuitively, chaotic behavior is often associated with disorder, but this is not correct. An example from physics may clarify the situation: consider a laminar flow of a fluid through a pipeline (which in a way is the analogue of a reaction at steady state). If the pressure difference over the pipe exceeds a critical value (associated with the Reynolds number), the flow may become turbulent (the analogue of a chaotic reaction). Note that in such turbulent flow the degree of order is relatively high, as large numbers of molecules move through circular patterns of macroscopic dimensions. Such a self-organizing system has a much lower entropy than in laminar flow, or in equilibrium when the flow stops. For an interesting discussion of nonlinear nonequilibrium phenomena, we refer to books by Prigogine [14] and by Gleick [15].

For chaotic or oscillating behavior, it is necessary that the mechanism contains an autocatalytic step:

$$A + X \xrightarrow{k_1} 2X \tag{2.197}$$

As molecule X catalyzes its own production, the reaction is called autocatalytic. If we couple this reaction to two additional steps

$$X + Y \xrightarrow{k_2} 2Y$$
$$Y \xrightarrow{k_3} E \tag{2.198}$$

we arrive at an oscillating reaction that has actually been used to model ecological systems, but is also of interest in chemistry.

To visualize the situation, one may take Y as a population of foxes, E as the foxes that have died, X as rabbits, and A as carrots. Rabbits live very well on carrots and the population would grow exponentially if there are sufficient carrots. The fox population starts to grow when the rabbit population is high, until there are more foxes than can be sustained by the rabbits. Famine sets in and the fox population diminishes, after which the rabbit population starts to grow again. In other words, X and Y are oscillating out of phase. It is essential that the mechanism contains

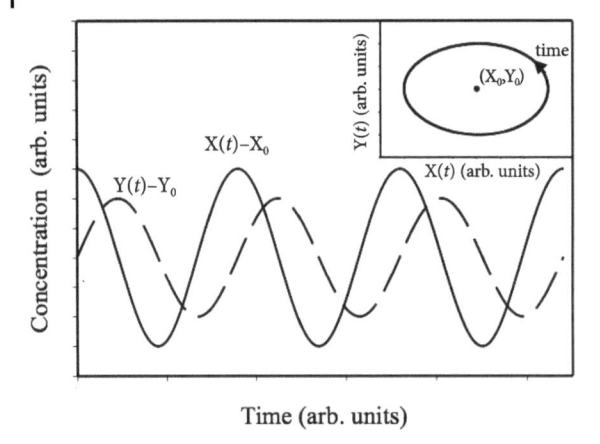

Figure 2.13 Concentrations of X and Y of the system (Eqs. (2.197) and (2.198)) are oscillating out of phase around their steady solution. The type of representation in the insert is sometimes referred to as a descriptor diagram.

autocatalytic steps, and that there is a continuous supply of reactant A, which keeps the system far away from equilibrium.

The system (Eqs. (2.197) and (2.198)) could well be a balanced ecosystem with steady state concentrations $[X_0] = k_3/k_2$ and $[Y_0] = k_1[A]/k_2$, as the reader may easily verify. However, one can also show (albeit with a little more computational effort) that far away from equilibrium solutions of the type

$$[X] = [X_0] + xe^{i\omega t} \; ; \quad \omega = \pm\sqrt{k_1 k_3 [A]} \; ; \quad i = \sqrt{-1} \qquad (2.199)$$

and similar for [Y] satisfy the rate equations as well. As $e^{i\omega t} = \cos \omega t + i \sin \omega t$, one sees that the solution oscillates around the steady state value, as indicated in Figure 2.13. For more details we refer the reader to Van Santen and Niemantsverdriet [3].

The best known and most investigated oscillating system in chemistry is probably the Belousov–Zhabotinsky reaction, discovered in Russia in 1958 and reported to the western world in 1970 [16]. Many biological processes (biological clocks, biorhythms, heartbeat) rely on oscillating reactions, all of which are critically dependent on a source of energy that enables the system to remain in its highly organized, out-of-equilibrium state.

Oscillations may also occur on catalyst surfaces. As an example, we briefly discuss the impressive work on the oxidation of CO on platinum surfaces by Ertl and coworkers [17, 18]. Here, the oscillation goes hand in hand with a structural reorganization of the surface (see also Chapter 5). Figure 2.14 shows the 'normal' (110) surface as well as the reconstructed surface, indicated by (1 × 2) as the unit cell of this structure is twice as large in one direction. Oscillations occur because O_2 is hardly dissociated by the reconstructed surface; the surface switches between the two structures at a critical coverage of CO. Referring to the reaction mechanisms summarized in Figure 2.14, when the surface is in the (1 × 1), that is, unrecon-

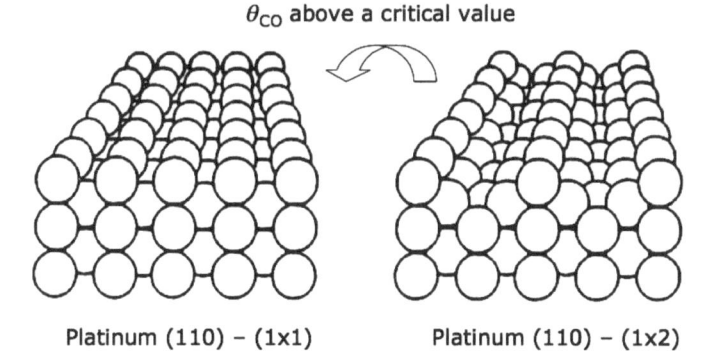

2.11 Entropy, Entropy Production, Auto Catalysis, and Oscillating Reactions

θ_{CO} above a critical value

Platinum (110) – (1×1)   Platinum (110) – (1×2)

$O_2 + 2*_{110}$ → $2\,O*_{110}$ (θ_O increases)	$O_2 + 2\#_{1\times 2}$ → $2\,O\#_{1\times 2}$ ($\theta_O \sim 0$)
$CO + *_{110}$ → $CO*_{110}$	$CO + \#_{1\times 2}$ → $CO\#_{1\times 2}$ (θ_{CO} increases)
$CO*_{110} + O*_{110}$ → $CO_2 + 2*_{110}$ (high rate, θ_{CO} decreases)	$CO\#_{1\times 2} + O\#_{1\times 2}$ → $CO_2 + 2\#_{1\times 2}$ (rate ~ 0)

θ_{CO} under a critical value

Figure 2.14 Reaction mechanisms for the CO oxidation on the (110) surface of platinum (left) and on the reconstructed surface (right). See text for explanation.

structed state, oxygen atoms are amply available, and they react with the CO to give CO_2.

However, as the CO coverage falls below a critical value, the surface switches to the less reactive (1 × 2) state, to which CO readily adsorbs, and reacts to remove the O atoms, which are *not* replenished on this surface, and the rate drops while the CO coverage builds up. As soon as the critical CO coverage is exceeded, the surface switches back to the original (1 × 1) state, and the cycle starts anew. The state of the surface can be monitored by low energy electron diffraction (explained in Chapter 4). Figure 2.15 shows that the CO_2 formation rate indeed changes in synchrony with the intensities of the diffraction features from the two surface structures. However instead of the entire surface switching from one state to the other, zones of alike structure move over the surface, setting up spectacular spatiotemporal patterns. Such patterns have not only been simulated but also imaged by advanced electron microscopes (Figure 2.16).

Oscillations such as in Figure 2.15 are quite regular and can be sustained for hours if the conditions are kept the same. Depending on the feed rate of reactants, which determines how far the system deviates from equilibrium, the oscillations may become more complex and develop into chaotic oscillations [19].

How relevant are these phenomena? First, many oscillating reactions exist and play an important role in living matter. Biochemical oscillations and also the inorganic oscillatory Belousov–Zhabotinsky system are very complex reaction net-

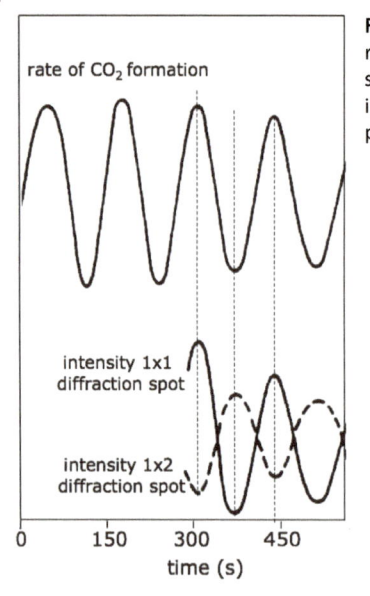

Figure 2.15 The rate of CO_2 formation in the $CO + O_2$ reaction of Pt(1110) oscillates synchronously with the surface reconstruction, from (1 × 1) to (1 × 2), shown in Figure 2.14 (reprinted from Eiswirth et al. [20] with permission of AIP Publishing).

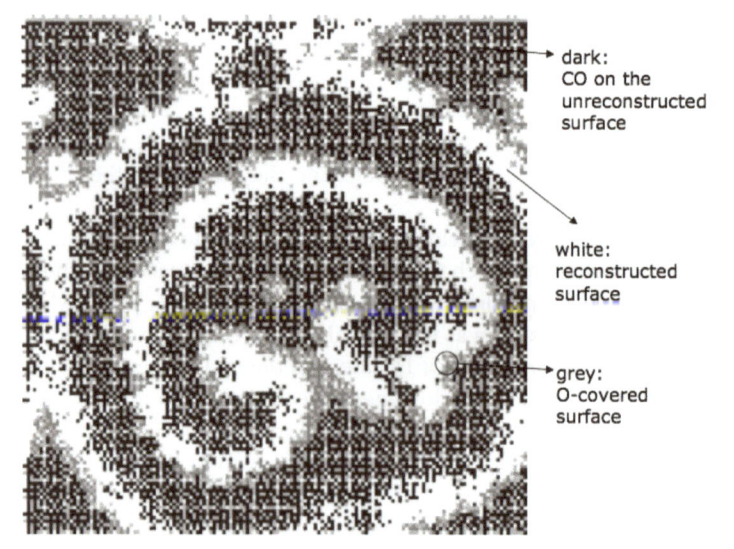

Figure 2.16 Computer simulation of spatiotemporal pattern formation in CO oxidation on a surface (adapted from Gelten et al. [21]).

works. Oscillating surface reactions though are much simpler and so offer convenient model systems to investigate the realm of nonequilibrium reactions on a fundamental level. Second, as mentioned above, the conditions under which nonlinear effects such as those caused by autocatalytic steps lead to uncontrollable situations, which should be avoided in practice. Hence, some knowledge about the subject is desired. Finally, the application of forced oscillations in some reactions may lead to better performance in favorable situations, for example, when a catalytic system alternates between conditions where the catalyst deactivates due to carbon deposition and conditions where this deposit is reacted away.

2.12
Kinetics of Enzyme-Catalyzed Reactions

Enzymes are highly specific catalysts in biological systems [22, 23]. They are proteins that consist of many amino acids coupled to each other by peptide bonds. The rather small enzyme insulin, for example, consists of 51 amino acids. The chain of amino acids folds into a defined 3D conformation, which is known in detail for many enzymes. Somewhere in this body is a functional group, for example, a carboxyl or an amine, which acts as the active site.

Amino acids contain two active groups, namely a carboxylic (–COOH) and an amino (–NH$_2$) group.

$$\begin{array}{c} H \quad O \\ | \quad \| \\ R-C-C-OH \\ | \\ NH_2 \end{array}$$

Twenty different amino acids are known. They combine to give proteins by forming an amide or peptide bond between the carbon from COOH and nitrogen from NH$_2$.

$$\begin{array}{cc} \begin{array}{c} H \quad O \\ | \quad \| \\ R-C-C-OH \\ | \\ NH_2 \end{array} & \begin{array}{c} H \quad O \\ | \quad \| \\ H_2N-C-C-OH \\ | \\ R' \end{array} \end{array}$$

$$\Downarrow$$

$$\begin{array}{c} H \quad O \quad\quad H \quad O \\ | \quad \| \quad\quad | \quad \| \\ R-C-C-N-C-C-OH \\ | \quad\quad\quad | \quad | \\ NH_2 \quad\quad H \quad R' \end{array}$$

Enzymes are catalytic proteins. Their active site can, for example, be a carboxylic or an amino group, embedded in a specific geometry. Several weak interactions (electrostatic, H-bonds, van der Waals) help in establishing the highly specific manner in which a substrate molecule binds to the active site, making enzymes the most efficient class of catalysts.

Enzymes can perform a multitude of reactions, though every enzyme usually only catalyzes very specifically the reaction of a single substrate. They are named after the reaction they catalyze, or the substrate with which they react, by adding the suffix -ase. Hence, there are oxidases, reductases, dehydrogenases, and hydrolases. The enzyme that catalyzes the decomposition of urea is called urease, and so on.

The kinetics of enzyme-catalyzed reactions resembles those of the heterogeneous reactions discussed in the previous sections. However, because in practice there are a few characteristic differences in how the equations are handled, we treat the enzymatic case separately.

The enzyme, E, acts by forming a complex with the reactant, S (commonly referred to as substrate in the language of this field), to a product, P, according to the following scheme:

$$S + E \underset{k_1^-}{\overset{k_1^+}{\rightleftharpoons}} ES \tag{2.200}$$

$$ES \overset{k_2}{\rightarrow} P + E \tag{2.201}$$

Because enzyme, substrate, and product are all in the same medium, we can conveniently work with concentrations. With the total enzyme concentration equal to $[E]_{tot}$, conservation of active species requires that

$$[E] + [ES] = [E]_{tot} \tag{2.202}$$

The rate of product formation follows straightforwardly from the reaction equation:

$$\frac{d[P]}{dt} = k_2[ES] \tag{2.203}$$

Unlike the case of a surface reaction, there is no need to specify the units of measurement because all reacting species are in the three-dimensional space of the reaction medium, and all relevant rates come in concentration per unit of time. The unknown concentration of the unoccupied enzymes follows by assuming that the reaction is at steady state:

$$\frac{d[ES]}{dt} = k_1^+[E][S] - (k_1^- + k_2)[ES] = 0 \tag{2.204}$$

which leads with Eq. (2.202) to

$$k_1^+[E]_{tot}[S] - k_1^+[ES][S] - (k_1^- + k_2)[ES] = 0 \tag{2.205}$$

or

$$[ES] = \frac{k_1^+[E]_{tot}[S]}{(k_1^- + k_2) + k_1^+[S]} \tag{2.206}$$

Substituting into Eq. (2.203) and introducing the Michaelis constant (K_M, Eq. (2.112)), we obtain

$$r = \frac{d[P]}{dt} = \frac{k_2[E]_{tot}[S]}{K_M + [S]} \quad \text{with} \quad K_M = \frac{k_1^b + k_2}{k_1^f} \tag{2.207}$$

This is the Michaelis–Menten expression (which dates back to 1913) for the rate of an enzymatic reaction.

In comparison with the case of a gas phase molecule that reacts in a monomolecular reaction on a solid catalyst, the reciprocal of the Michaelis constant takes the place of the equilibrium constant of adsorption in the Langmuir–Hinshelwood equations.

If the substrate concentration is very high, the rate reaches its maximum:

$$r_{max} = k_2 [E]_{tot} \quad \text{for} \quad [S] \gg K_M \tag{2.208}$$

and the enzyme is used very efficiently. Because k_2 equals $r_{max}/[E]_{tot}$, it is often referred to as the turnover frequency and, hence, it is also often referred to as k_{cat}. If, on the other hand, the substrate concentration is much smaller than K_M, the rate is given by

$$r = \frac{k_2}{K_M} [E]_{tot} [S] \quad \text{for} \quad [S] \ll K_M \tag{2.209}$$

and most of the enzyme is free. The ratio k_2/K_M, called specificity constant, is a suitable quantity for comparing the enzyme's specificity for different substrates. When an enzyme can convert two molecules, A and B, the rates of the two conversions relate as $(k_2/K_M)_A : (k_2/K_M)_B$. There is a practical upper limit to the value of k_2/K_M of about 10^9 mol L^{-1} s^{-1}, set by the rate of diffusion of substrate molecules to the enzyme through the solution. Hence, enzymatic reaction systems approaching this upper limit come close the perfection.

As an example, we mention the enzyme catalase, which catalyzes the decomposition of H_2O_2 to H_2O and O_2 at a turnover number of $k_{cat} = 10^7$ s^{-1} and a high specificity constant of $k_{cat}/K_M = 4 \times 10^8$ mol L^{-1} s^{-1}. Such activities are orders of magnitude higher than those of heterogeneous catalysts. Another important point to consider is that of control. As Figure 2.17 shows, when the enzymes are almost saturated the rate hardly changes with the concentration of the substrate, implying that the rate of product formation cannot be controlled by [S]. Of course, control is optimally possible in the low substrate concentration regime. Hence, in cases where substrate control of the rate is important, the reaction should ideally proceed in the region of [S] between 5 and 10 K_M.

To use the Michaelis–Menten rate expression in practice, it is convenient to rearrange it to the form

$$\frac{1}{r} = \frac{1}{k_2 [E]_{tot}} + \frac{K_M}{k_2 [E]_{tot} [S]} = \frac{1}{r_{max}} + \frac{K_M}{k_2 [E]_{tot} [S]} \tag{2.210}$$

Hence, a plot of $1/r$ versus $1/[S]$ is linear with a slope $K_M/k_2 [E]_{tot}$ and an intercept $1/r_{max}$. Such a graph shown in Figure 2.18 is known as a Lineweaver–Burk plot.

As with any reaction, temperature has an important effect on the rate of an enzymatic reaction, albeit the range of interest is limited. For each enzyme an optimum temperature exists (37 °C for reactions in human beings). At high temperatures, the activity decreases due to thermal denaturation of the protein constituting the enzyme.

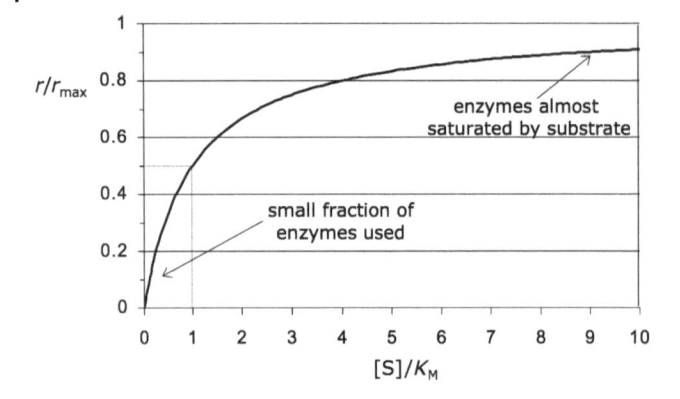

Figure 2.17 Normalized rate for an enzyme-catalyzed reaction as a function of $[S]/K_M$.

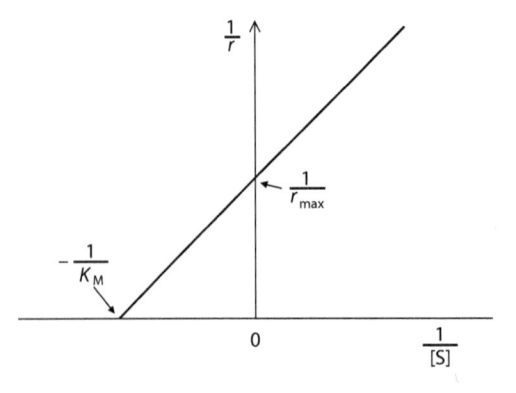

Figure 2.18 The Lineweaver–Burk plot.

The pH of the reaction medium is also an important variable as it controls the charge state of the enzyme and, thus, its conformation. Optimum values exist for every combination of enzyme and substrate.

Although we shall restrict our discussion here to the very simple unimolecular isomerization or decomposition of a single substrate, we need to mention the effect that inhibitors have on the rate.

Inhibitors are species that bind to enzymes, modifying their activity. Competitive inhibitors bind at the same site as the substrate binds; this is analogous to competitive adsorption in heterogeneous catalysis. The reaction scheme becomes

$$S + E \underset{k_1^+}{\overset{k_1^-}{\rightleftharpoons}} ES \overset{k_2^+}{\longrightarrow} P + E \tag{2.211}$$

$$I + E \underset{k_i^+}{\overset{k_i^-}{\rightleftharpoons}} EI \tag{2.212}$$

Noncompetitive inhibitors also exist. These bind to other sites on the enzyme, thereby modifying the latter and its activity.

Competitive inhibition is important in biological control mechanisms, for instance, if the product assumes the role of an inhibitor. The enzyme invertase catalyzes the hydrolysis of sucrose into glucose and fructose. As glucose is a competitive inhibitor, it ensures that the reaction does not proceed too far.

An interesting practical application of competitive inhibition is the use of ethanol to treat methanol poisoning in people (who may have wanted to try a cheaper type of alcohol). Although methanol itself is not particularly toxic, it is broken down in the liver by the enzyme alcohol dehydrogenase to formaldehyde and formic acid, two highly toxic compounds that can cause blindness and even death after the consumption of only small amounts of methanol. However, as alcohol dehydrogenase binds more strongly to ethanol than to methanol, patients who have ingested methanol are administered ethanol at a dose that saturates the enzyme. In the meantime, the methanol intoxication can be treated. The same treatment also works in cases where ethylene glycol has been ingested. Although all this represents useful knowledge, we trust that our readers will never have to rely on its practical use.

References

1 Boudart, M. and Djega-Mariadassou G. (1984) *Kinetics of Heterogeneous Catalytic Reactions*, Princeton University Press, Princeton.

2 Cortright, R.D. and Dumesic, J.A. (2001) Kinetics of heterogeneous catalytic reactions: Analysis of reaction schemes. *Advances in Catalysis*, **46**, 161–264, (eds B.C. Gates and H. Knözinger).

3 Van Santen R.A. and Niemantsverdriet J.W. (1995) *Chemical Kinetics and Catalysis*, Plenum, New York.

4 Laidler K.J. (1987) *Chemical Kinetics*, HarperCollinsPublishers, Inc., New York.

5 Chase M.W., Davies C.A., Downey J.R., Frurip D.J., McDonald R.A. and Syverud A.N. (1985) Janaf Thermochemical Tables, 3rd edn. *Journal of Physical and Chemical Reference Data*, **14**, 1–926.

6 Chase M.W., Davies C.A., Downey J.R., Frurip D.J., McDonald R.A. and Syverud A.N. (1985) Janaf Thermochemical Tables, 3rd edn. *Journal of Physical and Chemical Reference Data*, **14**, 927–1856.

7 Rod T.H. and Nørskov J.K. (2002) The surface science of enzymes. *Surface Science*, **500**, 678–698.

8 Atkins P.W. (1998) *Physical Chemistry*, Oxford University Press, Oxford.

9 Wijngaarden, R.J., Kronberg A. and Westerterp K.R. (1998) *Industrial Catalysis: Optimizing Catalysts and Processes*, Wiley-VCH Verlag GmbH, Weinheim.

10 Murzin D.Y. (2013) *Engineering Catalysis*, De Gruyter, Berlin.

11 Davis, M.E. and Davis R.J. (2003) *Fundamentals of Chemical Reaction Engineering*, McGraw-Hill, New York.

12 Engel T. and Ertl G. (1978) Molecular-beam investigation of catalytic oxidation of CO on Pd (111). *Journal of Chemical Physics*, **69**, 1267–1281.

13 Bowker M., Guo Q.M., Li Y.X. and Joyner R.W. (1993) Structure sensitivity in CO oxidation over rhodium. *Catalysis Letters*, **18**, 119–123.

14 Prigogine I. (1980) *From Being to Becoming: Time and Complexity in the Physical Sciences*, WH Freeman, San Francisco.

15 Gleick J. (1987) *Chaos, Making a New Science*, Viking Penguin, New York.

16 Zaikin A.N. and Zhabotinsky A.M. (1970) Concentration wave propagation in 2-dimensional liquid-phase self-oscillating system. *Nature*, **225**, 535–537.

17 Imbihl R. and Ertl G. (1995) Oscillatory kinetics in heterogeneous catalysis. *Chemical Reviews*, **95**, 697–733.
18 Ertl G. (2008) Reactions at surfaces: From atoms to complexity (Nobel lecture). *Angewandte Chemie – International Edition*, **47**, 3524–3535.
19 Cobden P.D., Siera J. and Nieuwenhuys B.E. (1992) Oscillatory reduction of nitric-oxide with hydrogen over Pt (100). *Journal of Vacuum Science and Technology A – Vacuum Surfaces and Films*, **10**, 2487–2494.
20 Eiswirth M., Moller P., Wetzl K., Imbihl R. and Ertl G. (1989) Mechanisms of spatial self-organization in isothermal kinetic oscillations during the catalytic co oxidation on Pt single-crystal surfaces, *Journal of Chemical Physics*, **90**, 510–521.
21 Gelten R.J., Jansen A.P.J., van Santen R.A., Lukkien J.J., Segers J.P.L. and Hilbers P.A.J. (1998) Monte Carlo simulations of a surface reaction model showing spatio-temporal pattern formations and oscillations. *Journal of Chemical Physics*, **108**, 5921–5934.
22 Buchholz K., Kasche V. and Bornscheuer U.T. (2012) *Biocatalysts and Enzyme Technology*, Wiley-Blackwell, London.
23 Bommarius A.S. and Riebel B.R. (2004) *Biocatalysis*, Wiley-VCH Verlag GmbH, Weinheim.

Chapter 3
Reaction Rate Theory

3.1
Introduction

This chapter will present theories that are capable of predicting the rate of a reaction, in particular the value of the pre-exponential factor [1–3]. In Chapter 2, we introduced the Arrhenius equation,

$$k = \nu e^{-E_a/k_B T} \tag{3.1}$$

as an entirely empirical expression for the rate constant of a reaction. We will discuss two formalisms, namely collision theory and transition state theory, which provide a sound basis for the Arrhenius equation and a framework in which to interpret the meaning of a particular value of a pre-exponential factor.

The Arrhenius equation corresponds to the situation sketched in Figure 3.1 for a reaction between two molecules. The two will only react if their joint potential energy on the moment of collision is sufficiently high to overcome the activation energy. If not, they will separate again and leave without reacting. This plausible picture, supported by the reasonable good description of the temperature dependence of many reactions, remains the background for the reaction rate theories in this chapter. However, to approach the problem without invoking the explicit form of the Arrhenius equation, we will distinguish between the height of the energy barrier, denoted ΔE, and the activation energy, E_a, which is an empirical

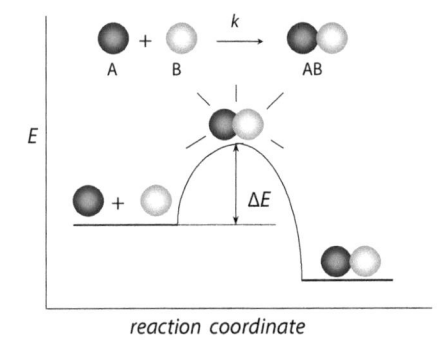

Figure 3.1 To react, the potential energy of the reaction complex must be higher than that of the barrier.

Concepts of Modern Catalysis and Kinetics, Third Edition. I. Chorkendorff and J.W. Niemantsverdriet.
© 2017 WILEY-VCH Verlag GmbH & Co. KGaA. Published 2017 by WILEY-VCH Verlag GmbH & Co. KGaA.

quantity. In Section 2.4, we described a general recipe for deriving the activation energy from experimental data. Here, we will apply the same procedure to derive the activation energy from the theoretical expressions for the reaction rate. We will see that E_a and ΔE are in general not equal, although the difference is small.

The two reaction rate theories discussed in this chapter are both based on the concept that reacting molecules acquire energy by collisions, thereby reaching an activated state (the top of the activation barrier) from which the product forms. Collision theory treats reacting molecules as billiard balls or hard spheres, for which only the kinetic energy upon collision determines whether the activation barrier can be crossed. Transition state theory includes the vibrations and rotations of the reacting molecules, which are heavily involved in the exchange of energy when molecules collide. These are also the degrees of freedom that are excited when a reacting complex passes over the activation barrier.

To understand how collision theory has been derived, we need to know the velocity distribution of molecules at a given temperature, as it is given by the Maxwell–Boltzmann distribution. To use transition state theory, we need the partition functions that follow from the Boltzmann distribution. Hence, we must devote a section of this chapter to statistical thermodynamics.

Collisions play a tremendously important role in stimulating reacting systems to cross the activation barrier. The theory of Lindemann emphasizes this and provides us with a method to describe the influence of the surrounding medium upon a chemical reaction. It gives important insight in the conditions under which the reaction rate theories discussed here are valid.

Finally, we will apply transition state theory and collision theory on some of the elementary surface reactions that are important in catalysis.

3.2
The Boltzmann Distribution and the Partition Function

Here, we briefly summarize statistical thermodynamics, as far as we need it in this chapter. For a thorough coverage of the subject we refer the reader to Atkins' *Physical Chemistry* [2].

Suppose we have a large ensemble of particles (electrons, atoms, molecules) with energy levels ε_i, with $i = 0, 1, 2, \ldots, N$, where N may also be infinitely large. A large number is mandatory since we want to derive meaningful averages. According to the Boltzmann distribution, the chance P_i that we find the system at a temperature T in a state with energy ε_i, is given by

$$P_i = \frac{e^{-\varepsilon_i/k_B T}}{\sum_{i=0}^{\infty} g_i e^{-\varepsilon_i/k_B T}} \tag{3.2}$$

This distribution follows automatically if we require that the entropy of a system with many members that is in equilibrium is at maximum. The denominator in the Boltzmann distribution ensures that the frequencies P_i are normalized and add up to unity, or 100%. This summation of states (Zustandssumme in German) is called a partition function:

$$q \equiv \sum_{i=0}^{\infty} g_i e^{-\varepsilon_i/k_B T} \qquad (3.3)$$

where g_i is the degeneracy factor, which ensures that all states are taking into account even if they have the same energy.

The partition function is a thermodynamic function of state which, through the sum over all the energy levels ε, contains all properties of the system. Although obviously correct, this statement may not at first sight appear very useful because large sums with many terms are awkward to work with. However, infinite mathematical series often yield simple expressions for their sums. In addition, if the energy levels become close enough for the energy to be handled as a continuous variable, the sum may be replaced by an integral that might yield a useful expression. Important properties of a system, such as average energy, free energy (chemical potential), and entropy follow straightforwardly from the partition function.

The average energy of the molecules follows by logarithmic differentiation:

$$\bar{\varepsilon} = k_B T^2 \frac{\partial \ln q}{\partial T} \qquad (3.4)$$

We shall not derive all formulas given, but this one is easy to prove as differentiation readily produces the definition of average energy in the corresponding distribution.

$$k_B T^2 \frac{\partial \ln q}{\partial T} = k_B T^2 \frac{1}{q} \frac{\partial q}{\partial T}$$

$$= k_B T^2 \frac{1}{\sum_{i=0}^{\infty} g_i e^{-\varepsilon_i/k_B T}} \sum_{i=0}^{\infty} \frac{\varepsilon_i}{k_R T^2} g_i e^{-\varepsilon_i/k_B T} \qquad (3.5)$$

$$= \frac{\sum_{i=0}^{\infty} \varepsilon_i g_i e^{-\varepsilon_i/k_B T}}{\sum_{i=0}^{\infty} g_i e^{-\varepsilon_i/k_B T}} \equiv \bar{\varepsilon}$$

Other thermodynamic quantities such as chemical potential and entropy also follow directly from the partition function, as we demonstrate later on. However, to illustrate what a partition function means, we will first discuss two relatively simple but instructive examples.

Example Partition Function of a Two-Level System

The first illustration of the concept of a partition function is that of a two-level system, for example, an electron in a magnetic field, with its spin either up or down (parallel or antiparallel to the magnetic field) (Figure 3.2). The ground state has energy $\varepsilon_0 = 0$ and the excited state has energy $\Delta\varepsilon$. By substituting these values into Eq. (3.3), we find the following partition function for this two-level system:

$$q = 1 + e^{-\Delta\varepsilon/k_B T} \qquad (3.6)$$

3 Reaction Rate Theory

At low temperatures, the system will be entirely in the ground state, and the partition function approaches 1 in the limit of $T \to 0$:

$$\lim_{T \to 0} q = 1 + \lim_{T \to 0} e^{-\Delta\varepsilon/k_B T} = 1 + 0 = 1 \tag{3.7}$$

At very high temperatures, however, the excited state will also be occupied. Entropy maximization requires that both levels be equally populated. We calculate the high-temperature limit of the partition function:

$$\lim_{T \to \infty} q = 1 + \lim_{T \to \infty} e^{-\Delta\varepsilon/k_B T} = 1 + 1 = 2 \tag{3.8}$$

Hence, at high temperatures the partition function equals the number of energy levels that are available. The average energy of the system follows by applying Eq. (3.4):

$$\bar{\varepsilon} = k_B T^2 \frac{\partial \ln(1 + e^{-\Delta\varepsilon/k_B T})}{\partial T} = \frac{\Delta\varepsilon \, e^{-\Delta\varepsilon/k_B T}}{1 + e^{-\Delta\varepsilon/k_B T}} \tag{3.9}$$

with the expected limits at zero and infinite temperature:

$$\lim_{T \to 0} \bar{\varepsilon} = 0; \quad \lim_{T \to \infty} \bar{\varepsilon} = 1/2 \Delta\varepsilon$$

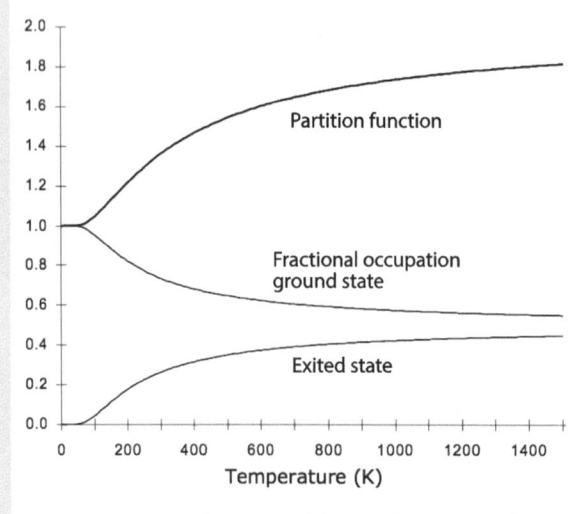

Figure 3.2 Partition function and fractional occupation for a two-level system.

Example   Partition Function of a System with an Infinite Number of Levels

We will now do the same for a system with an infinite number of equidistant energy levels, separated by $\Delta\varepsilon$. For such a system, the partition function becomes the following mathematical series with a well-known sum:

$$q = \sum_{i=0}^{\infty} e^{-i\Delta\varepsilon/k_B T} = \frac{1}{1 - e^{-\Delta\varepsilon/k_B T}} \qquad (3.10)$$

The reader may verify that the limiting values of the partition function for low and high temperature are 1 and ∞, respectively. The average energy becomes

$$\bar{\varepsilon} = k_B T^2 \frac{\partial \ln\left(1/(1 - e^{-\Delta\varepsilon/k_B T})\right)}{\partial T} = \frac{\Delta\varepsilon\, e^{-\Delta\varepsilon/k_B T}}{1 - e^{-\Delta\varepsilon/k_B T}} \qquad (3.11)$$

with the expected limiting values of zero and infinity for low and high temperatures.

The above two examples illustrate that the value of the partition function is an indicator for how many of the energy levels are occupied at a particular temperature. At $T = 0$, where the system is in the ground state, the partition function has the value $q = 1$. In the limit of infinite temperature, entropy demands that all states are equally occupied and the partition function becomes equal to the total number of energy levels.

3.3
Partition Functions of Atoms and Molecules

Partition functions are very important in estimating equilibrium constants and rate constants in elementary reaction steps. Therefore, we shall take a closer look at the partition functions of atoms and molecules. Motion, or translation, is the only degree of freedom that atoms have. Molecules also possess internal degrees of freedom, namely vibration and rotation.

3.3.1
The Boltzmann Distribution

In the following, we consider a system of distinguishable particles (note that a collection of atoms or molecules would be *in*distinguishable; an issue that only matters for the configurational entropy). We assume that the different degrees of freedom are independent and that the total energy of the system can be written as the sum of the energies of the individual members. Our system consists of N particles in total, distributed over N_i particles with energy ε_i, all adding up to the total energy E_{tot}. Again, N must be a very large number, a condition usually fulfilled in most practical cases. The probability for finding particle i with the energy ε_i is P_i:

$$P_i = \frac{N_i}{N}; \quad \sum_i N_i = N; \quad \sum_i P_i = \sum_i \frac{N_i}{N} = 1 \tag{3.12}$$

The total energy E_{tot} is given by

$$E_{\text{tot}} = \sum_i \varepsilon_i N_i; \quad \sum_i P_i \varepsilon_i = \bar{\varepsilon} = \frac{E_{\text{tot}}}{N} \tag{3.13}$$

where $\bar{\varepsilon}$ is the average energy per particle. How will this system fulfill the two stated constraints on particle number N and energy content E_{tot} and at the same time maximize the entropy S of the system? According to Boltzmann, the entropy is given by a constant k_B (Boltzmann's constant) times the logarithm of the number of ways (W) in which we can distribute the N particles over i states each containing N_i particles. Hence, Boltzmann's expression for the entropy, which is also written on his gravestone in Vienna, is

$$S_{\text{tot}} = k_B \ln(W) \tag{3.14}$$

where

$$W = \frac{N!}{\prod_i N_i!} \tag{3.15}$$

The quantity S_{tot} is typically referred to as the configurational entropy. It can easily be expressed in terms of the probability P_i by using Stirling's approximation for large N:

$$\ln(N!) \approx N \ln(N) - N \tag{3.16}$$

leading to

$$S = \frac{S_{\text{tot}}}{N} = -k_B \sum_i P_i \ln(P_i) \tag{3.17}$$

There is a standard mathematical tool for solving the problem of maximizing a quantity that has to satisfy constraints, namely the method of *Lagrange undetermined multipliers*. It boils down to finding the maximum of the function $f(P_i)$, which contains the entropy and the constraints weighted by (yet unknown) constants λ_1 and λ_2, the so-called Lagrange multipliers:

$$f(P_i) = S = S(P_i) - \lambda_1 \left(\sum_i P_i - 1 \right) - \lambda_2 \left(\sum_i \varepsilon_i P_i - \bar{\varepsilon} \right) \tag{3.18}$$

The extremum of this function is found by differentiation:

$$\frac{\partial f(P_i)}{\partial P_i} = \frac{\partial \left(S(P_i) - \lambda_1 \sum_i P_i - \lambda_2 \sum_i \varepsilon_i P_i \right)}{\partial P_i} = 0 \tag{3.19}$$

resulting in maximum entropy for

$$P_i = e^{-\left(\frac{\lambda_1}{k_B}+1\right)} e^{-\frac{\lambda_2}{k_B}\varepsilon_i} \tag{3.20}$$

By using the condition that the sum of all the probabilities must be unity, we immediately obtain

$$\sum_i P_i = 1 = \sum_i e^{-\left(\frac{\lambda_1}{k_B}+1\right)} e^{-\frac{\lambda_2}{k_B}\varepsilon_i} = e^{-\left(\frac{\lambda_1}{k_B}+1\right)} \sum_i e^{-\frac{\lambda_2}{k_B}\varepsilon_i} \tag{3.21}$$

from which it is easily seen that

$$e^{\left(\frac{\lambda_1}{k_B}+1\right)} = \sum_i e^{-\frac{\lambda_2}{k_B}\varepsilon_i} \tag{3.22}$$

We have now expressed λ_1 in terms of λ_2 but we still need to determine what λ_2 is. We can do this by relating the above equation to something we know from thermodynamics. If we do so (see e.g., [2, 3]), we find that λ_2 equals $1/T$ and that the above quantity is simply the partition function q. For the interested reader, we justify choosing λ_2 equal to $1/T$ in the following section.

3.3.1.1 Justification for Equating λ_2 with 1/T

It is instructive to compare Eq. (3.22) with the situation of a particle in a box because this automatically yields a useful expression for the partition function of translation (Figure 3.3).

If we consider the system to be a monatomic gas in a one-dimensional box (which can easily be generalized to three dimensions), the energy levels of the atom are given by quantum mechanics as

$$\varepsilon_i = \frac{i^2 h^2}{8ml^2} = i^2 c \tag{3.23}$$

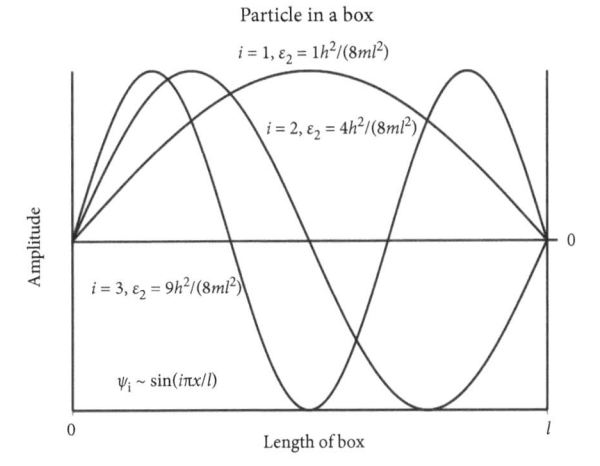

Figure 3.3 Possible solutions to a particle in a potential box with infinitely high walls.

where h is Planck's constant, m is the mass of the atom, and l is the length of the box. This result follows from the fact that the atom has to be described by a simple wave function that equals zero at the walls of the box. We can now calculate a value known from thermodynamics, namely the average energy in one dimension $\bar{\varepsilon} = \frac{k_B T}{2}$. Thus, by replacing the sum by an integral (this is allowed since the energy levels are very close), we obtain

$$q = e^{\left(\frac{\lambda_1}{k_B}+1\right)} = \sum_i e^{-\frac{\lambda_2}{k_B}\varepsilon_i} \cong \int_0^\infty e^{-\frac{\lambda_2 c i^2}{k_B}} di = \frac{1}{2}\sqrt{\frac{\pi k_B}{\lambda_2 c}} = \sqrt{\frac{2\pi k_B m l^2}{h^2 \lambda_2}} \quad (3.24)$$

$$\bar{\varepsilon} = \frac{k_B T}{2} = \sum_i P_i \varepsilon_i = \frac{\sum_i \varepsilon_i e^{-\lambda_2 \varepsilon_i/k_B}}{q} \cong \frac{\int_0^\infty i^2 c e^{-\frac{\lambda_2 c i^2}{k_B}} di}{\sqrt{\frac{2\pi k_B m l^2}{h^2 \lambda_2}}} = \frac{\frac{1 k_B}{2\lambda_2}\sqrt{\frac{2\pi k_B m l^2}{h^2 \lambda_2}}}{\sqrt{\frac{2\pi k_B m l^2}{h^2 \lambda_2}}} = \frac{1 k_B}{2\lambda_2}$$

indicating that $\lambda_2 = 1/T$ is simply the reciprocal temperature. This leaves us with the following expression for the probability of finding a particle in the state i

$$P_i = \frac{e^{-\frac{\varepsilon_i}{k_B T}}}{\sum_i e^{-\frac{\varepsilon_i}{k_B T}}} = \frac{e^{-\frac{\varepsilon_i}{k_B T}}}{q} \quad (3.25)$$

where q is the translational partition function in one dimension:

$$q_{\text{trans}} = \frac{l\sqrt{2\pi k_B m T}}{h} \quad (3.26)$$

3.3.2
Maxwell–Boltzmann Distribution of Velocities

Often, we will be interested in how the velocities of the molecules are distributed. Therefore, we need to transform the Boltzmann distribution of energies into the Maxwell–Boltzmann distribution of velocities, thereby changing the variable from energy to velocity or, rather, momentum p_x (not to be confused with pressure). If the energy levels are very close (as they are in the classic limit), we can replace the sum by an integral:

$$\sum_i P_i = 1 = \int_{-\infty}^{\infty} f(p_x) dp_x = \int_0^\infty \frac{e^{-\varepsilon_i/k_B T}}{q} di = \int_0^\infty \frac{e^{-\varepsilon_i/k_B T}}{\frac{l\sqrt{2\pi k_B m T}}{h}} di \quad (3.27)$$

and we obtain

$$\int_{-\infty}^{\infty} f(p_x) dp_x = \int_0^\infty \frac{e^{-i^2 p^2/2mk_B T}}{\sqrt{2\pi k_B m T}} 2p\, di = \int_{-\infty}^{\infty} \frac{e^{-p_x^2/2mk_B T}}{\sqrt{2\pi k_B m T}} dp_x \quad (3.28)$$

giving us the Maxwell–Boltzmann distribution in one dimension:

$$f(p_x)\,dp_x = \frac{e^{\frac{-p_x^2}{2mk_BT}}}{\sqrt{2\pi m k_B T}}\,dp_x \tag{3.29}$$

This distribution is readily generalized to three dimensions where it takes the form

$$f(p_x, p_y, p_z) = f(p_x)f(p_y)f(p_z) \tag{3.30}$$

in Cartesian coordinates or

$$f(v) = 4\pi \left(\frac{m}{2\pi k_B T}\right)^{3/2} v^2 e^{-mv^2/2k_B T} \tag{3.31}$$

in spherical coordinates, with v the magnitude of the velocity. In fact, the Maxwell–Boltzmann velocity distribution follows as a consequence of having a system with the constraints of a fixed number of particles at a certain total energy, for which the entropy is maximized.

3.3.3
Total Partition Function of a System

For a system of distinguishable particles, the total partition function of the system is the product of all the individual partition functions, that is,

$$Q = q^N \quad \text{distinguishable particles} \tag{3.32}$$

However, if the particles are indistinguishable, as the atoms in a gas, the number of possible configurations is significantly reduced and the partition function for an ensemble of atoms or molecules is therefore

$$Q = \frac{q^N}{N!} \quad \text{or} \quad Q = \prod_i \frac{q_i^{N_i}}{N_i!} \quad \text{indistinguishable particles} \tag{3.33}$$

if we have a mixture of noninteracting gases of type i.

Consider a particle (a molecule or atom) of which the different degrees of freedom are independent and the energy is simply the sum of the energies contained in the different degrees of freedom. We can then write the partition function as the product of the partition functions for the various degrees of freedom. For an atom, this is rather trivial:

$$q = q_{\text{trans}}\, q_{\text{elec}}\, q_{\text{nucl}} \tag{3.34}$$

where q_{trans} describes the partition function for the translational degrees of freedom, while q_{elec} and q_{nucl} are the partition functions related to the electrons and the nuclei of the atom. For molecules, this is complicated due to the internal degrees of freedom related to vibrations and rotations. Here, the partition function will have additional terms like

$$q = q_{\text{trans}}\, q_{\text{rot}}\, q_{\text{vib}}\, q_{\text{elec}}\, q_{\text{nucl}} \tag{3.35}$$

Several relevant physical parameters can be extracted for this partition function, such as

$$\mu = -k_B T \left(\frac{\partial \ln(Q)}{\partial N} \right)_{V,T} \tag{3.36}$$

$$E = k_B T^2 \left(\frac{\partial \ln(Q)}{\partial T} \right)_{N,V} \tag{3.37}$$

$$p = k_B T \left(\frac{\partial \ln(Q)}{\partial V} \right)_{N,T} \tag{3.38}$$

$$S = \frac{\partial}{\partial T}(k_B T \ln(Q)_{N,V}) = k_B \ln(Q) + k_B T \left(\frac{\partial \ln(Q)}{\partial T} \right)_{N,V} \tag{3.39}$$

where p here refers to the pressure. To determine the overall partition function, we need to understand the partition function for the individual molecules or atoms.

3.3.3.1 Translational Partition Function

Starting with the partition function of translation, consider a particle of mass m moving in one dimension x over a line of length l with velocity v_x. Its momentum equals $p_x = mv_x$ and its kinetic energy $\varepsilon_x = p_x^2/2m$. The coordinates available for the particle x, p_x in phase space can be divided into small cells each of size h which is Planck's constant. Since the division is so incredibly small, we can replace the sum with integration over phase space in x and p_x and so calculate the partition function. By normalizing with the size of the cell h the expression becomes

$$q_{\text{trans}} = \frac{1}{h} \int_0^l dx \int_{-\infty}^{\infty} dp_x e^{-p_x^2/2mk_B T} = l \frac{(2\pi m k_B T)^{1/2}}{h} \tag{3.40}$$

This is the translational partition function for any particle of mass m, moving over a line of length l, in one dimension. Please note that this result is exactly the same as we calculated from quantum mechanics for a particle in a one-dimensional box. For a particle moving over an area A on a surface, the partition function of translation is

$$q_{\text{trans}}^{2D} = A \frac{2\pi m k_B T}{h^2} \tag{3.41}$$

and for a particle free to move inside a volume V:

$$q_{\text{trans}}^{3D} = V \frac{(2\pi m k_B T)^{3/2}}{h^3} \tag{3.42}$$

Hence, we conclude that the translational partition function of a particle depends on its mass, the temperature, the dimensionality, as well as the dimensions of the space in which it moves. As a result, translational partition functions may be large numbers. The translational partition function is conveniently calculated per volume, which is the quantity used, for example, when the equilibrium conditions are determined, as we shall see later. The partition function can conveniently be written as

$$q_{\text{trans}}^{3D} = V\frac{(2\pi m k_B T)^{3/2}}{h^3} = \frac{V}{\Lambda^3} \tag{3.43}$$

where Λ correspond to the thermal wavelength (the de Broglie wavelength) of the particle in one dimension. It is worth noting that the Boltzmann distribution is only valid when $V/\Lambda^3 \gg 1$, meaning that the wavelength of the particle must be much smaller than the length of the container in which it is confined. This requirement will be fulfilled for all systems encountered in this book. Exceptions only occur under extreme conditions, for example, at very low temperatures (i.e., for He at a few K where it becomes a super fluid) or at extremely high pressures as in stars. Under such extreme conditions either the Bose–Einstein or the Fermi–Dirac distribution need to be applied, depending on whether the particles have integral or half-order spin. For further details, we refer the reader to textbooks on statistical mechanics [4].

The average kinetic energy of a particle in space follows if we apply Eqs. (3.4) or (3.37) to Eq. (3.42):

$$\bar{\varepsilon}_{\text{trans}} = k_B T^2 \frac{\partial}{\partial T} \ln\left[V\frac{(2\pi m k_B T)^{3/2}}{h^3}\right] = \frac{3}{2}k_B T \tag{3.44}$$

which is the expected result. In a classical system, each dimension contributes $1/2 k_B T$ to the kinetic energy.

3.3.3.2 Vibrational Partition Function

Molecules also possess internal degrees of freedom, namely vibration and rotation. The vibrational energy levels in the harmonic oscillator approximation of a vibration with frequency $h\nu$ are given by

$$\varepsilon_i = \left(i + \frac{1}{2}\right)h\nu \tag{3.45}$$

Note that the zero of energy is now the bottom of the potential, and the ground state – the lowest occupied level – lies $1/2h\nu$ higher. As partition functions are usually given with respect to the lowest occupied state, we shift the zero of energy upward by $1/2h\nu$ to obtain

$$\varepsilon_i' = ih\nu = \varepsilon_i - \frac{1}{2}h\nu \tag{3.46}$$

The partition function with respect to the lowest occupied state, thus, becomes

$$q_{\text{vib}}' = \sum_{i=0}^{\infty} e^{-ih\nu/k_B T} = \frac{1}{1 - e^{-h\nu/k_B T}} \tag{3.47}$$

whereas the partition function relative to the bottom of the potential energy curve is

$$q_{\text{vib}} = \sum_{i=0}^{\infty} e^{-\left(i+\frac{1}{2}\right) h\nu/k_\text{B} T} = \frac{e^{-\frac{1}{2}h\nu/k_\text{B}T}}{1 - e^{-h\nu/k_\text{B}T}} \tag{3.48}$$

Both expressions are important, as we will see later. The value of the vibrational partition function q'_{vib} is often close to unity, unless the vibration has such a low frequency that $h\nu \ll k_\text{B}T$, in which case we can use the approximation

$$e^{\pm x} \approx 1 \pm x \tag{3.49}$$

and the partition function approaches the classical limit:

$$q'_{\text{vib,class}} \approx \frac{k_\text{B}T}{h\nu}, \quad \text{for } h\nu \ll k_\text{B}T \tag{3.50}$$

Note that a diatomic molecule in the gas phase has only one vibration but as soon as it adsorbs on the surface it acquires several more modes, some of which may have quite low frequencies. The total partition function of vibration then becomes the product of the individual partition functions:

$$q_{\text{vib}} = \prod_i \frac{e^{-\frac{1}{2}h\nu_i/k_\text{B}T}}{1 - e^{-h\nu_i/k_\text{B}T}} \tag{3.51}$$

By applying Eqs. (3.4) or (3.36), we can calculate the average vibrational energy of a molecule with N vibrational modes, and we do this with respect to the zero of the vibrational potentials, implying that we will include all zero-point energies:

$$\begin{aligned} \bar{\varepsilon}_{\text{vib}} &= k_\text{B}T^2 \frac{\partial}{\partial T} \ln \prod_{i=0}^{N} \frac{e^{-\frac{1}{2}h\nu_i/k_\text{B}T}}{1 - e^{-h\nu_i/k_\text{B}T}} \\ &= \sum_{i=0}^{N} \left(\frac{h\nu_i}{e^{h\nu_i/k_\text{B}T} - 1} + \frac{1}{2}h\nu_i \right) \\ &\approx \sum_{i=0}^{N} \frac{1}{2} h\nu_i \end{aligned} \tag{3.52}$$

The approximation is valid for every mode with a not-too-low vibrational frequency, otherwise the full term for this mode needs to be used.

3.3.3.3 Rotational (and Nuclear) Partition Function

Finally, the rotational partition function of a diatomic molecule follows from the quantum mechanical energy level scheme:

$$\varepsilon_j = \frac{j(j+1)h^2}{8\pi^2 I} \tag{3.53}$$

in which j is the rotational quantum number, and I the moment of inertia,

$$I = \mu r^2 \tag{3.54}$$

Table 3.1 The symmetry factor for different symmetry groups and examples of molecules belonging to them.

Molecular symmetry	σ	Types of molecules
C_1, C_i, and C_s	1	CO, CHFClBr, *meso*-tartaric acid, and CH_3OH
C_2, C_{2v}, and C_{2h}	2	H_2, H_2O_2, H_2O, and *trans*-dichloroethylene
C_{3v} and C_{3h}	3	NH_3, and planar $B(OH)_3$

with μ the reduced mass

$$\frac{1}{\mu} = \frac{1}{m_1} + \frac{1}{m_2} \tag{3.55}$$

and m_1 and m_2 the masses of the atoms in the diatomic molecule. Inserting the expression for the energy into that for the partition function, and taking the degeneracy of rotational levels into account, yields

$$\begin{aligned} q_{\text{rot}} &= \frac{1}{\sigma} \sum_{j=0}^{\infty} (2j+1) e^{-\frac{j(j+1)h^2}{8\pi^2 I k_B T}} \\ &\approx \frac{1}{\sigma} \int_0^{\infty} d(j(j+1)) e^{-\frac{j(j+1)h^2}{8\pi^2 I k_B T}} \\ &= \frac{1}{\sigma} \frac{8\pi^2 I k_B T}{h^2} \quad \text{for} \quad T > \frac{h^2}{8\pi^2 I k_B} \end{aligned} \tag{3.56}$$

The final expression is the classical limit, valid above a certain critical temperature, which, however, in practical cases is low (i.e., 85 K for H_2, 3 K for CO). For a homonuclear or a symmetric linear molecule, the factor σ equals 2, while for a heteronuclear molecule $\sigma = 1$ (Table 3.1). This symmetry factor stems from the indistinguishable permutations the molecule may undergo due to the rotation and actually also involves the nuclear partition function. The symmetry factor can be estimated directly from the symmetry of the molecule.

The average energy of a rotating molecule follows immediately by logarithmic differentiation:

$$\bar{\varepsilon}_{\text{rot}} = k_B T^2 \frac{\partial}{\partial T} \ln \left(\frac{1}{\sigma} \frac{8\pi^2 I k_B T}{h^2} \right) = k_B T \tag{3.57}$$

For completeness, we also give the general formula for a larger molecule with moments of inertia I_A, I_B, and I_C along the three principal axes:

$$q_{\text{rot}} = \frac{1}{\sigma} \left(\frac{8\pi^2 k_B T}{h^2} \right)^{3/2} \sqrt{\pi I_A I_B I_C} \tag{3.58}$$

As an exercise, we leave the reader to show that the average energy of a diatomic molecule in the gas phase at temperatures where only the vibrational ground state is populated equals $5k_B T/2$. What is it at high temperatures?

3.3.3.4 Electronic and Nuclear Partition Functions

The electronic partition function is often simply unity since the energy separation between excited states is usually very large compared to $k_B T$. But it is not always the case – a well-known example is NO, which has an unpaired electron in its highest occupied molecular orbital. As a consequence, the orbital momentum can assume two orientations with respect to the axis (clockwise and anticlockwise circulation). In addition, there are two orientations for the spin of the electron, giving four different configurations with a relatively small difference in energy between 2 and 4. All this results in the electronic partition function of the NO molecule varying between 2 and 4, depending on the temperature.

Usually we would choose the separate atoms in their ground state as the zero energy. The electronic partition function is then

$$q_{el}(T) = \omega_{e0} e^{-\frac{\varepsilon_0}{k_B T}} + \omega_{e1} e^{-\frac{\varepsilon_1}{k_B T}} + \ldots \tag{3.59}$$

where ω_{ei} denotes the degeneracy of state ε_i, and $i > 0$ refers to excited electronic states. The ground state of a system if we, for example, consider a diatomic molecule is usually

$$\varepsilon_0 = -D_0 = -D_e + \frac{h\nu}{2} \tag{3.60}$$

as illustrated in Figure 3.4. In this case, the energy gained by forming the molecule from the two atoms is D_0 and is equal to the dissociation energy, since we should remember that the molecule has a zero-point vibrational energy equal to $1/2 h\nu$.

The nuclear partition function does not usually contribute to the partition function and can therefore be taken as unity. We shall ignore this contribution in the following.

Finally, an overview of the formulae for partition functions is given in Table 3.2.

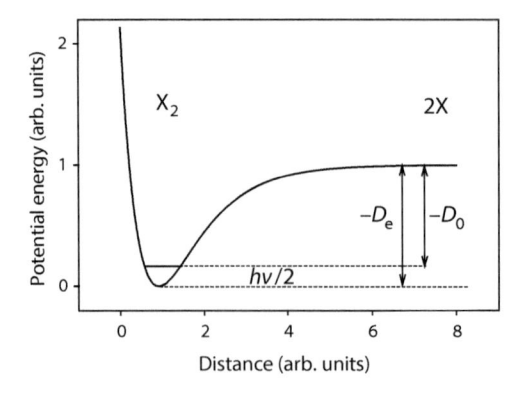

Figure 3.4 Potential energy diagram for dissociation of a diatomic molecule.

Table 3.2 Summary of the most important results of statistical mechanics for diatomic gases.

Partition functions

$$P_i = \frac{e^{-\varepsilon_i/k_B T}}{\sum_{i=0}^{\infty} e^{-\varepsilon_i/k_B T}}$$

$$\bar{\varepsilon} = k_B T^2 \frac{\partial \ln q}{\partial T}$$

$$\mu = -k_B T \ln q$$

$$q = \sum_{i=0}^{\infty} e^{-\varepsilon_i/k_B T}$$

$$s = \frac{\partial k_B T \ln q}{\partial T}$$

$$K(T) = \prod_i q^{\nu_i}$$

Partition functions of a diatomic molecule (per degree of freedom)

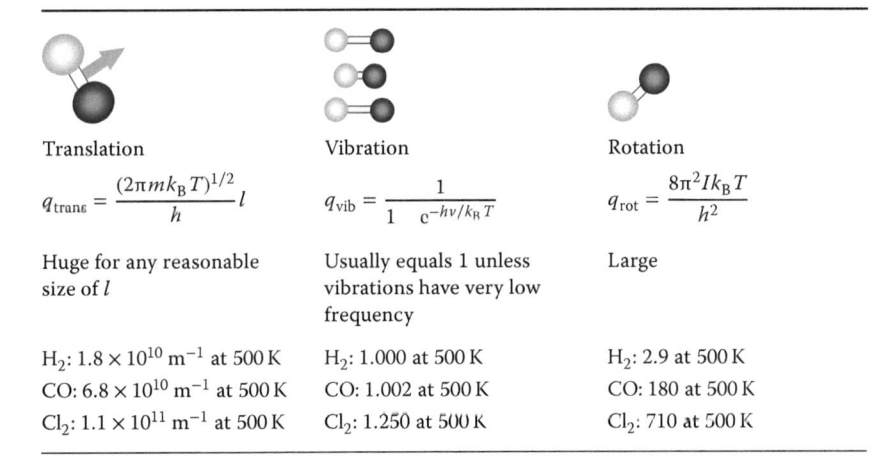

Translation	Vibration	Rotation
$q_{trans} = \frac{(2\pi m k_B T)^{1/2}}{h} l$	$q_{vib} = \frac{1}{1 - e^{-h\nu/k_B T}}$	$q_{rot} = \frac{8\pi^2 I k_B T}{h^2}$
Huge for any reasonable size of l	Usually equals 1 unless vibrations have very low frequency	Large
H_2: 1.8×10^{10} m^{-1} at 500 K	H_2: 1.000 at 500 K	H_2: 2.9 at 500 K
CO: 6.8×10^{10} m^{-1} at 500 K	CO: 1.002 at 500 K	CO: 180 at 500 K
Cl_2: 1.1×10^{11} m^{-1} at 500 K	Cl_2: 1.250 at 500 K	Cl_2: 710 at 500 K

3.4 Molecules in Equilibrium

In Chapter 2 we discussed both chemical equilibrium and equilibrium constants. We shall now return to the chemical reactions and see how equilibrium constants can be determined directly from the partition functions of the molecules participating in the reaction. Consider the following reaction which was described in Chapter 2:

$$\nu_A A + \nu_B B \underset{k^-}{\overset{k^+}{\rightleftharpoons}} \nu_C C + \nu_D D \quad (3.61)$$

The criterion for chemical equilibrium is that the Gibbs free energy is at its minimum:

$$\frac{dG}{d\xi} = -\nu_A \mu_A - \nu_B \mu_B + \nu_C \mu_C + \nu_D \mu_D = \sum_i \nu_i \mu_i = 0 \quad (3.62)$$

By using Eq. (3.35), we find the chemical potential directly from the partition function:

$$\mu_i = -k_B T \left(\frac{\partial \ln(Q_i)}{\partial N_i} \right) = -k_B T \left(\frac{\partial \ln\left(\frac{q_i^{N_i}}{N_i!}\right)}{\partial N_i} \right) \approx -k_B T \ln\left(\frac{q_i}{N_i}\right) \quad (3.63)$$

where we have used Stirling's approximation, Eq. (3.16). Substituting Eq. (3.63) into Eq. (3.62), we obtain

$$\prod_i \left(\frac{q_i}{N_i}\right)^{\nu_i} = 1 = \left(\frac{q_C}{N_C}\right)^{\nu_C} \left(\frac{q_D}{N_D}\right)^{\nu_D} \left(\frac{q_A}{N_A}\right)^{-\nu_A} \left(\frac{q_B}{N_B}\right)^{-\nu_B} \quad (3.64)$$

Since we have assumed that we are dealing with an ideal gas at pressure p_i, we may use

$$N_i = \frac{p_i V}{k_B T} \quad (3.65)$$

and, hence, we can rearrange the equation into

$$\frac{\left(\frac{q_C}{V}\right)^{\nu_C} \left(\frac{q_D}{V}\right)^{\nu_D}}{\left(\frac{q_A}{V}\right)^{\nu_A} \left(\frac{q_B}{V}\right)^{\nu_B}} (k_B T)^{\nu_C + \nu_D - \nu_A - \nu_B} = \frac{p_C^{\nu_C} p_D^{\nu_D}}{p_A^{\nu_A} p_B^{\nu_B}} \quad (3.66)$$

Since q_i are proportional to V (this stems from the translational partition function), $\frac{q_i}{V}$ only depends on T. Nevertheless, if

$$\nu_C + \nu_D - \nu_A - \nu_B \neq 0 \quad (3.67)$$

the equilibrium constant will have dimensions of pressure. We therefore normalize the pressures by $p_0 = 1$ bar such that the equilibrium constant becomes dimensionless. Hence, we have an expression for the equilibrium constant in terms of the molecular partition functions q_i:

$$K(T) = \frac{\left(\frac{q_C}{V}\right)^{\nu_C} \left(\frac{q_D}{V}\right)^{\nu_D}}{\left(\frac{q_A}{V}\right)^{\nu_A} \left(\frac{q_B}{V}\right)^{\nu_B}} \left(\frac{k_B T}{p_0}\right)^{\nu_C + \nu_D - \nu_A - \nu_B} = \frac{\left(\frac{p_C}{p_0}\right)^{\nu_C} \left(\frac{p_D}{p_0}\right)^{\nu_D}}{\left(\frac{p_A}{p_0}\right)^{\nu_A} \left(\frac{p_B}{p_0}\right)^{\nu_B}} \quad (3.68)$$

Thus, given sufficient detailed knowledge of the internal energy levels of the molecules participating in a reaction, we can calculate the relevant partition functions, and then the equilibrium constant from Eq. (3.68). This approach is applicable in general: determine the partition function, then estimate the chemical potentials of the reacting species and the equilibrium constant can be determined. A few examples will illustrate this approach.

Example

It is instructive to illustrate the relation between the partition function and the equilibrium constant with a simple, entirely hypothetical example. Consider the equilibrium between an ensemble of molecules A and B, each with energy levels as indicated in Figure 3.5. The ground state of molecule A is the zero of energy; hence, the partition function of A will be

$$q_A = \sum_{i=0}^{1} e^{-i\Delta\varepsilon_A/k_BT} = 1 + e^{-\Delta\varepsilon_A/k_BT} \tag{3.69}$$

The partition function of molecule B does not have the same zero energy in its ground state, which we therefore have to correct for by adjusting the energy level with u

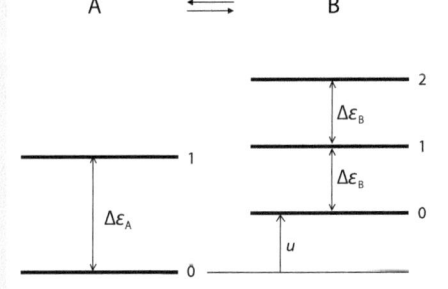

$$q_B = \sum_{i=0}^{2} e^{-(u+i\Delta\varepsilon_B)/k_BT} = e^{-u/k_BT} \sum_{i=0}^{2} e^{-i\Delta\varepsilon_B/k_BT} \tag{3.70}$$

Figure 3.5 Energy level scheme of two hypothetical molecules in equilibrium.

By applying

$$\sum_i v_i \mu_i = 0 \quad \text{and}$$

$$\mu_i = -k_B T \left(\frac{\partial \ln(Q_i)}{\partial N_i}\right) = -k_B T \left(\frac{\partial \ln\left(\frac{q_i^{N_i}}{N_i!}\right)}{\partial N_i}\right) \approx -k_B T \ln\left(\frac{q_i}{N_i}\right) \tag{3.71}$$

we obtain the following expression for the equilibrium constant:

$$\frac{N_A}{N_B} = K = \frac{q_B}{q_A} = \frac{e^{-u/k_BT}(1 + e^{-\Delta\varepsilon_B/k_BT} + e^{-2\Delta\varepsilon_B/k_BT})}{1 + e^{-\Delta\varepsilon_A/k_BT}} \tag{3.72}$$

In the low-temperature limit, we find $K = 0$ and $q_A = 1$, indicating that the system is entirely in the ground state of A, as this yields the lowest possible (free) energy. At high temperatures, though, we find $K = 3/2$, because all levels will be equally occupied. This example nicely illustrates how energy dominates at low temperature, but entropy takes over at high temperature.

Example   Equilibrium Constant of H_2, N_2, and O_2 Dissociation

Here we consider the possible dissociation of H_2 by heating it to a sufficiently high temperature. We assume ideal gas behavior and consider the reactions

$$H_2 \rightleftharpoons 2H \tag{3.73}$$

and calculate the partial pressure of atomic hydrogen for a constant total pressure at $T = 1000$, 2000, and 3000 K. If the pressure is kept constant at $p_{tot} = 1$ bar, then in the equilibrium situation $n_{H_2} = n_{initial} - \xi$, while $n_H = 2\xi$, leaving us with $n_{final} = n_{initial} + \xi$ (where ξ is the extent of reaction):

$$p_{H_2} = \frac{n_{initial} - \xi}{n_{initial} + \xi} p_{tot}, \quad p_H = \frac{2\xi}{n_{initial} + \xi} p_{tot} \tag{3.74}$$

At equilibrium, we have $\sum_i v_i \mu_i = 0$, implying that

$$-\mu_{H_2} + 2\mu_H = 0 \tag{3.75}$$

As

$$\mu_i = -k_B T \left(\frac{\partial \ln(Q_i)}{\partial N_i} \right) = -k_B T \left[\frac{\partial \ln\left(\frac{q_i^{N_i}}{N_i!}\right)}{\partial N_i} \right] \approx -k_B T \ln\left(\frac{q_i}{N_i}\right) \tag{3.76}$$

and

$$\frac{N_H^2}{N_{H_2}} = \frac{q_H^2}{q_{H_2}} \quad \text{and} \quad N_i = \frac{p_i V}{k_B T} \tag{3.77}$$

we obtain the equilibrium constant:

$$\frac{\left(\frac{p_H}{p_0}\right)^2}{\left(\frac{p_{H_2}}{p_0}\right)} = \left(\frac{k_B T}{p_0}\right) \frac{\left(\frac{q_H}{V}\right)^2}{\frac{q_{H_2}}{V}} = K(T) \tag{3.78}$$

By introducing Eq. (3.74), we find

$$\xi = \frac{n_{initial}}{\sqrt{1 + \frac{4P_{tot}}{K(T)}}} \quad \text{and} \quad p_H = \frac{2p_{tot}}{1 + \sqrt{1 + \frac{4p_{tot}}{K(T)}}} \tag{3.79}$$

3.4 Molecules in Equilibrium

We have in Eq. (3.80) expressed the equilibrium constant in terms of the relevant partition functions, which must be calculated:

$$K(T) = \left(\frac{k_B T}{p_0}\right) \frac{\left(\frac{q_H^{trans}}{V} q_H^{el}\right)^2}{\frac{q_{H_2}^{trans}}{V} q_{H_2}^{vib} q_{H_2}^{rot} q_{H_2}^{el}} \tag{3.80}$$

Note that this is the equilibrium constant for one hydrogen molecule dissociating into two atoms. Had we defined the process as being half a molecule of H_2 forming one H atom we would have had an equilibrium constant that was the square root of the one given by Eq. (3.80).

The translational partition functions at $T = 1000$ K are easily calculated for both species as

$$\frac{q_{trans}}{V} = \frac{(2\pi m_x k_B T)^{3/2}}{h^3} \tag{3.81}$$

leading to

$$\frac{q_H^{trans}}{V} = \frac{(2 \cdot 3.141 \cdot 1.66 \times 10^{-27} \cdot 1.38 \times 10^{-23} \cdot 1000)^{3/2}}{(6.63 \times 10^{-34})^3} = 6.01 \times 10^{30} \tag{3.82}$$

$$\frac{q_{H_2}^{trans}}{V} = \frac{2^{3/2} q_H^{trans}}{V} = 1.70 \times 10^{31} \tag{3.83}$$

The electronic partition function presents no difficulty for H_2 since the excited electronic configurations are well separated from the ground state. Hence, at the temperatures considered here, we do not have to take electronic excitations into account, resulting in $q_{H_2}^{el} = 1$. For the hydrogen atom, the situation is very similar. The ground state H_{1s} has one unpaired electron, implying it is doubly degenerated, leading to the term $^2S_{1/2}$. The first excited electronic states would be H_{2p}, $^2P_{1/2}$ or H_{2s}, $^2S_{1/2}$ which both lie 10.4 eV above the ground state. Thus,

$$q_H^{el} = \omega_{e1} + \omega_{e2} e^{\frac{-\Delta E_{12}}{k_B T}} + \cdots \cong \omega_{e1} = 2$$

where ω_{ei} refers to the degeneration.

The vibrational partition function is

$$q_{vib} = \prod_i \frac{e^{-1/2 h\nu_i/k_B T}}{1 - e^{-h\nu_i/k_B T}} \tag{3.84}$$

H_2 only has one vibrational mode at 4400.39 cm^{-1} and hence $\frac{h\nu}{k_B} = 6332$ K and

$$q_{H_2}^{vib} = \frac{e^{-1/26342/T}}{1 - e^{-6342/T}} = 0.042 \quad \text{at} \quad T = 1000 \text{ K} \tag{3.85}$$

To calculate the rotational partition function for the molecule, we need to be careful and check whether the assumption under which Eq. (3.56) has been derived is valid.

$$q_{H_2}^{rot} = \frac{1}{\sigma}\frac{8\pi^2 I k_B T}{h^2} \quad \text{for} \quad T \gg \frac{h^2}{8\pi^2 I k_B} \equiv \Theta_r \quad (3.86)$$

Since for molecular H_2 $\Theta_r = 87.6$ K, there is no problem for the temperatures considered here, although at temperatures well below 1000 K one would need to consider the sum instead of the integral in Eq. (3.56). For $T = 1000$ K we may safely use the approximation Eq. (3.86) and we find

$$q_{H_2}^{rot} \cong \frac{1}{\sigma}\frac{T}{\Theta_r} = \frac{1}{2}\frac{1000}{87.6} = 5.71 \quad (3.87)$$

in which we have used $\sigma = 2$ for a homonuclear molecule such as H_2.

Choosing the separate atoms as the zero energy, the electronic partition function of the hydrogen molecule is

$$q_{H_2}^{el} = \omega_{e1} e^{\frac{D_e}{k_B T}} + \cdots \cong 1 e^{\frac{D_e}{k_B T}} \quad (3.88)$$

The factor D_e can either be determined from the dissociation energy and the ground state vibration energy or from thermodynamic data. The heat of formation of H atoms from H_2 molecules can be found in the literature, but some care should be exercised considering the total energy content of H atoms and H_2 molecules at standard conditions.

The heat of formation an H atom is $\Delta H° = 218.0$ kJ/mol at 298 K (standard conditions) per H atom. There would be no problem at 0 K since the heat of formation would simply be half of the dissociation energy D_0, and D_e would be given by $D_e = D_0 + h\nu/2$, corresponding to $T = 0$ in Figure 3.6. However, under standard conditions, both the molecules and the atoms possess energy, and in different amounts, due to differences in internal degrees of freedom. With

$$H = E + pV = k_B T^2 \left(\frac{\partial \ln(Q)}{\partial T}\right)_{N,V} + k_B T \quad (3.89)$$

we find

$$H_H = E + pV = k_B T^2 \left(\frac{\partial \ln(Q_H)}{\partial T}\right)_{N,V} + k_B T = \frac{3}{2} k_B T + k_B T \quad (3.90)$$

and

$$H_{H_2} = E + pV = k_B T^2 \left(\frac{\partial \ln(Q_{H_2})}{\partial T}\right)_{N,V} + k_B T$$
$$= \frac{3}{2} k_B T + k_B T + \frac{h\nu}{2} + k_B T - D_e \quad (3.91)$$

3.4 Molecules in Equilibrium

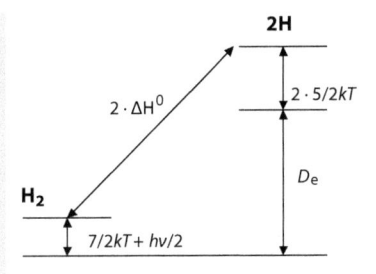

Figure 3.6 Detailed energy diagram for determining the dissociation energy D_e of H_2, from thermodynamic data.

The molecule has $3/2\, k_B T$ from translational energy, $k_B T$ from the term pV, $k_B T$ from the two rotational degrees of freedom, and then the zero-point vibrational energy. The atom has only contributions from translational energy and the pV term:

$$2H_H = H_{H_2} + 2\Delta H° \tag{3.92}$$

$$\frac{7}{2} k_B T + \frac{h\nu}{2} + 2\Delta H° - D_e = 2\frac{5}{2} k_B T \tag{3.93}$$

which leads to

$$D_e = -3/2 k_B T + \frac{h\nu}{2} + 2\Delta H° = 458.70 \, \text{kJ/mol} \tag{3.94}$$

We can now calculate both the equilibrium constant and the partial pressure of atomic hydrogen for the different temperatures. For $T = 1000$ K, we obtain

$$K_{H_2}(T=1000\,K) = \left(\frac{k_B T}{P_0}\right) \frac{\left(\frac{q_H^{trans}}{V} q_H^{el}\right)^2}{\frac{q_{H_2}^{trans}}{V} q_{H_2}^{vib} q_{H_2}^{rot} q_{H_2}^{el}}$$

$$= \frac{(1.38 \times 10^{-25})(6.01 \times 10^{30} \cdot 2)^2}{1.70 \times 10^{31} \cdot 0.042 \cdot 5.86 \cdot 9.11 \times 10^{23}}$$

$$= 5.24 \times 10^{-18} \tag{3.95}$$

Having demonstrated the manner to calculate K_{H_2} it is now straightforward to establish K_{H_2} and P_H at the various temperatures (Table 3.3).

Table 3.3 Equilibrium constants for the dissociation of H_2, N_2, and O_2 and the partial pressures of the atoms at different temperatures calculated from fundamental data given in Table 3.4.

T(K)	$K_{H_2}(T)$	p_H/p_0	$K_{N_2}(T)$	p_N/p_0	$K_{O_2}(T)$	p_O/p_0
298	5.81×10^{-72}	2.41×10^{-36}	6.35×10^{-160}	2.52×10^{-80}	6.13×10^{-81}	7.83×10^{-41}
1000	5.24×10^{-18}	2.29×10^{-9}	2.55×10^{-43}	5.05×10^{-22}	4.12×10^{-19}	6.42×10^{-10}
2000	3.13×10^{-6}	1.76×10^{-3}	2.23×10^{-18}	1.80×10^{-9}	1.22×10^{-5}	3.49×10^{-3}
3000	1.77×10^{-3}	1.72×10^{-1}	1.01×10^{-9}	3.18×10^{-5}	5.04×10^{-1}	5.01×10^{-1}

Table 3.4 Data used for calculating the fraction of the molecules dissociated. Atomic hydrogen is doubly degenerate in its ground state and the molecule is not degenerate. Similar assumptions are made for the nitrogen molecule. For oxygen, special care has to be taken as the molecular ground state is triply degenerate and the atomic state has three close-lying states that are 5-, 3-, and 1-fold degenerate; these levels are separated by 1901 and 2718 J/mol where the 5-fold degenerate state is the ground state.

Molecule	Vibrational frequency (cm^{-1})	Rotational frequency (cm^{-1})	$D_e = D_0 + h\nu/2$ (kJ/mol)
H$_2$	4400.39	60.864	458.7
N$_2$	2358.07	1.9987	956.8
O$_2$	1580.36	1.4457	504.1

Only at the very highest temperatures is it possible to obtain any appreciable amounts of atomic hydrogen. We have performed a similar calculation for nitrogen and oxygen for reference (Tables 3.3 and 3.4). Since the N$_2$ molecule is very stable (only CO is more stable), we obtain a D_e that is more than twice that of hydrogen, implying that it is practically impossible to dissociate N$_2$ in the gas phase. Hence, it is not feasible to produce ammonia by simply heating the gases N$_2$ and H$_2$ since the equilibrium of the reaction greatly disfavors ammonia production at high temperatures required to dissociate the reactants. Try to calculate the equilibrium constant for ammonia using the partition functions for the relevant molecules.

3.5
Collision Theory

If we assume that molecules can be considered as billiard balls (hard spheres) without internal degrees of freedom, then the probability of reaction between, say, A and B depends on how often a molecule A meets a molecule B, and also if during this collision sufficient energy is available to cross the energy barrier that separates the reactants, A and B, from the product, AB. Hence, we need to calculate the collision frequency for molecules A and B.

We will call ρ_A and ρ_B the number of molecules A and B per unit volume, respectively. Because the probability of collision is obviously related to the size of the molecules, we also define the diameters of the effective spheres that represent A and B, namely d_A and d_B.

According to Trautz and Lewis, who gave the first treatment of reaction rates in terms of the kinetic theory of collisions in 1916–1918 [1], the rate of collisions (not yet reaction) between the spheres A and B is

$$r_{\text{coll}} = k_{\text{coll}} \rho_A \rho_B = \frac{\pi d^2 \overline{u}}{\sigma_{AB}} \rho_A \rho_B; \quad \sigma_{AB} = 1 + \delta_{AB} \quad (3.96)$$

3.5 Collision Theory

where πd^2 is an effective cross section for collision, with $d = (d_A + d_B)/2$ the average between the two diameters, σ_{AB} is a symmetry factor to prevent double counting if A and B are identical molecules ($\delta_{AB} = 1$ if A = B, and 0 if A ≠ B) and $\bar{u}$ is the average relative velocity between the colliding molecules.

The effect of double counting is most easily seen in the following calculation. Suppose that the density of molecules is $\rho_A = \rho_B = 10$ and that A and B are identical. Consequently, the number of collisions between A and B is

$$r_{ccoll,AB} = \frac{\pi d^2 \bar{u}}{\sigma_{AB}} \rho_A \rho_B = \pi d^2 \bar{u} \cdot 100 \tag{3.97}$$

If we now take B equal to A, we have $\rho_A = 20$. The total number of collisions in the volume does not change and becomes:

$$r_{coll,AA} = \frac{\pi d^2 \bar{u}}{\sigma_{AA}} \rho_A^2 = \frac{\pi d^2 \bar{u}}{2} \cdot 20^2 = \pi d^2 \bar{u} \cdot 200 \tag{3.98}$$

that is, twice as many as we had collisions between A and B. However, in counting the total number of collisions in the AB mixture we need to include the AA and BB collisions amounting to

$$r_{coll,AA} + r_{coll,BB} = \frac{\pi d^2 \bar{u}}{\sigma_{AA}} \rho_A^2 + \frac{\pi d^2 \bar{u}}{\sigma_{BB}} \rho_B^2 = \pi d^2 \bar{u} \cdot 100 \tag{3.99}$$

so that the total number of collisions in the AB mixture is with $200\pi d^2 \bar{u}$ the same as in the volume containing twice the amount of A and no B. Had we forgotten to take the symmetry numbers into account, then the number of collisions in the volume with $\rho_A = 20$ would have been too high.

Figure 3.7 clarifies expression (3.96): the molecule A, along with corresponding area $1/4\pi d_A^2$, moves towards a second molecule B with a relative velocity u between the two. If the center of B lies within a cylinder with cross section π^2 and length $u\Delta t$, there will be a collision within time Δt.

The relative velocity between the molecules not only determines whether A and B collide, but also if the kinetic energy involved in the collision is sufficient to surmount the reaction barrier. Velocities in a mixture of particles in equilibrium are distributed according to the Maxwell–Boltzmann distribution in spherical coordinates:

$$f(v) = 4\pi \left(\frac{m}{2\pi k_B T}\right)^{3/2} v^2 e^{-mv^2/2k_B T} \tag{3.100}$$

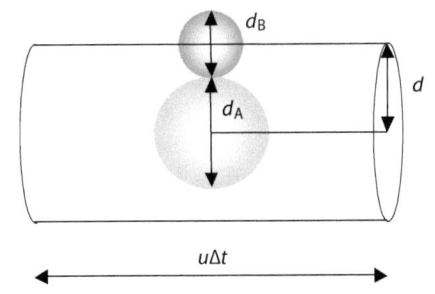

Figure 3.7 Collisions between molecules A and B occur if the center of molecule B lies within a cylinder with radius $d = (d_A + d_B)/2$ and length $u\Delta t$.

The same is true for the distribution of relative velocities, provided one replaces the mass with the reduced mass μ (defined in Eq. (3.55)) of the two molecules:

$$f(u) = 4\pi \left(\frac{\mu}{2\pi k_B T}\right)^{3/2} u^2 e^{-\mu u^2/2k_B T} \qquad (3.101)$$

We calculate the averages of the absolute and relative velocities in the Maxwell–Boltzmann distribution:

$$\bar{x} = \int_0^\infty x f(x) \, dx = \int_0^\infty x 4\pi \left(\frac{\mu}{2\pi k_B T}\right)^{3/2} x^2 e^{-\mu x^2/2k_B T} \, dx$$

$$= \sqrt{\frac{2k_B T}{\mu}} \frac{4}{\sqrt{\pi}} \int_0^\infty y^3 e^{-y^2} \, dy; \quad y = \sqrt{\frac{\mu}{2k_B T}} x \qquad (3.102)$$

leading to

$$\bar{v} = \left(\frac{8k_B T}{\pi m}\right)^{1/2}; \quad \bar{u} = \left(\frac{8k_B T}{\pi \mu}\right)^{1/2} \qquad (3.103)$$

which is often also used in a form expressing the masses in molar weights, and the energy per mol:

$$\bar{v} = \left(\frac{8RT}{\pi M}\right)^{1/2}; \quad \bar{u} = \left(\frac{8RT}{\pi \hat{\mu}}\right)^{1/2} \qquad (3.104)$$

With the expression for the average velocity, we arrive at the following expression for the rate of collisions between A and B per unit volume:

$$r_{\text{coll}} = \frac{\pi d^2}{\sigma_{AB}} \left(\frac{8k_B T}{\pi \mu}\right)^{1/2} \rho_A \rho_B; \quad \sigma_{AB} = 1 + \delta_{AB} \qquad (3.105)$$

Let us consider a 1 : 1 gas mixture of hydrogen and nitrogen at a total pressure of 1 bar at 300 K. We shall assume that the average molecular diameter is 3 Å or 3×10^{-10} m. Then the collision number is

$$r_{\text{coll}} = \pi d^2 \left(\frac{8k_B T}{\pi \mu}\right)^{1/2} \frac{P_{N_2} P_{H_2}}{(k_B T)^2} = 7.6 \times 10^{34} \text{ m}^{-3} \text{ s}^{-1} \qquad (3.106)$$

The average velocities of molecules (at $T = 300$ K) are on the order of 500 m s^{-1} (476 m s^{-1} for N_2 and 1781 m s^{-1} for H_2) and collision rates for a liter of gas at atmospheric pressures are of the order 10^{31} per second. Note that the collision rate depends strongly on density, through the terms ρ_A and ρ_B, but only weakly on temperature, as $T^{1/2}$.

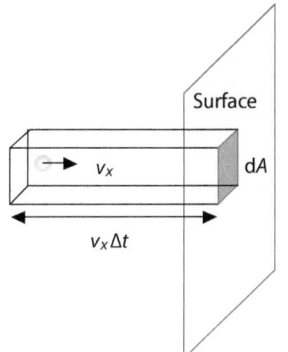

Figure 3.8 All molecules with velocity v_x perpendicular to the surface in the box with volume $v_x \Delta t\, dA$ will hit the surface within the area dA.

The Rate of Surface Collisions

As we are particularly interested in surface reactions and catalysis, we will calculate the rate of collisions between a gas and a surface. For a surface of area A (see Figure 3.8), the molecules that will be able to hit this surface must have a velocity component orthogonal to the surface v_x. All molecules with velocity v_x, given by the Maxwell–Boltzmann distribution $f(v_x)$ in Cartesian coordinates, at a distance $v_x \Delta t$ orthogonal to the surface will collide with the surface. The product $v_x \Delta t A = V$ defines a volume and the number of molecules therein with velocity v_x is $f(v_x)V(v_x)\rho$ where ρ is the density of molecules. By integrating over all v_x from 0 to infinity, we obtain the total number of collisions in time interval Δt on the area A. Since we are interested in the collision number per time and per area, we calculate

$$\begin{aligned}
r_{\text{coll-surf}} &= \frac{1}{\Delta t A} \int_0^\infty f(v_x) V(v_x) \rho\, dv_x = \rho \int_0^\infty v_x f(v_x)\, dv_x \\
&= \frac{p}{k_B T} \int_0^\infty v_x \sqrt{\frac{m}{2\pi k_B T}} e^{-\frac{m v_x^2}{2 k_B T}}\, dv_x \qquad (3.107) \\
&= \frac{p}{k_B T} \bar{v}_x = \frac{p}{k_B T} \sqrt{\frac{k_B T}{2\pi m}} = \frac{p}{\sqrt{2\pi m k_B T}}
\end{aligned}$$

where we have assumed that the gas behaves ideally, that is, $\rho = p/k_B T$.

The number of surface collisions at $p = 1$ bar and $T = 300$ K is, thus, $r_{\text{coll-surf}} = 1.08 \times 10^{28}$ m^{-2} s^{-1} for hydrogen and 2.88×10^{27} m^{-2} s^{-1} for nitrogen. Since there are typically 1.5×10^{19} surface atoms/m^2, a surface atom will on average be hit a billion times per second under ambient conditions. This, however, does not necessarily mean that the gas molecule reacts, particularly not if the reaction is an activated process.

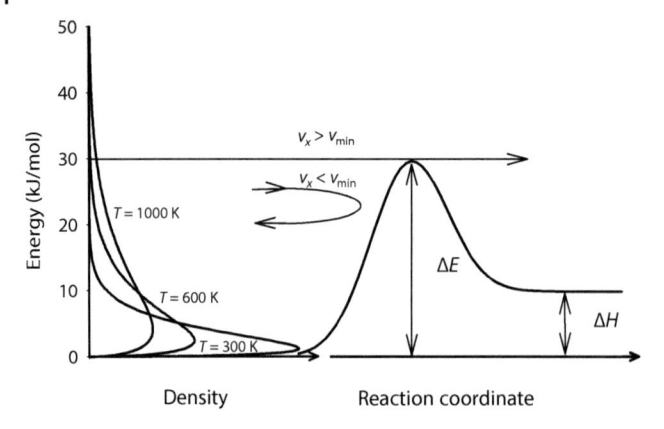

Figure 3.9 Molecules only react if they have sufficient energy to overcome the energy barrier ΔE. The height of the barrier can be transformed into a minimum velocity u_{min}. In the above case, $\Delta E = 30$ kJ/mol and only for a gas at $T = 1000$ K or higher can a substantial number of molecules overcome the barrier.

3.5.1
Reaction Probability

Returning to the general case of reaction between two molecules, we now know their rate of collision – a tremendously large number, of the order of 10^{31} L^{-1} s^{-1}. Now we need to determine the probability that the colliding molecules react. The condition for reaction is that the collision energy in the reaction coordinate along u (the line connecting the reacting species) is higher than the energy of the barrier to reaction, as shown in Figure 3.9. Hence, we need to calculate the fraction of collisions, P, for which the relative kinetic energy, $1/2\mu u^2$, is larger than ΔE. With u_{min} corresponding to the velocity for which $1/2\mu u_{min}^2 = \Delta E$, the fraction of reactive collisions becomes:

$$P = \frac{\int_{u_{min}}^{\infty} u f(u) \, du}{\int_0^{\infty} u f(u) \, du}$$

$$= \frac{\int_{u_{min}}^{\infty} 4\pi \left(\frac{\mu}{2\pi k_B T}\right)^{3/2} u^3 e^{-\mu u^2/2k_B T} \, du}{\int_0^{\infty} 4\pi \left(\frac{\mu}{2\pi k_B T}\right)^{3/2} u^3 e^{-\mu u^2/2k_B T} \, du} \quad (3.108)$$

$$= e^{-\Delta E/k_B T}$$

and we arrive at the expression for the rate of reaction per unit volume between two molecules, A and B, in the collision theory:

$$r_{react} = \frac{\pi d^2}{\sigma_{AB}} \left(\frac{8k_B T}{\pi \mu}\right)^{1/2} e^{-\Delta E/k_B T} \rho_A \rho_B; \quad \sigma_{AB} = 1 + \delta_{AB} \quad (3.109)$$

The rate is thus the number of collisions between A and B – a very large number – multiplied by the reaction probability, which may be a very small number. For example, if the energy barrier corresponds to 100 kJ/mol, the reaction probability is only 3.5×10^{-11} at 500 K. Hence, only a very small fraction of all collisions leads to product formation. In a way, reaction is a rare event! For examples of the application of collision theory see Laidler [1].

Expression (3.109) appears to be similar to the Arrhenius expression, but there is an important difference. In the Arrhenius equation, the temperature dependence is in the exponential only, whereas in the collision theory we find a $T^{1/2}$ dependence in the pre-exponential factor. We shall see later that transition state theory predicts even stronger dependences on T.

The corresponding activation energy is obtained by logarithmic differentiation:

$$E_{a,\,coll} \equiv k_B T^2 \frac{\partial}{\partial T} \ln r_{coll} = \Delta E + 1/2 k_B T \tag{3.110}$$

Hence, to write the rate in the form of the Arrhenius equation, we replace the energy barrier ΔE by the activation energy $\Delta E + 1/2 k_B T$, which means that the pre-exponential factor contains an additional factor $e^{1/2}$:

$$v_{coll} = \frac{e^{1/2} \pi d^2}{\sigma_{AB}} \left(\frac{8 k_B T}{\pi \mu} \right)^{1/2} ; \quad \sigma_{AB} = 1 + \delta_{AB} \tag{3.111}$$

In experimental practice, we usually ignore the temperature dependence of the prefactor and extract the activation energy by making an Arrhenius plot, as discussed in Chapter 2. The consequence of collision theory, however, is that a curved plot, rather than a straight line, will result if the activation energy is of the same order of $k_B T$.

How useful is the rate expression derived from collision theory for describing adsorption? For cases in which adsorption is not activated, that is, $E_a = 0$, the collision frequency describes in essence the rate of impingement of a gas on a surface. This is an upper limit for the rate of adsorption. In general, the rate of adsorption is lower because the molecules must, for example, (1) interact inelastically with the surface, (2) have a specific orientation, or (3) hit on a specific site on the surface unit cell in order to adsorb. Therefore, an efficiency coefficient is introduced, called the *sticking coefficient*, which describes the probability that a molecule becomes adsorbed upon collision with the surface. In the case of activated adsorption (e.g., dissociative adsorption of N_2 and CH_4 on many metals), the exponential term is usually included in the sticking coefficient. For inactivated adsorption, the collision theory can only account for sticking by the opportunistic introduction of an improvised correction coefficient. Transition state theory offers a more sensible description of the sticking coefficient, as we shall see.

3.5.2
Fundamental Objection against Collision Theory

Although collision theory provides a useful formalism to estimate an upper limit for the rate of reaction, it possesses the great disadvantage that it is not capable of describing the free energy changes of a reaction event, since it only deals with the individual molecules and does not take the ensemble into consideration. As such, the theory is essentially in conflict with thermodynamics. This becomes immediately apparent if we derive equilibrium constants on the basis of collision theory.

Consider the equilibrium

$$A + B \underset{k^+}{\overset{k^-}{\rightleftharpoons}} C + D \qquad (3.112)$$

Applying collision theory to the forward and the reverse reaction, and taking the ratio, we obtain the equilibrium constant:

$$K = \frac{k^+}{k^-} = \frac{\pi d_{AB}^2 \left(\frac{8k_B T}{\pi \mu_{AB}}\right)^{1/2} e^{-\Delta E/k_B T}}{\pi d_{CD}^2 \left(\frac{8k_B T}{\pi \mu_{CD}}\right)^{1/2} e^{-(\Delta E - \Delta H)/k_B T}} = \frac{d_{AB}^2 \mu_{CD}^{1/2}}{d_{CD}^2 \mu_{AB}^{1/2}} e^{-\Delta H/k_B T} \qquad (3.113)$$

Hence, we find a relation between K and the enthalpy of the reaction, instead of the free energy, and the expression for the equilibrium is in conflict with equilibrium thermodynamics, in particular with Eq. (3.32) of Chapter 2, since the prefactor can not be related to the change of entropy of the system. Hence, collision theory is not in accordance with thermodynamics.

3.6
Activation of Reacting Molecules by Collisions: The Lindemann Theory

A question that intrigued several kineticists around 1920 was the following. For bimolecular reactions of the type A + B = Products, collision theory gave at least a plausible conceptual picture: if the collision between A and B is sufficiently vigorous, the energy barrier separating reactants and products can be crossed. How, though, can one explain the case of monomolecular elementary reaction, for example, an isomerization, such as cyclopropane to propylene, or the decomposition of a molecule such as sulfuryl, SO_2Cl_2, into SO_2 and Cl_2? Highly original, though not always realistic, explanations were suggested. For instance, Jean Perrin proposed in 1919 that the walls of the reaction vessel radiated a kind of reaction energy that enabled the reaction. Frederick Lindemann, the later Lord Cherwell and Minister of Defense in Churchill's Second World War cabinet, was strongly against Perrin's "radiation theory of chemical action" and in 1921 he proposed an alternative explanation, which is still generally accepted today [1].

According to the Lindemann–Christiansen hypothesis, formulated independently by both scientists in 1921, all molecules acquire and lose energy by collisions with surrounding molecules. This is expressed in the simplified form of the Lindemann mechanism, in which we use an asterisk to indicate a highly energetic or activated molecule, which has sufficient energy to cross the barrier towards the product side, and M is a molecule from the surroundings; M may be from the same type as A:

$$A + M \underset{k^+}{\overset{k^-}{\rightleftharpoons}} A^* + M \tag{3.114}$$

$$A^* \xrightarrow{k_2^+} P \tag{3.115}$$

Application of the steady state approximation to the energized intermediate A* gives the concentration of this elusive species:

$$\frac{d[A^*]}{dt} = k_1^+[A][M] - k_2^+[A^*] - k_1^-[A^*][M] \equiv 0 \tag{3.116}$$

$$[A^*] = \frac{k_1^+[A][M]}{k_2^+ + k_1^-[M]} \tag{3.117}$$

and hence the rate of reaction becomes

$$\frac{d[P]}{dt} = \frac{k_1^+ k_2^+[A][M]}{k_2^+ + k_1^-[M]} \tag{3.118}$$

In the normal pressure regime, where the number of collisions is very high (according to the previous section 10^{31} s^{-1} L^{-1} at 1 bar), the denominator will be dominated by the second term, and hence we find the "normal" result that the overall rate is first order in the concentration of A. However, at low enough pressures, the first term of the denominator becomes dominant. Suppose that A and M are the same species, then the rate of the unimolecular reaction from A to P becomes

$$\frac{d[P]}{dt} = \frac{k_1^+ k_2^+[A][A]}{k_2^+ + k_1^-[A]} \approx k_1^+[A]^2 \quad \text{for} \quad k_1^-[A] \ll k_2^+ \tag{3.119}$$

that is, second order in [A]. Hence, according to the Lindemann theory a truly unimolecular reaction does not exist because collisions with surrounding molecules are needed to bring the reacting molecule in a sufficiently energetic state that it becomes capable of crossing the energy barrier.

Note that if we nevertheless write the rate equation of a unimolecular reaction in the form of a first order reaction, $r = k[A]$, and perform kinetic measurements over a large range of pressures, we will find a rate constant that falls off at low pressures. Such behavior has been reported for several isomerization and decomposition reactions.

3.7
Transition State Theory

Building on the Lindemann theory described above, Henry Eyring, and independently also M.G. Evans and Michael Polanyi, developed around 1935 a theory for the rate of reaction that is still used, namely the transition state theory [1].

In qualitative terms, the reaction proceeds via an activated complex, the transition state, located at the top of the energy barrier between reactants and products. Reacting molecules are activated to the transition state by collisions with surrounding molecules. Crossing the barrier is only possible in the forward direction. The reaction event is described by a single parameter, called the reaction coordinate, which is usually a vibration. The reaction can, thus, be visualized as a journey over a potential energy surface (a mountain landscape) where the transition state lies at the saddle point (the col of a mountain pass).

The situation is represented by the diagram in Figure 3.10 and by the following reaction scheme

$$R \rightleftharpoons R^{\#} \longrightarrow P \tag{3.120}$$

which shows that passage over the barrier can proceed in the forward direction only. We assume that R is fully equilibrated with $R^{\#}$, with equilibrium constant $K^{\#}$. Next we describe the reaction in terms of a suitable reaction coordinate. If the reaction is a dissociation of a diatomic molecule, then the reaction coordinate would be the stretching vibration between the two atoms in the molecule because activation of this vibration would weaken the bond between them. The rate of reaction from the transition state to the product is taken as the frequency of the reaction coordinate, $h\nu$. A little thought tells us that this is plausible as if the bond of our dissociating molecule is stretched further than in the transition state, it will break. The rate at which the transition state attempts to achieve this equals the frequency of vibration.

Hence, the rate of the overall reaction becomes

$$\frac{d[P]}{dt} = \nu \left[R^{\#}\right] = \nu K^{\#}[R] = \nu \frac{q'^{\#}}{q}[R] \tag{3.121}$$

We express the equilibrium constant in the partition functions of both the reactant and the transition state, and we take the partition function of the reaction coordinate separately:

$$\frac{d[P]}{dt} = \nu q_\nu \frac{q^{\#}}{q}[R] = \nu \frac{e^{-\frac{1}{2}\frac{h\nu}{k_B T}}}{1-e^{-\frac{h\nu}{k_B T}}} \frac{q^{\#}}{q}[R] \tag{3.122}$$

Because the frequency of a weakly bonded vibrating system is relatively small, that is, $k_B T \gg h\nu$, we may approximate its partition function by the classical limit $\frac{k_B T}{h\nu}$, and arrive at the rate expression in the transition state theory:

$$\frac{d[P]}{dt} = \frac{k_B T}{h} \frac{q^{\#}}{q}[R] = k_{TST}[R] \tag{3.123}$$

$$k_{TST} \equiv \frac{k_B T}{h} \frac{q^{\#}}{q} \tag{3.124}$$

where we should keep in mind that the reaction coordinate is excluded from the partition function of the transition state $q'^{\#}$; in fact it is accounted for by the factor

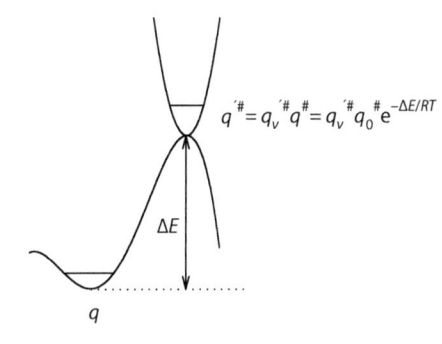

Figure 3.10 The common zero energy is with respect to the electronic ground state of q. The electronic ground state of $q^{\#}$ is, thus, shifted ΔE upwards and $q_0^{\#}$ refers to this new zero point.

$$q^{'\#} = q_v^{'\#} q^{\#} = q_v^{'\#} q_0^{\#} e^{-\Delta E/RT}$$

$k_B T/h$. Notice this is not strictly correct since it is somewhat arbitrary to discuss a vibration in a potential that gets weaker as we approach the transition state in the reaction coordinate. At the transition state, the potential actually vanishes and the particle can be considered to be freely moving in this one dimension.

The partition functions $q^{\#}$ and q in Eq. (3.124) are given with respect to the same zero of energy, which is not convenient. The energy difference between the ground state and the transition state is the energy barrier, ΔE, used in collision theory. By referring the transition state to its own electronic ground state (see Figure 3.10), a more convenient $q_0^{\#}$ is defined. Incorporating it in our rate expression, we automatically obtain an expression that bears great resemblance to the Arrhenius equation:

$$k_{\text{TST}} = \frac{k_B T}{h} \frac{q_0^{\#}}{q} e^{-\Delta E/k_B T} \tag{3.125}$$

Thermodynamic Form of the Transition State Rate Expression

Expression (3.123) is often also written in the form

$$k_{\text{TST}} = \frac{k_B T}{h} K^{\#} \tag{3.126}$$

with the provision that the reaction coordinate is excluded from the equilibrium constant or, in other words, that the transition state is fully equilibrated with the reactant except for the degree of freedom representing the reaction coordinate. Inserting the relation between the equilibrium constant and the Gibbs free energy, we find

$$k_{\text{TST}} = \frac{k_B T}{h} e^{-\Delta G_0^{\#}/RT} = \frac{k_B T}{h} e^{\Delta S_0^{\#}/R} e^{-\Delta H_0^{\#}/RT} \tag{3.127}$$

Because the thermodynamic quantities are commonly expressed in kJ/mol, we have replaced the Boltzmann constants by the gas constant, R. We calculate the activation energy (in kJ/mol) by applying the definition

$$E_{a,\text{TST}} = RT^2 \frac{\partial}{\partial T} \ln k_{\text{TST}} = \Delta H_0^{\#} + RT \tag{3.128}$$

If we choose to write the rate in the Arrhenius form as

$$k_{\text{TST}} = \nu_{\text{TST}} e^{-\frac{E_{a,\text{TST}}}{RT}} \tag{3.129}$$

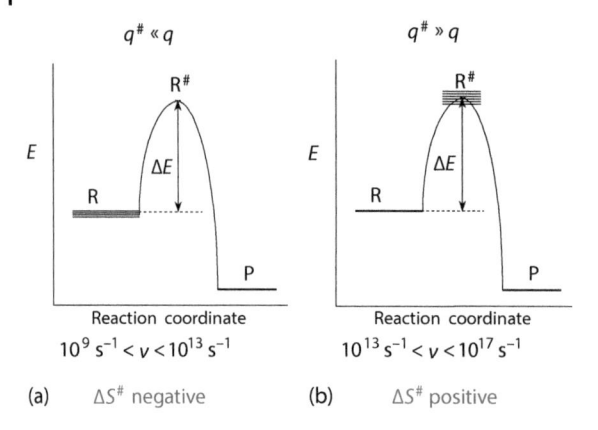

Figure 3.11 (a) A tight transition state possesses lower entropy than the ground state of the reactants; therefore, the pre-exponential factor will be lower than the standard value of $ek_BT/h \approx 10^{13}$ s^{-1}. (b) A loose transition state, on the other hand, has higher entropy and consequently the pre-exponential factor is higher.

the pre-exponential factor becomes

$$\nu_{TST} \equiv e \frac{k_B T}{h} e^{\Delta S_0^\#/R} \tag{3.130}$$

As expression (3.130) shows, a gain in entropy in going from the reactants to the transition state results in a large pre-exponential factor and increases the rate of reaction. This situation corresponds to partition functions of vibration, rotation, and/or translation that are higher in the transition state than in the ground state of the reactant; this is often called a loose transition state (see Figure 3.11). An alternative viewpoint is that the transition state provides more energy levels that can be occupied at the given temperature than the ground state does.

Of course, the converse situation, in which the entropy of the transition state is lower than that of the ground state of the reactant, can also occur (Figure 3.11). In this case, one speaks of a tight transition state; tight, because rotations, vibration or motion of the activated complex are more restricted than in the ground state of the reactant. The dissociation of molecules on a surface provides an example that we shall discuss in the next section.

The neutral situation, in which the entropy of the transition state does not notably differ from that of the reactant(s) in the ground state yields the standard pre-exponential factor $h\nu = ek_B T/h$ which is on the order of 10^{13} s^{-1}. The influence of the entropy term gives rise to a range of possibilities for the pre-exponential factor from about 10^9 to 10^{17} s^{-1}. If a measured value falls outside this range, one should wonder if the step is elementary or if perhaps only a small number of the reactants participate in the reaction. In the chapters on surface reactions, we will encounter several situations where this is the case.

Example

An example application of transition state theory is the gas-phase dissociation of a diatomic molecule such as CO

$$CO \rightleftharpoons CO^{\#} \longrightarrow C + O \tag{3.131}$$

The reaction coordinate is the C–O stretch vibration of the CO molecule in its transition state. The rate constant is

$$k_{\text{TST}} = \frac{k_B T}{h} \frac{q_{\text{trans}}^{3D,\#} q_{\text{rot}}^{\#}}{q_{\text{trans}}^{3D} q_{\text{vib}} q_{\text{rot}}} e^{-\Delta E/k_B T} \tag{3.132}$$

where the vibrational frequency of the transition state is omitted because it is the reaction coordinate. Realizing that the translational partition functions, depending on mass, temperature, and available volume per molecule, are equal in the ground state and the transition state, while the vibrational frequency of the CO molecule is large enough (2143 cm^{-1}) that q'_{vib} does not appreciably deviate from 1, we have

$$q_{\text{vib}} = \frac{e^{-\frac{h\nu}{2k_B T}}}{1 - e^{-\frac{h\nu}{k_B T}}} \cong e^{-\frac{h\nu}{2k_B T}} \tag{3.133}$$

As the rotational partition function for a diatomic molecule such as CO is

$$q_{\text{rot}}(CO) = \frac{8\pi^2 k_B T}{h^2} I; \quad I = \mu r_{\text{C-O}}^2 \tag{3.134}$$

we can simplify the expression for the rate constant of CO dissociation in the gas phase to

$$k_{\text{TST}} = \frac{k_B T}{h} \left(\frac{r_{\text{C-O}}^{\#}}{r_{\text{C-O}}}\right)^2 e^{-\left(\Delta E - \frac{h\nu}{2}\right)/k_B T} \tag{3.135}$$

This leaves us with an activation energy equal to

$$E_a = \Delta E + k_B T - \frac{h\nu}{2} \tag{3.136}$$

and a pre-exponential factor

$$\nu = \frac{k_B T e}{h} \left(\frac{r_{\text{C-O}}^{\#}}{r_{\text{C-O}}}\right)^2 \tag{3.137}$$

Example

Consider a simple gas phase reaction where CO reacts with O_2:

$$CO + O_2 \rightleftharpoons CO_2 + O \tag{3.138}$$

for which the transition state is assumed to consist of a OCOO complex. We will determine the temperature dependence of the pre-exponential factor for this reaction on the assumption that the transition state is linear and (for simplicity) that all the vibrational energies are much lower than $k_B T$ ($q_{vib} \propto k_B T/(h\nu)$).

The rate constant is written according to transition state theory as

$$k_{TST} = \frac{k_B T}{h} \frac{q'^{\#}_{OCOO}}{q'_{O_2} q'_{CO}} e^{\frac{-\Delta E}{k_B T}} = \nu_{TST} e^{\frac{-\Delta E}{k_B T}} \tag{3.139}$$

where all vibrational zero-point energies have been lumped into ΔE. As all three molecules are linear, we only have two rotational modes for each. There are three translational and $3N - 5$ vibrational modes. Note that since the molecules considered here are relatively simple, we do not have to consider phenomena such as internal rotations, which may further complicate the situation. Table 3.6 summarizes all temperature-dependent terms.

We obtain the temperature dependence of the pre-exponential factor by inserting the values of Table 3.6 in Eq. (3.139) for the prefactor:

Table 3.5 The number of degrees of freedom in translation, rotation, and vibrations of the reacting molecules and the transition state in the gas-phase reaction of CO and O_2 and the temperature dependence these modes contribute to the partition function. Note that one of the modes of the transition state complex is the reaction coordinate, so that only six vibrational modes are listed.

Species	Modes q_{trans}	Modes q_{rot}	Modes q_{vib}	Resulting q
CO	$3T^{3/2}$	$2T$	$1T$	$T^{7/2}$
O_2	$3T^{3/2}$	$2T$	$1T$	$T^{7/2}$
OCOO	$3T^{3/2}$	$2T$	$6T^6$	$T^{17/2}$

$$\nu_{TST} \propto T \frac{T^{17/2}}{T^{7/2} T^{7/2}} = T^{5/2} \tag{3.140}$$

The reader may now wish to verify that the activation energy calculated by logarithmic differentiation obtains a contribution $5k_B T/2$ in addition to ΔE, whereas the pre-exponential needs to be multiplied by a factor $e^{5/2}$ in order to properly compare Eq. (3.139) with the Arrhenius equation. Although the prefactor turns out to have a rather strong temperature dependence, the deviation of a $\ln k$ versus $1/T$ Arrhenius plot from a straight line will be small if the activation energy is not too small.

3.8
Transition State Theory of Surface Reactions

In this section, we derive rate expressions for elementary surface reactions, such as adsorption, desorption, and dissociation, in general terms. Specific cases will be dealt with in later chapters. The expressions give important insight into what experimentally determined sticking coefficients and pre-exponential factors mean. We limit ourselves to systems where the adsorbates are randomly distributed on the surface and we ignore interactions between adsorbed species, that is, we employ the mean-field approximation, which works well in the low coverage limit if there are no or only repulsive interactions between the absorbed species. With attractive interactions, however, the absorbed species tend to agglomerate into islands even at low coverage. Such situations are much more complex and call for treatment on the basis of Monte Carlo simulations.

3.8.1
Adsorption of Atoms

In the adsorption of atoms and molecules, we need to distinguish between two cases: direct or indirect adsorption. The direct adsorption process is one in which the particle collides with the surface and stays at the point of impact as an adsorbed species. In the much more common indirect mode, the particle first adsorbs in a weakly bound precursor state, in which it moves freely over the surface, until after some time it forms a bond with an adsorption site.

3.8.1.1 Indirect Adsorption

For the case of indirect atomic adsorption, the transition state is that of two-dimensional motion over the surface:

$$A + * \rightleftharpoons A*^{\#}_{\text{mobile}} \tag{3.141}$$

After some time in the mobile precursor state, the atom finds a free site and forms a true chemical bond with the surface:

$$A*^{\#}_{\text{mobile}} \longrightarrow A* \tag{3.142}$$

The system we consider consists of a volume of gas containing N_g gas atoms interacting with a surface containing M adsorption sites or $N_0 = \frac{M}{A}$ adsorption sites per area. In terms of transition state theory, the rate of reaction is then

$$\frac{dN_{A*}}{dt} = \nu N^{\#} = \nu K^{\#} N_g \tag{3.143}$$

Here, we use the convenient notation # for the transition state $A*^{\#}_{2D}$. As we already have seen, it is customary to introduce the concept of surface coverage when dealing with reactions on surfaces. The coverage of $A*$ is given by

$$\theta_{A*} = \frac{N_{A*}}{M} \tag{3.144}$$

3 Reaction Rate Theory

resulting in

$$\frac{d\theta_{A*}}{dt} = \frac{dN_{A*}}{M\,dt} = \nu\frac{K^{\#}N_g}{M} = \nu\frac{K^{\#}}{M}\frac{V}{k_B T}p_A \equiv k_{TST}p_A \qquad (3.145)$$

where k_{TST} relates the interesting quantities of coverage of $A*$ on, and pressure of A above, the surface. Note that we have assumed ideal gas behavior $N_g = \frac{pV}{k_B T}$.

We now need an expression for the equilibrium constant between the gas phase and the transition state complex. The reaction coordinate is again the (very weak) vibration between the atom and the surface. There are no other vibrations parallel to the surface because the atom is moving freely in two dimensions. The relevant partition functions for the atoms in the gas phase and in the transition state are

$$Q_{gas} = \frac{q_g^{N_g}}{N_g!}; \quad Q_{\#} = \frac{q_{\#}^{N_{\#}}}{N_{\#}!} \qquad (3.146)$$

where N_{gas} and $N_{\#}$ are the numbers of atoms in the two states and

$$q_g = q_{trans}^{3D}; \quad q_{\#} = q_v^{\#} q_{trans}^{2D} \qquad (3.147)$$

Since we assume the gas phase to be in equilibrium with the transition state, we may use the result from statistical mechanics that $\mu_{gas} = \mu_{\#}$, in which $\mu = -k_B T \frac{\delta \ln(Q)}{\delta N}$.

This leads to

$$K^{\#} = \frac{N_{\#}}{N_g} = \frac{q_v^{\#} q_{trans}^{2D}}{q_{trans}^{3D}} \qquad (3.148)$$

and the uptake rate can now be expressed as

$$\begin{aligned}\frac{d\theta_A}{dt} &= \frac{\nu N^{\#}}{M} = \frac{\nu K^{\#}}{M}\frac{V p_A}{k_B T} = \frac{\nu q_v^{\#} q_{trans}^{2D}}{M\, q_{trans}^{3D}}\frac{V p_A}{k_B T} \\ &\cong \frac{k_B T}{Mh}\frac{A(2\pi m k_B T)/h^2}{V(2\pi m k_B T)^{3/2}/h^3}\frac{V p_A}{k_B T} \\ &= \frac{A p_A}{M\sqrt{2\pi m k_B T}} = \frac{p_A}{N_0\sqrt{2\pi m k_B T}}\end{aligned} \qquad (3.149)$$

Hereby, we obtain the important result

$$\frac{d\theta_{A*}}{dt} = \frac{p_A}{N_0\sqrt{2\pi m k_B T}} \qquad (3.150)$$

which, if multiplied with the number of sites per area N_0, gives us the number of atoms adsorbed per area

$$N_0\frac{d\theta_{A*}}{dt} = \frac{M}{A}\frac{d\theta_{A*}}{dt} = \frac{1}{A}\frac{dN_{A*}}{dt} = r_{coll.surface} = \frac{p_A}{\sqrt{2\pi m k_B T}} \qquad (3.151)$$

which is equivalent to the formula derived earlier from collision theory, Eq. (3.107). This is expected because atoms are essentially hard spheres without internal degrees of freedom. Also, in deriving Eq. (3.106) we did not enforce any limitations on the motion of the atoms in the plane of surface after collision.

Under the conditions described above, all atoms hitting the surface will be adsorbed into the precursor state. However, atoms that are already absorbed may block sites for new incoming atoms; thus, the above model is in principle only valid for low coverages. For higher coverages, we have to consider the detailed behavior of the atoms adsorbed into the precursor state. Often, such precursors are very weakly adsorbed and the atoms may either react further into a chemisorbed state (whereby they are localized at an adsorption site) or they desorb. If a weakly adsorbed state exists on top of already adsorbed atoms, it is called an extrinsic precursor (contrary to the intrinsic precursor on the clean surface), and coverage will have no effect on the rate of adsorption. If such an extrinsic precursor state is not available, the adsorption rate will be proportional to the free surface area. The model breaks down, however, at high coverage, since one cannot have a free two-dimensional gas on a limited and fragmented surface area.

3.8.1.2 Direct Adsorption

In the direct adsorption process, the ground state of the atom is in the three-dimensional space above the surface (we apologize for the somewhat counterintuitive terminology of using ground state for a molecule above the surface). The transition state is the fully immobilized atom on the surface site, which is not allowed to move around

$$A + * \rightleftharpoons A*^{\#}_{\text{immobile}} \qquad (3.152)$$

$$A*^{\#}_{\text{immobile}} \longrightarrow A* \qquad (3.153)$$

The reaction coordinate that describes the adsorption process is the vibration between the atom and the surface. Strictly speaking, the adsorbed atom has three vibrational modes, one perpendicular to the surface, corresponding to the reaction coordinate, and two parallel to the surface. Usually these two vibrations – also called frustrated translational modes – are very soft meaning that $k_B T \gg h\nu$. Associative (nondissociative) adsorption furthermore usually occurs without an energy barrier, and we will therefore assume that $\Delta E = 0$. Hence, we can now write the transition state expression for the rate of direct adsorption of an atom via this transition state, applying the same method as used above for the indirect adsorption.

The gas-phase partition function Q_{gas} of the atom is the same; however, since the atoms are immediately immobilized on a two-dimensional surface, we need to take the configuration of the adsorbed atoms into account in the transition state. Again we consider a surface containing M sites each with an area of a^2. The density of sites per area is $N_0 = \frac{M}{A} = \frac{1}{a^2}$. The M sites are not necessarily free as some could be occupied already; thus, the number of free sites will be M' and $\theta_* = M'/M = (1 - \theta_A)$. If we have $N_{\#}$ atoms adsorbed on these sites (we again write # for the transition state $A^{+\#}_{immobile}$), the partition function for this system is given by

$$Q^{\#} = \frac{M'!}{N_{\#}!(M'-N_{\#})!}(q^{\#})^{N_{\#}} \qquad (3.154)$$

Setting the chemical potentials equal, $\mu_g = \mu_{\#}$, and applying Stirling's formula $\ln(N!) \cong N* \ln(N) - N$, we obtain after some manipulations

$$K^{\#} = \frac{N_{\#}}{N_g} = (M'-N_{\#})\frac{q^{\#}}{q_g} \quad \text{or} \quad K^{\#} = M(\theta_* - \theta_{\#})\frac{q^{\#}}{q_g} \qquad (3.155)$$

where we introduce the coverage $\theta_{\#} = \frac{N_{\#}}{M}$ of species in the transition state. In general, $\theta_{\#}$ is $\ll \theta_*$ and can be neglected, that is, $\theta_* - \theta_{\#} \cong \theta_*$. We can now easily derive the Langmuir isotherm, to which we shall return to later. Using Eq. (3.145), we find

$$\frac{d\theta_A}{dt} = \nu\frac{N^{\#}}{M} = \frac{\nu K^{\#}}{M}\frac{Vp_A}{k_BT} = \frac{\nu(M'-N_{\#})}{M}\frac{q_{\#}}{q_g}\frac{Vp_A}{k_BT}$$

$$= \nu(\theta_* - \theta_{\#})\frac{q_v^{\#} q_{2D\text{-vib}}^{\#}}{q_{\text{trans}}^{3D}}\frac{Vp_A}{k_BT}$$

$$\cong \frac{Vp_A\theta_*}{h}\frac{q_{2D\text{-vib}}^{\#}}{q_{\text{trans}}^{3D}}$$

$$\cong \frac{Vp_A\theta_*}{h}\frac{\left(\frac{k_BT}{h\nu_{2D}}\right)^2 h^3}{V(2\pi m k_B T)^{3/2}} e^{\frac{-(\Delta E + h\nu_{2D\text{-vib}})}{k_BT}}$$

$$= \frac{p_A\theta_*}{\sqrt{2\pi m k_B T}}\frac{\left(\frac{k_BT}{h\nu_{2D}}\right)^2 h^2}{2\pi m k_B T} e^{\frac{-(\Delta E + h\nu_{2D\text{-vib}})}{k_BT}} \qquad (3.156)$$

$$= \frac{N_0 p_A\theta_*}{N_0\sqrt{2\pi m k_B T}}\frac{\left(\frac{k_BT}{h\nu_{2D}}\right)^2 h^2}{2\pi m k_B T} e^{\frac{-(\Delta E + h\nu_{2D\text{-vib}})}{k_BT}}$$

$$= \frac{\theta_* p_A S_0(T)}{N_0\sqrt{2\pi m k_B T}} = \frac{(1-\theta_A) p_A S_0(T)}{N_0\sqrt{2\pi m k_B T}}$$

Here, we have utilized Eq. (3.147) and assumed that the electronic ground state of the transition state has been raised by ΔE (to refer partition functions to the transition state's own ground state) and $q_{2D\text{-vib}}^{\#}$ is referred with respect to the bottom of the potential, as in Figure 3.10. Expression (3.156) shows that the adsorption rate per area is the collision number for that area times a factor $S_0(T)$, the so-called sticking coefficient, which must always be smaller than one. The sticking coefficient describes which fraction of the incident atoms were successful in reaching the adsorbed state.

Thus, the rate of adsorption becomes

$$\frac{d\theta_A}{dt} = \frac{S_0(T)p_A}{N_0\sqrt{2\pi m k_B T}}(1 - \theta_A) \tag{3.157}$$

where

$$S_0(T) = \frac{Mq_{2D\text{-vib}}^{\#}}{q_{trans}^{2D}} = \frac{q_{2D\text{-vib}}^{\#}}{\frac{q_{trans}^{2D}}{M}} = \frac{q_{2D\text{-vib}}^{\#}}{q_{trans\text{-unitcell}}^{2D}} \tag{3.158}$$

and

$$q_{trans\text{-unitcell}}^{2D} = \frac{q_{trans}^{2D}}{M} = \frac{A}{M}\frac{2\pi m k_B T}{h^2} = a^2\frac{2\pi m k_B T}{h^2} \tag{3.159}$$

is the translational partition function in a unit cell of area a^2.

To get a feeling for the value of a sticking coefficient for direct adsorption, we substitute some typical numbers, such as $a = 2.5$ Å, $m = 40 \cdot 1.66 \times 10^{-27}$ kg-Ar, $T = 300$ K, $\Delta E = 0$, and $h\nu = 40$ cm^{-1} into Eq. (3.157), and find (for argon)

$$S_0(T) = \frac{q_{2D\text{-vib}}^{\#}}{q_{trans\text{-unitcell}}^{2D}} = \frac{\left(\frac{k_B T}{h\nu_{2D}}\right)^2 h^2}{a^2 2\pi m k_B T} e^{-\frac{(\Delta E + h\nu_{2D\text{-vib}})}{k_B T}}$$

$$= \left(\frac{k_B T}{h\nu_{2D}}\right)^2 e^{-\frac{h\nu_{2D\text{-vib}}}{k_B T}} 4.0 \times 10^{-3} = 0.09 \tag{3.160}$$

This model breaks down when $h\nu$ becomes small in comparison to $k_B T$, as the potential becomes so soft that the atoms will no longer be located at a specific site but diffuse over the surface. In this intermediate regime the atoms move around in a potential

$$V(x, y) = \frac{1}{2}V_0\left(2 - \cos\left(\frac{2\pi x}{a}\right) - \cos\left(\frac{2\pi y}{a}\right)\right) \tag{3.161}$$

which, however, complicates things to a degree that is outside the scope of this treatment. Eventually, if the potential is sufficiently weak or the temperature high enough, the atoms will travel freely, corresponding to the situation of a two-dimensional gas. The vibrational partition function $q_{2D\text{-vib}}^{\#}$ then needs to be converted into a two-dimensional translational partition function, that is,

$$\left(\frac{k_B T}{h\nu_{2D}}\right)^2 e^{-\frac{h\nu_{2D\text{-vib}}}{k_B T}} \longrightarrow \frac{a^2 2\pi m k_B T}{h^2} \tag{3.162}$$

and we are back to the formulae for the case of indirect adsorption. From the above calculations, it is easily seen that S_0 varies from 1 (when the precursor is free to move in two dimensions) down to 10^{-3} (when the transition state is completely frozen, or immobilized) dependent on the nature of the transition state, even if $\Delta E = 0$. If the adsorption is activated, that is, $\Delta E > 0$, the sticking coefficient can be much smaller. Furthermore, in the case of molecules adsorbing on a surface, there may also be steric hindrance, so that the activation energy depends on the orientation of the molecule with respect to the atoms of the surface, and on the exact position of impact in the unit cell of the surface since the latter is not uniform. Hence, there are several good reasons to expect low sticking coefficients if the process proceeds through a direct and activated process. We note that the sticking coefficient S_0 for atoms can conveniently be approximated by

$$S_0(T) = \frac{q^{\#}_{\text{2D-vib}}}{q^{\text{2D}}_{\text{trans-unitcell}}} = \frac{q^{0\#}_{\text{2D-vib}}}{q^{\text{2D}}_{\text{trans-unitcell}}} e^{-\frac{\Delta E}{k_B T}} \qquad (3.163)$$

$$= \frac{q''^{0\#}_{\text{2D-vib}}}{q^{\text{2D}}_{\text{trans-unitcell}}} e^{-\frac{\Delta E - \frac{2h\nu_{\parallel}}{2}}{k_B T}} = S_0^0 e^{-\frac{\Delta E_{\text{act}}}{k_B T}}$$

Here, $q^{0\#}_{\text{2D-vib}}$ refers to the electronic ground state of the transition state and $q''^{0\#}_{\text{2D-vib}}$ to the vibrational ground state of the transition state. We have combined the two zero-point vibrations with ΔE into an effective activation energy ΔE_{act}. We shall later explain how this important quantity can be measured.

3.8.2
Adsorption of Molecules

The description of molecular adsorption is very similar to that of atoms, provided we account for the molecules possess internal degrees of freedom. Hence, we need to consider how these degrees change in going from the gas phase to the transition state of adsorption. The most general form for the rate constant of adsorption in the transition state theory is

$$\frac{d\theta_{A*}}{dt} = \frac{dN_{A*}}{M\,dt} = \frac{\nu N^{\#}}{M} = \nu \frac{K^{\#} N_g}{M} = \nu \frac{K^{\#}}{M} \frac{V}{k_B T} p_A \qquad (3.164)$$

The reaction coordinate is the vibration between the molecule and the surface, and is not include in the vibrational partition function of the transition state. Again we can distinguish two extremes.

3.8.2.1 Precursor-Mediated or Indirect Adsorption
If the molecule adsorbs via a physisorbed precursor state in which it is free to move across the surface, while rotating and vibrating (with possibly modified frequencies and rotational modes), we obtain

$$K^{\#} = \frac{N}{N_g} = \frac{q^{\#}}{q_{\text{gas}}} \qquad (3.165)$$

$$\frac{d\theta_A}{dt} = \frac{\nu N^{\#}}{M} = \frac{\nu V p_A K^{\#}}{k_B T M} = \frac{V p_A}{M h} \frac{q_{trans}^{2D} q_{rot}^{\#} q_{vib}^{\#}}{q_{trans}^{3D} q_{rot}^{gas} q_{vib}^{gas}}$$

$$= \frac{p_A}{N_0 \sqrt{2\pi m k_B T}} \frac{q_{rot}^{\#} q_{vib}^{\#}}{q_{rot}^{gas} q_{vib}^{gas}} \quad (3.166)$$

$$= \frac{p_A}{N_0 \sqrt{2\pi m k_B T}} S_0(T)$$

Thus,

$$\frac{d\theta_{A*}}{dt} = \frac{p_A S_0(T)}{N_0 \sqrt{2\pi m k_B T}} \quad (3.167)$$

in which we recognize the rate constant of impingement for collisions between the gas molecules and a ratio that we may again interpret as the sticking coefficient, that is, the probability that a molecule stays at the surface upon collision. Naturally if no changes occur in the internal coordinates $q_{rot}^{\#} = q_{rot}^{gas}$ and $q_{vib}^{\#} = q_{vib}^{gas}$, the sticking coefficient will become unity since

$$S_0(T) = \frac{q_{rot}^{\#} q_{vib}^{\#}}{q_{rot}^{gas} q_{vib}^{gas}} \quad (3.168)$$

Hence, according to the transition state theory, adsorption becomes more likely if the molecule in the mobile physisorbed precursor state retains its freedom to rotate and vibrate as it did in the gas phase. Of course, this situation corresponds to minimal entropy loss in the adsorption process. In general, the transition from the gas phase into confinement in two dimensions will always be associated with a loss in entropy and the sticking coefficient is normally smaller than unity.

3.8.2.2 Direct Adsorption

The other extreme is direct adsorption, in which the molecule lands immediately at its final adsorption site without the possibility to move over the surface. In this case, the only degrees of freedom the molecule has in the transition state are vibrational, among which the vibration between the molecule and the surface represents the reaction coordinate. This leaves us with the following expression, which immediately indicates that the rate constant is small:

$$\frac{d\theta_A}{dt} = \frac{\nu N^{\#}}{M} = \frac{\nu K^{\#}}{M} \frac{V p_A}{k_B T} = \frac{(\theta_* - \theta_{\#}) V p_A}{h} \frac{q'^{\#}}{q_{trans}^{3D} q_{rot}^{gas} q_{vib}^{gas}} \quad (3.169)$$

$$= \frac{S_0(T) p_A (\theta_* - \theta_{\#})}{N_0 \sqrt{2\pi m k_B T}} = \frac{S_0(T) p_A}{N_0 \sqrt{2\pi m k_B T}} (1 - \theta_A)$$

where we have used $\theta_* - \theta_{\#} \cong \theta_* = 1 - \theta_A$ and Eq. (3.155) for the equilibrium constant; $q'^{\#}$ is the partition function for the transition state, from which the reaction coordinate has been excluded. This leaves us with the following expression for the rate of direct adsorption:

$$\frac{d\theta_A}{dt} = \frac{p_A}{N_0\sqrt{2\pi m k_B T}} S_0(T)(1-\theta_A) \tag{3.170}$$

where

$$S_0(T) = \frac{M q'^{\#}}{q_{\text{trans}}^{2D} q_{\text{rot}}^{\text{gas}} q_{\text{vib}}^{\text{gas}}} = \frac{q'^{\#}}{q_{\text{trans-unitcell}}^{2D} q_{\text{rot}}^{\text{gas}} q_{\text{vib}}^{\text{gas}}} \tag{3.171}$$

Clearly, the sticking coefficient for the direct adsorption process is small since a considerable amount of entropy is lost when the molecule is "frozen in" on an adsorption site. In fact, adsorption of most molecules occurs via a mobile precursor state. Nevertheless, direct adsorption does occur, but it is usually coupled with the activated dissociation of a highly stable molecule. An example is the dissociative adsorption of CH_4, with sticking coefficients of the order 10^{-8}–10^{-6}. In this case, the sticking coefficient not only contains the partition functions but also an exponential activation energy term since the transition electronic ground state is increased by ΔE as illustrated in Figure 3.10:

$$q'^{\#} = q'^{\#0} e^{-\frac{\Delta E}{k_B T}} = q_{\text{vib}}^{\#0} q_{\text{frus-rot}}^{\#0} q_{\text{frus-trans}}^{\#0} e^{-\frac{\Delta E}{k_B T}} \tag{3.172}$$

The indices ^{0} refer to the fact that the partition function are given with respect to the bottom of the potential of the transition state and not that of the ground state, whereby we obtain the exponential factor. The partition function of the transition state can conveniently be split up into three terms. The first, $q_{\text{vib}}^{\#0}$, describes all the vibrations of the molecule in the transition state except the one that serves as the reaction coordinate. The factor $q_{\text{frus-rot}}^{\#0}$ describes the vibration due to the fact that the molecule may have a preferred orientation on the surface and is no longer capable of rotating. Instead the rotation becomes frustrated, as indicated in Figure 3.12 for a diatomic molecule such as CO. The last term $q_{\text{frus-trans}}^{\#0}$ describes the frustrated translational modes parallel to the surface (Figure 3.12). Again, if we know the ground state precisely, we can subtract the zero-point vibrational energies to arrive at ΔE_{act}, a quantity that can be measured.

In general, a nonlinear molecule with N atoms has three translational, three rotational, and $3N-6$ vibrational degrees of freedom in the gas phase, which reduce to three frustrated vibrational modes, three frustrated rotational modes, and $3N-6$ vibrational modes, minus the mode which is the reaction coordinate. For a linear molecule with N atoms, there are three translational, two rotational, and $3N-5$ vibrational degrees of freedom in the gas phase, and three frustrated vibrational modes, two frustrated rotational modes, and $3N-5$ vibrational modes, minus the reaction coordinate, on the surface. Thus, the transition state for direct adsorption of a CO molecule consists of two frustrated translational modes, two frustrated rotational modes, and one vibrational mode. In this case, the third frustrated translational mode vanishes since it is the reaction coordinate. More complex molecules may also have internal rotational levels, which further complicate the picture. It is beyond the scope of this book to treat such systems.

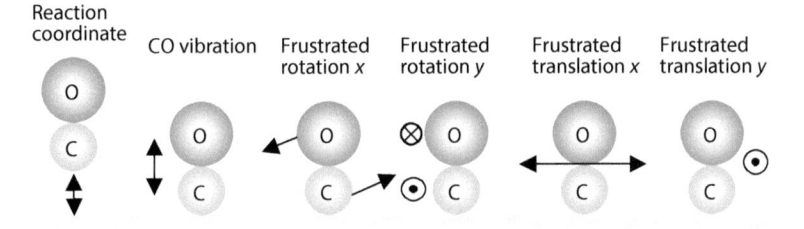

Figure 3.12 CO adsorbed on a surface possesses five modes of vibration. Note that the surface is not necessarily symmetric in x and y, and hence the vibrations in the x and y direction do not have to be degenerate.

3.8.3
Reaction between Adsorbates

In the following, we consider a surface with adsorbed atoms or molecules that react. We will leave out the details of the internal coordinates of these adsorbed species, but note that their partition functions can be found using the schemes presented above. Let us assume that species A reacts with B to form an adsorbed product AB via an activated complex $AB^{\#}$:

$$A* + B* \rightleftharpoons AB**^{\#} \tag{3.173}$$

$$AB**^{\#} \longrightarrow AB* + * \tag{3.174}$$

Again we assume that the species are distributed randomly over the surface. The equilibrium constant for this system is given by

$$K^{\#} = \frac{\theta_{AB^{\#}}}{\theta_A \theta_B} = \frac{q_{AB^{\#}}}{q_A q_B} \tag{3.175}$$

The rate of the surface reaction becomes

$$\frac{d\theta_{AB}}{dt} = \nu_{AB^{\#}} \theta_{AB^{\#}} = \nu_{AB^{\#}} K^{\#} \theta_A \theta_B$$
$$= \frac{\nu_{AB^{\#}} k_B T}{h \nu_{AB^{\#}}} \frac{q'^{\#}_{AB^{\#}}}{q_A q_B} \theta_A \theta_B \equiv k^+ \theta_A \theta_B \tag{3.176}$$

where $q'^{\#}$ refers to the fact that the reaction coordinate has been omitted from the partition function. If we also refer the partition function of the transition state to its electronic ground state, we can extract the shift in energy between the two ground states $\Delta E_{A+B-AB^{\#}}$ and write the k^+ in a more convenient manner

$$k^+ = \frac{k_B T}{h} \frac{q'^{\#}_{AB^{\#}}}{q_A q_B} = \frac{k_B T}{h} \frac{q'^{0\#}_{AB^{\#}}}{q_A q_B} e^{-\frac{\Delta E_{A+B-AB^{\#}}}{k_B T}} = k_0^+ e^{-\frac{\Delta E_{A+B-AB^{\#}}}{k_B T}} \tag{3.177}$$

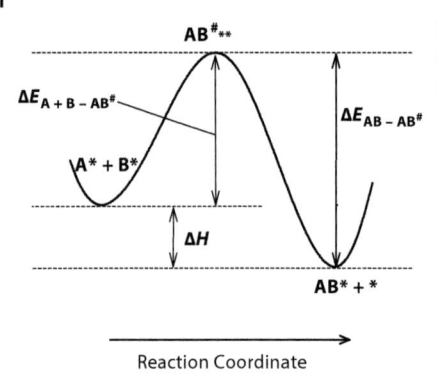

Figure 3.13 Potential energy diagram for the forward and reverse reaction A∗ + B∗ ⇌ AB∗ + ∗.

The reverse of reaction (3.173) and (3.174) is the dissociation of AB on the surface. It is easy to see that had we considered the reaction

$$AB* + * \rightleftharpoons AB^{\#}** \tag{3.178}$$

we would have obtained an equilibrium constant $K^{\#} = \frac{\theta_{AB^{\#}}}{\theta_* \theta_{AB}} = \frac{q_{AB^{\#}}}{q_{AB}}$ and the rate constant

$$\frac{d\theta_{AB}}{dt} = -\nu_{AB^{\#}} \theta_{AB^{\#}} = -k^{-} \theta_{AB} \theta_* \tag{3.179}$$

with

$$k^{-} = \frac{k_B T}{h} \frac{q'^{\#}_{AB^{\#}}}{q_{AB}} = \frac{k_B T}{h} \frac{q'^{0\#}_{AB^{\#}}}{q_{AB}} e^{-\frac{\Delta E_{AB-AB^{\#}}}{k_B T}} = k_0^{-} e^{-\frac{\Delta E_{AB-AB^{\#}}}{k_B T}} \tag{3.180}$$

Figure 3.13 illustrates the corresponding potential energy diagram.

If we assume that both sides of the transition state are in equilibrium with the transition state we find

$$\frac{d\theta_{AB^{\#}}}{dt} = k^{+} \theta_A \theta_B - k^{-} \theta_{AB} \theta_* = 0 \tag{3.181}$$

leading to

$$K_{eq} \equiv \frac{k^{+}}{k^{-}} = \frac{q_{AB}}{q_A q_B} e^{\frac{\Delta E_{AB-AB^{\#}} - \Delta E_{A+B-AB^{\#}}}{k_B T}} = e^{\frac{\Delta S}{k_B}} e^{-\frac{\Delta H}{k_B T}} \tag{3.182}$$

Note how the partition function for the transition state vanishes as a result of the equilibrium assumption and the equilibrium constant is determined, as it should be, by the initial and final state only. This result will prove to be useful when we consider more complex reactions. If several steps are in equilibrium, and we express the overall rate in terms of partition functions, many terms cancel. However, if there is no equilibrium, we can use the above approach to estimate the rate, provided we have sufficient knowledge of the energy levels in the activated complex to determine the relevant partition functions.

3.8.4
Desorption of Molecules

The opposite of adsorption, desorption, represents the end of the catalytic cycle. It is also the basis of temperature programmed desorption (TPD), an important method to studying the heats of adsorption and reactions on surface, so that the activation energies and pre-exponential factors of desorption can be measured. Hence, it is important to have theoretical expressions for the rates of desorption to compare with empirical data.

The transition state of a desorbing molecule can either be mobile, resembling the precursor state of indirect adsorption, or immobile, implying that the molecule only possesses vibrational modes in the transition state. However, for desorption at elevated temperatures, the adsorbed molecule may also be mobile. An example is the desorption of a diatomic molecule such as CO which, starting from an immobilized ground state but desorbing via a mobile transition state, may result in a pre-exponential factor well above the standard value of $ek_B T/h$ (as we shall derive in general terms in the following).

An immobile ground state does *not* imply that the molecule can not diffuse to neighboring sites. It can do so by hopping, which is considered as a separate reaction possibility, characterized by its own rate constant. Hence, we treat the immobile ground state of the molecule as the situation in which it resides on a certain site. The molecule has the option of two competing reactions, hopping to a neighboring site or desorbing to the gas phase. The opposite of the immobile ground state is the mobile ground state, where the molecule behaves as a two-dimensional gas.

We write the desorption reaction as

$$AB* \rightleftharpoons AB*^{\#} \tag{3.183}$$

$$AB*^{\#} \longrightarrow AB + * \tag{3.184}$$

The molecule reaches the transition state and from there it desorbs into the gas phase. To evaluate the rate constant, we use the same procedure as above: write down the partition functions for the participating species, equalize the chemical potentials, and find an expression for the number of molecules in the transition state. Since it is much more practical to do this in terms of coverage, we immediately obtain

$$K^{\#} = \frac{\theta_{AB^{\#}}}{\theta_{AB}} = \frac{q_{AB}^{\#}}{q_{AB}} \tag{3.185}$$

$$-\frac{d\theta_{AB}}{dt} = \nu_{AB^{\#}}\theta_{AB^{\#}} = \nu_{AB^{\#}}K^{\#}\theta_{AB} \cong \frac{k_B T}{h}\frac{q_{AB}^{\prime\#}}{q_{AB}}\theta_{AB}$$

$$= \frac{k_B T}{h}\frac{q_{AB}^{\prime 0\#}}{q_{AB}} e^{-\frac{\Delta E}{k_B T}}\theta_{AB} \tag{3.186}$$

where the negative sign shows that the reaction removes AB molecules from the surface. Here, we again refer $q'^{\#}_{AB}$ and $q'^{0\#}_{AB}$ to, respectively, the partition function of the transition state without the reaction coordinate and the transition partition function with respect to the electronic ground state of the transition state. Thus, ΔE is again the energy difference between the ground state and the transition state.

Usually the desorption process is written as

$$-\frac{d\theta_{AB}}{dt} = \nu e^{-\frac{E_a}{k_B T}} \theta_{AB} \tag{3.187}$$

which corresponds according to the definition of Arrhenius energy in Eq. (2.48) of Chapter 2 to

$$E_a = \Delta E + k_B T; \quad \nu = \frac{k_B T}{h} \frac{q'^{0\#}_{AB}}{q_{AB}} \tag{3.188}$$

This is not strictly correct, since often the vibrational terms are extracted out of $h\nu$ and included in the activation energy, which then deviates slightly from ΔE or E_a as given by Eq. (3.188).

Note that if the ratio of partition functions in Eq. (3.189) leaves a nonvanishing temperature dependence, it generates additional terms $k_B T$ in E_a and factors e in $h\nu$. Consequently, the prefactor ν strongly depends on the nature of the transition state.

Often the adsorbed species are bound rather strongly and can be considered immobile at the bottom of a vibrational well. The transition state may, however, have several possibilities, being, for example, a precursor that is highly mobile in two dimensions. Thus, depending on the values of the partition functions of rotation and translation, the pre-exponential factor for desorption may become significantly larger than the standard value of $k_B T/h \approx 10^{13}$ s^{-1}. Figure 3.14 summarizes a number of desorption processes with the corresponding range of pre-exponential factors. Table 3.6 contains several experimentally determined desorption parameters of CO from different surfaces. Most of the pre-exponential factors for CO desorption can be rationalized in terms of the loose transition state that we just sketched.

3.9
Summary

We finish this chapter by repeating some of the most important results. If we have detailed knowledge of the energy levels for an ensemble of particles (remember that statistical mechanics always operates on the basis of large numbers), it is possible to derive the partition function for the system and so calculate important macroscopic values such as energy, pressure, entropy, and in particular the chemical potential of the system.

Adsorbed state	Transition state	Desorbed state	Preexponential factor
	Mobile	○	~10^{15} s^{-1}
	Immobile	○	~10^{13} s^{-1}
	Mobile	○—●	~10^{14-16} s^{-1}
	Immobile	○—●	~10^{13} s^{-1}

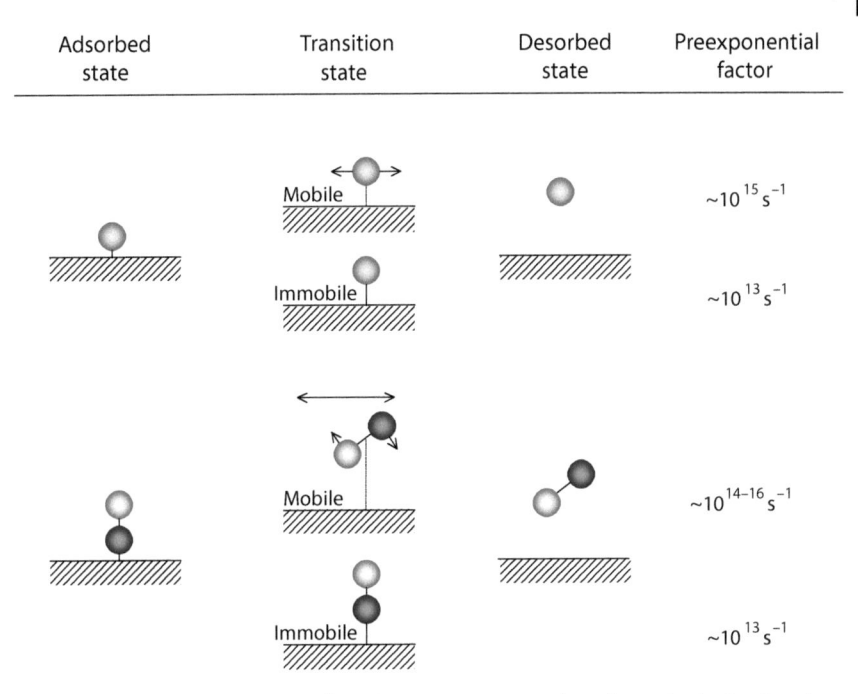

Figure 3.14 Microscopic pictures of the desorption of atoms and molecules via mobile and immobile transition states. If the transition state resembles the ground state, we expect a prefactor of desorption on the order of 10^{13} s^{-1}. If the adsorbates are mobile in the transition state, the prefactor increases by one or two orders of magnitude. For desorbing molecules, free rotation in the transition state increases the prefactor even further. The prefactors are roughly characteristic for atoms such as C, N, and O and molecules such as N_2, CO, NO, and O_2 (reproduced from [7] with permission of Wiley-VCH).

For two systems in chemical equilibrium, we can calculate the equilibrium constant from the ratio of partition functions by requiring the chemical potentials of the two systems to be equal.

Even if a system is not in chemical equilibrium, it is possible to calculate at which rate it is approaching equilibrium if we have sufficiently detailed knowledge of the energies involved in the transition state (so that it is possible to calculate the partition functions – the crucial step). However, computational chemistry has advanced to a level that good estimates of reaction rates can almost be obtained routinely. We will discuss examples in Chapter 6.

By applying the machinery of statistical thermodynamics, we have derived expressions for the adsorption, reaction, and desorption of molecules on and from a surface. The rate constants can in each case be described as a ratio between partition functions of the transition state and the reactants. Below, we summarize the most important results for elementary surface reactions. In principle, all the important constants involved (prefactors and activation energies) can be calculated from the partitions functions. These are, however, not easily obtainable and, where possible, experimentally determined values are used.

Table 3.6 Experimental activation energies and pre-exponential factors for CO and NO desorbing from a range of clean and well-defined single crystals. All data were obtained in the low coverage regime (Zhdanov et al. [5, 6]).

System	Prefactor (s^{-1})	Activation energy (kJ/mol)
CO/Co(0001)	10^{15}	118
CO/Ni(111)	10^{15}	130
CO/Ni(111)	10^{17}	155
CO/Ni(111)	10^{15}	126
CO/Ni(100)	10^{14}	130
CO/Cu(100)	10^{14}	67
CO/Ru(0001)	10^{16}	160
CO/Rh(111)	10^{14}	134
CO/Pd(111)	10^{14}	143
CO/Pd(111)	10^{15}	147
CO/Pd(100)	10^{16}	160
CO/Pd(211)	10^{14}	147
CO/Ir(110)	10^{13}	155
CO/Pt(111)	10^{14}	134
NO/Pt(111)	10^{15}	126
NO/Pt(111)	10^{16}	113

Adsorption

$$AB + * \longrightarrow AB* \tag{3.189}$$

$$\frac{d\theta_{AB}}{dt} = \frac{S_0(T) p_{AB}}{N_0 \sqrt{2\pi m k_B T}} \theta_* \tag{3.190}$$

$$S_0(T) = \frac{q'_{AB^\#}}{q^{2D}_{\text{trans-unitcell}} q^{\text{gas}}_{\text{rot}} q^{\text{gas}}_{\text{vib}}} = S_0^0 e^{-\frac{\Delta E_{\text{act}}}{k_B T}} \tag{3.191}$$

Reaction

$$A* + B* \longrightarrow AB* + * \tag{3.192}$$

$$\frac{d\theta_{AB^\#}}{dt} = k^+ \theta_A \theta_B \qquad k^+ = \frac{k_B T}{h} \frac{q'^\#_{AB^\#}}{q_A q_B} = k_0^+ e^{-\frac{\Delta E_{\text{act}-A+B}}{k_B T}} \tag{3.193}$$

$$\frac{d\theta_{AB^\#}}{dt} = -k^- \theta_{AB} \theta_* \qquad k^- = \frac{k_B T}{h} \frac{q'^\#_{AB^\#}}{q_{AB}} = k_0^- e^{-\frac{\Delta E_{\text{act}-AB}}{k_B T}} \tag{3.194}$$

Desorption

$$\text{AB}* \longrightarrow \text{AB} + * \tag{3.195}$$

$$\frac{d\theta_{AB}}{dt} = \frac{-k_B T}{h} \frac{q'^{\#}_{AB}}{q_{AB}} \theta_{AB} \tag{3.196}$$

$$\frac{d\theta_{AB}}{dt} = -\nu e^{-\frac{\Delta E_{\text{desorption}}}{k_B T}} \theta_{AB} \quad \nu = \frac{k_B T}{h} \frac{q'^{0\#}_{AB}}{q_{AB}} \tag{3.197}$$

Equilibrium

If we assume equilibrium between adsorption and desorption, we find the Langmuir isotherm.

$$\text{AB} + * \rightleftharpoons \text{AB}* \tag{3.198}$$

$$\frac{d\theta_{AB}}{dt} = \frac{S_0(T) p_{AB}}{N_0 \sqrt{2\pi m k_B T}} \theta_* - \nu e^{-\frac{\Delta E_{\text{desorption}}}{k_B T}} \theta_{AB} = 0 \tag{3.199}$$

$$\theta_{AB} = \frac{K_{eq} p_{AB}}{K_{eq} p_{AB} + 1} \tag{3.200}$$

where

$$K_{eq} = \frac{S_0(T) e^{\frac{\Delta E_{\text{desorption}}}{k_B T}}}{N_0 \nu \sqrt{2\pi m k_B T}} \tag{3.201}$$

References

1 Laidler, K.J. (1987) *Chemical Kinetics*, HarperCollinsPublishers Inc., New York.
2 Atkins, P.W. (1998) *Physical Chemistry*, Oxford University Press, Oxford.
3 Van Santen, R.A. and Niemantsverdriet J.W. (1995) *Chemical Kinetics and Catalysis*, Plenum, New York.
4 McQuarrie, D.A. (1976) *Statistical Mechanics*, Harper & Row, New York.
5 Zhdanov, V.P., Pavlicek, J., and Knor, Z. (1988) Preexponential factors for elementary surface processes. *Catalysis Reviews – Science and Engineering*, **30**, 501–517.
6 Zhdanov, V.P. (1991) *Elementary Physicochemical Processes on Solid Surfaces*, Plenum Press, New York.
7 Niemantsverdriet, J.W. (2007) *Spectroscopy in Catalysis, an Introduction*, Wiley-VCH Verlag GmbH, Weinheim.

Chapter 4
Catalyst Characterization

4.1
Introduction

Characterization is an important field in catalysis. Spectroscopy, microscopy, diffraction, and methods based on adsorption and desorption or bulk reactions (reduction, oxidation) all offer tools to investigate the nature of an active catalyst. With such knowledge, we hope to understand catalysts better, so that we can improve them or even design new catalysts.

In Chapter 1, we already emphasized that the properties of a heterogeneous catalyst surface are determined by its composition and structure on the atomic scale. Hence, from a fundamental point of view, the ultimate goal of catalyst characterization should be to examine the surface atom by atom under the reaction conditions in which the catalyst operates, that is, *in situ*. However, a catalyst often consists of small particles of metal, oxide, sulfide on a support material. Chemical promoters may have been added to the catalyst to optimize its activity and/or selectivity, and structural promoters may have been incorporated to improve the mechanical properties and stabilize the particles against sintering. As a result, a heterogeneous catalyst can be quite complex. Moreover, the state of the catalytic surface generally depends on the conditions under which it is used.

Therefore, in many fundamentally oriented studies the complex catalyst is replaced by a simplified model, which is better defined. Such models may range from supported particles from which all promoters have been removed, via well-defined particles deposited on planar substrates, to single crystals (Figure 4.1). With the latter, we are in the domain of surface science, where a wealth of informative techniques is available that do not work on technical catalysts.

In industry, the emphasis is mainly on developing an active, selective, stable, and mechanically robust catalyst. To accomplish this, tools are needed which identify those structural properties that discriminate efficient from less efficient catalysts. All information in pursuit of this achievement is therefore highly desirable. Empirical relationships between those factors that govern catalyst composition (e.g., particle size and shape, and pore dimensions) and those that determine catalytic performance are extremely useful in catalyst development, although they do not

Concepts of Modern Catalysis and Kinetics, Third Edition. I. Chorkendorff and J.W. Niemantsverdriet.
© 2017 WILEY-VCH Verlag GmbH & Co. KGaA. Published 2017 by WILEY-VCH Verlag GmbH & Co. KGaA.

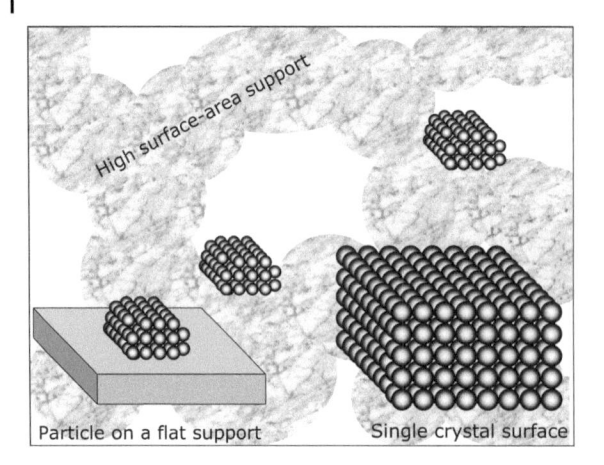

Figure 4.1 Supported catalyst, consisting of small particles on a high surface area carrier such as silica or alumina, along with two simplified model systems, which in general offer much better opportunities for characterization at the molecular level.

always give fundamental insight into how the catalyst operates on the molecular level.

Several spectroscopic, microscopic and diffraction techniques are used to investigate catalysts. As Figure 4.2 illustrates, such techniques are based on some type of excitation (arrows pointing inward in Figure 4.2) to which the catalyst responds (symbolized by the arrows point outward). For example, irradiating a catalyst with X-ray photons generates photoelectrons, which are employed in X-ray photoelectron spectroscopy (XPS) – one of the most useful characterization tools. One can also heat a spent catalyst and look at what temperatures reaction intermediates and products desorb from the surface (temperature-programmed desorption, TPD).

Figure 4.3 shows some statistics on the use of characterization techniques in catalysis. In this chapter, we briefly introduce the most important of these meth-

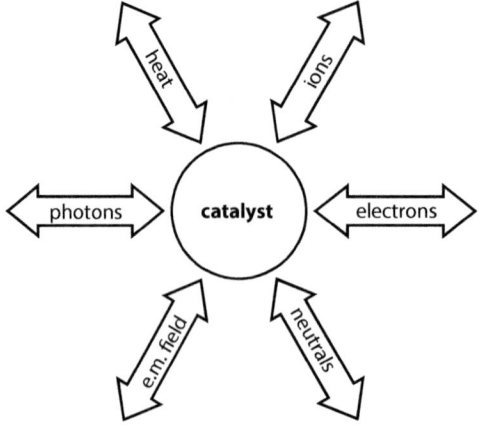

Figure 4.2 Catalyst characterization techniques: the circle represents the sample under study, the inward arrows denote excitation processes, and the outward arrows indicate how the information should be extracted.

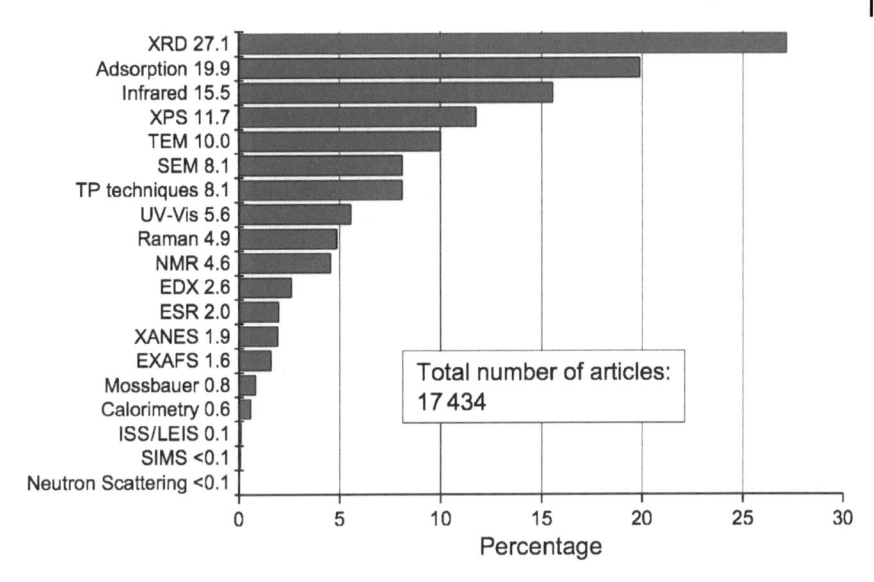

Figure 4.3 Percentage of papers in which the listed techniques were used, out of the 17 434 articles published between 2009 and 2013 by ten journals in the field of catalysis (Journal of Catalysis, Applied Catalysis A and B, Catalysis Today, Catalysis Letters, Catalysis Communications, ACS Catalysis, ChemCatChem, Topics in Catalysis, and Journal of Molecular Catalysis A).

ods and illustrate their use. Further examples can be found in subsequent chapters and in more specialized books [1–4].

4.2
X-ray Diffraction (XRD)

X-ray diffraction is one of the oldest and most frequently applied techniques in catalyst characterization. It is used to identify crystalline phases inside catalysts by means of lattice structural parameters and to obtain an indication of particle size [5].

X-ray diffraction occurs in the elastic scattering of X-ray photons by atoms in a periodic lattice. The scattered monochromatic X-rays that are in phase give constructive interference. Figure 4.4 illustrates how diffraction of X-rays by crystal planes allows one to derive lattice spacings using the Bragg relation:

$$n\lambda = 2d \sin\theta; \quad n = 1, 2, \ldots \quad (4.1)$$

where λ is the wavelength of the X-rays, d is the distance between two lattice planes, θ is the angle between the incoming X-rays and the normal to the reflecting lattice plane, and n is an integer called order of the reflection.

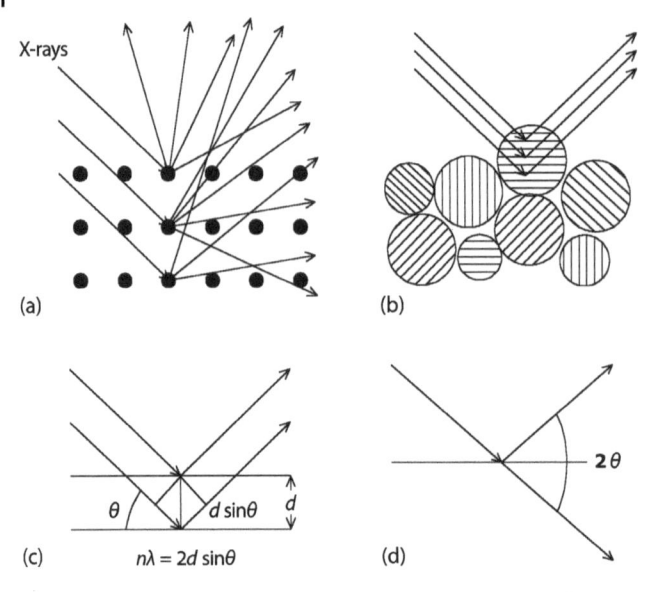

Figure 4.4 X-rays scattered by atoms in an ordered lattice interfere constructively in directions given by Bragg's law. The angles of maximum intensity enable one to calculate the spacings between the lattice planes and allow furthermore for phase identification. Diffractograms are measured as a function of the angle 2θ. When the sample is a polycrystalline powder, the diffraction pattern is formed by a small fraction of the particles only. Rotation of the sample during measurement enhances the number of particles that contribute to diffraction (reprinted from [1] with permission of Wiley-VCH).

If one measures the angles, 2θ, under which constructively interfering X-rays leave the crystal, the Bragg relation (Eq. (4.1)), gives the corresponding lattice spacings, which are characteristic for a certain compound.

With powdered samples, an image of diffraction lines occurs because a small fraction of the powder particles will be oriented such that, by chance, a certain crystal plane is at the right angle θ to the incident beam for constructive interference (see Figure 4.4).

In catalyst characterization, diffraction patterns are mainly used to identify the crystallographic phases that are present in the catalyst. Figure 4.5 gives an example where XRD has been used to study the reduction of two Co/SiO_2 catalysts, one with small crystallites of a few nanometers and the other with much larger crystallite of about 28 nm [6]. The XRD patterns of the latter clearly reveal how the initially oxidic catalyst consists of Co_3O_4, which during reduction in H_2 at temperatures around 150–160 °C converts to CoO while metallic cobalt starts to form at 180–200 °C. However, the XRD patterns of the small-particle catalyst show little detail beyond the features of the silica support (broad peaks at low angles). Only after reduction at high temperatures of 400 °C do we see broad peaks in the range where contributions from metallic cobalt are expected, but the peaks are much less distinct than for the large particle sample.

4.2 X-ray Diffraction (XRD)

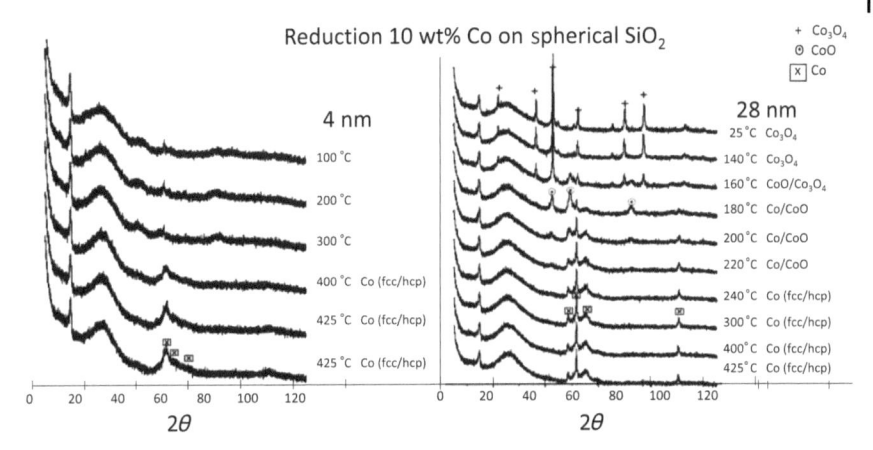

Figure 4.5 *In situ* X-ray diffraction analysis of the reduction of a spherical Co/SiO$_2$ model catalyst with two particle sizes. The results can be compared with the XANES (X-ray Absorption Near-Edge Spectroscopy) data of Figure 4.14 and the TPR (Temperature programmed Reaction) data of Figure 4.23 (reproduced from Saib *et al.* [6] with permission from Elsevier).

The example reveals that X-ray diffraction has an important limitation: clear diffraction peaks are only observed when the sample possesses sufficient long-range order. The advantage of this limitation is that the width (or rather the shape) of diffraction peaks carries information on the dimensions of the reflecting planes. Diffraction lines from perfect crystals are very narrow, but for crystallite sizes below 100 nm, line broadening occurs due to incomplete destructive interference in scattering directions where the X-rays are out of phase.

The Scherrer formula relates crystal size to line width:

$$\langle L \rangle = \frac{K\lambda}{\beta \cos\theta} \tag{4.2}$$

in which $\langle L \rangle$ is a measure for the dimension of the particle in the direction perpendicular to the reflecting plane, λ is the X-ray wavelength, β is the peak width, θ is the angle between the beam and the normal on the reflecting plane, and K is a constant (often taken as 1). X-ray line broadening provides a quick but not always reliable estimate of the particle size. Better procedures to determine particle sizes from X-ray diffraction are based on lineprofile analysis with Fourier transform methods.

One of the great advantages of using X-rays is their penetrating power, such that XRD can be used to study catalysts under realistic conditions in specially designed *in situ* reactors. This enables one to monitor solid-state reactions such as reduction, oxidation, and sulfidation that play a role in the activation of catalysts. In this respect, the use of synchrotron radiation as a source for XRD offers excellent opportunities, as collection times for diffractograms become appreciably shorter. Moreover, the high intensity of the radiation gives data of much better signal-to-noise ratio and, as a consequence, patterns with broad peaks from small particles can be determined with much better accuracy.

The strength of XRD for catalyst characterization is that it gives clear and unequivocal structure information on particles that are sufficiently large, along with an estimate of their size, and it can reveal this information under reaction conditions. The limitation of XRD is that it can not detect particles that are either too small or amorphous. Hence, one can never be sure that there are no other phases present than the ones detected with XRD. In particular, the surface, where catalytic activity resides, is invisible in standard XRD.

4.3
X-ray Photoelectron Spectroscopy (XPS)

XPS is among the most frequently used techniques in catalysis. It yields information on (1) the elemental composition, (2) the oxidation state of the elements, and (3) in favorable cases, the dispersion of one phase over another [1, 7].

XPS is based on the photoelectric effect: an atom absorbs a photon of energy $h\nu$ so that a core or valence electron with binding energy E_b is ejected with kinetic energy (Figure 4.6):

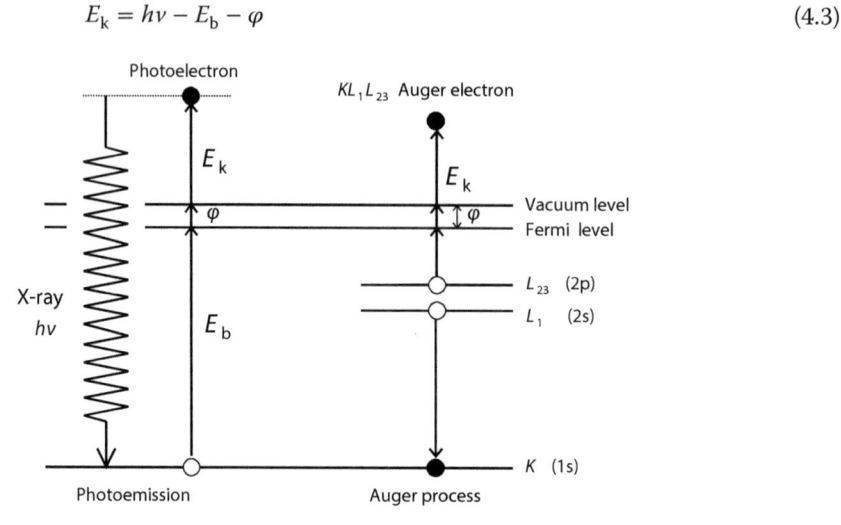

$$E_k = h\nu - E_b - \varphi \tag{4.3}$$

Figure 4.6 Photoemission and the Auger process. Left: An incident X-ray photon is absorbed and a photoelectron emitted. Measurement of its kinetic energy allows one to calculate the binding energy of the photoelectron. The atom becomes an unstable ion with a hole in one of the core levels. Right: The excited ion relaxes by filling the core hole with an electron from a higher shell. The energy released by this transition is taken up by another electron, the Auger electron, which leaves the sample with an element-specific kinetic energy. In Auger spectroscopy, a beam of energetic (2–5 keV) electrons creates the initial core holes (reprinted from [1] with permission of Wiley-VCH).

where E_k is the kinetic energy of the photoelectron, h is Planck's constant, ν is the frequency of the exciting radiation, E_b is the binding energy of the photoelectron with respect to the Fermi level of the sample, and φ is the work function of the spectrometer.

Routinely applied X-ray sources use Mg K_α ($h\nu = 1253.6\,\text{eV}$) and Al K_α ($h\nu = 1486.3\,\text{eV}$) radiation, implying that the kinetic energies of the photoelectrons fall roughly in the range of 0–1 keV. At these energies, electrons travel no more than a few atomic distances through the solid. Figure 4.7 shows that the mean free path of electrons, λ, is limited to less than 2 nm for such energies. Optimum surface sensitivity ($\lambda \approx 0.5\,\text{nm}$) is achieved with electrons at kinetic energies in the range 50–200 eV, where almost half of the photoelectrons come from the outermost layer. Measuring XPS at a synchrotron offers the opportunity to choose the excitation energy which yields optimal surface sensitivity while at the same time spectra of high energy resolution are obtained. This application of XPS is more common in surface science than in catalyst characterization though [7].

In XPS, one measures the intensity of photoelectrons $N(E)$ as a function of their kinetic energy. By using Eq. (4.3), one converts kinetic energy into binding energy, which is usually the property appearing on the x-axis of a spectrum. Binding energies of electrons are fully characteristic of the element from which the photoelectron originates. Figure 4.8 shows the XPS spectrum of an alumina-supported rhodium catalyst, prepared by impregnating the support with $RhCl_3$ in water. Peaks due to Rh, Cl, Al, O, and C, owing to an ever-present contamination by hydrocarbons, are readily assigned by consulting binding energy tables.

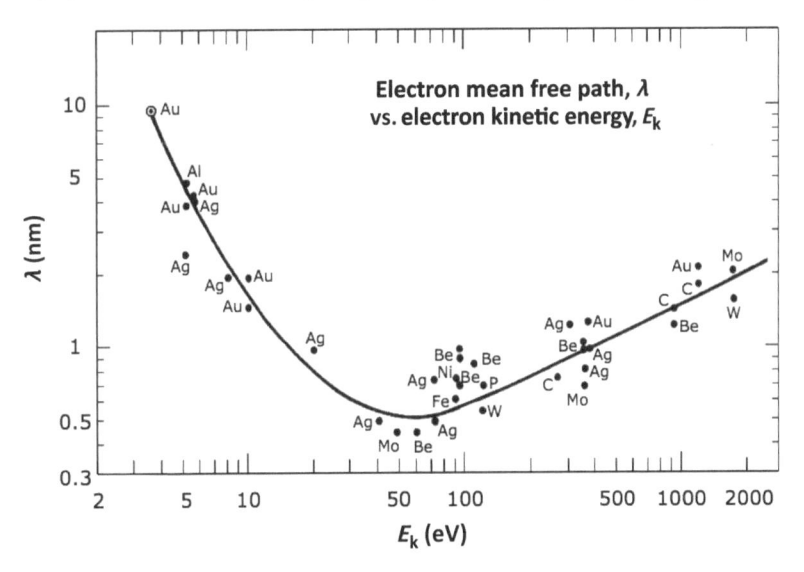

Figure 4.7 The mean free path of an electron depends on its kinetic energy and determines how much surface information it carries. Optimum surface sensitivity is obtained with electrons in the 25–200 eV range (adapted from [1] and [8]).

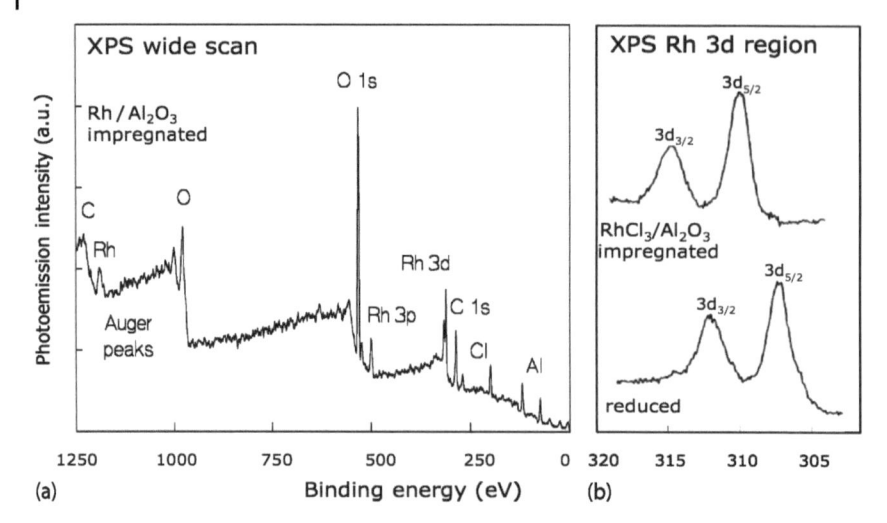

Figure 4.8 XPS wide-scan spectrum of a Rh/Al$_2$O$_3$ model catalyst prepared by impregnating Al$_2$O$_3$ with a solution of RhCl$_3$ in water. The photoelectron and Auger peaks (a) are given, along with a region of interest from the Rh 3d spectrum of the fresh and the reduced catalyst (b), showing the sensitivity of the binding energy for the oxidation state of rhodium (reprinted with permission from Borg et al. [9], copyright 1992, American Vacuum Society).

In addition to the expected photoelectron peaks, the spectrum in Figure 4.8 also exhibits peaks due to Auger electrons. These originate because the atom from which the photoelectron has left is a highly excited ion, with a hole in one of its inner shells. This ion relaxes according to the scheme on the right of Figure 4.6. One readily sees that Auger electrons have fixed *kinetic* energies that are independent of the energy that created the initial core hole. Nevertheless, Auger peaks are plotted on the binding energy scale, which has of course no physical significance. The main peak of the O KVV Auger signal in Figure 4.8 has a kinetic energy of about 500 eV, but appears at a binding energy of about 986 eV because the spectrum was taken with Al K_α X-rays of 1486 eV. Auger peaks can be recognized by recording the spectrum at two different X-ray energies: XPS peaks appear at the same binding energies while Auger peaks will shift on the binding energy scale. This is the main reason why X-ray sources often contain a dual anode of Mg and Al.

Photoelectron peaks are labeled according to the quantum numbers of the level from which the electron originates. An electron with orbital momentum l (0, 1, 2, 3, … indicated as s, p, d, f, …) and spin momentum s has a total momentum $j = l + s$. As the spin may be up ($s = +1/2$) or down ($s = -1/2$), each level with $l \geq 1$ has two sublevels, with an energy difference called the spin-orbit splitting. Thus, the Rh 3d level gives two photoemission peaks, $3d_{5/2}$ (with $l = 2$ and $j = 2 + 1/2$) and $3d_{3/2}$ ($l = 2$ and $j = 2 - 1/2$). Auger electrons, however, are labeled according to the terminology commonly used in X-ray spectroscopy. An Auger electron labeled KLM originates from a transition with the initial core hole in the

K shell, which is filled by an electron from the L shell, whereas the Auger electron is emitted from the M shell.

Spin-orbit splittings and binding energies of a particular electron level increase with increasing atomic number. The intensity ratio of the peaks from a spin-orbit doublet is determined by the multiplicity of the corresponding levels, equal to $2j + 1$. Hence, the intensity ratio of the $j = 5/2$ and $j = 3/2$ components of the Rh 3d doublet is 6 : 4 or 3 : 2. Thus, photoelectron peaks from core levels come in pairs – doublets – except for s levels, which normally give a single peak.

Binding energies are not only element specific but also contain chemical information because the energy levels of core electrons depend slightly on the chemical state of the atom. Chemical shifts are typically in the range of 0–3 eV. In general, the binding energy increases with increasing oxidation state and for a fixed oxidation state with the electronegativity of the ligands. To appreciate the meaning of a binding energy, it is necessary to consider final state effects. In practice, we use XPS data as if they were characteristic for the atoms as they are before the photoemission event takes place. This is not correct, however, as photoemission data represent a state from which an electron has just left. Consequently, the binding energy of a photoelectron contains information both on the state of the atom before photo ionization (the initial state) and on the core-ionized atom left behind after the emission of an electron (the final state). Fortunately, it is often qualitatively correct to interpret binding energy shifts as those in Figure 4.8 in terms of initial state effects. The charge potential model elegantly explains the physics behind such binding energy shifts, by means of the formula

$$E_b^i = kq_i + \sum_j \frac{q_j}{r_{ij}} + E_b^{ref} \qquad (4.4)$$

in which E_b^i is the binding energy of an electron from an atom i, q_i the charge on the atom, k is a constant, q_j the charge on a neighboring atom j, r_{ij} is the distance between atoms i and j, and E_b^{ref} is a suitable energy reference. The first term in Eq. (4.4) indicates that the binding energy increases with increasing positive charge on the atom from which the photoelectron originates. In ionic solids, the second term counteracts the first because the charge on a neighboring atom will have the opposite sign. Because of its similarity to the lattice potential in ionic solids, the second term is often referred to as the Madelung sum.

Figure 4.8b illustrates how the binding energy is sensitive for the oxidation state of rhodium in Rh/Al_2O_3 catalysts. The Rh 3d XPS spectrum of the freshly impregnated catalyst reveals that the $Rh3d_{5/2}$ binding energy is 310 eV, a value characteristic for trivalent rhodium (as in $RhCl_3$). After reduction in hydrogen, the binding energy of the rhodium has decreased to 307.4 eV, indicative of metallic rhodium. Hence, XPS reveals both that rhodium is present in the catalyst and the oxidation state in which it occurs.

An unavoidable experimental problem with techniques based on detection of charged particles such as electrons and ions is that electrically insulating samples may charge up during measurement. The potential the sample acquires is deter-

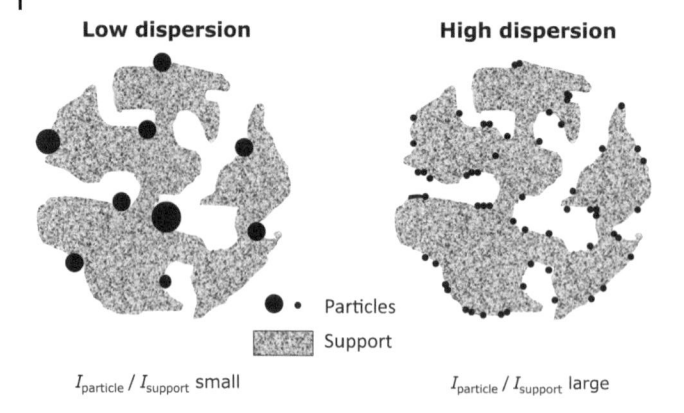

Figure 4.9 The XPS intensity ratio of the signals from particles and the support, I_P/I_S, reflects the dispersion of the particles over the support (reproduced from [1] with permission of Wiley-VCH).

mined by the balance of photoelectrons leaving the sample, the current through the sample holder towards the sample, and the flow of Auger and secondary electrons from the source window onto the sample. Due to positive charge on the sample, all XPS peaks in the spectrum shift by the same amount, to higher binding energies. This is easily corrected if the sample contains an element in a state for which the binding energy is known. In SiO_2-supported catalysts, for example, one uses the binding energy of the Si 2p electrons, which should be 103.4 eV. If nothing else is available, one can use the C 1s binding energy (284.6 eV) of ever-present carbon contamination.

Whereas correction of shifts due to homogenous charging are usually readily corrected, the broadening of peaks due to inhomogeneous charging of samples is more problematic, as it results in decreased resolution and a lower signal-to-noise ratio. The use of a flood gun, which sprays the sample with low energy electrons, and sample mounting techniques in which powders are pressed in indium foil may alleviate the charging problems.

Because XPS is a surface sensitive technique, it recognizes how well particles are dispersed over a support. Figure 4.9 schematically shows two catalysts with the same quantity of supported particles but with different dispersions. When the particles are small, almost all atoms are at the surface, and the support is largely covered. In this case, XPS measures a high intensity I_P from the particles, but a relatively low intensity I_S for the support. Consequently, the ratio I_P/I_S is high. For poorly dispersed particles, I_P/I_S is low. Thus, the XPS intensity ratio I_P/I_S reflects the dispersion of a catalyst on the support. Several models have been reported that derive particle dispersions from XPS intensity ratios, frequently with success. Hence, XPS offers an alternative determination of dispersion for catalysts that are not accessible to investigation by the usual techniques for particle size determination, such as electron microscopy and hydrogen chemisorption.

In conclusion, XPS is among the most frequently used techniques in characterizing catalysis. It readily provides the composition of the surface region and also reveals information on both the oxidation state of metals and the electronegativity of its ligands. XPS can also provide insight into the dispersion of particles over supports, which is particularly useful if the more common techniques employed for this purpose, such as electron microscopy or hydrogen chemisorption, cannot discriminate between support and active phase.

4.4
X-ray Absorption Spectroscopy (EXAFS and XANES)

EXAFS (extended X-ray absorption fine structure) and XANES (X-ray absorption near-edge spectroscopy) are, like XPS, based on the absorption of X-rays and the creation of photoelectrons [2, 10, 11]. However, whereas in XPS the photoelectrons are collected to analyze their kinetic energy, EXAFS deals with the ways these electrons are scattered by neighboring atoms, which is visible in interference effects in the X-ray absorption spectrum. The technique yields detailed information on the distance, number, and type of neighbors of the absorbing atom and, thus, gives insight into local structure on the subnanometer length scale. The XANES spectrum is limited to the absorption edge and gives valuable information on phase compositions by comparison with spectra of known reference compounds. As X-rays have high penetrating power, EXAFS and XANES offer attractive *in situ* opportunities.

4.4.1
Extended X-ray Absorption Fine Structure (EXAFS)

Although known since the 1920s, the phenomenon was only exploited as an analytical tool when bright, tunable sources of X-rays at synchrotrons became available in the 1970s. Figure 4.10 illustrates the EXAFS phenomenon. First, we consider the X-ray absorption spectrum of a free atom, which has an electron with binding energy E_b. If we irradiate this atom with X-rays of energy $h\nu$, absorption takes place when $h\nu \geq E_b$ and the electron leaves the atom with kinetic energy $E_k = h\nu - E_b$. The X-ray absorption spectrum shows a series of edges corresponding to the binding energies of all electrons present in the atom. Fine structure arises if neighbors surround the atom. In this case the photoelectron, having both particle and wave character, can be scattered back from a neighboring atom (see Figure 4.10). Because of its wave character, the outgoing and the backscattered electron interfere. Depending on the wavelength of the electron, the distance between emitting and scattering atom, as well as the shift in phase caused by the scattering event, the two waves enhance or destroy each other. As a result, the cross section for X-ray absorption is modulated by the interference between the photoelectron waves, such that it is enhanced at energies where constructive interference occurs. As indicated schematically in Figure 4.10, the X-ray absorp-

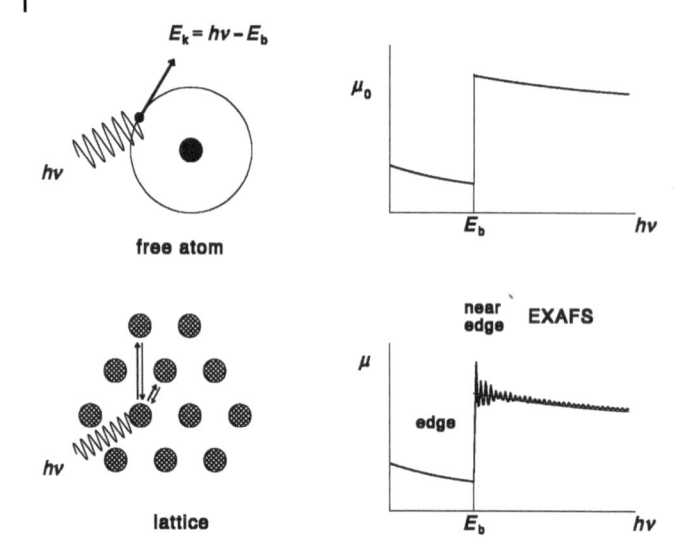

Figure 4.10 Absorption of X-rays as a function of photon energy $E = hv$ by a free atom and by atoms in a lattice. The fine structure, due to the interference of waves associated with the outgoing and backscattered photoelectrons, represents the EXAFS function (reproduced from [1] with permission of Wiley-VCH).

tion spectrum exhibits fine structure, which extends to several hundred electron volts above the absorption edge. The absorption around the edge arises from electrons with low kinetic energies that interact with valence electrons. This near-edge range in the spectrum is often referred to as NEXAFS or XANES.

It is good to have a feeling for how distances, coordination numbers, and concentrations affect the EXAFS spectrum: the intensity of the wiggles goes up if the number of neighbors increases, the number of oscillations depends inversely on interatomic distances (as in any scattering or diffraction experiment), and the step height of the edge is proportional to the concentration of atoms in the sample.

However, mathematics is essential to explain how structural data are derived from EXAFS. The EXAFS function, $\chi(k)$, is extracted from the X-ray absorption spectrum in Figure 4.10 by removing the approximately parabolic background and the step, that is, the spectrum of the free atom. As in any scattering experiment, it is customary to express the signal as a function of the wavenumber k, rather than of energy. The relation between k and the kinetic energy of the photoelectron is

$$k = \frac{2\pi}{h}\sqrt{2m_e E_k} = \frac{2\pi}{h}\sqrt{2m_e(hv - E_b)} \tag{4.5}$$

where k is the wavenumber of the photoelectron, h is Planck's constant, m_e is the mass of an electron, E_k is the kinetic energy of the photoelectron, v is the X-ray frequency, and E_b is the binding energy of the photoemitted electron. For those unfamiliar with wave vectors, $hk/2\pi$ is the momentum of a wave quantum, whereas $\sqrt{(2m_e E_{kin})} = m_e v$ is the classical momentum of the electron when considered as a particle.

In a mono-atomic solid, the EXAFS function $\chi(k)$ is the sum of the scattering contributions of all atoms in neighboring coordination shells:

$$\chi(k) = \sum_j A_j(k) \sin\left[2kr_j + \varphi_j(k)\right] \tag{4.6}$$

where $\chi(k)$ is the EXAFS function, j labels the coordination shells around the electron-emitting atom, $A_j(k)$ is the amplitude, equal to the scattering intensity due to the jth coordination shell, r_j is the distance between the central atom and atoms in the jth shell, and $\varphi(k)$ is the total phase shift.

Each coordination shell contributes a sine function multiplied by an amplitude, which contains the number of neighbors in a coordination shell, N_j, as the most desirable information:

$$A_j(k) = N_j \frac{e^{-2r_j/\lambda(k)}}{kr_j^2} S_o^2(k) F_j(k) e^{-2k^2 \sigma_j^2} \tag{4.7}$$

Other terms include the exponential attenuation of electrons traveling through the solid, a correction for relaxation effects in the emitting atom (S_o), the effect of lattice vibrations in the last exponential term, and finally a very important quantity, F_j, which is the backscattering factor of atoms in the jth shell. The dependence of the backscattering factor on energy is characteristic for an element. Hence, via $F_j(k)$ one can often identify the scattering atom. Figure 4.11 illustrates this for the

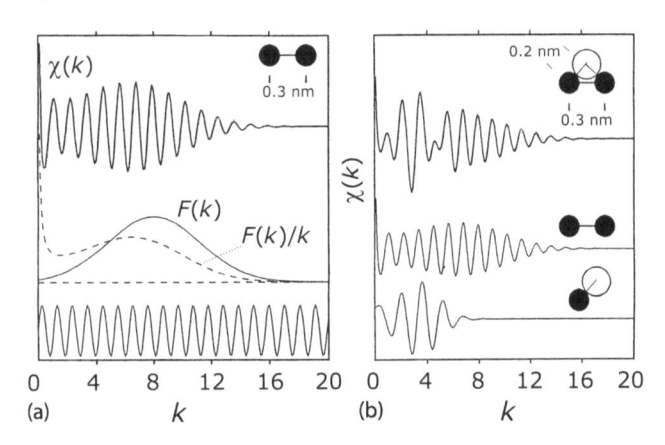

Figure 4.11 (a) Simulated EXAFS spectrum of a dimer such as Cu_2, showing that the EXAFS signal is the product of a sine function and a backscattering amplitude $F(k)$ divided by k, as expressed by Eq. (4.6) and (4.7). Note that $F(k)/k$ remains visible as the envelope around the EXAFS signal $\chi(k)$. (b) The Cu EXAFS spectrum of a cluster such as Cu_2O is the sum of a Cu–Cu and a Cu–O contribution. Fourier analysis is the mathematical tool to decompose the spectrum into the individual Cu–Cu and Cu–O contributions. Note the different backscattering properties of Cu and O, revealed in the envelope of the individual EXAFS contributions. For simplicity, phase shifts have been ignored in the simulations (reproduced from [1] with permission of Wiley-VCH).

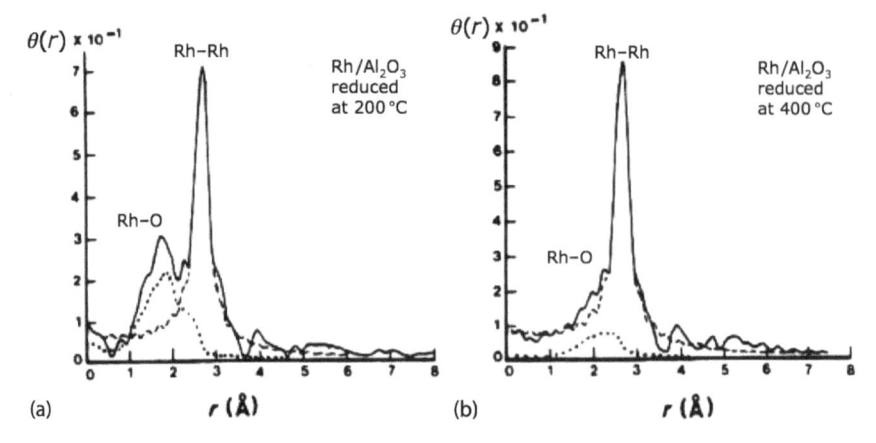

Figure 4.12 Rh-EXAFS Fourier transforms of Rh/Al$_2$O$_3$ catalysts after reduction at (a) 200 °C and (b) 400 °C showing a dominant contribution from Rh nearest neighbors at 0.27 nm and contributions from oxygen neighbors in Rh$_2$O$_3$ and in the metal support interface (reprinted with permission from Koningsberger et al. [12]; copyright 1985 American Chemical Society).

simple cases of a metal dimer and a metal oxide trimer. Clearly, the backscattering factor determines the shape of the EXAFS contribution from a specific neighbor. Note how $F_j(k)$ is recognizable in the envelope of the oscillating function. As Figure 4.11 schematically shows, the backscattering function for oxygen is limited to a smaller range of k values than that of the metal.

The essence of analyzing an EXAFS spectrum is to recognize all sine contributions in $\chi(k)$. The obvious mathematical tool to achieve this is Fourier analysis. The argument of each sine contribution in Eq. (4.6) depends on k (which is known), on r (to be determined), and on the phase shift $\varphi(k)$. The latter is a characteristic property of the scattering atom in a certain environment and is best derived from the EXAFS spectrum of a reference compound for which all distances are known. The EXAFS information becomes accessible if we convert it to a radial distribution function, $\theta_n(r)$, by means of Fourier transformation:

$$\theta_n(r) = \frac{1}{\sqrt{2\pi}} \int_{k_{min}}^{k_{max}} k^n \chi(k) e^{2ikr} \, dk \qquad (4.8)$$

where n is an integer, chosen to emphasize light ($n = 1$) or heavier scatterers ($n = 2$ or 3). The function $\theta_n(r)$ represents the probability of finding an atom at a distance r, modified by the two r-dependent terms in the amplitude, which progressively decrease the intensity of distant shells.

Figure 4.12 shows Fourier-transformed EXAFS data of a highly dispersed ($H/M = 1.2$) Rh/Al$_2$O$_3$ catalyst under hydrogen, after reduction at 200 and 400 °C. The Fourier transforms show the presence of Rh neighbors at 0.27 nm and oxygen neighbors at shorter distances. The data analysis indicates that the catalyst reduced at 200 °C still contains RhO contributions characteristic of Rh$_2$O$_3$, which are attributed to unreduced particles. However, the Rh–O contribution in the radial distribution function of the fully reduced catalyst is due to contact between rhodium atoms and the oxygen atoms of the alumina support. In this way, EXAFS can give information on the nature of metal–support interactions.

An important requirement for meaningful EXAFS data on catalysts is that the particles are monodisperse, such that the average environment, which determines the EXAFS signal, is the same throughout the entire catalyst. For multicomponent catalysts, the technique has the drawback that data analysis becomes progressively more complicated and time consuming with an increasing number of constituent atoms, and that considerable expertise is required to avoid ambiguities. However, EXAFS of optimized and monodisperse catalysts, for which the data analysis is carried out with care, provides unique structure information on the scale of interatomic distances!

4.4.2
X-ray Absorption Near-Edge Spectroscopy (XANES)

The near-edge region of the spectrum, see Figure 4.10, includes the area just before to just after the edge, and it contains valuable information on the phases present in a sample. First, the edge position bears similar information as a photoelectron peak in XPS, and hence it is sensitive to oxidation state and electronegativity of neighboring atoms. Second, the precise shape of the edge region is also sensitive to the environment of the absorbing atom. Figure 4.13 illustrates this with the XANES spectra of a few cobalt reference compounds. XANES spectra are often analyzed by fitting linear combinations of reference spectra, although deeper analysis of the spectral shapes is certainly possible.

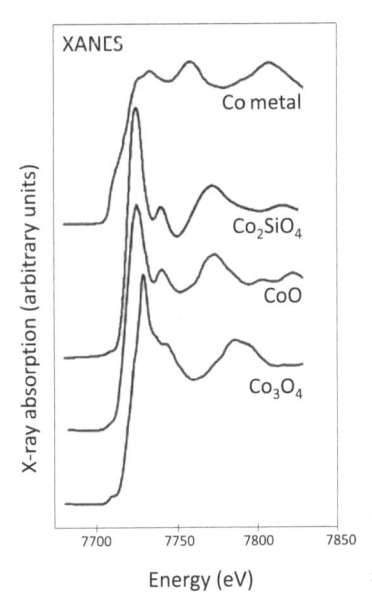

Figure 4.13 XANES spectra of cobalt reference compounds (reproduced from Saib et al. [6] with permission from Elsevier).

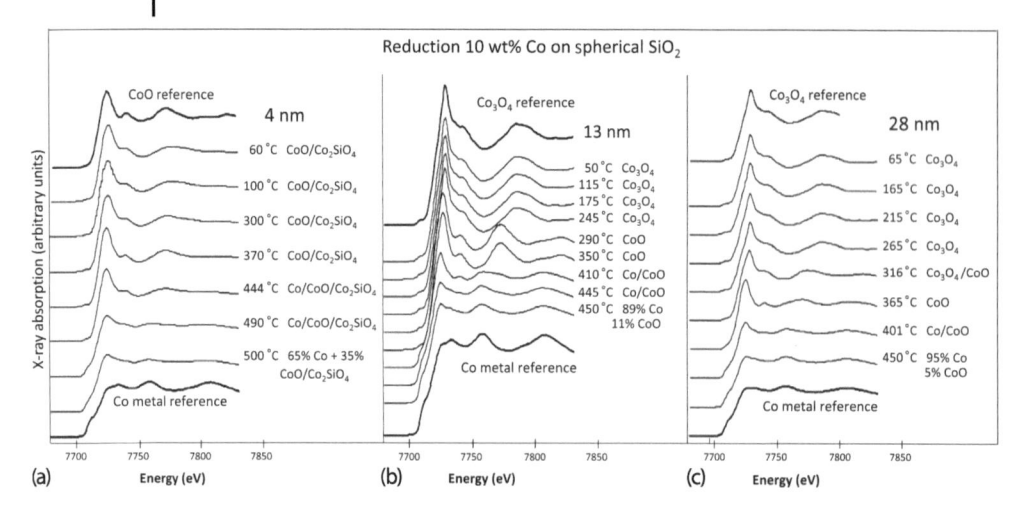

Figure 4.14 *In situ* XANES spectra of the reduction of 10 wt% Co/SiO$_2$ spherical model catalysts with large (Co 28 nm) (a), medium-sized (Co 13 nm) (b), and small (Co 4 nm) (c) cobalt crystallites following calcination at 250 °C in air. The results can be compared with the XRD results of Figure 4.5 and the TPR data of Figure 4.23 (reproduced from Saib *et al.* [6] with permission from Elsevier).

The example on reduction of supported cobalt catalysts of different particle size in Figure 4.14 shows how the initial cobalt oxide converts into metallic cobalt. Note that considerably higher temperatures are needed to reduce the small particles than to reduce the large particles. It is interesting to compare these measurements with the XRD patterns in Figure 4.5, and the TPR spectra of Figure 4.23, which we discuss later in this chapter. Where XRD relies on structures that are coherently structured over domains of several nanometers, XANES samples the X-ray absorption of each cobalt atom, and reveals information on its local environment. As a result XANES has become very popular for the study of catalysts. However, a synchrotron is needed to collect the data.

4.5
Electron Microscopy

Electron microscopy is a rather straightforward technique to determine the size and shape of supported particles [13–15]. Electrons have characteristic wavelengths of less than 1 Å, and come close to monitoring atomic detail. Figure 4.15 summarizes what happens when a primary electron beam of energy between 100 and 400 keV hits a sample:

- Depending on sample thickness, a fraction of the electrons passes through the sample without suffering energy loss. As the attenuation of the beam depends on density and thickness, the transmitted electrons form a two-dimensional projection of the sample.

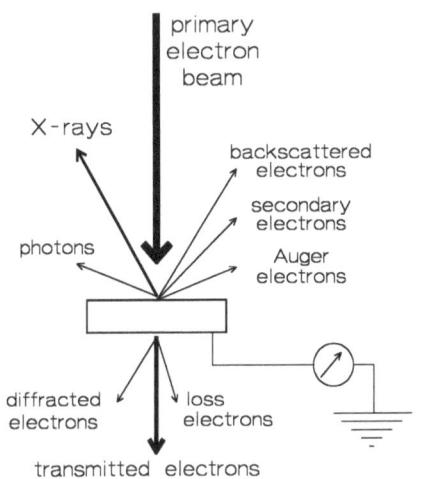

Figure 4.15 Interaction between the primary electron beam and the sample in an electron microscope leads to a number of detectable signals (reproduced from [1] with permission of Wiley-VCH).

- Electrons are diffracted by particles if these are favorably oriented towards the beam, enabling one to obtain crystallographic information.
- Electrons can collide with atoms in the sample and be scattered back; backscattering becomes more effective as the mass of the atom increases. If a region of the sample contains heavier atoms (e.g., Pt) than the surroundings, it can be distinguished due to a higher yield of backscattered electrons.
- Auger electrons and X-rays are formed in the relaxation of core-ionized atoms, as discussed in the sections on Auger Electron Spectroscopy (AES) and XPS. Both carry element-specific information and can be used for compositional analysis.
- Electrons excite characteristic transitions in the sample, which can be studied by analyzing the energy loss suffered by the primary electrons in the electron energy loss spectroscopy or EELS mode. When coupled with scanning, this leads to elemental composition maps of samples, with the opportunity to obtain oxidation state information in favorable cases as well.
- Many electrons lose energy in a cascade of consecutive inelastic collisions. Most secondary electrons emitted by the sample undergo their last loss process in the surface region.
- The emission of a range of photons from UV to infrared, called cathodoluminescence, is mainly caused by recombination of electron–hole pairs in the sample.

Thus, the interaction of the primary beam with the sample provides a wealth of information on morphology, crystallography, and chemical composition. Using transmission electron microscopy to make a projection of the sample density is a routine way to study particle sizes in catalysts.

Figure 4.16 schematically shows how transmission and scanning electron microscopy (TEM and SEM) work. A TEM instrument is similar to an optical microscope if one replaces optical by electromagnetic lenses. In TEM, a primary

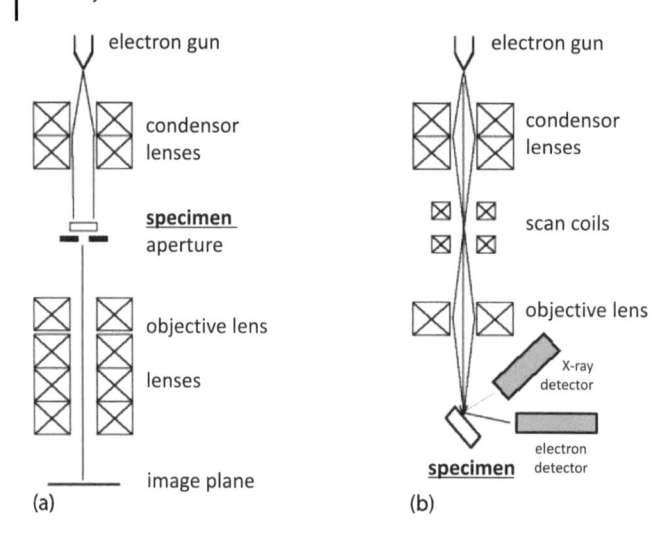

Figure 4.16 Schematic set up of an electron microscope (a) in the transmission (TEM) and (b) the scanning (SEM) mode. The SEM instrument also contains an X-ray detector for composition analysis (adapted from [1]).

electron beam of high energy and high intensity passes through a condenser to produce parallel rays that impinge on the sample. As the attenuation of the beam depends on the density and the thickness of the sample, the transmitted electrons form a two-dimensional projection of the sample mass, which is subsequently magnified by the electron optics to produce a so-called bright field image. The dark field image is obtained from the diffracted electron beams, which are slightly off-angle from the transmitted beam. Typical operating conditions of a TEM instrument are 100–200 keV electrons, 10^{-6} mbar vacuum, 0.5 nm resolution, and a magnification of 3×10^5 to 10^6.

Scanning electron microscopy (SEM) involves rastering a narrow electron beam over the surface and detecting the yield of either secondary or backscattered electrons as a function of the position of the primary beam. Contrast is caused by the orientation: parts of the surface facing the detector appear brighter than parts of the surface with their surface normal, pointing away, from the detector. The secondary electrons have mostly low energies (about 5–50 eV) and originate from the surface region of the sample. Backscattered electrons come from deeper and carry information on the composition of the sample because heavy elements are more efficient scatterers and appear brighter in the image.

Dedicated SEM instruments have a resolution of about 5 nm. The main difference between SEM and TEM is that SEM sees contrast due to the topology and composition of a surface, whereas the electron beam in TEM projects all information on the mass it encounters in a two-dimensional image, which, however, is of subnanometer resolution.

As illustrated by Figure 4.15, an electron microscope offers additional possibilities for analyzing the sample. Diffraction patterns (spots from a single-crystal particle and rings from a collection of randomly oriented particles) enable one

Cu/ZnO
under H₂   H₂O : H₂ = 1 : 3   5% CO in H₂

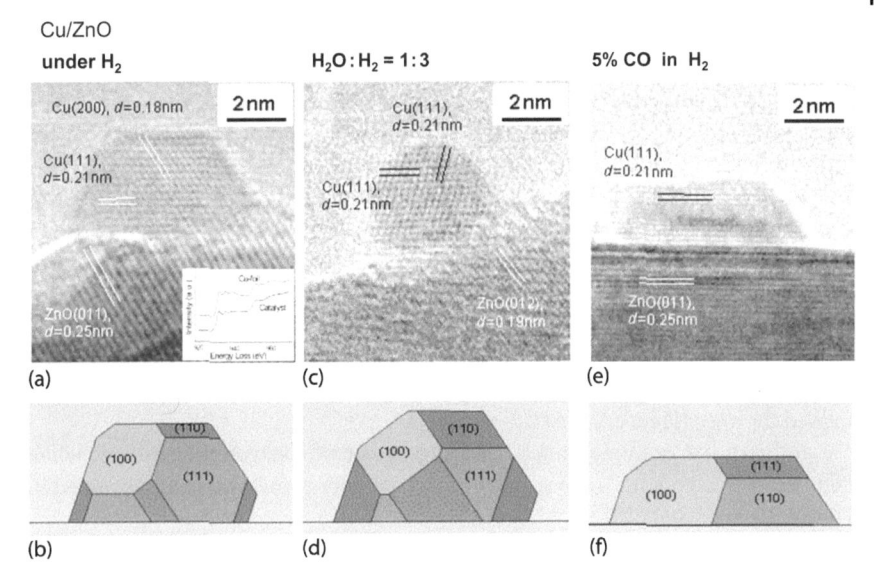

Figure 4.17 Atomically resolved TEM images of a Cu/ZnO model catalyst in various gas environments together with the corresponding Wulff construction of the Cu particle. (a,b) Cu nanocrystal faceted by (100), (110), and (111) surfaces; the TEM image was recorded at 1.5 mbar of H₂ at 220 °C with the electron beam parallel to the [011] zone-axis of copper. The insert shows EELS data at the Cu $L_{2,3}$-edge of the catalyst and a Cu foil, both reduced *in situ*; (c,d) metallic Cu nanocrystal in a mixture of H₂ : H₂O = 3 : 1 at a total pressure of 1.5 mbar at 220 °C; (e,f) Cu nanocrystal in a 5 mbar mixture of 5% CO in H₂ at 220 °C (from [16]; reprinted with permission from AAAS).

to identify crystallographic phases as in XRD. Emitted X-rays are characteristic for an element and allow for determination of the chemical composition of a selected part of the sample. This technique is referred to as energy-dispersive X-ray analysis (EDX).

Transmission electron microscopy is one of the techniques most often used for the characterization of catalysts. In general, detection of supported particles is possible provided that there is sufficient contrast between particles and support – a limitation that may impede applications of TEM on well-dispersed supported oxides. The determination of particle sizes or of distributions therein is now a routine matter, although it rests on the assumption that the size of the imaged particle is truly proportional to the size of the actual particle and that the detection probability is the same for all particles, independent of their dimensions.

Although experiments involving electrons generally have to be performed in a vacuum, recent experimental developments have made it possible to study the behavior of catalysts under low pressures of a gas. This is important, as the morphology of small particles often depends critically on the gaseous environment. Figure 4.17 shows *in situ* TEM images of a model catalyst for methanol synthesis, consisting of copper particles on a zinc oxide substrate. The high-resolution TEM microscope was modified with a cell enabling the study of samples under gas at pressures up to 20 mbar and temperatures up to 700–1000 °C [14, 16].

Figure 4.17 illustrates that the shape and the morphology of the copper particles vary dynamically (and reversibly) with the gas composition. Such changes in morphology are driven by the interface energy between the copper particle and the ZnO surface. When under hydrogen or a mixture of water and hydrogen, the copper particles tend to form facetted half-spheres while the particles spread over the ZnO surface in the CO-containing atmosphere. The latter is believed to partially reduce the ZnO surface, thereby giving it a higher surface free energy. According to the Wulff construction (a procedure for predicting particle morphology on the basis of surface and interface energy considerations discussed in the next chapter), one expects a larger contact area between Cu and ZnO, as is indeed the case (Figure 4.17e,f).

Another great example of the resolving power of electron microscopy, when coupled with elemental analysis in the electron energy loss mode, is discussed in Figure 9.7.

4.6
Mössbauer Spectroscopy

Mössbauer spectroscopy is a specialist characterization tool in catalysis [1, 17]. Nevertheless, it has yielded essential information on a number of important catalysts, such as the iron catalyst for ammonia and Fischer–Tropsch synthesis, as well as the CoMoS hydrotreating catalyst. Mössbauer spectroscopy provides the oxidation state, the internal magnetic field, and the lattice symmetry of a limited number of elements such as iron, cobalt, tin, iridium, ruthenium, antimony, platinum, and gold, and can be applied *in situ*.

The Mössbauer effect, discovered by Rudolf L. Mössbauer in 1957, can in short be described as the recoil-free emission and resonant absorption of gamma radiation by nuclei. In the case of iron, the source consists of ^{57}Co, which decays with a half-life of 270 days to an excited state of ^{57}Fe (natural abundance in iron: 2%). The latter, in turn, decays rapidly to the first excite state of this isotope. The final decay generates a 14.4 keV photon and a very narrow natural linewidth of the order of neV.

The absorber contains iron with nuclei in the ground state. These nuclei could in principle absorb the radiation from the source if their nuclear levels were separated by exactly the same energy difference as in the source. This is in general not the case because nuclear levels are shifted or split by hyperfine interactions between the nucleus and its environment. To cover all possible transitions between the shifted and split nuclear levels, one must be able to vary the energy of the source radiation. This is done by using the Doppler effect. If the source has a velocity v towards the absorber, the energy of the gamma radiation is given by

$$E(v) = E_\circ \left(1 + \frac{v}{c}\right) \tag{4.9}$$

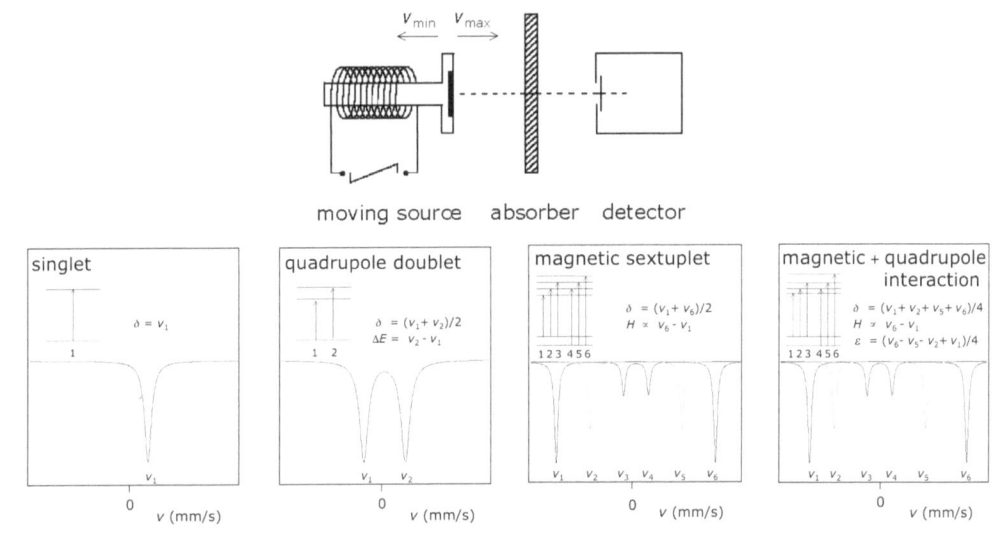

Figure 4.18 To cover all possible transitions in the absorbing nucleus, the energy of the source radiation is modulated by using the Doppler effect, such that the emitted radiation has an energy $E(v) = E_o(1 + v/c)$. For ^{57}Fe the required velocities fall in the range −1 to +1 cm/s. In Mössbauer emission spectroscopy, the sample under investigation is the source, and a single line absorber is used to scan the emission spectrum. The spectra shown are the three most common types observed in iron-containing catalysts. Also indicated are the corresponding nuclear transitions and derivation of the Mössbauer parameters from the spectra (reproduced from [1] with permission of Wiley-VCH).

in which E_o is the energy of the stationary source and c the velocity of light. A velocity range of −10 to +10 mm s^{-1} is sufficient to cover all shifts in iron.

Gamma rays involved in nuclear transitions are quanta with an extremely narrow range of energy. It is this spectral precision that enables the detection of the so-called hyperfine interactions, the minute changes in the nuclear energy levels that result from a change in chemical state or in magnetic or electric surroundings of the atom to which the nucleus belongs. The three hyperfine interactions are illustrated in Figure 4.18.

1. The isomer shift, δ, arises from the Coulomb interaction between the positively charged nucleus and the negatively charged s electrons and is, thus, a measure for the s electron density at the nucleus, yielding useful information on the oxidation state of the iron in the absorber. An example of a single line spectrum is face-centered cubic (fcc) iron, as in stainless steel or in many alloys with noble metals.
2. The electric quadrupole splitting is caused by the interaction of the electric quadrupole moment with an electric field gradient. The latter can be caused by asymmetrically distributed electrons in incompletely filled shells of the atom itself and by charges on neighboring ions. The splitting occurs only in the excited level (Figure 4.18) and is proportional to the magnitude of the electric field gradient at the nucleus. The spectrum contains two peaks, the so-called

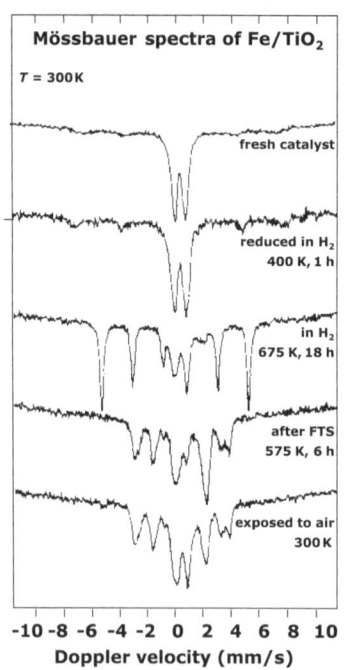

Figure 4.19 Mössbauer spectra give detailed information on the state of iron in a TiO$_2$-supported iron catalyst after different treatments. Here, reduction was in H$_2$ at 675 K for 18 h. *FTS* Fischer–Tropsch synthesis (reproduced from Van der Kraan *et al.* [18] with permission of Elsevier).

quadrupole doublet. This type of spectrum is often observed with highly dispersed iron(III) oxides.

3. Magnetic hyperfine splitting, or the socalled Zeeman effect, arises from the interaction between the nuclear magnetic dipole moment and the magnetic field H at the nucleus. This interaction splits both nuclear levels and removes all degeneracy. From the eight possible transitions only six are allowed (Figure 4.18) and the spectrum contains six equidistant peaks, called a sextet or sextuplet. The separation between the peaks in the spectrum is proportional to the magnetic field at the nucleus. The best-known example is that of metallic or body-centered cubic (bcc) iron with a magnetic hyperfine field of 33 T (330 kOe).

4. If the nucleus feels both a magnetic field and an electric field gradient, and the electric quadrupole interaction is small, then the excited levels shift further and make the sextet asymmetrical, as observed in the spectrum of Fe$_2$O$_3$.

The Mössbauer effect can only be detected in the solid state because the absorption and emission events must occur without energy losses due to recoil effects. The fraction of the absorption and emission events without exchange of recoil energy is called the recoilless fraction, f. It depends on temperature and on the energy of the lattice vibrations: f is high for a rigid lattice, but low for surface atoms.

Figure 4.19 shows how Mössbauer spectroscopy reveals the identity of the major iron phases in a supported iron catalyst after different treatments. The top spectrum belongs to a fresh Fe/TiO$_2$ catalyst, that is, after impregnation and drying. The quadrupole doublet has an isomer shift corresponding to iron in the ferric or Fe^{3+} state. After reduction in H$_2$ at 675 K, the catalyst consists mainly of metallic iron, as evidenced by the sextet, along with some unreduced iron, which gives rise to two doublet contributions of Fe^{2+} and Fe^{3+} in the center. The overall degree of iron reduction, as reflected by the relative area under the bcc ion sextet, is high. Fischer–Tropsch synthesis at 575 K in CO and H$_2$ converts the metallic iron into the Hägg carbide, Fe$_5$C$_2$. The unreduced iron is mainly present as Fe^{2+}. Exposure of the carburized catalyst to air at room temperature leaves most of the carbide phase unaltered but oxidizes ferrous iron to ferric iron.

The example is typical for many applications of Mössbauer spectroscopy in catalysis: a catalyst undergoes a certain treatment, then its Mössbauer spectrum is measured *in situ* at room temperature. However, if the catalyst contains highly dispersed particles, the measurement of spectra at cryogenic temperatures becomes advantageous as the recoil-free fraction of surface atoms increases substantially at temperatures below 300 K. Second, spectra of small particles which behave superparamagnetically at room temperature become magnetically split and, therefore, more informative at cryogenic temperatures.

In Mössbauer emission spectroscopy, one prepares the catalyst with the radioactive source element (e.g., ^{57}Co) and uses a suitable moving single-line absorber of ^{57}Fe to record the spectra. In this way, one can also study Co-containing catalysts, although strictly speaking the information concerns the iron in the catalyst that forms by the ^{57}Co → ^{57}Fe decay process.

The great advantage of Mössbauer spectroscopy is that it can be applied *in situ*. The major limitation of the technique is that it can only be applied to a couple of elements, among which iron and tin are the easiest to study.

4.7
Ion Spectroscopy: SIMS, LEIS, RBS

Although used considerably less than XPS, ion spectroscopy yields highly specific information on the surface of catalysts. When a beam of low energy ions (e.g., 0.5–5 keV Ar$^+$) impinges on a surface, the ions can penetrate the sample and lose their energy in a series of inelastic collisions with the target atoms. This may lead to the emission of secondary ions and neutrals from the surface, which is the basis of secondary ion mass spectrometry (SIMS). Another possibility is that the ions scatter elastically from the surface atoms, as in low-energy ion scattering (LEIS) or, when high-energy ions are used, from atoms deeper in the sample (Rutherford backscattering) [1, 19].

Secondary ion mass spectrometry (SIMS) is by far the most sensitive surface technique, but also the most difficult to quantify. When a surface is exposed to a beam of ions (Ar$^+$, 0.5–5 keV), energy is deposited in the surface region of the sample by a collisional cascade. Some of the energy will return to the surface and stimulate the ejection (desorption) of atoms, ions, and multiatomic clusters. In

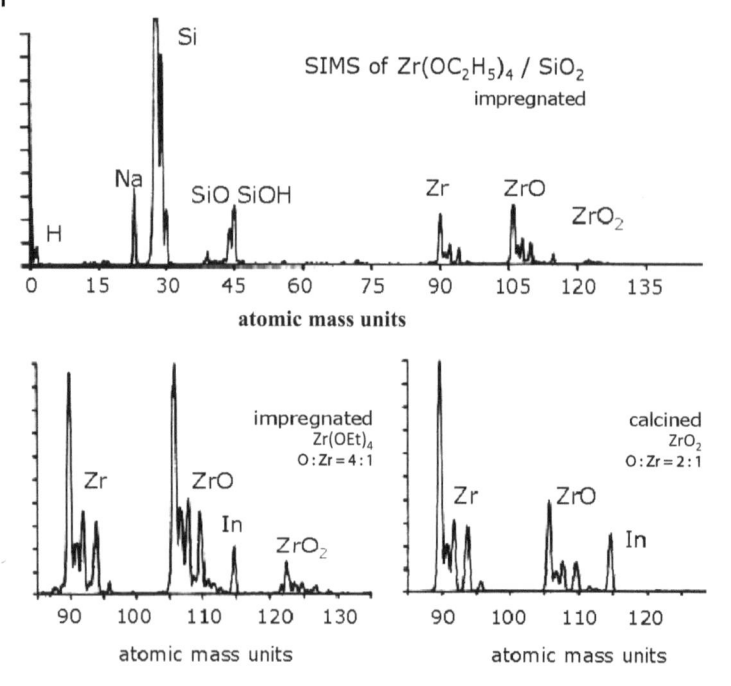

Figure 4.20 SIMS spectra of a ZrO₂/SiO₂ catalyst after impregnation from Zr(OC₂H₅)₄ and after calcination in air. Note how the ZrO/Zr intensity ratio reflects the oxygen : zirconium stoichiometry, 4 : 1 in the ethoxy and 2 : 1 in the oxide (reproduced from Meijers et al. [20] with permission of Elsevier).

SIMS, positive or negative secondary ions are detected directly with a quadrupole mass spectrometer.

Figure 4.20 shows the positive SIMS spectrum of a silica-supported zirconium oxide catalyst precursor, freshly prepared by a condensation reaction between zirconium ethoxide and the hydroxyl groups of the support. Note the simultaneous occurrence of single ions (H⁺, Si⁺, Zr⁺) and molecular ions (SiO⁺, SiOH⁺, ZrO⁺, ZrOH⁺, ZrO₂⁺). The isotope pattern of zirconium is also clearly visible. Isotopes are important in the identification of peaks because all peak intensity ratios must agree with the natural abundances. In addition to the peaks expected from zirconia on silica mounted on an indium foil, the spectrum of Figure 4.20 also contains peaks from Na⁺, K⁺, and Ca⁺. This is typical for SIMS: sensitivities vary over several orders of magnitude and elements such as the alkalis are detected when present in trace amounts.

SIMS is strictly speaking a destructive technique. In the dynamic mode, used for making concentration depth profiles, several tens of monolayers are removed per minute. In the static limit, which is used to study adsorbed molecules, the rate of removal corresponds to less than one monolayer per hour, implying that the surface structure does not change during the measurement. In this case, the molecular ion fragments are indicative of the chemical structure on the surface.

The advantages of SIMS are its high sensitivity (detection limit of ppm for certain elements), its ability to detect hydrogen, and the emission of molecular fragments that often bear tractable relationships with the parent structure on the surface. Disadvantages are that secondary ion formation is a poorly understood phenomenon and that quantification is often difficult. A major drawback is the matrix effect: secondary ion yields of one element can vary tremendously with its chemical environment. This matrix effect and the elemental sensitivity variation of five orders of magnitude across the periodic table make quantitative interpretation of SIMS spectra of technical catalysts extremely difficult.

In spite of quantitation problems, SIMS can give useful information if one uses appropriate reference materials. The spectra in Figure 4.20 indicate that the relative intensities of the ZrO_2^+, ZrO^+, and Zr^+, contain information on the chemical environment of the zirconium: the zirconium ethoxide (O : Zr = 4 : 1) shows higher intensities of the ZrO_2^+ and ZrO^+ signals than the calcined ZrO_2 (O : Zr = 2 : 1). This information is interpreted by comparing the spectra of the catalysts with reference spectra of ZrO_2 and zirconium ethoxide [20].

In low energy ion scattering (LEIS, also called ion scattering spectroscopy, ISS), a beam of noble gas ions with energy of a few keV scatters elastically from a solid surface [19]. The energy of the outgoing ion is determined by energy and momentum conservation and reveals the mass of the target atom from which it scattered. When the ion energy detector is placed at an angle of 90° with respect to the incident beam, one has

$$\frac{E}{E_o} = \frac{M_{at} - M_{ion}}{M_{at} + M_{ion}} \qquad (4.10)$$

in which E_o is the energy of the incident ion, E its energy after scattering, M_{ion} the mass of the incident ion, and M_{at} the mass of the atom with which the ion collides. Hence, the energy spectrum of the backscattered ions is equivalent to a mass spectrum of the surface atoms. Figure 4.21 shows the LEIS data of a multicomponent oxidation catalyst. In agreement with the expression for the energy, the spectra illustrate that the primary ions lose less energy to heavy atoms than to light atoms.

Two factors determine the intensity of the scattered beam: the scattering cross section for the incident ion–target atom combination and the neutralization probability of the ion in its interaction with the solid. It is the latter quantity that makes LEIS surface sensitive: 1 keV He ions have a neutralization probability of about 99% in passing through one layer of substrate atoms. Hence, the majority of ions that reach the detector must have scattered off the outermost layer. At present, there is no simple theory to describe the scattering cross section and the neutralization probability. However, satisfactory calibration procedures by use of reference samples exist. The fact that LEIS provides quantitative information on the outer layer composition of multicomponent materials makes it an extremely powerful tool for the characterization of catalysts. Nevertheless, the technique remains a specialty in the field.

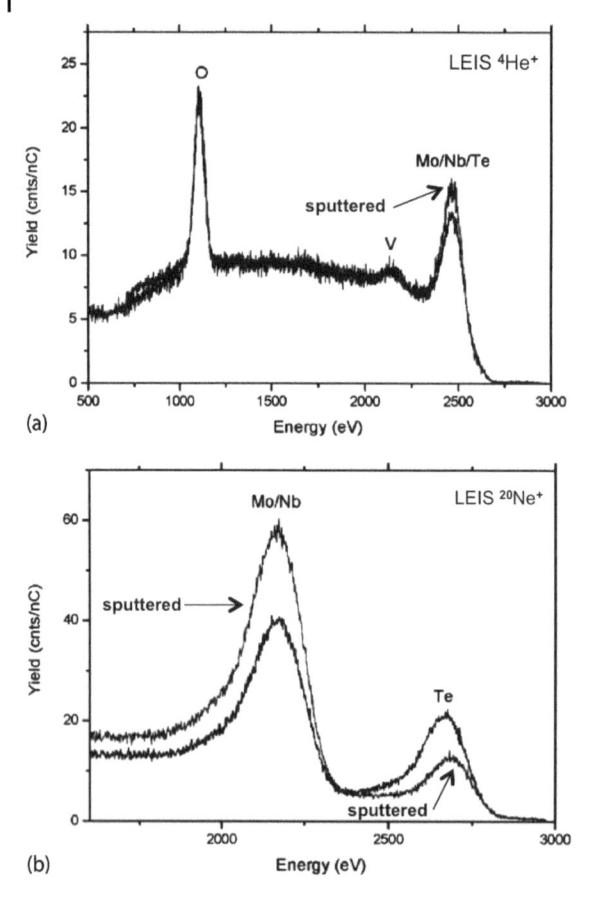

Figure 4.21 Low energy ion scattering (LEIS) of a Mo-V-Te-Nb-O catalyst for oxidation and ammoxidation reactions, illustrating the effect of the incident ion. According to Eq. (4.10), mass discrimination of heavier elements in the target increases when the mass of the incident ion M_{ion} is higher, although elements with $M_{at} < M_{ion}$ would escape detection. Hence, the Mo/Nb/Te signals cannot be distinguished in the $^4He^+$ scattering while Te is clearly resolved in the $^{20}Ne^+$ scattering experiment. The latter also demonstrates how the effect of light sputtering can be used to identify which elements enrich at the surface, in this case Te (reproduced from Guliants et al. [21] with permission of the American Chemical Society).

Rutherford backscattering, RBS, is the high energy equivalent of LEIS. Here, one uses a primary beam of high energy H^+ or He^+ ions (1–5 MeV), which scatter from the nuclei of the atoms in the target. A fraction of the incident ions is scattered back and is subsequently analyzed for energy. As in LEIS, the energy spectrum represents a mass spectrum, but this time it is characteristic for the interior of the sample. The technique is not common in the characterization of technical catalysts, but has been successfully applied on planar model catalysts [1].

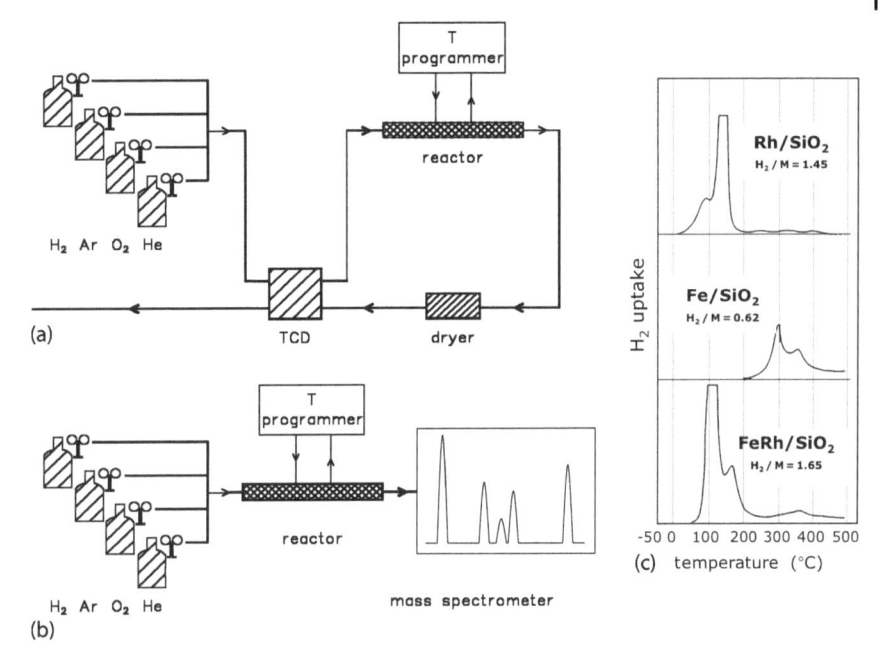

Figure 4.22 Experimental set-ups for temperature-programmed (TP) reduction, oxidation, and desorption. (a) The reactor is inside the oven, the temperature of which can be increased linearly with time. Gas consumption by the catalyst is monitored by the change in thermal conductivity of the gas mixture; it is essential to remove traces of water and other contaminants, etc. because these would affect the thermal conductivity measurement. (b) A TP apparatus equipped with a mass spectrometer. (c) TPR patterns of silica-supported Rh, Fe, and Fe-Rh catalysts, which had been previously calcined to ensure that all metals are oxidized at the start of the measurement (adapted from [1, 24]).

4.8
Temperature-Programmed Reduction, Oxidation, and Sulfidation

Temperature-programmed reaction (TPR) methods form a class of techniques in which a chemical reaction is monitored while the temperature increases linearly in time [1, 3, 22, 23]. Several forms are in use: temperature-programmed reduction, oxidation, and sulfidation. The instrumentation for temperature-programmed investigations is relatively simple. The reactor, charged with catalyst, is controlled by a processor, which heats the reactor at a linear rate of typically 0.1 to 20 °C min^{-1}. A thermal conductivity detector or, preferably, a mass spectrometer measures the composition of the outlet gas.

TPR provides useful information on the temperatures needed for the complete reduction of a catalyst. Figure 4.22 shows the TPR patterns of silica-supported iron and rhodium catalysts. There are three things to note. First, the difference in reduction temperature between the noble metal rhodium and the nonnoble metal iron. Second, the area under a TPR or TPO (temperature programmed oxidation) curve represents the total hydrogen consumption and is commonly expressed in

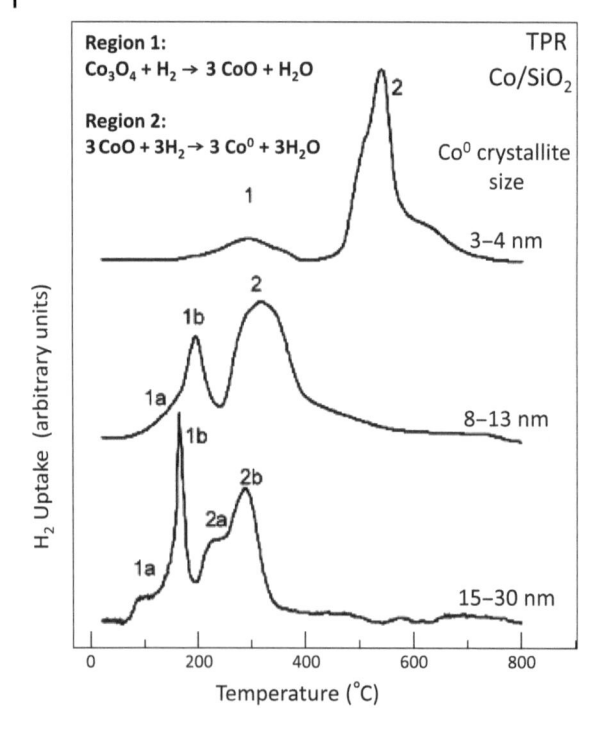

Figure 4.23 Temperature-programmed reduction of three silica-supported cobalt catalysts with different particle sizes, along with an assignment of the chemical reactions to the peaks, based on the XRD and XANES data discussed in Figures 4.5 and 4.14 (reproduced from Saib et al. [6] with permission from Elsevier).

moles of H_2 consumed per mol of metal atoms, H_2/M. The ratios of almost 1.5 for rhodium in Figure 4.22c indicate that rhodium was present as Rh_2O_3. For iron, the H_2/M ratios are significantly lower, indicating that this metal is only partially reduced. Third, the TPR pattern of the bimetallic catalyst shows that the bimetallic combination is reduced at lower temperatures than the iron catalyst, indicating that rhodium catalyzes the reduction of the less noble iron. This is evidence that rhodium and iron are well mixed in the fresh catalyst. The reduction mechanism is as follows: as soon as rhodium becomes metallic it dissociates hydrogen; atomic hydrogen migrates to iron oxide in contact with metallic rhodium and instantaneously reduces the oxide.

Earlier in this chapter, we discussed the reduction of silica supported cobalt catalysts at different temperatures, as studied by X-ray diffraction and absorption spectroscopy (Figures 4.5 and 4.14, respectively). Figure 4.23 shows the TPR spectra of the three catalysts with different particle sizes [6]. It is instructive to compare the different reduction regimes in the TPR spectra with the spectroscopic data of XANES and the structural information from XRD. In this way, the various peaks in the TPR can be attributed to the reductions of Co_3O_4 to CoO and of CoO to Co metal. The TPR clearly reveals that reduction of the catalyst with the high disper-

4.8 Temperature-Programmed Reduction, Oxidation, and Sulfidation | 157

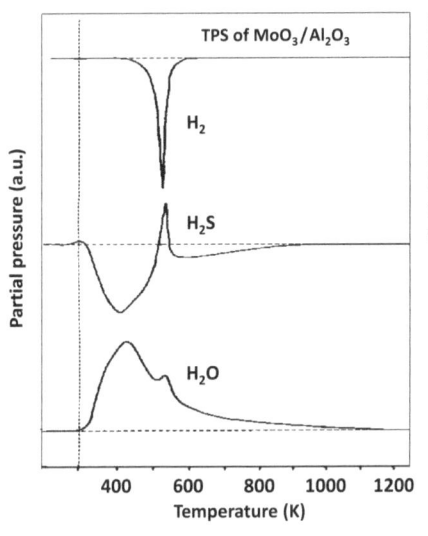

Figure 4.24 Temperature-programmed sulfidation (TPS) of MoO_3/Al_2O_3 catalysts in a mixture of H_2S and H_2, showing the consumption of these gases and the production of H_2O as a function of temperature. Note that H_2S evolves from the catalyst around 500 K, which is attributed to the hydrogenation of elementary sulfur (reproduced from Arnoldy et al. [25] with permission of Elsevier).

sion occurs slower, which is probably due to the higher fraction of cobalt at the interface with the silica.

Catalysts used for hydrodesulfurization (HDS) and hydrodenitrogenation (HDN) of heavy oil fractions are largely based on alumina-supported molybdenum or tungsten, to which cobalt or nickel is added as a promoter. Since the catalysts are active in the sulfided state, activation is carried out by treating the oxidic catalyst precursor in a mixture of H_2S and H_2 (or by exposing the catalyst to the sulfur-containing feed). The function of hydrogen is to prevent the decomposition of the relatively unstable H_2S to elemental sulfur, which would otherwise accumulate on the surface of the catalyst. Figure 4.24 shows the TPS (Temperature Programmed Sulfidation) patterns of MoO_3/Al_2O_3; negative peaks mean that the corresponding gas is consumed, positive that it is produced. Below 400 K, H_2S is consumed and H_2O produced while no uptake of H_2 is detected, suggesting that the prevailing reaction at low temperatures is the exchange of sulfur for oxygen.

Around 500 K, the catalyst consumes H_2, as shown by the sharp peak while simultaneously H_2S and some additional H_2O are produced, which indicates that the catalyst has taken up too much sulfur at lower temperatures, which is now released in the form of H_2S. At higher temperatures, the catalyst continues to exchange oxygen for sulfur until all molybdenum is present as MoS_2. TPS has proven very useful in the study of the sulfidation of MoO_3 as well as Co and Ni promoted catalysts.

4.9
Infrared Spectroscopy

The most common application of infrared spectroscopy in catalysis is to identify adsorbed species and to study the way in which these species are chemisorbed on the surface of the catalyst [2, 26]. Sometimes infrared spectra of adsorbed probe molecules such as CO and NO give valuable information on adsorption sites on a catalyst. We will first summarize the theory behind infrared absorption.

Molecules possess discrete levels of rotational and vibrational energy. Transitions between vibrational levels occur by absorption of photons with frequencies v in the infrared range (wavelength 1–1000 μm, wave numbers 10 000–10 cm^{-1}, energy differences 1240–1.24 meV). The C–O stretch vibration, for example, is at 2143 cm^{-1}. For small deviations of the atoms in a vibrating diatomic molecule from their equilibrium positions, the potential energy $V(r)$ can be approximated by that of the harmonic oscillator

$$V(r) = \frac{1}{2}k(r - r_{eq})^2 \tag{4.11}$$

in which k is the force constant of the vibrating bond, r the distance between the two atoms, and r_{eq} the distance at equilibrium. The corresponding vibrational energy levels are

$$E_n = \left(n + \frac{1}{2}\right)hv; \quad v = \frac{1}{2\pi}\sqrt{\frac{k}{\mu}}; \quad \frac{1}{\mu} = \frac{1}{m_1} + \frac{1}{m_2} \tag{4.12}$$

in which $n = 0, 1, 2, \ldots$, h is Planck's constant, v the frequency, m_1 and m_2 the masses of the vibrating atoms, and μ the reduced mass. Allowed transitions in the harmonic approximation are those for which the vibrational quantum number changes by one unit. A general selection rule for the absorption of a photon is that the dipole moment of the molecule must change during the vibration.

The harmonic approximation is only valid for small deviations of the atoms from their equilibrium positions. The most obvious shortcoming of the harmonic potential is that the bond between to atoms can not break. With physically more realistic potentials, such as the Lennard-Jones or the Morse potential, the energy levels are no longer equally spaced and vibrational transitions with $\Delta n > 1$ are no longer forbidden. Such transitions are called overtones. The overtone of gaseous CO at 4260 cm^{-1} (slightly less than 2×2143 cm^{-1} = 4286 cm^{-1}) is an example.

The simple harmonic oscillator picture of a vibrating molecule has important implications. First, knowing the frequency, one can immediately calculate the force constant of the bond. Note from Eq. (4.11) that k, as coefficient of r^2, corresponds to the curvature of the interatomic potential and not primarily to its depth, the bond energy. However, as the depth and the curvature of a potential usually change hand in hand, the infrared frequency is often taken as an indicator for the strength of the bond. Second, isotopic substitution can be useful in the assignment of frequencies to bonds in adsorbed species because frequency shifts

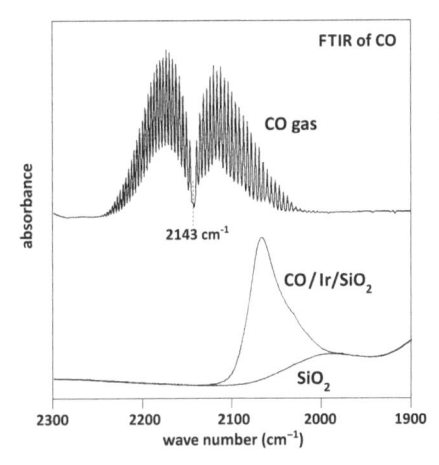

Figure 4.25 The infrared spectrum of gas phase CO shows rotational fine structure, which disappears upon adsorption, as shown by the spectrum of CO adsorbed on an Ir/SiO$_2$ catalyst (adapted from [1]).

due to isotopic substitution (of, for example, D for H in adsorbed ethylene, or OD for OH in methanol) can be predicted directly.

Molecules in the gas phase have rotational freedom, and the vibrational transitions are accompanied by rotational transitions. For a rigid rotor that vibrates as a harmonic oscillator, the expression for the available energy levels is

$$E = (n + 1/2)h\nu + \frac{h^2}{8\pi^2 I} j(j+1) \tag{4.13}$$

where j is the rotational quantum number and I the moment of inertia of the molecule. Here, a third selection rule applies: $\Delta j = \pm 1$. Note that because of this third selection rule the purely vibrational transition is not observed in the gas phase. Instead, two branches of rotational side bands appear (see Figure 4.25 for CO).

Upon adsorption, the molecule loses its rotational freedom and so only the vibrational transition is observed (but at a different frequency, see Figure 4.25). For CO, three factors contribute to this shift:

- Mechanical coupling of the C–O molecule to the heavy substrate increases the C–O frequency by some 20–50 cm^{-1}.
- Interaction between the C–O dipole and its image in the conducting, polarizable metal weakens the C–O frequency by 25–75 cm^{-1}.
- Formation of a chemisorption bond between C–O and the substrate alters the distribution of electrons over the molecular orbitals and weakens the C–O bond.

Thus, it is not strictly correct to interpret the frequency difference between adsorbed and gas phase C–O in terms of chemisorption bond strength only.

Infrared frequencies are characteristic for certain bonds in molecules and they can often be used to identify chemisorbed species on surfaces. The infrared spectrum of CO or NO can sometimes also be used to recognize sites on the surface of a catalyst, as the following example shows.

Sulfided Mo and Co-Mo catalysts, used in hydrotreating reactions, contain Mo as MoS_2. This compound has a layer structure consisting of sandwiches, each of a Mo layer between two S layers. The chemical activity of MoS_2 is associated with the edges of the sandwich where the Mo is exposed to the gas phase; the basal plane of the S^{2-} anions is largely unreactive. The infrared spectrum of NO on a sulfided Mo/Al_2O_3 catalyst (Figure 4.26) shows two peaks at frequencies that agree with those observed in organometallic clusters of Mo and NO groups. NO on sulfided Co/Al_2O_3 gives also two infrared peaks (Figure 4.26 (ii)), but at different frequencies, as observed on MoS_2. These results suggest that NO can be used as a probe to titrate the number of Co and Mo sites in the $Co-Mo/Al_2O_3$ catalyst. Figure 4.26 (iii) confirms that this idea works. Moreover, comparison of the intensities of the NO/Mo infrared signals on Mo/Al_2O_3 and $Co-Mo/Al_2O_3$ reveals that the presence of Co decreases the number of Mo atoms that are accessible to NO. This means that Co most probably decorates the edges of the MoS_2 sandwiches because the edges constitute the adsorption sites for NO [27].

Transmission infrared spectroscopy is very popular for studying the adsorption of gases on supported catalysts or for studying the decomposition of infrared active catalyst precursors during catalyst preparation. Infrared spectroscopy is an *in situ* technique that is applicable in transmission or diffuse reflection mode on real catalysts.

4.10
Surface Science Techniques

Beyond the characterization methods discussed so far, surface science offers an impressive array of experimental techniques to investigate surfaces. Unfortunately, many are only applicable to well-defined macroscopic single crystals (see also Figure 4.1). In the next chapter, we tell more about the structure of surfaces. Due to the complexity of catalysts, we are often forced to use models to investigate the surface properties of catalyst particles, for example, if we are interested in bonding modes, ordering phenomena, and reactivity aspects of adsorbed species. By using single crystals of metals or oxides, and by utilizing the many methods that surface science offers, many important questions can be addressed. Such information has often proved relevant for understanding technical catalysts. It is beyond the scope of this book to treat all surface science methods that are available and the reader is referred to relevant textbooks [28–31]. Surface sensitive characterization techniques such as XPS, IR, SIMS and ISS, which are also applicable to supported catalysts, have already been introduced and are often used on single crystals as well. Here, we complete our overview by describing a few techniques that are used to study the structure of surfaces: low energy electron diffraction (LEED), scanning tunneling microscopy (STM), and atomic force microscopy (AFM).

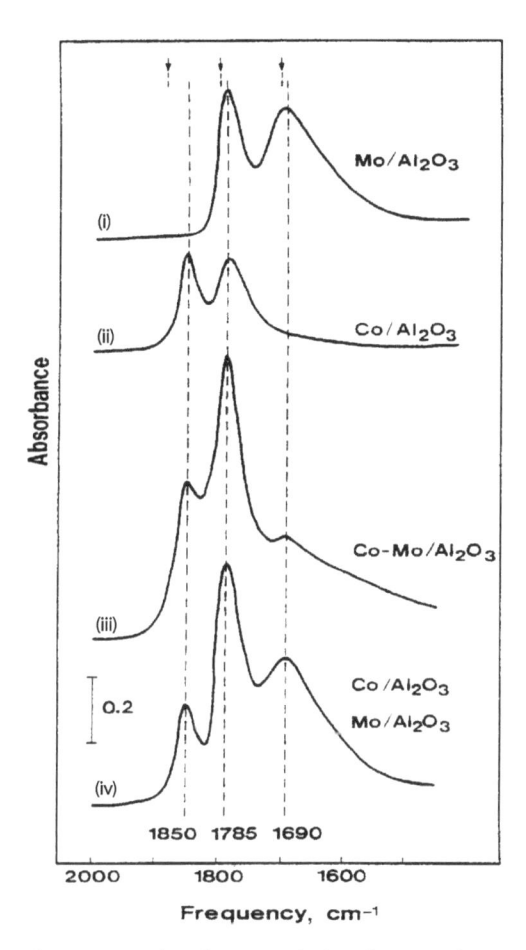

Figure 4.26 Infrared spectra of NO probe molecules on sulfided Mo, Co, and Co-Mo hydrodesulfurization catalysts. The peak assignments are supported by the Infra Red (IR) spectra of organometallic model compounds. These spectra allow for a quantitative titration of Co and Mo sites in the Co-Mo catalyst (reproduced from Topsøe and Topsøe [27] with permission of Elsevier).

4.10.1
Low Electron Energy Diffraction (LEED)

LEED is the surface analogue of X-ray diffraction. As the name indicates, the major difference is that one uses electrons instead of X-rays. As electrons of low kinetic energy (40–200 eV) do not penetrate very far into the material without losing energy, the elastically reflected electrons carry only information on the outermost layers of the surface (see Figure 4.7).

The LEED experiment relies on the duality of electrons, which have both particle and wave character. Electrons of primary energy, E_p, somewhere in the min-

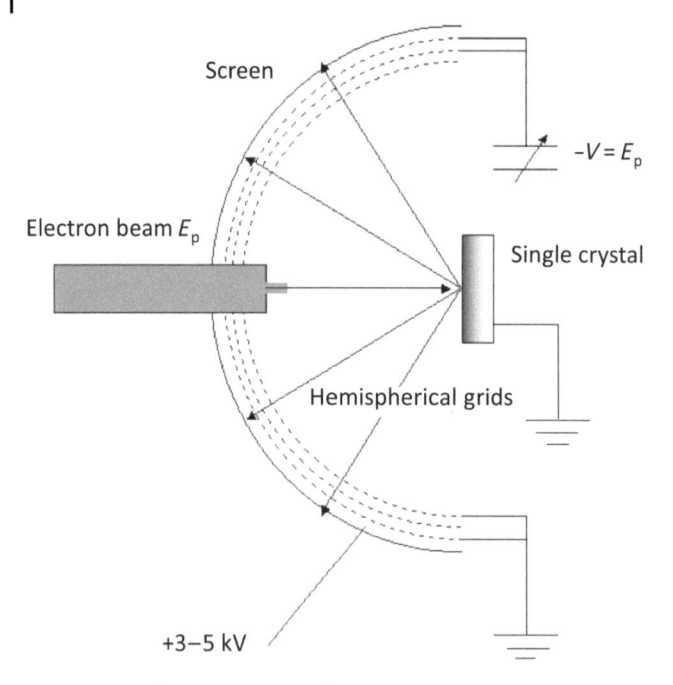

Figure 4.27 Schematic drawing of LEED optics.

imum of the mean-free path curve (Figure 4.7) possess a wavelength, λ, that is comparable with the distance between atoms in a lattice:

$$\lambda = \frac{h}{p} \quad \text{and} \quad p = \hbar k = \sqrt{2m_e E_p} \tag{4.14}$$

where p is the momentum of the electron, h is Planck's constant, k the wave number, and m_e the mass of the electron. The equation can be rearranged to

$$\lambda(\text{Å}) = \sqrt{\frac{150.4}{E_p(\text{eV})}} \tag{4.15}$$

Since the wavelength is of the order of lattice distances, electrons that are scattered elastically undergo constructive and destructive interference (as with X-rays in XRD). The backscattered electrons form a pattern of spots on a fluorescent screen from which the symmetry and structure of the surface may be deduced.

Figure 4.27 shows the experimental setup. A beam of electrons with energy E_p from a relative simple electron gun mounted in the center of the LEED optics is directed towards the surface. The crystal is positioned in the center of four concentric hemispherical grids. The inner grid is at ground potential, whereas the two middle grids are kept at a negative potential, slightly below E_p, such that only electrons that have been scattered elastically are allowed to pass. Between the last grid

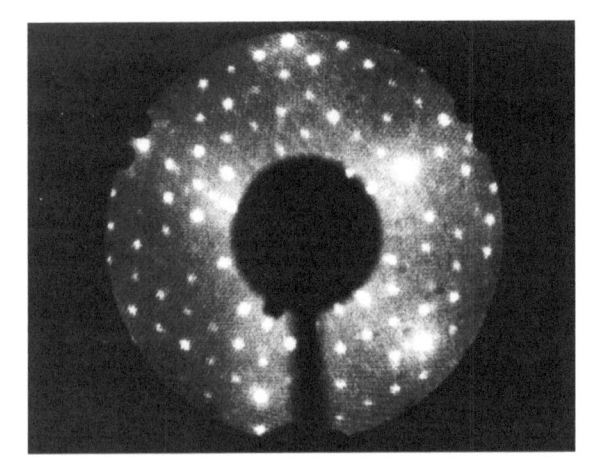

Figure 4.28 LEED pattern obtained from N-atoms on Fe(111) at $E_p = 42$ eV. The surface is estimated to have a coverage of 0.96 ML nitrogen and is reconstructed into a 5 × 5 overlayer structure. The dark shadow in the middle of the picture is due to the electron gun (reproduced from Alstrup et al. [32] with permission of Elsevier).

and an outer fluorescent screen, there is a potential difference between 2 and 5 kV to accelerate the electrons toward the fluorescent screen. The latter is transparent such that the spot pattern can be observed from the rear without being obstructed by the sample manipulator. The spots can then be photographed or recorded with a CCD camera. More elaborate setups have been developed for spot profile analysis LEED (SPALEED) where the detailed intensity distribution inside a spot is measured to reveal details on the long-range structure of the surface.

Figure 4.28 shows an example of a LEED picture that was obtained after exposing a clean single crystal of Fe(111) to nitrogen under conditions where N_2 dissociates and the surface becomes saturated with N atoms [32]. The surface is believed to reconstruct significantly to accommodate these high coverages.

The principle of a LEED experiment is shown schematically in Figure 4.29. The primary electron beam impinges on a crystal with a unit cell described by vectors a_1 and a_2. The (00) beam is reflected directly back into the electron gun and can not be observed unless the crystal is tilted. The LEED image is congruent with the reciprocal lattice described by two vectors, a_1^* and a_2^*. The kinematic theory of scattering relates the reciprocal lattice vectors to the real-space lattice through the following relations

$$a_1^* \equiv 2\pi \frac{a_2 \times z}{a_1 \cdot a_2 \times z} \qquad (4.16)$$

$$a_2^* \equiv 2\pi \frac{a_1 \times z}{a_1 \cdot a_2 \times z} \qquad (4.17)$$

in which z is a unit vector perpendicular to the surface. Hence, by analyzing the LEED pattern one can deduce the symmetry and structure of the clean or covered

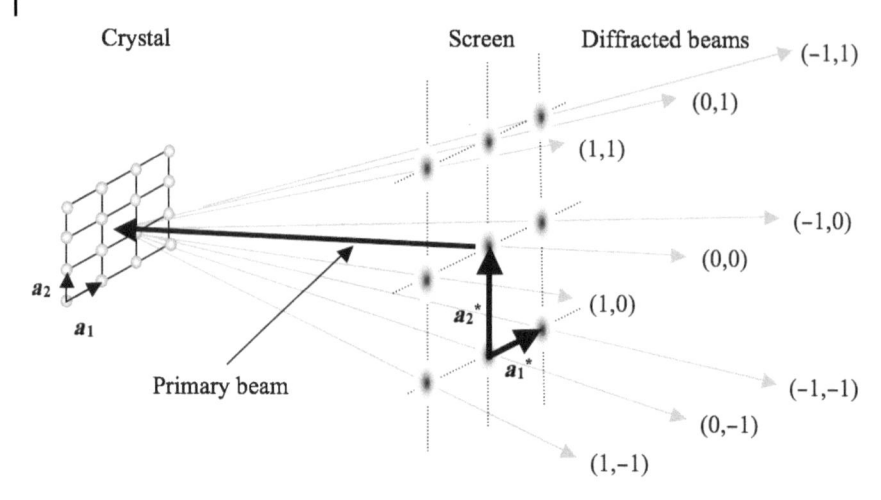

Figure 4.29 Schematic drawing of the LEED experiment on a single crystal with a unit cell given by vectors a_1 and a_2. The LEED pattern corresponds to the reciprocal lattice described by vectors a_1^* and a_2^*.

surface. Note, however, that in the case of an ordered overlayer LEED only yields the size of the unit cell for the overlayer–substrate combination and not where the adsorbates are actually positioned relative to the substrate atoms. If the latter is desired, more elaborate methods must be used. For example, the intensity of the spots formed on the screen is a function of the primary energy. Thus, intensity versus E_p data (called $I-V$ curves) can be compared with predictions based on an assumed structure. In this way one may deduce the accurate position of both adsorbate and substrate atoms. One should be aware that the LEED picture is an average of all ordered structures over the area hit by the electrons. Also, LEED only probes structures if they show order over a sufficiently large area – typically at least $30 \times 30 \,\text{Å}^2$.

4.10.2
Scanning Probe Microscopy

The most recent developments in determining the surface structure are the scanning tunneling microscope (STM) and the scanning or atomic force microscope (SFM or AFM) [33–35]. These techniques are capable of imaging the local surface topography with atomic resolution. The general concept behind scanning probe microscopy is that a sharp tip is rastered across a surface by piezoelectric translators while a certain property reflecting the interaction between the tip and the surface is monitored. As a result, scanning probe microscopy yields local information.

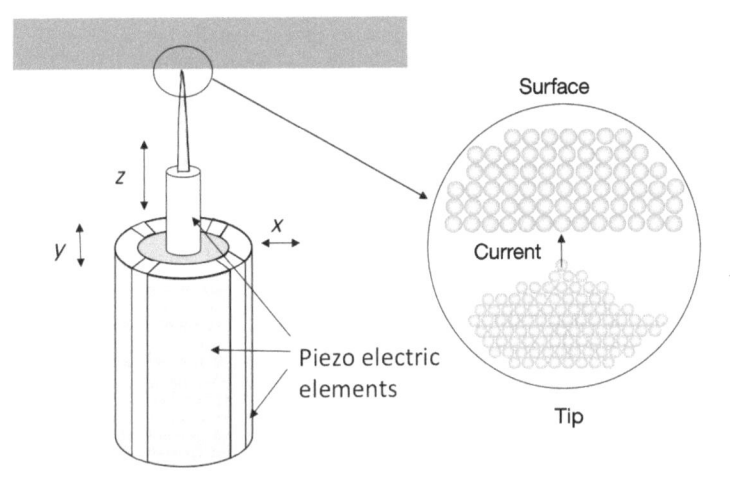

Figure 4.30 Schematic drawing of an STM. Notice that the tip is rather blunt on the atomic scale with a curvature of roughly 100 Å. The scanning motion in the x and y direction is enabled by a cylinder divided in four piezo elements while the z motion is enabled by the cylindrical piezo holding the tip.

4.10.2.1 Scanning Tunneling Microscopy (STM)

The STM method was developed by Binnig and Rohrer, who received the Nobel Prize for their invention. STM is the most mature of the scanning probe methods. A sharp tip (curvature of the order 100 Å) is brought close to a surface and a low potential difference (the bias voltage) is applied between sample and tip.

Classically, there should not be a current between the sample and the tip, but as the distance becomes below 0.5 nm quantum mechanics takes over and electrons may tunnel through the gap, giving rise to a tunnel current on the order of 1 nA, which can be measured. The experimental setup is shown schematically in Figure 4.30.

STM can be performed in several modes, that is, constant current or constant distance while scanning the tip over a desired surface area. The constant current mode is the most common. Here, the current is kept constant by varying the tip–surface distance in a feedback loop. The tip is mounted on a piezoelectric tube, which can expand or contract depending on the potential difference. The coarse approach of the tip to the surface is dealt with by applying a so-called inchworm, which is also based on piezoelectric elements. This inchworm allows the tip to be moved several mm per minute while the much more accurate z motion during scanning is carried out by a tube which can extract or expand over about 100 nm with a resolution of 0.001 nm! Similarly, scanning in the x–y-plane is obtained by applying potential differences on a tube that is split in four parts along the axis. Such an arrangement allows movements between less than an Å to a few μm. The entire system is kept small to obtain high resonance frequencies and is mounted on a vibrationally damped set-up to eliminate external disturbances as much as possible. To obtain a typical STM picture, the tip may scan 256 lines with 256 points for each line. The scan rate is very fast so that it usually takes between 1 and 10 s to obtain an image. By acquiring subsequent pictures, it is possible to make movies of dynamical processes on the surface, such as diffusion or chemical reaction.

The tip is usually made of tungsten, at least for ultrahigh vacuum applications. Other less reactive materials, such as Pt or Pt/Ir alloys, may be applied for studies under ambient conditions. Recipes are available for preparing sharp tips, which are necessary if surfaces are not flat. For perfect single crystal surfaces, however, relatively blunt tips prepared by cutting a tungsten wire by pliers may also work, since as long as just one atom sticks out further than the rest it will conduct basically all the current. STM provides excellent pictures of conducting surfaces in real space and it often allows one to deduce the positions of the atoms and adsorbates.

The extremely favorable resolution is due to the tunneling phenomenon that is possible if empty electron states of the surface overlap with filled states at the tip, or vice versa. Thus, what is depicted in an STM experiment is *not* the atom but merely the density of states around the Fermi level.

In a quantum mechanical treatment, the tunnel current is given as a function of distance d between tip and surface as

$$I(d) \propto e^{-1.025 d(\text{Å})\sqrt{\varphi(eV)}} \tag{4.18}$$

where φ is the work function. Crucially, the current depends exponentially on the distance between tip and surface, mainly because the overlap between empty and filled states is determined by the tails of the respective wave functions out of the surface and the tip into the vacuum. This explains why it is possible to prepare a satisfactory tip. If one atom on the tip protrudes by only 1 Å above the other atoms, then Eq. (4.18) predicts that it carries already 90% of the current. Expression (4.18) also explains why atomic resolution is possible in STM.

Figure 4.31 shows an example where STM recognizes the individual metal atoms in an alloy, thus, revealing highly important structural information on the atomic level. The technique does not require a vacuum and can in principle be applied under *in situ* conditions (even in liquids). Unfortunately, STM only works on well-defined, planar, and conducting surfaces such as metals and semiconductors, and not on oxide-supported catalysts. For the latter surfaces, atomic force microscopy offers better perspectives.

4.10.2.2 The Atomic Force Microscope (AFM)

Atomic force microscopy (AFM) or, as it is also called, scanning force microscopy (SFM) is based on the minute but detectable forces – of the order of nanonewtons – between a sharp tip and atoms in the surface [1, 35]. The tip is mounted on a flexible arm, called cantilever, and is positioned at a subnanometer distance from the surface. If the sample is scanned under the tip in the x–y-plane, it feels the attractive or repulsive force from the surface atoms and hence it is deflected in the z-direction. The deflection can be measured with a laser and photo detectors

Figure 4.31 STM image of a PtRh(100) surface. Although the bulk contains equal amounts of each element, the surface consists of 69% of platinum (dark) and 31% of rhodium (bright), in agreement with the expected surface segregation of platinum on clean Pt-Rh alloys in ultrahigh vacuum. The black spots are due to carbon impurities. It is seen that platinum and rhodium have a tendency to cluster in small groups of the same elements (reproduced from Wouda et al. [36] with permission of Elsevier).

as indicated schematically in Figure 4.32. Atomic force microscopy can be applied in two ways.

In the contact mode, the tip is within a few ångstrøms of the surface, and the interaction between them is determined by the interactions between the individual atoms in the tip and on the surface.

The second mode of operation is the noncontact mode, in which the distance between the tip and sample is much larger, between 2 and 30 nm. In this case, one describes the forces in terms of the macroscopic interaction between bodies. Magnetic force microscopy, in which the magnetic domain structure of a solid can be imaged, is an example of noncontact mode operation.

A third mode, which has recently become the standard for work on surfaces that are easily damaged, is in essence a hybrid between contact and noncontact mode and is sometimes called the tapping mode. In this case, the cantilever is brought into oscillation such that the tip just touches the surface at the maximum deflection towards the sample. When the oscillating cantilever approaches the maximum deflection, it starts to feel the surface and the oscillation becomes damped, which is detected by the electronics and used as the basis for monitoring the topography when the sample is scanned. In tapping mode, shear forces due to dragging the tip horizontally along the surface ("scratching") are avoided while

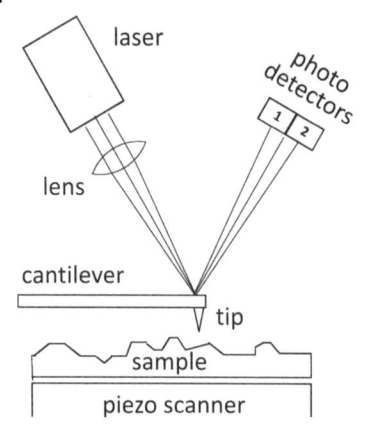

Figure 4.32 Experimental set-up for atomic force microscopy. The sample is mounted on a piezo-electric scanner and can be positioned with a precision better than 0.01 nm in the x, y, and z directions. The tip is mounted on a flexible arm, the cantilever. When the tip is attracted or repelled by the sample, the deflection of the cantilever/tip assembly is measured as follows: a laser beam is focused at the end of the cantilever and reflected to two photodiodes, numbered 1 and 2. If the tip bends towards the surface, photodiode 2 receives more light than 1, and the difference in intensity between 1 and 2 is a measure of the deflection of the cantilever and, thus, for the force between the sample and the tip (reproduced from [1] with permission of Wiley-VCH).

forces in a perpendicular direction are greatly reduced. It has become the favored way of imaging small particles on planar substrates used as models for catalysts.

Figure 4.33 shows the noncontact AFM image, measured in tapping mode, of small polyethylene deposits grown on a planar Cr/SiO_2 model catalyst. As the polymer is a soft material, such images can only be obtained in noncontact mode, which nowadays represent the standard mode of operation.

The image is always a convolution of the topography of the surface and that of the tip, and the one with the least steep features determines the image. Flat surfaces are scanned with the conventional pyramidal tips, which have a wide opening angle and are relatively blunt. If surfaces contain features that are sharper than

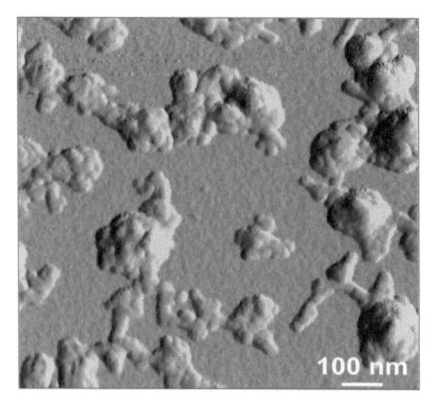

Figure 4.33 AFM image of polyethylene islands on a planar $Cr/SiO_2/Si(100)$ model catalyst (courtesy P.C. Thüne, TU Eindhoven).

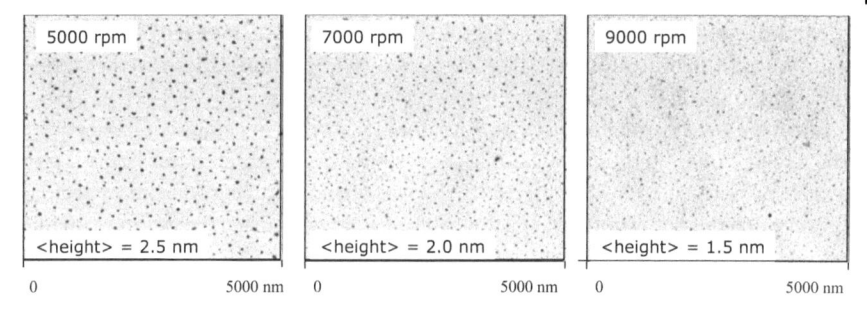

Figure 4.34 AFM images of rhodium particles deposited by spin-coating impregnation on a flat SiO$_2$ on Si(100) substrate particle, after reduction in hydrogen (adapted from [37]).

the tip, one images the tip shape rather than the surface topography. Figure 4.34 shows AFM images of rhodium particles on a planar silica substrate. The particles, with average diameters of a few nanometers only, appear to be larger because the size of the tip determines the resolution of the image. One nevertheless obtains correct information on the height of the particles. As long as the tip can reach the substrate, one measures the correct height difference between particle apex and substrate level.

In the example of Figure 4.34, rhodium particles have been deposited by spin-coating impregnation of a SiO$_2$/Si substrate with an aqueous solution of rhodium trichloride. After drying, the particles were reduced in hydrogen. The images show samples prepared at three different rotation speeds in the spin coating process, but with concentrations adjusted such that each sample contains about the same amount of rhodium atoms. The particles prepared at high rotation speeds are smaller, which is attributed to the shorter time during which the solvent evaporates and the rhodium salt deposits. This favors nucleation over growth, leading to many small deposits of rhodium precursor material. Reduction in hydrogen produces the small metal particles that have been imaged in Figure 4.34. Hence, even if the morphology in directions parallel to the surface is not correctly reproduced, AFM determines the height of features on flat parts of the substrates with high precision.

4.11
Concluding Remarks

In this chapter, we have limited ourselves to the most common techniques in catalyst characterization. Of course, there are several other methods available, such as nuclear magnetic resonance (NMR), which is very useful in the study of zeolites, electron spin resonance (ESR) and Raman spectroscopy, which may be of interest for certain oxide catalysts. Also, all of the more generic tools from analytical chemistry, such as elemental analysis, UV-Vis spectroscopy, atomic absorption, calorimetry, thermogravimetry, etc. are often used on a routine basis.

Methods based on adsorption, used to determine the surface area of the entire catalyst or of the metallic particles, are also very important. These are discussed in Chapter 5.

We have already mentioned that fundamental studies in catalysis often require the use of single crystals or other model systems. As catalyst characterization in academic research aims to determine the surface composition on the molecular level under the conditions where the catalyst does its work, one can in principle adopt two approaches. The first is to model the catalytic surface, for example, with that of a single crystal. By using the appropriate combination of surface science tools, the desired characterization on the atomic scale is certainly possible in favorable cases. However, although one may be able to study the catalytic properties of such samples under realistic conditions (pressures of 1 atm or higher), most of the characterization is necessarily carried out in ultrahigh vacuum (UHV), and not under reaction conditions.

The second approach is to study real catalysts with *in situ* techniques such as infrared and Mössbauer spectroscopy, EXAFS, and XRD, under reaction conditions, or, as is more often done, under a controlled environment after quenching of the reaction. The *in situ* techniques, however, are not sufficiently surface specific to yield the desired atom-by-atom characterization of the surface. At best they determine the composition of the particles.

The dilemma is thus: investigations of real catalysts under relevant conditions by *in situ* techniques gives little information on the surface of the catalyst while techniques that are surface sensitive can often only be applied on model surfaces under vacuum. Bridging the gap between UHV and high pressures and between the surfaces of single crystals and of real catalysts is, therefore, an important issue in catalysis.

Finally, while many techniques undoubtedly reveal usable information on catalysts, the information required from a fundamental point of view can often not be obtained. In this situation, it is a good strategy to combine all techniques that tell us at least something about the catalyst. The catalysis literature contains several examples where this approach has been remarkably successful.

References

1 Niemantsverdriet, J.W. (2007) *Spectroscopy in Catalysis, an Introduction*, Wiley-VCH Verlag GmbH, Weinheim.
2 Weckhuysen, B.M. (2004) *In-Situ Spectroscopy of Catalysts*, American Scientific Publishers, San Diego.
3 Che, M. and Vedrine, J.C. (Eds.) (2012) *Characterization of Solid Materials and Heterogeneous Catalysts – From Structure to Surface Reactivity*, Wiley-VCH Verlag GmbH, Weinheim.
4 Rodriguez, J.A., Hanson, J.C., and Chupas, P.J. (Eds.) (2013) *In-Situ Characterization of Heterogeneous Catalysts*, John Wiley & Sons, Inc., Hoboken.
5 Schloegl, R. (2009) X-ray Diffraction: A Basic Tool for Characterization of Solid Catalysts in the Working State, in *Advances in Catalysis*, Vol 52 (eds B.C. Gates, H. Knözinger), pp. 273–338.
6 Saib, A.M., Borgna, A., van de Loosdrecht, J., van Berge, P.J., Geus, J.W., and Niemantsverdriet, J.W. (2006) Preparation and characterisation of spherical CO/SiO_2 model catalysts with well-

defined nano-sized cobalt crystallites and a comparison of their stability against oxidation with water. *Journal of Catalysis*, **239**, 326–339.
7 Knop-Gericke, A., Kleimenov, E., Haevecker, M., Blume, R., Teschner, D., Zafeiratos, S., Schloegl, R., Bukhtiyarov, V.I., Kaichev, V.V., Prosvirin, I.P., Nizovskii, A.I., Bluhm, H., Barinov, A., Dudin, P., and Kiskinova, M. (2009) X-ray Photoelectron Spectroscopy for Investigation of Heterogeneous Catalytic Processes, in *Advances in Catalysis*, Vol 52 (eds B.C. Gates, H. Knözinger), pp. 213–272.
8 Somorjai, G.A. and Li, Y. (2010) *Introduction to Surface Chemistry and Catalysis*, John Wiley & Sons, Inc., Hoboken.
9 Borg, H.J., Van den Oetelaar, L.C.A., Van Ijzendoorn, L.J., and Niemantsverdriet, J.W. (1992) Surface-chemistry of catalyst preparation studied by using flat alumina model supports. *Journal of Vacuum Science & Technology A – Vacuum Surfaces and Films*, **10**, 2737–2741.
10 Fernandez-Garcia, M. (2002) XANES analysis of catalytic systems under reaction conditions. *Catalysis Reviews – Science and Engineering*, **44**, 59–121.
11 Bare, S.R. and Ressler, T. (2009) Characterization of Catalysts in Reactive Atmospheres by X-ray Absorption Spectroscopy, in *Advances in Catalysis*, Vol 52 (eds B.C. Gates, H. Knözinger), pp. 339–465.
12 Koningsberger, D.C., Van Zon, J., Van't Blik, H.F.J., Visser, G.J., Prins, R., Mansour, A.N., Sayers, D.E., Short, D.R., and Katzer, J.R. (1985) An extended X-ray absorption fine-structure study of rhodium oxygen bonds in a highly dispersed Rh/Al_2O_3 catalyst. *Journal of Physical Chemistry*, **89**, 4075–4081.
13 Thomas, J.M. and Gal, P.L. (2004) Electron Microscopy and the Materials Chemistry of Solid Catalysts, in *Advances in Catalysis*, Vol 48 (eds B.C. Gates, H. Knözinger), pp. 171–227.
14 Hansen, P.L., Helveg, S., and Datye, A.K. (2006) Atomic-scale Imaging of Supported Metal Nanocluster Catalysts in the Working State, in *Advances in Catalysis*, Vol 50 (eds B.C. Gates, H. Knözinger), pp. 77–95.
15 Wagner, J.B., Cavalca, F., Damsgaard, C.D., Duchstein, L.D.L., and Hansen, T.W. (2012) Exploring the environmental transmission electron microscope. *Micron*, **43**, 1169–1175.
16 Hansen, P.L., Wagner, J.B., Helveg, S., Rostrup-Nielsen, J.R., Clausen, B.S., and Topsøe, H. (2002) Atom-resolved imaging of dynamic shape changes in supported copper nanocrystals. *Science*, **295**, 2053–2055.
17 Millet, M.M.J. (2007) Mössbauer Spectroscopy in Heterogeneous Catalysis, in *Advances in Catalysis*, Vol 51 (eds B.C. Gates, H. Knözinger), pp. 309–350.
18 Van der Kraan, A.M., Nonnekens, R.C.H., Stoop, F., and Niemantsverdriet, J.W. (1986) Characterization of $FeRu/TiO_2$ and Fe/TiO_2 catalysts after reduction and Fischer-Tropsch synthesis by Mössbauer-spectroscopy. *Applied Catalysis*, **27**, 285–298.
19 Brongersma, H.H., Draxler, M., de Ridder, M., and Bauer, P. (2007) Surface composition analysis by low-energy ion scattering. *Surface Science Reports*, **62**, 63–109.
20 Meijers, A., De Jong, A.M., Van Gruijthuijsen, L.M.P., and Niemantsverdriet, J.W. (1991) Preparation of zirconium-oxide on silica and characterization by X-ray photoelectronspectroscopy, secondary ion mass-spectrometry, temperature programmed oxidation and infrared-spectroscopy. *Applied Catalysis*, **70**, 53–71.
21 Guliants, V.V., Bhandari, R., Hughett, A.R., Bhatt, S., Schuler, B.D., Brongersma, H.H., Knoester, A., Gaffney, A.M., and Han, S. (2006) Probe molecule chemisorption-low energy ion scattering study of surface active sites present in the orthorhombic Mo-V-(Te-Nb)-O catalysts for propane (amm)oxidation. *Journal of Physical Chemistry B*, **110**, 6129–6140.
22 Falconer, J.L. and Schwarz, J.A. (1983) Temperature-programmed desorption and reaction – applications to supported

catalysts. *Catalysis Reviews – Science and Engineering*, **25**, 141–227.
23 Reiche, M.A., Maciejewski, M., and Baiker, A. (2000) Characterization by temperature programmed reduction. *Catalysis Today*, **56**, 347–355.
24 Van't Blik, H.F.J. and Niemantsverdriet, J.W. (1984) Characterization of bimetallic FeRh/SiO$_2$ catalysts by temperature programmed reduction, oxidation and Mössbauer spectroscopy. *Applied Catalysis*, **10**, 155–162.
25 Arnoldy, P., Van den Heijkant, J.A.M., De Bok, G.D., and Moulijn, J.A. (1985) Temperature-programmed sulfiding of MoO$_3$/Al$_2$O$_3$ catalysts. *Journal of Catalysis*, **92**, 35–55.
26 Busca, G. (2014) *Heterogeneous Catalytic Materials, Solids State Chemistry, Surface Chemistry and Catalytic Behavior*, Elsevier, Amsterdam.
27 Topsøe, N.Y. and Topsøe, H. (1983) Characterization of the structures and active-sites in sulfided Co-Mo/Al$_2$O$_3$ and Ni-Mo/Al$_2$O$_3$ catalysts by NO chemisorption. *Journal of Catalysis*, **84**, 386–401.
28 Ertl, G. and Kueppers, J. (1985) *Low Energy Electrons and Surface Chemistry*, Wiley-VCH Verlag GmbH, Weinheim.
29 Feldman, L.C. and Mayer, J.W. (1986) *Fundamentals of Surface and Thin Film Analysis*, North-Holland, Amsterdam.
30 Woodruff, D.P. and Delchar, T.A. (1986) *Modern Techniques of Surface Science*, Cambridge University Press, Cambridge.
31 Kolasinski, K. (2012) *Surface Science: Foundations of Catalysis and Nanoscience*, John Wiley & Sons, Inc., Hoboken.
32 Alstrup, I., Chorkendorff, I., and Ullmann, S. (1997) The interaction of nitrogen with the (111) surface of iron at low and at elevated pressures. *Journal of Catalysis*, **168** 217–234.
33 Lauritsen, J.V. and Besenbacher, F. (2006) Model catalyst surfaces investigated by scanning tunneling microscopy, in *Advances in Catalysis*, Vol 50 (eds B.C. Gates, H. Knözinger), Elsevier, 97–147.
34 Wintterlin, J. (2000) Scanning Tunneling Microscopy Studies of Catalytic Reactions, in *Advances in Catalysis*, Vol 45 (eds B.C. Gates, H. Knözinger), 131–206.
35 Wiesendanger, R. (1994) *Scanning Probe Microscopy and Spectroscopy*, Cambridge University Press, Cambridge.
36 Wouda, P.T., Nieuwenhuys, B.E., Schmid, M., and Varga, P. (1996) Chemically resolved STM on a PtRh(100) surface. *Surface Science*, **359**, 17–22.
37 Niemantsverdriet, J.W., Engelen, A.F.P., A.M. de Jong, Wieldraaijer, W., and Kramer, G.J. (1999) Realistic surface science models of industrial catalysts. *Applied Surface Science*, **144/145**, 366–374.

Chapter 5
Solid Catalysts

5.1
Requirements of a Successful Catalyst

Catalysts can be metals, oxides, sulfides, carbides, nitrides, acids, salts, virtually any type of material [1–4]. Solid catalysts come in a multitude of different forms and can be loose particles, or small particles on a support. The support can be a porous powder, such as aluminum oxide particles, or a large monolithic structure, such as the ceramics used in the exhaust systems of cars. Clays or zeolites can also be catalysts [5].

The preparation of catalysts is a mixture of art and science, but most of all much experience [3, 6]. Although the underlying chemistry is largely known, many catalyst preparation recipes are so complicated that it is not possible to write a complete scheme of chemical reactions in detail.

Suppose that our tests in the laboratory have yielded a formulation with an excellent activity in terms of turnover per active site for a certain reaction, and a fabulous selectivity towards the desired product. Will this substance be a successful catalyst in an industrial application? Not necessarily. It will have to be developed into a material with the following properties:

- High activity per unit of volume in the eventual reactor.
- High selectivity towards the desired product at the conversion levels used in the eventual reactor, and the lowest possible selectivity to byproducts that generate waste problems.
- Sufficiently long lifetime with respect to deactivation.
- Possibility to regenerate, particularly if deactivation is fast.
- Reproducible preparation.
- Sufficient thermal stability against sintering, structural change, or volatilization inside the reaction environment (e.g., when steam is a byproduct of the reaction).
- High mechanical strength with respect to crushing (e.g., under the weight of the catalyst bed or during the shaping process).
- High attrition resistance (resistance to mechanical wear).

Concepts of Modern Catalysis and Kinetics, Third Edition. I. Chorkendorff and J.W. Niemantsverdriet.
© 2017 WILEY-VCH Verlag GmbH & Co. KGaA. Published 2017 by WILEY-VCH Verlag GmbH & Co. KGaA.

Catalysts are generally developed for a special process, that is, for a *certain reaction* in a *certain reactor* under *certain conditions*. Mass and heat transport phenomena put their own requirements on the morphology and heat-conducting properties of the catalyst. For example, it will be essential that the catalyst can be processed into shaped particles, such as extrudates or tablets, to avoid build up of pressure gradients over the catalyst bed. If our catalyst scores favorably on all of the points above but it cannot be brought into a form that allows operation in a reactor, it will never be used in practice.

Finally, factors such as originality (is our catalyst sufficiently original for a new patent or is it covered under an existing patent owned by the competition?) and economics have to be considered before our catalyst will be adopted.

In principle, any heterogeneous catalytic reaction can be considered to occur at an active site or an ensemble, where the elementary reactions take place. The specific activity of such a site is scientifically of great importance when comparing, for example, different metals or geometric configurations. We define the turn over frequency (TOF) as the number of reactant molecules that are converted over this site per second. Note that the TOF is not necessarily a good practical measure of catalyst activity. Industry is more interested in the activity per unit volume of catalyst. Having a high dispersion is important, but also if the particles provide a high number of the desired sites.

Since catalytic reactions often run at elevated pressure, the size of the reactor is an important parameter for the capital investment needed. In this context it is also important to consider the space–time product, expressing how much product the reactor can produce per unit time per unit volume. It makes little sense to have a catalyst that fully converts the reactant only if the space velocity is zero. Here, the space velocity is the ratio between the gas flow and the volume of the catalyst. For testing catalysts, the space velocity is chosen so high that one approaches the zero conversion limit, but for running an industrial process we want optimal trade-off between production and approach to equilibrium with the smallest amount of catalyst. Catalyst cost is of course also an important consideration and may actually result in using a less active metal because the optimum metal is scarce and expensive. Figure 5.1 shows the price of some of the most often used noble metals. Strong fluctuations occur as these metals are object of speculation. Several of the metals shown are also scarce and sometimes even a rumor that a new catalyst has been developed that may require substantial amounts of one of the less common metals may easily influence the market. Of course, metals such as nickel or iron are so inexpensive that developing alternative catalysts for, for example, the ammonia synthesis (iron based) or the steam reforming process (nickel based) is an almost impossible challenge.

Catalyst activity per unit volume is not the only interesting parameter. Also durability and selectivity form important considerations. It is not interesting to have a high conversion of reactants if it leads to a wide range of different products. This would result in expensive separation procedures for isolating the relevant products. Thus, we also want catalysts that are selective for the desired product. The selectivity of a catalyst leading to two products A and B is defined as

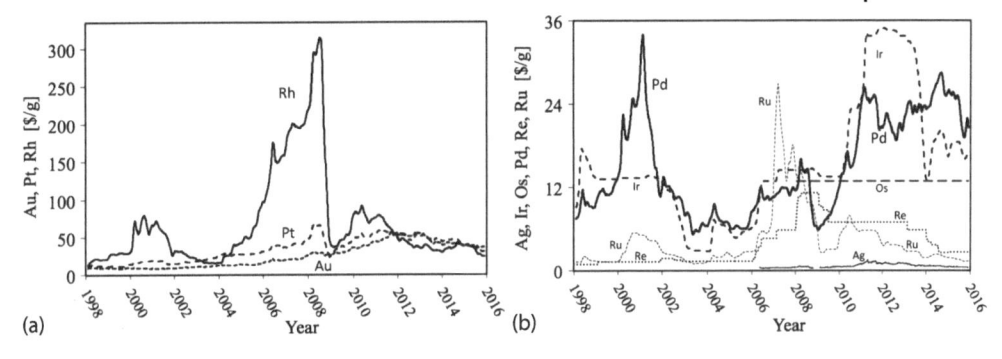

Figure 5.1 Price development of selected metals over the last 18 years. Note the rapid oscillations, illustrating that certain metals are subject to speculation (data from apps.catalysts.basf.com).

A/(A + B) × 100% for species A. Thus, an ideal catalyst operates at high conversion and high selectivity at the same time.

This is a major challenge for the development of the so-called dream reactions where one, for example, wants to convert methane directly into methanol by oxidation with oxygen. Looking at the reaction scheme it may look straight forward to do so, since one just needs to add one oxygen atom into the methane. Moreover, such a process is strongly exothermic (for details see Section 8.2.6.1). However, the problem is that the full oxidation into carbon dioxide and water is much more exothermic. Catalysts that can perform highly selective (i.e., 90%) oxidation of methane into methanol do exist, but they only operate at low conversion, making the product yield too low to be of interest for practical applications. Hence, direct methane oxidation to methanol remains a dream reaction.

5.2
The Structure of Metals, Oxides, and Sulfides and Their Surfaces

5.2.1
Metal Structures

The most important metals for catalysis are those of the groups VIII and I-B of the periodic system. Three crystal structures are important, face-centered cubic (fcc: Ni, Cu, Rh, Pd, Ag, Ir, Pt, Au), hexagonally close-packed (hcp: Co, Ru, Os), and body-centered cubic (bcc: Fe). Figure 5.2 shows the unit cell for each of these structures. Note that the unit cells contain 4, 2, and, 6 atoms for the fcc, bcc, and hcp structure, respectively. Many other structures, however, exist when considering more complex materials such as oxides, sulfides, etc., which we shall not treat here. Before discussing the surfaces that the metals expose, we mention a few general properties.

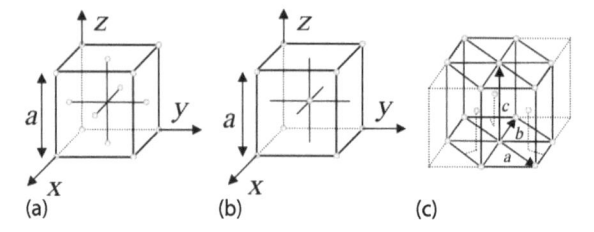

Figure 5.2 Unit cells of the (a) face-centered cubic (fcc), (b) body-centered cubic (bcc), and (c) hexagonally closed packed (hcp) lattices.

If the lattices are viewed as close-packed spheres, the fcc and the hcp lattices have the highest density, possessing about 26% empty space. Each atom in the interior has 12 nearest neighbors, or in different words, an atom in the interior has a coordination number of 12. The bcc lattice is slightly more open and contains about 32% of empty space. The coordination of a bulk atom inside the bcc lattice is 8.

5.2.2
Surface Crystallography of Metals

Surface crystallography is nothing but the two-dimensional analogue of bulk crystallography, in which we consider the structure of the different planes on which the atoms in the three-dimensional crystal reside. We limit ourselves to the most often encountered structures for metals as shown in Figure 5.2 For other structures, we refer to textbooks on solid state physics [7–9].

5.2.2.1 Crystal Planes
The structure and geometry of a surface play a dominant role with respect to its reactivity in adsorption and catalysis [10, 11]. It is therefore always necessary to specify which structure we are dealing with and, hence, it is important to have a notation that describes the various surfaces in a unique manner. A crystal surface is described by a vector normal to it, given by

$$H = hx + ky + lz \tag{5.1}$$

The surface is then indicated by the set $h, k,$ and l between parentheses (hkl), often in combination with the metals, such as in Cu(100) or Pt(111). In case of a (100) surface, the atoms reside in a plane parallel to the y–z-plane. Negative numbers are given by a bar over the value as in Cu(0$\bar{1}$0). Note that permutations like Cu(100), Cu(010), Cu(001), Cu(0$\bar{1}$0), and Cu($\bar{1}$00) all describe surfaces of the same structure as shown in Figure 5.3.

The three basal planes of cubic crystals are (100), (110), and (111). The respective cross sections are shown in Figure 5.3. The positions of the surface atoms that appear by applying these cuts on the unit cells of Figure 5.2 are shown in Figure 5.4.

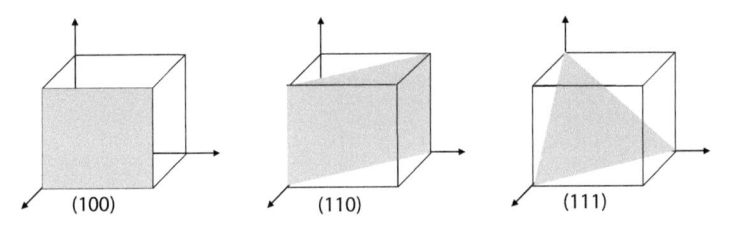

5.2 The Structure of Metals, Oxides, and Sulfides and Their Surfaces

Figure 5.3 Basal planes formed by cutting the unit cells of the simple crystal structures.

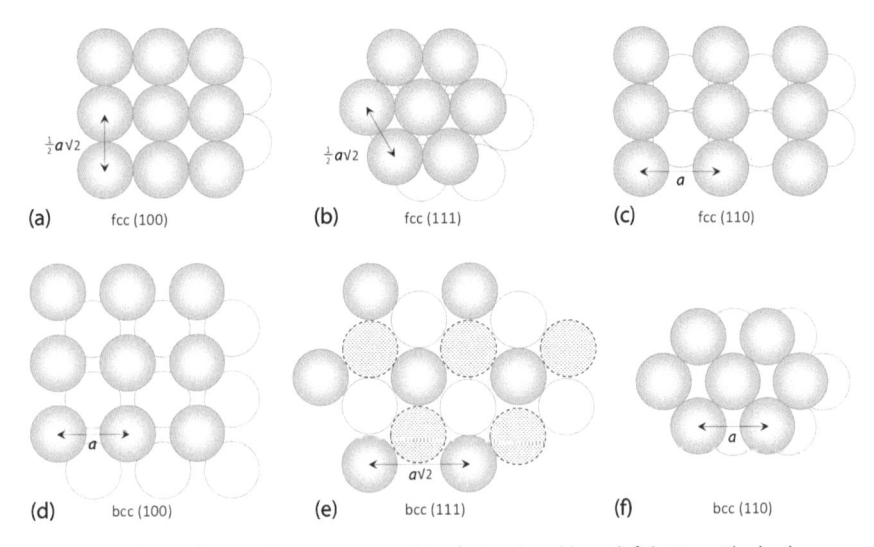

Figure 5.4 Basal plane surface structures of the fcc (a–c) and bcc (d–f) lattices. The broken lines indicate atoms in the second layer. Interatomic distances are given in terms of the lattice constant a.

An important parameter for surface reactivity is the density of atoms in the surface. The general rule of thumb is that the more open the surface, the more reactive it is. We return to this effect in much more detail in Chapter 6. Note that (110) is the most open basal plane of an fcc crystal, whereas (111) exhibits the closest packing. For bcc crystals the order is the opposite, that is, (111) is the most open and (110) the most packed.

The difference between the fcc and hcp structure is best seen if one considers the sequence of close-packed layers. For fcc lattices, this is the (111) plane (see Figures 5.2 and 5.4), for hcp lattices the (001) or more precise, the (0001) plane. The geometry of the atoms in these planes is exactly the same. Both lattices can now be built up by stacking close-packed layers on top of each other. If one places the atoms of the third layer directly above those of the first, one obtains the hcp structure. The sequence of layers in the ⟨001⟩ direction is, for example, *ababab*. In the fcc structure, it is every atom of the *fourth* layer that is above an atom of the first layer. The sequence of layers in the ⟨111⟩ direction is *abcabcabc…*

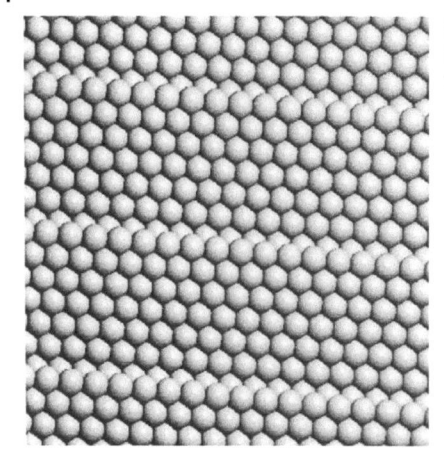

Figure 5.5 Model of the fcc (775) surface. Note that it basically consists of a (111) surface with a step on each sixth row.

More complicated surface structures contain steps or even kinks may also be described in this notation as $n(h_t, k_t, l_t) \times (h_s, k_s, l_s)$, which indicates that the surface contains n rows of atoms forming a (h_t, k_t, l_t) terrace, and one step of (h_s, k_s, l_s) structure. The notation is also applicable to surfaces containing steps and kinks, which result in a rough step. Thus, a (775) surface is equivalent to $6(111) \times (11\bar{1})$. In this manner, all sorts of surfaces may be constructed. Of course, the question is whether these are stable or break up into facetted structures.

Single crystals with specific surface orientations can be used to model real catalysts. These model catalysts are typically 5–10 mm in diameter and 1–3 mm thick; they can be cut from single crystal rods. Nowadays, most metals are commercially available in this form. A particular surface structure may be prepared by spark cutting a macroscopic single crystal normal to the desired direction. The crystal is usually oriented with accuracies better than 0.5° with the help of Laue X-ray diffraction. The surface is then polished with decreasing grain-sized polish paste until all scratches have been removed and a mirror-like surface is obtained. Sometimes, in particular for soft materials such as Cu, an additional electropolish is needed to remove damaged layers. Finally, the crystal is mounted in the ultrahigh vacuum (UHV) chamber (base pressures of 10^{-10} mbar or less to ensure that the crystal is not contaminated by adsorbates) and is cleaned by repeated cycles of sputtering and annealing. Preparing a clean surface is often tedious and may sometimes take several months. The cleaning procedure varies from element to element and, usually, effective and successful recipes can be found in the literature. The fcc (775) structure shown in Figure 5.5 is easily prepared from a (111) rod by cutting it under an angle of 8.47° with the normal to the (111) surface, where the angle is calculated from

$$\cos(\varphi) = \frac{\mathbf{a}}{|\mathbf{a}|} \frac{\mathbf{b}}{|\mathbf{b}|} = \frac{(111)}{\sqrt{(1+1+1)}} \cdot \frac{(775)}{\sqrt{49+49+25}} = \frac{19}{\sqrt{369}} \Rightarrow$$

$$\varphi = 8.47°$$

(5.2)

The reactivity of a surface depends on the number of unsaturated bonds, an unsaturated bond being what is left from a former bond with a neighboring metal atom that had to be broken in order to create the surface. Thus, we need to know the number of missing neighbors, denoted by Z_s, of an atom in each surface plane. One can infer from Figure 5.2 that an atom in the fcc (111) or hcp (001) surface has six neighbors in the surface and three below, but misses the three neighbors above that were present in the bulk: $Z_s = 3$. An atom in the fcc (100) surface, on the other hand, has four neighbors in the surface, four below, and misses four above the surface: $Z_s = 4$. Hence, an atom in the fcc (100) surface has one more unsaturated bond than an atom in the fcc (111) surface and is, therefore, slightly more reactive. Using the same reasoning, it is obvious that the atoms at the monoatomic steps in Figure 5.5 are much more reactive than the atoms in the (111) terraces. We will see later that such sites may enhance the reactivity by many orders of magnitude.

5.2.2.2 Adsorbate Sites

Definitions of the most common adsorption sites are shown in Figure 5.6 They are named on-top site, bridge sites (long or short bridge), and hollow sites, which may be threefold or fourfold in character. In case of threefold adsorption on the fcc (111) surface, it is also necessary to distinguish between hcp and fcc sites, having an atom just below the site or not.

The number of adsorption sites on a surface per unit or area follows straightforwardly from the geometry. Consider, for example, adsorption on a fourfold hollow site on the fcc (100) surface. The number of available sites is simply the number of unit cells with area $(1/2a\sqrt{2})^2$ per m^2, where a is the lattice constant of the fcc lattice. Note that the area of the same (100) unit cell on a bcc (100) surface is just a^2, a being again the lattice constant of the bcc lattice.

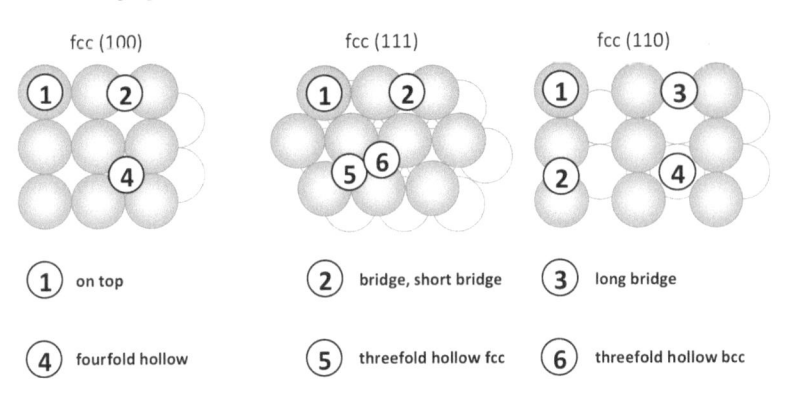

Figure 5.6 Adsorption sites on various fcc surfaces.

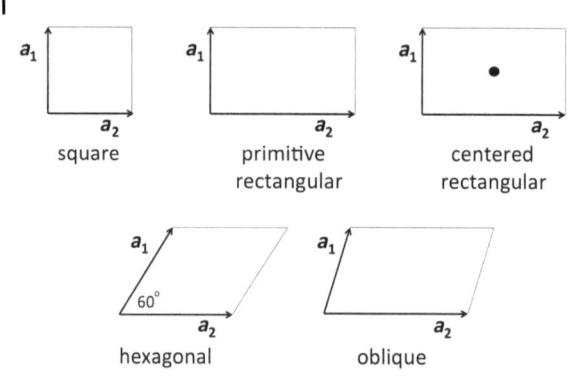

Figure 5.7 The five Bravais lattices of surfaces.

5.2.2.3 The Two-Dimensional Lattice

The common periodic structures displayed by surfaces are described by a two-dimensional lattice. Any point in this lattice is reached by a suitable combination of two basis vectors. Two unit vectors describe the smallest cell in which an identical arrangement of the atoms is found. The lattice is then constructed by moving this unit cell over any linear combination of the unit vectors. These vectors form the Bravais lattices, which is the set of vectors by which all points in the lattice can be reached.

In two dimensions, five different lattices exist, see Figure 5.7. One recognizes the hexagonal Bravais lattice as the unit cell of the cubic (111) and hcp (001) surfaces, the centered rectangular cell as the unit cell of the bcc and fcc (110) surfaces, and the square cell as the unit cell of the cubic (100) surfaces. Translation of these unit cells over vectors $h\boldsymbol{a}_1 + k\boldsymbol{a}_2$, in which h and k are integers, produces the surface structure.

In many cases the number of occupied adsorption sites is not equivalent to the number of unit cells. Often, repulsive interactions between adsorbed species prevents filling of all sites and adsorption may only be possible if all neighboring sites are unoccupied. Adsorbate structures are described according to how the new unit cell of adsorbate and substrate relate to the original unit cell of the substrate. Figure 5.8 shows a few examples along with the nomenclature used.

The unit cell of the p(2 × 2) adsorbate structure in Figure 5.8 is twice as large in both directions as the unit cell of the substrate and, hence, the structure is called p(2 × 2), where the p stands for primitive. The coverage corresponds to 0.25 monolayers, (commonly abbreviated to ML).

The Wood notation, as this way of describing surface structures is called, is adequate for simple geometries. However, for more complicated structures it fails, and one uses a 2 × 2 matrix which expresses how the vectors $\boldsymbol{a}_1$ and $\boldsymbol{a}_2$ of the substrate unit cell transform into those of the overlayer.

Figure 5.8 also shows the $\sqrt{2} \times \sqrt{2}(R45°)$ structure of S on Ni(100). The unit cell of the adsorbate is rotated over 45°. The structure is also referred to as c(2×2),

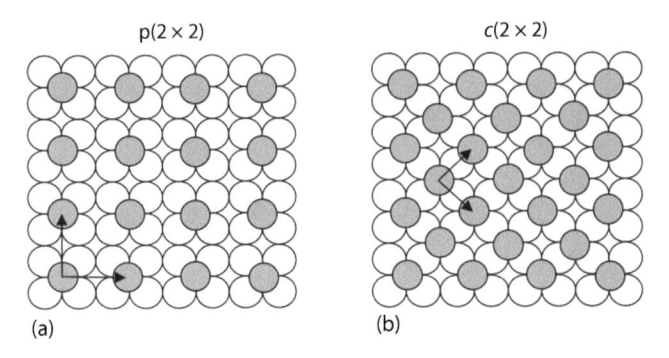

Figure 5.8 Structure of 2 × 2 and $\sqrt{2} \times \sqrt{2}\,(R45°)$ or $c(2 \times 2)$ adsorbate layers, which, for instance, are observed for sulfur atoms on a Ni(100) surface.

as it can be obtained by placing an additional atom in the center of the p(2 × 2) unit cell. The adsorbate coverage in the $c(2 \times 2)$ structure corresponds to 0.5 ML.

The saturation coverage depends on the adsorbate. For the Ni(100) surface, sulfur adsorbs to a maximum of 0.5 ML, whereas the saturation coverages of CO and H atoms are 0.58 and 1.0 ML, respectively. Note that the Wood notation does not specify where the adsorbate binds. This should be investigated experimentally, that is, by diffraction of scanning tunneling microscopy. Atoms such as carbon, nitrogen, oxygen, and sulfur all prefer bonding in sites of high coordination and will therefore typically be found in four- and threefold positions on the surface (see also Figure 5.6). Often, the energy gain involved in optimizing bonding around the adsorbate is so large that the whole surface may rearrange to achieve the high coordination.

A well-known example of adsorbate-induced surface reconstruction is that of carbon on the (100) surface of nickel [12]. Even though this surface already offers a fourfold coordination to the carbon atom in unreconstructed form, additional energy is gained by the so-called clock–anticlock reconstruction shown in Figure 5.9. Of course, it costs energy to rearrange the surface nickel atoms, but this investment is more than compensated by the higher bonding energy the carbon atoms obtain, and hence the total energy is lowered. Also the Ni(111) and the Ni(110) surfaces, which do not offer this type of sites, reconstruct such that carbon atoms obtain a high coordination.

Adsorbates on surfaces may attract or repel each other to some extent. If, for instance, the interaction is attractive, the adsorbates will tend to form islands that grow with increasing coverage until the whole surface is covered, while the repulsive interaction will lead to a random coverage until relatively high coverage where close packing of the adsorbates can no longer be avoided. All this has been a main topic of surface science for many years and a substantial database has been established on how different adsorbates are localized on the surface and how they rearrange the surface (see, for example, Somorjai and Li [10]).

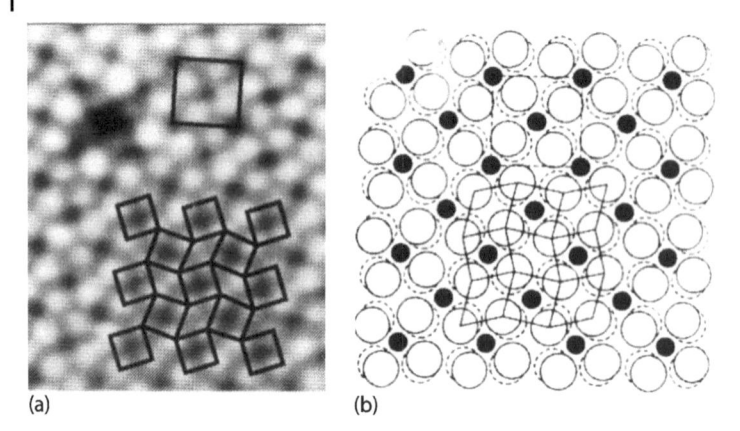

Figure 5.9 (a) STM picture of the so-called clock reconstruction that occurs when 0.5 ML of carbon is adsorbed on Ni(100). (b) Hard sphere model of the reconstruction. Dotted rings indicate the unreconstructed positions of the Ni surface atoms (reproduced with permission from Klink et al. [12]; copyright 1993 by the American Physical Society).

5.2.3
Oxides and Sulfides

Bulk structures of oxides are best described by assuming that they are made up of positive metal ions (cations) and negative O ions (anions) [13]. Locally the major structural feature is that cations are surrounded by O ions and oxygen by cations, leading to a bulk structure that is largely determined by the stoichiometry. The O^{2-} ions are in almost all oxides, larger than the metal cation. It does not exist in isolated form but is stabilized by the surrounding positive metal ions.

Figure 5.10 shows the different bulk terminations of MgO in the cubic rock salt structure. The (100) surface is by far the most stable, and MgO particles usually show only (100) facets. Note that there are two different (111) surfaces, namely those terminated by magnesium or by oxygen. Such surfaces possess a net dipole moment and are called polar. The (100) and (110) surfaces of MgO contain equal amounts of Mg and O; these are neutral or nonpolar.

The cations in transition metal oxides often occur in more than one oxidation state. Molybdenum oxide is a good example, as the Mo cation can be in the 6+, 5+, and 4+ oxidation state. Oxide surfaces with the cation in the lower oxidation state are usually more reactive than those in the highest oxidation state. Such ions can engage in reactions that involve changes in valence state.

Cations at the surface possess Lewis acidity, that is, they behave as electron acceptors. The oxygen ions behave as proton acceptors and are, thus, Brønsted bases. According to Brønsted's concept of basicity, species capable of accepting a proton are called a base, while a Brønsted acid is a proton donor. In Lewis' concept, every species that can accept an electron is an acid, while electron donors, such as molecules possessing electron lone pairs are bases. Hence, a Lewis base

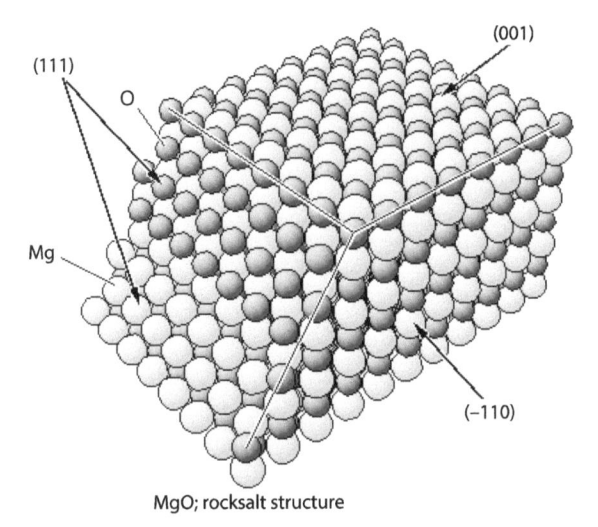

Figure 5.10 Hypothetical particle of cubic MgO, exhibiting three crystallographic surfaces. The (100) and (110) surfaces are nonpolar, implying that Mg and O ions are present in equal numbers, but the polar (111) surface can be terminated either by Mg cations or O anions. In practice, MgO crystals predominantly show (100) terminated surfaces.

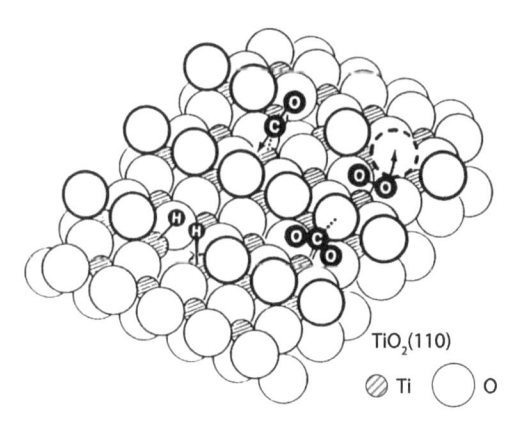

Figure 5.11 Defects consisting of oxygen vacancies constitute adsorption sites on a TiO$_2$(110) surface. Note how CO binds with its lone-pair electrons on a Ti ion (a Lewis acid site). O$_2$ dissociating on a defect furnishes an O atom that locally repairs the defect. CO$_2$ may adsorb by coordinating to an O atom, thus, forming a carbonate group (reproduced with permission from Göpel et al. [14]; copyright 1983 by the American Physical Society).

is in practice equivalent to a Brønsted base. However, the concepts of acidity are markedly different.

Densely packed oxide surfaces, such as MgO(100), are largely inactive, but defects, particularly those associated with oxygen vacancies, provide sites where adsorbates may bind strongly. Figure 5.11 shows the adsorption of different molecules on defects in a TiO$_2$ surface.

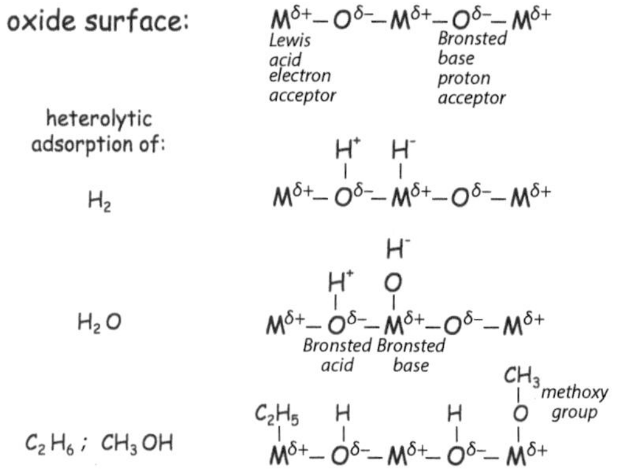

oxide surface:

$M^{\delta+}-O^{\delta-}-M^{\delta+}-O^{\delta-}-M^{\delta+}$
Lewis acid — electron acceptor
Bronsted base — proton acceptor

heterolytic adsorption of:

H_2

$\quad\quad H^+ \quad H^-$
$\quad\quad\ |\quad\ \ |$
$M^{\delta+}-O^{\delta-}-M^{\delta+}-O^{\delta-}-M^{\delta+}$

H_2O

$\quad\quad\quad\quad H^-$
$\quad H^+ \quad O$
$\quad\ |\quad\quad |$
$M^{\delta+}-O^{\delta-}-M^{\delta+}-O^{\delta-}-M^{\delta+}$
Bronsted acid — Bronsted base

C_2H_6; CH_3OH

$\quad\quad\quad\quad\quad\quad\quad\quad CH_3$ methoxy group
$\quad C_2H_5 \quad H \quad\quad H \quad O$
$\quad\ |\quad\quad |\quad\quad\quad |\quad\ |$
$M^{\delta+}-O^{\delta-}-M^{\delta+}-O^{\delta-}-M^{\delta+}$

Figure 5.12 Many gases dissociate heterolytically on oxide surfaces.

Many gases adsorb only after heterolytic dissociation; Figure 5.12 illustrates a few cases. Hydrogen interacts only weakly with most oxide surfaces. If the surface is sufficiently reactive, heterolytic splitting occurs. Water dissociates readily into H^+ and OH^- on almost all oxide surfaces. This leads to two different hydroxyl groups, a terminal one with basic character on the cation and a bridging M–(OH)–M group with acidic character. In hydrocarbons or alcohols, the C–H and O–H bonds usually break readily, while C–C bonds are more difficult to activate. For more information we refer the reader to an excellent book by Henrich and Cox [13].

Sulfides play an important role in hydrotreating catalysis. Whereas oxides are ionic structures, in which cations and anions preferably surround each other to minimize the repulsion between ions of the same charge, sulfides have largely covalent bonds; as a consequence, there is no repulsion which prevents sulfur atoms forming mutual bonds and, hence, the crystal structures of sulfides differ, in general, greatly from those of oxides.

The most important sulfides are MoS_2 and WS_2, which possess a layered structure. The metal atoms occupy trigonal-prismatic holes in the plane between the hexagonally arranged sulfur layers, as shown in Figure 5.13. The layers are shifted with respect to each other, making it necessary to include two layers of MoS_2 in the unit cell. This structure is denoted by $2H-MoS_2$. In catalysts, one often finds particles consisting of single layers of MoS_2 and WS_2, as well as stacks of such particles. The reactivity of the sulfides is associated with the edges, the basal plane consisting of sulfur is largely inactive. Edges can be sulfur terminated, or metal terminated, and in addition there may be sulfur vacancies, of which Figure 5.13 shows two types.

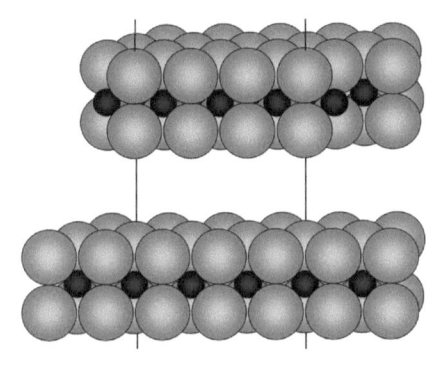

Figure 5.13 MoS$_2$ seen along the [010] direction, showing the arrangement of two slabs in the so-called 2H–MoS$_2$ (2H because there are two slabs in the unit cell). The upper slab shows two types of defects caused by sulfur vacancies.

5.2.4
Surface Free Energy

It costs energy to create a surface, because bonds have to be broken. Thus, in going from a piece of matter to two smaller pieces, the total energy of the system increases: the surface free energy is always positive.

The surface free energy γ is related to the cohesive energy of the solid, ΔH_{coh}, and to the number of bonds between an atom and its nearest neighbors that had to be broken to create the surface:

$$\gamma = \Delta H_{coh} \frac{Z_s}{Z} N_s \tag{5.3}$$

in which Z_s is the number of missing nearest neighbors of a surface atom, Z is the coordination number of a bulk atom, and N_s is the density of atoms in the surface.

Metals possess the highest surface free energies, of the order of 1.5–3 J m^{-2}, while values for ionic solids and oxides are much lower, roughly between 0.2 and 0.5 J m^{-2}. Hydrocarbons have among the lowest surface free energies, between 0.01 and 0.03 J m^{-2}.

Expressions such as Eq. (5.3) form the basis of so-called 'broken bond' calculations. Using typical numbers for an fcc metal (ΔH_{coh} = 500 kJ mol^{-1}, Z_s/Z = 0.25, N_s = 10^{15} atoms cm^{-2}), we obtain a surface free energy of approximately 2 J m^{-2}. Thus, the surface free energy of metals goes up and down with the cohesive energy, implying that transition metals with half-filled d-bands have the largest surface free energy, as we will see later.

Minimization of the surface free energy is the driving force behind a number of surface processes and phenomena. We mention a few:

- Surfaces are always covered by a substance that lowers the surface free energy of the system. Metals are usually covered by a monolayer of hydrocarbons, oxides, and often also by water (OH groups).

Minority element	Bulk	Minority element	Bulk
Ru, Ir	Co	Fe, Co	Cu
Fe, Ru, Rh, Ir, Pt	Ni	Ru, Rh, Ir	Pd
Ni, Pt	Cu	Fe, Ru, Ir	Ag
Fe, Co	Rh	Ru, Ir	Au
Fe, Co, Ni, Cu	Pd		
Co, Ni, Cu, Rh, Pd, Pt	Ag		
Fe, Co, Ru	Ir		
Fe, Co, Ni, Cu, Ru, Rh	Pt		
Fe, Co, Ni, Cu, Rh, Pd, Pt	Au		

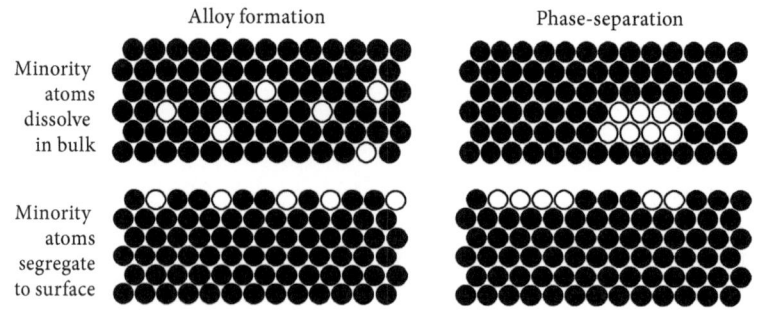

Alloy formation   Phase-separation

Minority atoms dissolve in bulk

Minority atoms segregate to surface

Minority element	Bulk	Minority element	Bulk
Co, Rh, Pd	Fe	Ni, Cu	Fe
Ag, Ir, Pt, Au	Fe	C	Co
Rh, Pd, Ag, Pt, Au	Co	Cu, Pd, Ag	Ru
Cu, Pd, Ag, Au	Ni	Cu, Pd, Ag	Rh
Ag, Au	Cu	Ag	Pd
Fe, Co, Ni, Pt, Au	Ru	Cu, Pd, Ag, Pt, Au	Ir
Ni, Pt	Rh	Ag, Au	Pt
Au	Pd		

Figure 5.14 Alloy formation and segregation in bimetallic systems with one of the metals present as a minority. The scheme qualitatively predicts whether two elements form a surface alloy or a solid solution. The results are valid in vacuum. As soon as an adsorbing gas is present, the component forming the strongest bond with the adsorbate will be enriched at the surface (for further details see Christensen et al. [15]).

- Clean, polycrystalline metals expose mostly their most densely packed surface because, to create this surface, the minimum number of bonds have to be broken (see Figure 5.4).
- Open surfaces such as fcc (110) often reconstruct to a geometry in which the number of neighbors of a surface atom is maximized.
- In alloys, the component with the lower surface free energy segregates to the surface, making the surface composition different from that of the bulk (Figure 5.14).

- Impurities in metals, such as C, O, or S, segregate to the surface because there they lower the total energy due to their lower surface free energy.
- Small metal particles on an oxidic support sinter at elevated temperatures because loss of surface area means a lower total energy. At the same time, a higher fraction of the support oxide with its lower surface free energy is uncovered. In oxidic systems, however, the surface free energy provides a driving force for spreading over the surface if the active phase has a lower surface free energy than the support.

5.3
Characteristics of Small Particles and Porous Material

Catalysis is a surface phenomenon; hence, efficient catalysts have a large surface area, implying that the active particles must be small. The properties of interest are the dispersion, that is, the fraction of atoms located at the surface of a particle, and the specific area, that is, the surface area per unit of weight. Small metal particles are often unstable and prone to sintering, particularly at temperatures where typical catalytic reactions are carried out. Therefore, most heterogeneous catalysts that are used in industry consist of particles inside the pores of an inert support. Silica, alumina, titania, magnesia, zinc oxide, zirconia, as well as carbon can all be used as support materials. The morphology of the particles is determined again by the surface energy, this time of the particles themselves and the substrate.

A catalyst nanoparticle will expose a number of facets, along with steps, kinks, adatoms, and defects. Moreover, the particle under reaction conditions is by no means static, but changes morphology all the time. Figure 5.15 shows a simulation of an fcc cobalt metal particle of approximately 4 nm in diameter, consisting of about 3000 atoms [16].

5.3.1
The Wulff Construction

How does a support affect the morphology of a particle on top of it? Which surface planes does the metal single crystal expose? The thermodynamically most stable configuration of such a small crystallites is determined by the free energy of the surface facets and the interface with the support, and can be derived by the so-called Wulff construction, which we demonstrate for a cross section through a particle–support assembly in two dimensions (Figure 5.16).

- The surface energy $\gamma(ijk)$ for each surface (ijk) is plotted in a polar plot such that the length of the vector is proportional to the surface energy.
- At the end of the vector, a surface plane is defined orthogonal to the vector.
- The inner envelope of these surfaces defines the equilibrium shape.

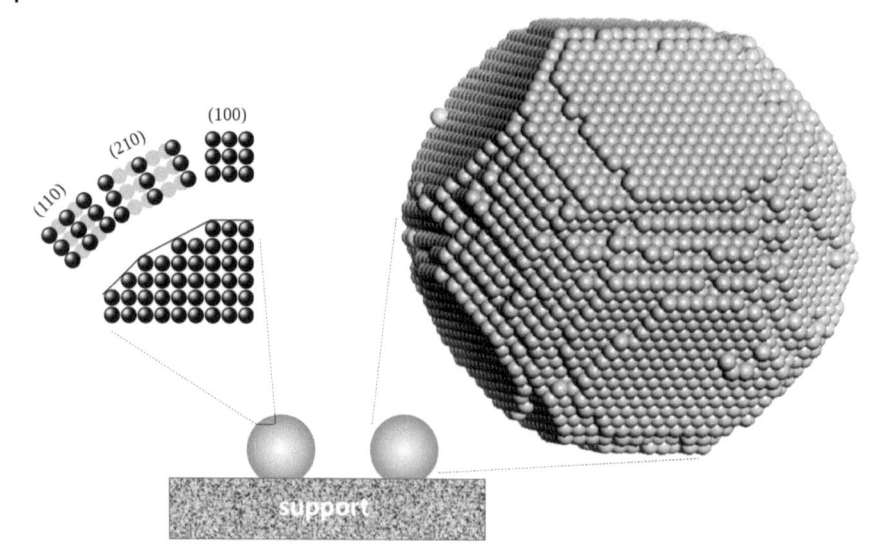

Figure 5.15 Catalyst particles show a variety of surface structures along with steps, kinks, adatoms and defects. The cobalt particle on the right has an fcc structure, contains 3027 atoms and has a diameter of about 4 nm (image of the particle reproduced from Van Helden et al., [16] with permission of Elsevier).

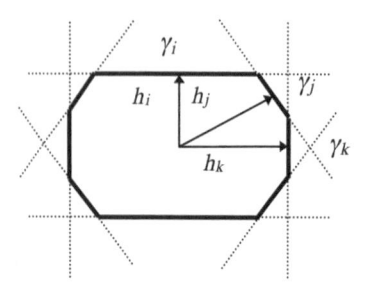

Figure 5.16 Wulff construction for a two-dimensional crystal where the surface energy has the following order: $\gamma_i < \gamma_j < \gamma_k$.

The construction relies on Wulff's assumption that the distance from the surface of a specific plane to the center of the crystallite is proportional to the surface energy, that is, $h_i \propto \gamma_i$. Thus, if we have a surface plane of small surface energy, its distance from the center of the crystallite will be small and this plane will then cut off all others and dominate the polyhedron.

For metals, the close-packed surfaces have, in general, the smallest surface free energy and, therefore, these surfaces dominate on small particles, for example, the (111) surfaces for the fcc and hcp metals, and the (110) surface for the bcc metals, although on iron particles the (100) surface is abundantly present. Surface free energies have been tabulated. To give an idea of how the values depend on crystal

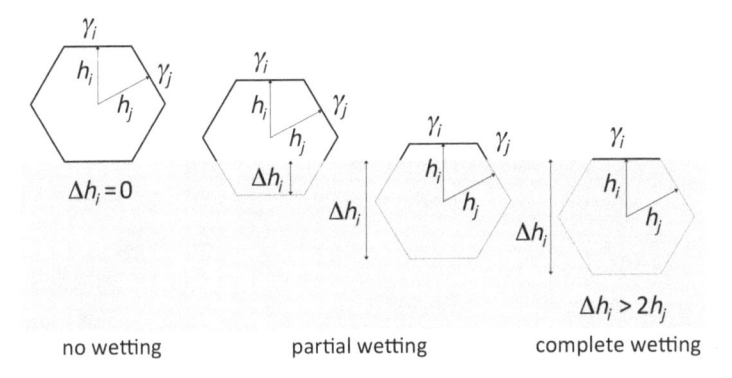

Figure 5.17 Two-dimensional Wulff construction that determines the equilibrium shape of the particle shown in Figure 5.16 when it is placed on an oxide surface. Note how the metal particle wets the surface when the interface energy β decreases (i.e., when a stable bond arises at the interface).

face, we list some values for palladium:

$$\gamma_{111} = 1.920\,\text{J}\,\text{m}^{-2}, \quad \gamma_{100} = 2.326\,\text{J}\,\text{m}^{-2}, \quad \gamma_{110} = 2.225\,\text{J}\,\text{m}^{-2}$$

As expected, the close-packed (111) surface of this fcc metal has the lowest surface free energy. On the basis of the simple broken bond model as expressed by Eq. (5.3), we would predict a value of $4/3 \times \gamma_{111} \approx 2560\,\text{J}\,\text{m}^{-2}$; the actual value is lower, which is in part explained by relaxation effects of the surface atoms that are not accounted for in (5.3).

In supported catalysts, particles are not free, as they would immediately minimize their free energy by merging to form large agglomerates. The oxide support hinders this process. The shape of the particle is to some extent also determined by the energy β associated with the interface between the metal particle and the oxide surface. The same scheme as above in Wulff's construction can be used by introducing a truncation of the vector pointing towards the surface $\Delta h_i = \frac{\beta h_i}{\gamma_i}$, where β is the adhesion energy at the surface (Figure 5.16). Figure 5.17 presents some examples of equilibrium shapes in three dimensions.

The interface energy β is difficult to measure. Only recently have sufficiently accurate determinations of equilibrium shape made it possible to extract reliable values. For example, the interface energy for small Pd particles on an Al_2O_3 surface was found to be $2.9\,\text{J}\,\text{m}^{-2}$. This value is significantly higher than the surface free energy of the Pd(111) surface, implying that this metal forms particles exhibiting (111) facets rather than spreading over the support. Indeed, STM measurements confirm that (111) facets dominate the surface [17].

One should be aware that the surface energy depends on the presence of adsorbates and hence the morphology of a particle may depend on the composition and pressure of the gas surrounding it. Particles in ultrahigh vacuum may therefore differ from those inside a catalytic reactor. Of course, the chemical state of a particle may also depend on the environment, for example, if an adsorbate oxidizes or reduces the particle. Such reactions can change the surface energies dras-

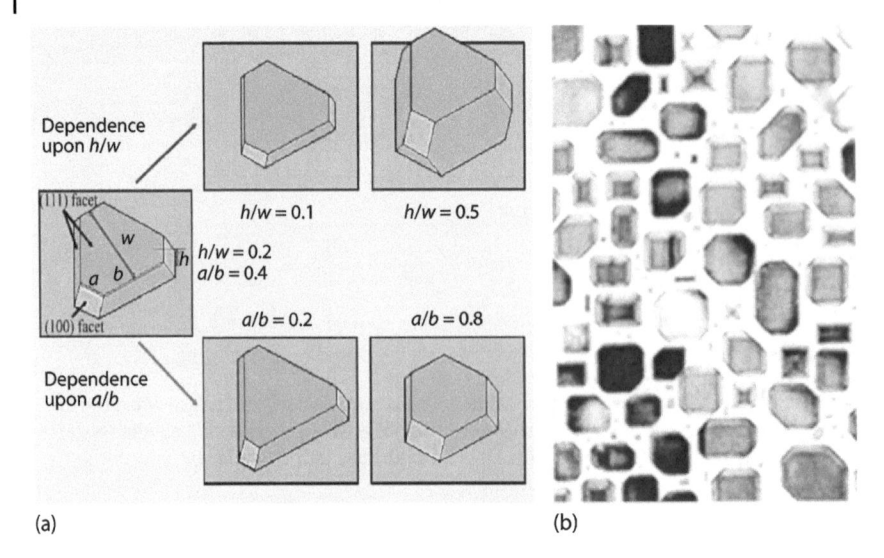

Figure 5.18 (a) Equilibrium polyhedrons of metal particles on a support for different combinations of free energies of surfaces and interfaces. (b) Transmission electron microscopy of structures obtained by depositing 5–10 ML of Pd on an MgO surface (TEM image reproduced from Henry [18] with permission of Elsevier).

tically and induce wetting and unwetting phenomena of the catalyst particles on the support. These phenomena have, for example, been observed in the Cu/ZnO methanol synthesis catalyst, see Figure 4.17.

5.3.2
The Pore System

Supports such as silica, alumina and carbon usually contain pores offering a high internal surface area. The pore system of a support is usually rather irregular in shape and contains macropores, due to the spaces between individual crystallites, with diameters on the order of 100 nm, and micropores with characteristic dimensions of 5–10 nm. A good support offers

- Controlled surface area and porosity
- Thermal stability
- High mechanical strength against crushing and attrition.

As surface area and pore structure are properties of key importance for any catalyst or support material, we will first describe how these properties can be measured. First, it is good to draw a clear borderline between roughness and porosity. If most features on a surface are deeper than they are wide, we call the surface porous (Figure 5.19). Although it is convenient to think about pores in terms of hollow cylinders, one should realize that pores may have all kinds of shapes. The pore system of zeolites consists of microporous channels and cages, whereas the pores of a silica gel support are formed by the interstices between spheres.

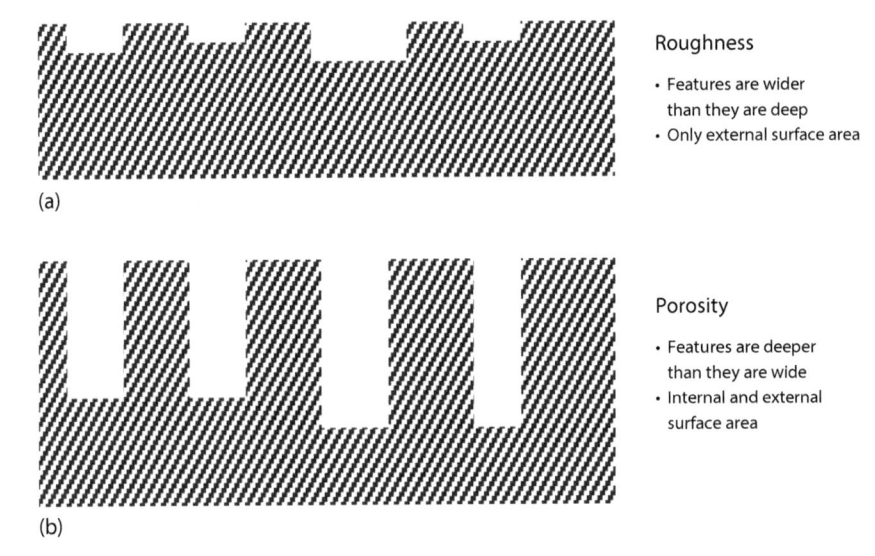

Figure 5.19 Schematic illustration of the difference between roughness (a) and porosity (b).

Alumina and carbon black, on the other hand, have platelet structures, resulting in slit-shaped pores. All support materials may contain micro-, meso-, and macropores (see the text box for definitions).

Adsorptive: gas to be adsorbed
Adsorbent: material on which gas adsorbs
Adsorbate: adsorbed gas

Micropore: pore width ≤ 2 nm
Mesopore: pore width 2–50 nm
Macropore: pore width ≥ 50 nm

5.3.3
The Surface Area

The principle underlying surface area measurements is simple: just physisorb an inert gas such as argon or nitrogen and determine how many molecules are needed to form a complete monolayer [2–4, 19]. As, for example, the N_2 molecule occupies $0.162\,nm^2$ at 77 K, the total surface area follows directly. Although this sounds straightforward, in practice molecules may adsorb beyond the monolayer to form multilayers. In addition, the molecules may condense in small pores. In fact, the narrower the pores are, the easier N_2 will condense in them. This phenomenon of capillary pore condensation, as described by the Kelvin equation, can be used to determine the types of pores and their size distribution inside a system. But first we need to know more about adsorption isotherms of physisorbed

5 Solid Catalysts

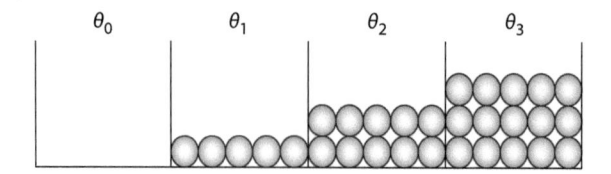

Figure 5.20 To derive the BET isotherm, the surface is divided into regions covered by *i* adsorbate layers; each region characterized a fractional coverage θ_i.

species. Thus, we will derive the isotherm of Brunauer, Emmett and Teller, usually called Brunauer–Emmet–Teller (BET) isotherm.

As illustrated in Figure 5.20, we divide the surface in areas which are uncovered, (fraction θ_0), are covered by a single monolayer (θ_1), two monolayers (θ_2), or by i layers (θ_i).

Suppose there are N_0 sites on the surface, then the number of atoms adsorbed is N_a

$$N_a = N_0 \sum_{i=0}^{\infty} i\theta_i \tag{5.4}$$

where we have the usual sum rule $\sum_{i=0}^{\infty} \theta_i = 1$.

If we now assume that this surface at temperature T is in equilibrium with a gas, then the adsorption rate equals the desorption rate. Since the atoms/molecules are physisorbed in a weak adsorption potential, there are no barriers and the sticking coefficient (the probability that a molecule adsorbs) is unity. This is not entirely consistent since there is an entropic barrier to direct adsorption on a specific site from the gas phase. Nevertheless, a lower sticking probability does not change the overall characteristics of the model. Hence, at equilibrium we have

$$\frac{d\theta_i}{dt} = F\theta_{i-1} - \nu e^{\frac{-E_d^i}{k_B T}} \theta_i - F\theta_i + \nu e^{\frac{-E_d^{i+1}}{k_B T}} \theta_{i+1}$$
$$\equiv F\theta_{i-1} - k_i \theta_i - F\theta_i + k_{i+1} \theta_{i+1} = 0 \tag{5.5}$$

where F is the incoming flux per site $F = \dfrac{P}{N_0 \sqrt{2\pi m k_B T}}$, and E_d^i is the desorption energy from layer i. Note that, due to the change in substrate from the first to the second layer, there may be a difference between E_d^1 and E_d^2. However, for $i = 2$ and higher, we consider desorption essentially as sublimation from a multilayer of gas, and hence $E_d^2 = E_d^{i>2}$ and $k_1 \neq k_2 = k_3 = \cdots = k_\infty$.

Writing the rate equations for the adsorption desorption equilibrium for each layer, we obtain

$$\frac{d\theta_0}{dt} = 0 = -F\theta_0 + k_1\theta_1 \Rightarrow \theta_1 = \frac{F}{k_1}\theta_0$$

$$\frac{d\theta_1}{dt} = 0 = F\theta_0 - k_1\theta_1 + k_2\theta_2 - F\theta_1 \Rightarrow \theta_2 = \frac{F}{k_1}\theta_1 = F^2\frac{\theta_0}{k_1 k_2}$$

$$\frac{d\theta_i}{dt} = 0 = F\theta_{i-1} - k_i\theta_i + k_{i+1}\theta_{i+1} - F\theta_i$$

$$\Rightarrow \theta_{i+1} = \frac{F}{k_i}\theta_i = F^{i+1}\frac{\theta_0}{k_1(k_2)^i} \quad (5.6)$$

The total coverage, θ_t, is

$$\theta_t = \frac{N_a}{N_0} = \sum_{i=1}^{\infty} i\theta_i = \sum_{i=1}^{\infty} i\left(\frac{F}{k_2}\right)^i \frac{k_2}{k_1}\theta_0$$

$$= \frac{k_2}{k_1}\theta_0 \sum_{i=1}^{\infty} i\left(\frac{F}{k_2}\right)^i = \frac{k_2}{k_1}\theta_0 \frac{\frac{F}{k_2}}{\left(1-\frac{F}{k_2}\right)^2} \quad (5.7)$$

where we have used $\sum_{i=1}^{\infty} ix^i = \frac{x}{(1-x)^2}$.

We have now expressed the total coverage as a function of rate constants and the incoming flux, but we still need to eliminate θ_0:

$$\sum_{i=0}^{\infty} \theta_i = 1 = \theta_0 + \sum_{i=1}^{\infty} \theta_i = \theta_0 + \sum_{i=1}^{\infty} \frac{k_2}{k_1}\theta_0\left(\frac{F}{k_2}\right)^i$$

$$= \theta_0\left(1 + \frac{k_2}{k_1}\sum_{i=1}^{\infty}\left(\frac{F}{k_2}\right)^i\right) = \theta_0\left[1 + \frac{k_2}{k_1}\frac{\left(\frac{F}{k_2}\right)}{1-\left(\frac{F}{k_2}\right)}\right] \quad (5.8)$$

Using

$$\sum_{i=1}^{\infty} x^i = \sum_{i=0}^{\infty} x^i - 1 = \frac{1}{1-x} - 1 = \frac{x}{1-x} \quad (5.9)$$

θ_0 becomes

$$\theta_0 = \left(1 + \frac{k_2}{k_1}\frac{\left(\frac{F}{k_2}\right)}{1-\left(\frac{F}{k_2}\right)}\right)^{-1} \quad (5.10)$$

and substitution in the Eq. (5.7) for θ_t leads to

$$\theta_t = \frac{\frac{k_2}{k_1}\frac{F}{k_2}}{\left(1-\frac{F}{k_2}+\frac{k_2}{k_1}\frac{F}{k_2}\right)\left(1-\frac{F}{k_2}\right)} = \frac{X\frac{P}{P_0}}{\left(1-\frac{P}{P_0}+X\frac{P}{P_0}\right)\left(1-\frac{P}{P_0}\right)} \quad (5.11)$$

where $\chi = \frac{k_2}{k_1}$ is the ratio of desorption rate constants for the second and first layer and $\frac{P}{P_0} = \frac{F}{k_2} = \frac{P}{k_2 N_0 \sqrt{2\pi m k_B T}}$ is the ratio of adsorption and desorption rate from the second layer and higher. For $\frac{P}{P_0} \to 1$, the total coverage $\theta_t \to \infty$, meaning that the adsorbate condenses into multilayers on the surface. This is the situation when the incoming flux equals the desorption rate from the multilayers, implying that P_0 is simply the equilibrium pressure of the condensed gas, a value that is tabulated in handbooks.

To proceed, our sample of unknown area is mounted in a small volume and cooled to low temperature (75 K if we use N_2). The equilibrium pressure (P_0) for N_2 at 75 K is 750 mbar. The amount of gas adsorbed is then measured as a function of the pressure and can conveniently be expressed in terms of the amount of gas adsorbed in one monolayer, that is,

$$\theta_t = \frac{N_a}{N_0} = \frac{\left(\frac{PV_a}{RT}\right)}{\left(\frac{PV_0}{RT}\right)} = \frac{V_a}{V_0} \tag{5.12}$$

By substituting this in the above expression for θ_t and by rearranging the equation we find

$$\frac{P}{V_a(P_0 - P)} = \frac{1}{\chi V_0} + \frac{(\chi - 1)}{\chi V_0}\frac{P}{P_0} \equiv \eta + \alpha \frac{P}{P_0} \tag{5.13}$$

Plotting $\frac{P}{V_a(P_0-P)}$ versus $\frac{P}{P_0}$ yields a straight line with slope $\alpha = \frac{\chi-1}{\chi V_0}$ crossing the y-axis at $\eta = \frac{1}{\chi V_0}$. The volume adsorbed in the first monolayer is found as $V_0 = \frac{1}{\alpha+\eta}$. The volume V_0 can be converted into the number of molecules absorbed by $N_0 = PV_0/k_B T$ and if we know how large an area each molecule occupies (A_0) then the total area, $A = N_0 A_0$, can be found.

Example Application of the BET method

The following volumes of adsorbed gas are found in a BET experiment of 0.2 g sample of catalyst:

Table 5.1 Pressure and corresponding volume of adsorbed N_2 on 0.2 g sample of catalyst.

P (mbar)	1.6	18	61	168	270
V (µL; standard conditions)	1200	1440	1640	2100	2500

Figure 5.21 shows a plot of the data in Table 5.1 according to Eq. (5.13). The slope and intercept result in a monolayer volume of $V_0 = 1616$ µL. Because each N_2 molecule occupies 0.16 nm^2 we can convert the monolayer volume into the total area of the catalyst: 6.4 m^2. Since this was for a sample of 0.2 g, the area per gram of catalyst corresponds to 32 m^2 g^{-1}.

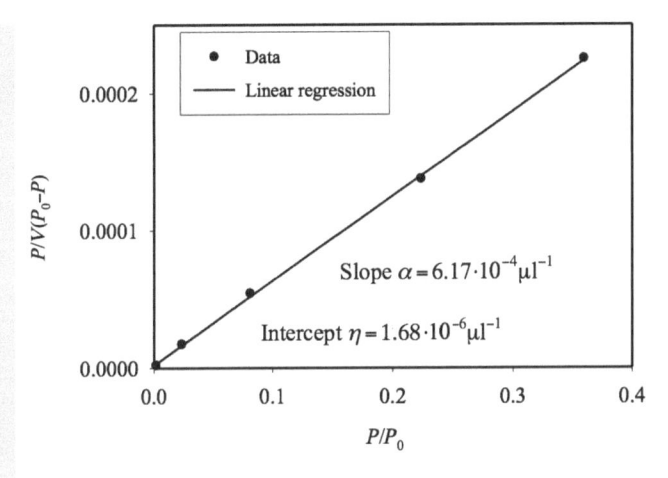

5.3 Characteristics of Small Particles and Porous Material

Figure 5.21 BET plot of the data in Table 5.1 used to determine α and η.

Slope $\alpha = 6.17 \cdot 10^{-4}\,\mu l^{-1}$

Intercept $\eta = 1.68 \cdot 10^{-6}\,\mu l^{-1}$

In this example, adsorption in the first layer is dominant, since the desorption rate from the second layer and higher is much larger than that from the first: $k_2/k_1 \cong 370$. Table 5.2 shows the coverages at a pressure of 61 mbar, when almost 90% of the total coverage is accounted for by the monolayer and only 8% of the coverage is due to multiple layers.

Table 5.2 Fraction of empty surface and N_2 coverages in 1, 2, 3, and 4 ML for the data in Table 5.1 and Figure 5.21 at a pressure of 61 mbar.

θ_0	θ_1	θ_2	θ_3	θ_4
0.030	0.892	0.072	0.0059	0.0005

The BET method has its limitations and several improvements exist, but these are beyond the scope of our treatment. We note that the BET isotherm is valid under the following assumptions:

- Dynamic equilibrium between adsorbate and adsorptive: the rate of adsorption and desorption in any layer are equal.
- In the first layer, molecules adsorb on equivalent adsorption sites.
- Molecules in the first layer constitute the adsorption sites for molecules in the second layer, and so on for higher layers.
- Adsorbate–adsorbate interactions are ignored.
- The adsorption–desorption conditions are the same for all layers but the first.
- The adsorption energy for molecules in the 2nd layer and higher equals the condensation energy.
- The multilayer grows to infinite thickness at saturation pressure ($P = P_0$).

For a supported metal catalyst, the BET method yields the total surface area of support and metal. If we do our measurements in the chemisorption domain, for

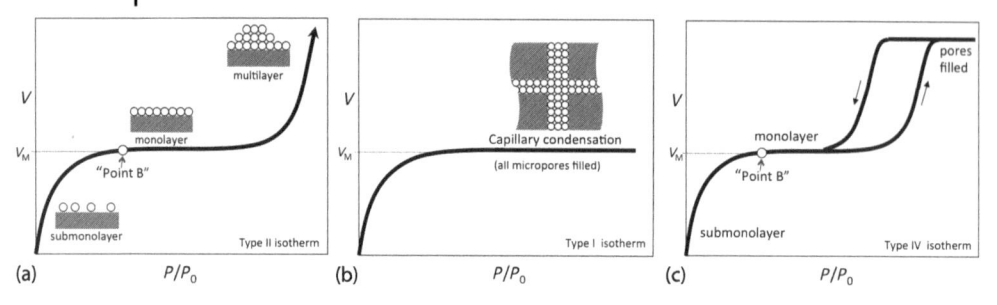

Figure 5.22 Schematic representation of several BET isotherms; (a) Type II is commonly observed for nonporous powders, (b) Type I for zeolites, and (c) Type IV for porous catalyst supports such as silica and alumina.

example with H_2 or CO at room temperature, adsorption is limited to the metallic phase, providing a way to determine the dispersion of the supported phase.

The BET area of a catalyst or of a catalyst support is one of the first properties one wants to know in catalyst development. All industrial laboratories and many academic laboratories possess equipment for measuring this property.

Figure 5.22 shows an idealized form of the adsorption isotherm for physisorption on a nonporous or macroporous solid. At low pressures, the surface is only partially occupied by the gas, until at higher pressures (point B on the curve), the monolayer is filled and the isotherm reaches a plateau. This part of the isotherm, from zero pressures to the point B, is equivalent with the Langmuir isotherm. At higher pressures a second layer starts to form, followed by unrestricted multilayer formation, which is in fact equivalent to condensation, that is, formation of a liquid layer. In the jargon of physisorption (approved by IUPAC), this is a Type II adsorption isotherm. If a system contains predominantly micropores, that is, a zeolite or an ultrahigh surface area carbon ($> 1000 \text{ m}^2 \text{ g}^{-1}$), multilayer formation is limited by the size of the pores.

Here, the phenomenon of capillary pore condensation comes into play. The adsorption on an infinitely extended, microporous material is described by the Type I isotherm of Figure 5.22. The plateau measures the internal volume of the micropores. For mesoporous materials, one will first observe the filling of a monolayer at relatively low pressures, as in a Type II isotherm, followed by build up of multilayers until capillary condensation sets in and puts a limit to the amount of gas that can be accommodated in the material. Removal of the gas from the pores will show a hysteresis effect: the gas leaves the pores at lower equilibrium pressures than at which it entered because capillary forces have to be overcome. This Type IV isotherm, shown in Figure 5.22, is characteristic for the common silica and alumina catalyst supports with specific surface areas of a few hundred $\text{m}^2 \text{ g}^{-1}$.

5.4
Catalyst Supports

Small metal particles are often unstable and prone to sintering, particularly at temperatures typical of catalytic reactions. Therefore, heterogeneous catalysts used in industry consist of relatively small particles stabilized in some way against sintering. This can be achieved by adding so-called structural promoters, or by applying particles inside the pores of an inert support. All kinds of materials that are thermally stable and chemically relatively inert can be used as supports, but alumina, silica, and carbon are the most common ones, with magnesia, titania, zirconia, zinc oxide, silicon carbide, and zeolites for particular applications. In special cases where an unrestricted flow of reacting gases is important, monoliths are applied. Table 5.3 gives an overview of supports used in catalysts for various reactions.

5.4.1
Silica

Silica is the support of choice for catalysts used in processes operated at relatively low temperatures (below about 300 °C), such as hydrogenations, polymerizations, or some oxidations. Its properties, such as pore size, particle size, and surface area, are easy to adjust to meet the specific requirements of applications. Compared to

Table 5.3 Example uses of supports in selected catalytic reactions.

Support	Catalytically active phase	Application
γ-Alumina, Al_2O_3	CoMoS, NiMoS, NiWS	Hydrotreating
	Pt, Pt-Re	Reforming
	Pt, Rh, Pd	Automotive exhaust cleaning
	Cu-ZnO	Methanol synthesis
	Cu-ZnO	Water–gas shift reaction
	Ni	Steam reforming
	TiO_2	Dehydration
	Pd, Pt, Ru, Rh	Hydrogenation
	Cr_2O_3, Pt	Dehydrogenation
	Pd	Dehydrochlorination
	$CuCl_2$	Oxychlorination
η-Alumina	Pt	Reforming, isomerization
α-Alumina	Ni	Steam reforming
	Ag	Epoxidation
Silica, SiO_2	CrO_x	Polymerization
	H_3PO_4	Hydration
	V_2O_5	Oxidation
Titania, TiO_2	V_2O_5	$DeNO_x$
Carbon	Pd, Pt	Hydrogenation

Table 5.4 Properties of shaped fumed silica supports (Jacobsen and Kleinschmit [20]).

Surface area of starting oxide	$m^2 g^{-1}$	50	130	200	300
Surface area of pelleted material	$m^2 g^{-1}$	45	110	160	230
Pore volume	$cm^3 g^{-1}$	0.6	0.8	0.75	0.8
Smallest pore diameter	nm	8	7	6	4
90% pores between	nm	10–40	10–40	10–30	7–25
Hardness	N	125	60	60	80

alumina, silica possesses lower thermal stability, and its propensity to form volatile hydroxides in steam at elevated temperatures limits its applicability as a support. Most silica supports are made by one of two different preparation routes: sol–gel precipitation to produce silica xerogels and flame hydrolysis to give so-called fumed silica.

In the sol–gel route, an alkaline (pH 12) sodium silicate solution, called water glass, is mixed with sulfuric acid. At low pH, the silicate ions convert into $Si(OH)_4$ monomer units which polymerize to colloidal silica particles and agglomerate into a hydrogel, a three-dimensional network of spheres. Sodium and sulfate are removed by washing. Drying at moderate temperatures (150–200 °C) produces a so-called silica xerogel, a highly porous network of interconnected silica spheres, which will absorb water from the ambient atmosphere equivalent to 5–10% of its own weight. This xerogel is insufficiently strong and lacks the macropores that are needed to prevent excessive pressure drops in reactor. Therefore, the material is milled to powders with the desired particle size, and mixed with binders and shaped into pellets or extrudates. The pore structure is primarily determined by the size of the colloidal silica spheres and the way in which these agglomerate in the hydrogel, but can still be subsequently affected adding acids or bases in the washing stage.

The second preparation route uses flame hydrolysis, a versatile way to produce all kinds of oxides with high specific surface areas. The advantages of fumed silica over xerogels are the better mechanical properties and higher purity of the former. The starting material to produce fumed silica is $SiCl_4$, which is volatilized in air, mixed with hydrogen, and fed into a flame reactor, where the mixture reacts in a complicated scheme to give an aerosol of nanometer-sized SiO_2 particles and HCl. The latter is removed from the surface of the silica by steam and air in a fluid bed. The thus-formed silica has a low density and needs to be compacted before it is pelleted or extruded into larger shapes. The original size and aggregation of the silica spheres can be controlled by several parameters such as the temperature of the flame, the H_2/O_2 ratio and the $SiCl_4$ content of the feed and the residence time in the flame reactor. Figure 5.23 gives a flow scheme of the process, and Table 5.4 lists some properties of fumed silica supports.

Silicas with surface areas up to $300 \, m^2 g^{-1}$ consist of dense particles with diameters of 7 nm and above, and do not contain micropores. Several companies produce fumed silica, Degussa's Aerosil® and Cabot's Cabosil® being among the best known.

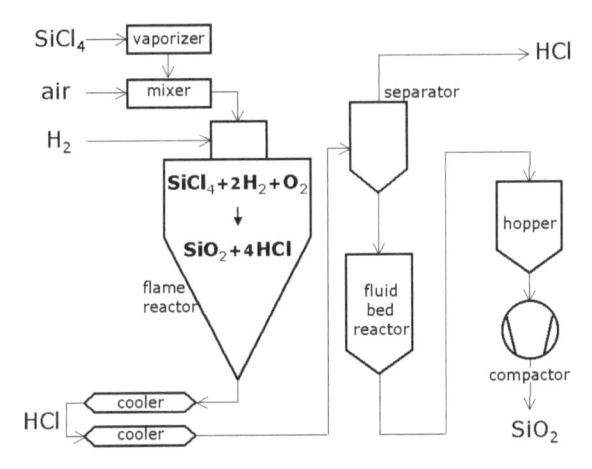

Figure 5.23 Scheme for production of silica (adapted from Jacobsen and Kleinschmit [20]).

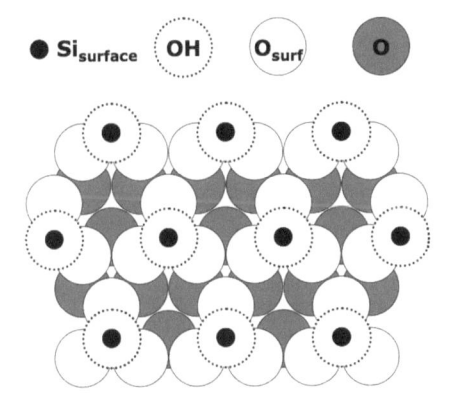

Figure 5.24 The surface of cristoballite is sometimes used as a structure model for silica.

Though silica supports are amorphous, the surface may exhibit some local order, such as that of the mineral β-cristoballite (Figure 5.24). The surfaces of silica supports contain OH groups at densities between 4 and 5.5 OH/nm^2; that of cristoballite is 4.55 OH/nm^2. Silica surfaces contain only terminal OH groups, that is, bound to a single Si atom. Heating leads to dehydroxylation, and at high temperatures only the isolated OH groups remain.

5.4.2
Alumina

Owing to its excellent thermal and mechanical stability and its rich chemistry, alumina is the most widely used support in catalysis. Although aluminum oxide exists in a variety of structures, only three phases are of interest, namely the nonporous, crystallographically ordered α-Al$_2$O$_3$, and the porous amorphous η- and γ-Al$_2$O$_3$.

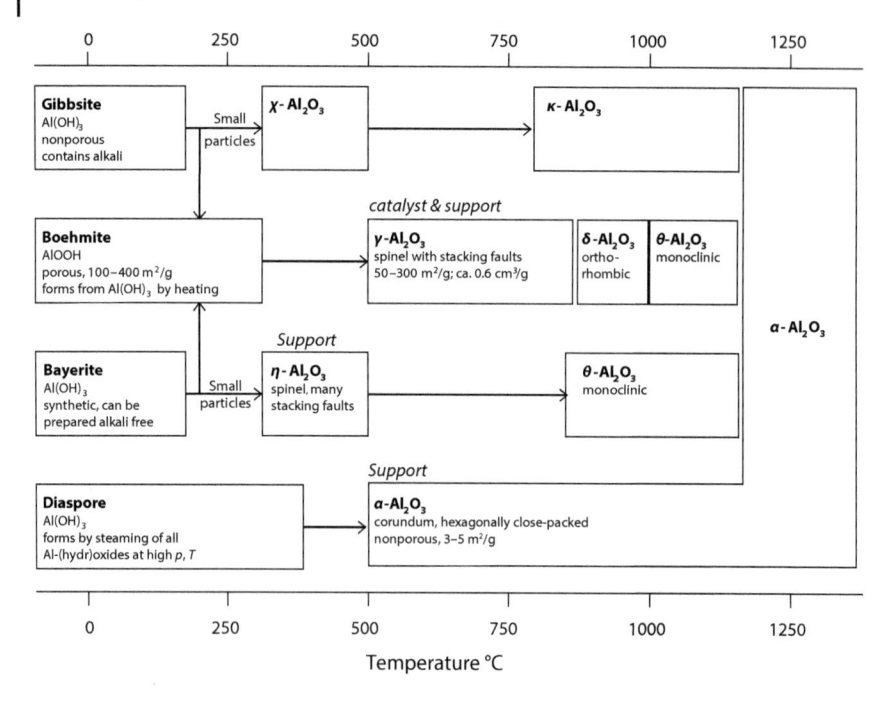

Figure 5.25 Alumina comes in many different forms, γ-Al$_2$O$_3$ being the most often used catalyst support.

The latter is also used as a catalyst by itself, for example, in the production of elemental sulfur from H$_2$S (the Claus process), the alkylation of phenol, or the dehydration of formic acid [21].

Alumina supports are made by thermal dehydration of Al(OH)$_3$ (gibbsite or bayerite) or AlOOH (boehmite) (see the scheme of Figure 5.25). Aluminum is found in nature in the form of bauxite (named after Les Baux, France), an ore consisting of aluminum hydroxides, silica, and other oxides. Depending on origin, the bauxite contains mainly gibbsite (Surinam, Ural), boehmite (France), or diaspore (Balkan). Aluminum is extracted from the ore by treating it with NaOH yielding a Na$_2$Al$_2$O$_4$ solution from which gibbsite is obtained by crystallization. As a consequence, gibbsite always contains alkali, which may be undesirable for some catalytic applications. Alkali-free aluminas are, therefore, prepared by hydrolysis of aluminum alcoholate, for example, Al(OC$_2$H$_5$)$_3$, which yields a gelatinous form of boehmite, also referred to as pseudoboehmite. Yet another preparation route starts from bayerite, formed by precipitation of Al(OH)$_3$ from aluminum salts to which ammonia is added in solution. Bayerite forms at high pH ($\approx$ 12), boehmite at more neutral pH values.

Alumina exists in several forms, among which only the α-form is crystalline. The others, called transitional aluminas, have spinel-like structures with different orders in which layers are stacked. In the spinel structure of, for example, MgAl$_2$O$_4$, oxygen ions are packed in a fcc manner, while Mg occupies tetrahedral and Al octahedral sites. As the stoichiometry of Al_2O_3 does not quite correspond to that of a true spinel, the transitional alumina structures are cation-deficient and contain many defects. Nevertheless, the X-ray diffractograms are characteristic of spinels.

As a support, γ-Al_2O_3 offers high surface areas (50–300 m² g⁻¹), mesopores of between 5 and 15 nm, pore volumes of about 0.6 cm³ g⁻¹, high thermal stability, and the ability to be shaped into mechanically stable extrudates and pellets. The thermal stability of γ-Al_2O_3 can be significantly improved by lanthanum oxide additives, which reduce the rate of sintering and retard the conversion into other phases of alumina. Its surface contains several hydroxyls, between 10 and 15 OH/nm², the linear ones being Brønsted bases (H⁺ acceptors) and the bridged ones Brønsted acids (H⁺ donors). After dehydroxylation, the surface develops Lewis acidity (electron acceptor) on the uncoordinated $Al^{\delta+}$ sites. Whereas cristoballite presented a realistic model for the surface of a silica support, such a convenient structure is not available for the alumina surface. The hydroxyl population is varied, exhibiting linear as well as two- and threefold bridged species, which can be identified by vibrational spectroscopy. In solution, alumina is a polyanion of positive charge at pH values below approximately 7 and negative at higher values, offering many possibilities to bind many ionic catalyst precursors.

The highly stable α-Al_2O_3 is used in high temperature applications, such as in steam reforming, or in cases where low surface areas are desired.

5.4.3
Carbon

Porous carbons are used as a support for noble metal catalysts in hydrogenation reactions of organic compounds, particularly in liquid media. Active carbons, containing functional groups at their surface, are usually prepared by pyrolysis of wood, coal, coconut shells, etc. in inert gas, CO_2, or steam at temperatures between 800 and 1500 °C. Activation of the surface is achieved by subsequent treatment in oxygen. Surface areas may be as high as 1500 m² g⁻¹ with micropores smaller than 1 nm. Treatment at higher temperatures leads to graphitic carbons of low surface area. The advantages of carbon are their relative chemical stability and the easy recoverability of the expensive noble metals from spent catalysts. Carbon is the support of choice for noble metals in low temperature hydrogenation catalysts, and also for liquid phase reactions.

5.4.4
Shaping of Catalyst Supports

The preparation routes for supports discussed above yield powders of small particles in the micro- to millimeter size range. Using this material in a fixed bed reactor of several meters would give a tremendous pressure drop over the bed. In rising bed or fluidized bed reactors, the catalyst would simply be blown out of the reactor. Hence, to apply catalysts in reactors, they need to be shaped into larger

Figure 5.26 Examples of the various forms of shaped catalysts (photo courtesy of Haldor Topsøe A/S).

bodies of sufficient mechanical strength (Figure 5.26). Densification of powder has the additional beneficial effect of enhancing the catalytic activity per volume of catalyst bed. As a rule of thumb, the density of a powder is increased by a factor of three by compression into a tablet. Shaping can be done with the support before it is loaded with active material, or afterwards with the impregnated catalyst.

The shape of the catalyst body influences the mass transport characteristics as well as the pressure drop in a catalyst bed, as the following example shows. If one compares a cylindrical pellet with a ring of similar length and diameter, the ring causes a 50% lower pressure drop and offers a 20% higher external surface in contact with the reacting gas phase than the pellet does.

Pellets or tablets (1.5–10 mm in diameter), rings (6–20 mm), and multichanneled pellets (20–40 mm in diameter and 10–20 mm high) are used when a high mechanical strength is required. They are produced by compressing a mixture of the support powder and several binders (kaolin clay, stearic acid) and lubricants (graphite) in a press.

In many cases, supports are shaped into simple cylinders (1–5 mm in diameter and 10–20 mm in length) in an extrusion process. The support powder is mixed with binders and water to form a paste that is forced through small holes of the desired size and shape. The paste should be sufficiently stiff such that the ribbon of extruded material maintains its shape during drying and shrinking. When dried, the material is cut or broken into pieces of the desired length. Extrusion is also applied to make ceramic monoliths as those used in automotive exhaust catalysts and in $DeNO_x$ reactors.

5.5
Preparation of Supported Catalysts

There are in principle two ways to make supported catalysts:

1. By coprecipitating the catalytically active component and the support to give a mixture that is subsequently dried, calcined (heated in air), and reduced to yield a porous material with a high surface area. This procedure is followed when materials are cheap and obtaining optimum catalytic activity per unit volume of catalyst is the main consideration.
2. By loading pre-existing support materials in the form of shaped bodies with the catalytically active phase by means of impregnation or precipitation from solution. This is the preferred method when catalyst precursors are expensive and the aim is to deposit the catalytically active phase in the form of nanometer-sized particles on the support. All noble metal catalysts are manufactured in this way.

We discuss illustrative examples of both methods. For general introductions into the subject we refer to De Jong [3] and to Regalbuto [22].

5.5.1
Coprecipitation

Solutions of salts of the catalytically active material and of the support are prepared, to which a precipitating agent is added, such as NaOH or $NaHCO_3$. As a result hydroxides or hydroxy salts precipitate and form a homogenous mixture that is filtered off. Removal of CO_2 and water during drying and calcination, and of oxygen during reduction, yields a porous catalyst. The process is difficult to control; it is essential to keep the solution homogeneous to allow the two components to precipitate simultaneously, and variations of the pH throughout the solution should be avoided. Examples of coprecipitated catalysts are Ni/Al_2O_3 used for steam reforming and $Cu\text{-}ZnO/Al_2O_3$ applied in the synthesis of methanol.

5.5.2
Impregnation, Adsorption, and Ion Exchange

Filling the pores of the support with a solution of the catalytically active element, after which the solvent is removed by drying, is a straightforward way to load a support with active material. However, in this process, various interactions are possible between the dissolved catalyst precursor and the surface of the support, which can be used to obtain a good dispersion of the active component over the support. To appreciate the importance of such interactions we need to take a closer look at the surface chemistry of hydroxylated oxides in solution.

Silica and alumina supports contain several types of hydroxyl groups, the main ones are indicated in Figure 5.27. Hydroxyl groups play a very important role in catalyst preparation because they represent anchoring sites where catalyst precur-

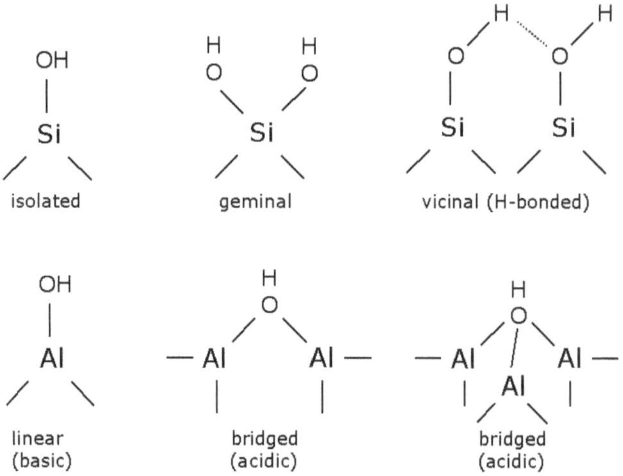

Figure 5.27 Examples of hydroxyl groups on silica and alumina.

sors can be attached to the support. Although those of silica are chemically indistinguishable, the OH groups on alumina differ significantly in chemical character. The linear hydroxyls on alumina have anionic (basic) character, as illustrated by their willingness to exchange with halogens such as Cl⁻ and F⁻ ions.

In water, the hydroxyls react with protons and OH⁻ groups, giving the surface ionic character. The following equilibrium reactions occur:

$$\text{Basic sites:} \quad \text{M–OH} + \text{H}^+ \rightleftharpoons \text{M–OH}_2^+ \tag{5.14}$$

$$\text{Acidic sites:} \quad \text{M–OH} + \text{OH}^- \rightleftharpoons \text{M–O}^- + \text{H}_2\text{O} \tag{5.15}$$

To an extent, the surface charges are determined by the pH of the solution and by the isoelectric point of the oxide, that is, the pH at which the oxide surface is neutral. The surface is negative at pH values below the isoelectric point and positive above it. Obviously, the charged state of the surface enables one to bind catalyst precursors of opposite charge to the ionic sites of the support.

For example, Pt/SiO_2 catalysts are conveniently made by impregnating a silica support with a basic solution (pH 8–9) of platinum tetraamine ions, $Pt(NH_3)_4^{2+}$ (dissolved as chloride). As the silica surface is negatively charged, the Pt-containing ions attach to the SiO⁻ entities and disperse over the surface. The pH should be kept below 9 because otherwise the silica surface starts to dissolve.

Negatively charged ions of platinum are used to prepare Pt/Al_2O_3 catalysts. An aqueous solution of hexachloroplatinic acid dissociates into H⁺ and $PtCl_6^{2-}$ ions, which react rapidly with the basic OH groups. If this preparation route is applied to preshaped alumina bodies, Pt will deposit preferentially on the outside, leading to a so-called egg-shell catalyst. As the basic OH groups also easily exchange with other anions, the principle of competitive adsorption can be utilized to affect the depth at which the platinum is deposited in the support. In the presence of acids that adsorb more strongly than hexachloroplatinic acid, such as oxalic or citric

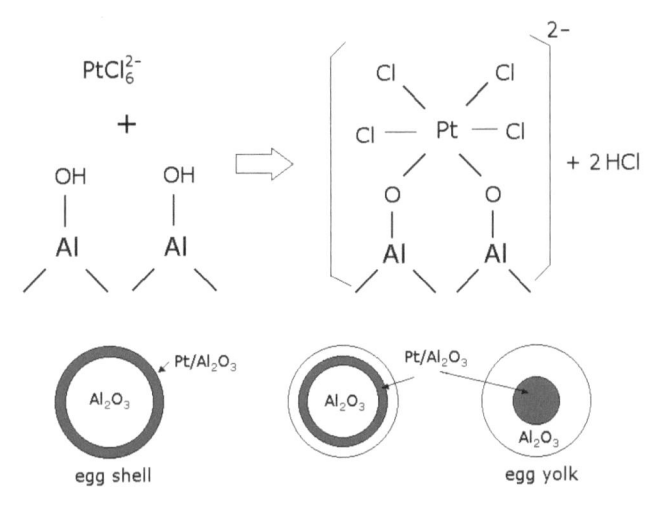

Figure 5.28 Examples of shell impregnations of support material with precious metals.

acid, one can even prevent platinum absorbing in the outer layer of the support, as indicated in Figure 5.28. Catalysts where the active material does not occur until a certain depth offer advantages when feed streams contain strongly adsorbing poisons, for example, lead in automotive exhaust gas.

In nonpolar solvents, for example, alcohols, the hydroxyls of the support can also be used to anchor alkoxy compounds to the surface in a condensation reaction, in which one alkoxy ligand reacts with the proton of the surface OH to give the corresponding alcohol, and the complex binds to the support. An example is the anchoring of zirconium ethoxide, $Zr(OC_2H_5)_4$, to silica by means of the reaction

$$Zr(OC_2H_5)_4 + Si-OH \longrightarrow Si-O-Zr(OC_2H_5)_3 + C_2H_5OH \tag{5.16}$$

which leads to a high dispersion of Zr over the support. Calcination in air is necessary to remove the ethoxy ligands and convert the zirconium to ZrO_2 [23].

5.5.3
Deposition Precipitation

The essence of this method is that small crystallites of metal hydroxide or carbonate precipitate from solution, preferably by heterogeneous nucleation at the interface between liquid and support. To this end, the support powder is suspended in the metal solution, and a base is added to raise the pH. As the pH must be homogeneous throughout the solution, efficient stirring is required, which, however, poses problems in large volumes, as well as in porous systems. A clever alternative is to add urea, $CO(NH_2)_2$, to the solution [3]. When, after thorough mixing, the temperature is raised to 70–90 °C, the urea hydrolyses slowly according to the reaction

$$CO(NH_2)_2 + 3H_2O \longrightarrow 2NH_4^+ + CO_2 + 2OH^- \quad (5.17)$$

and as a result the pH rises homogeneously throughout the solution, and at certain pH value nucleation starts at the support. This method has been used to prepare supported vanadium, molybdenum, manganese, iron, nickel, and copper catalysts with a homogenous distribution throughout the entire support. After precipitation, the solid is filtered off, washed, dried, and shaped. Finally, the catalyst is calcined and further activated by reduction or sulfidation (see below).

5.6
Unsupported Catalysts

Multicomponent catalysts, particularly those based on mixtures of oxides, can be prepared by fusion in an electric arc furnace. This is only possible for oxides that conduct electric current at high temperatures. Various oxides, for example, iron oxide (magnetite, Fe_3O_4), alumina, and potassium oxide for an ammonia synthesis catalyst, mix well in their molten state. After cooling, the catalyst is crushed into fragments, and the particles of the desired size are sieved out. Larger particles are recycled to the crusher, smaller particles go back in the furnace. Fused catalysts leave the furnace as nonporous solids, but develop porosity during reduction. For the ammonia synthesis catalyst, the solid mainly consists of Fe_3O_4 with roughly 1% K_2O and 2.5% Al_2O_3. This is an example of a doubly promoted catalyst since it is electronically promoted by potassium and structurally promoted by alumina. If CaO is also added it is triply promoted. During reduction in the synthesis gas, the oxygen in the Fe_3O_4 will be removed as water, while the other oxides are not so easily reduced. The result is a porous structure of metallic iron, which is the active part of the ammonia catalyst.

5.7
Zeolites

Zeolites form a unique class of oxides consisting of microporous, crystalline aluminosilicates that can be found in nature or can be synthesized [5, 19]. The zeolite framework is very open and contains channels and cages where cations, water, and adsorbed molecules may reside and react. The specific absorption properties of zeolites are used in detergents, toothpaste, and desiccants, whereas their acidity makes them attractive catalysts.

Zeolites were discovered in 1756 by the Swedish mineralogist Axel Frederick Cronstedt, who found out that the mineral stilbite lost significant amounts of water when heated. The word zeolite stems from the Greek language and means boiling stone. It took almost two centuries before zeolites received the attention of chemists. In the 1930s, Richard Barrer investigated the synthesis of zeolites

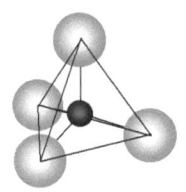

Figure 5.29 SiO$_4$ tetrahedra form the basic building blocks of zeolites.

and in 1948 he succeeded in making the first synthetic zeolite without counterpart in nature. Around that time, Milton discovered the synthesis of zeolite A. Nowadays, new zeolites and associated materials are still being discovered in laboratories worldwide.

Zeolites are used in various applications such as household detergents, desiccants, and as catalysts. In the middle of the 1960s, Rabo and coworkers at Union Carbide and Plank and co-workers at Mobil demonstrated that faujasitic zeolites were very interesting solid acid catalysts. Since then, a wealth of zeolite-catalyzed reactions of hydrocarbons has been discovered. For fundamental catalysis, they offer the advantage that the crystal structure is known and that the catalytically active sites are thus well defined. The fact that zeolites possess well-defined pore systems in which the catalytically active sites are embedded in a defined way gives them some similarity to enzymes.

Natural zeolites may bear the name of the mineral (mordenite, faujasite, ferrierite, silicalite) or sometimes that of the discoverer, for example, Barrerite after Prof Barrer, or the place where they were found, for example, Bikitaite from Bikita, Zimbabwe. Synthetic zeolites are usually named after the industry or university where they were developed, for example, VPI comes from Virginia Polytechnic Institute, and ZSM stands for Zeolite Socony Mobil.

With over 600 currently known zeolites and new ones discovered every year, it is useful to have a general classification of structures endorsed by the IUPAC. In this system each structure has three letters, for example FAU for faujasites, MFI for ZSM-5, and MOR for mordenite. Within a given structure, there can still be many different zeolites, as the composition may vary.

5.7.1
Structure of a Zeolite

Basically, zeolites consist of SiO$_4$ and AlO$_4$ tetrahedra (Figure 5.29), which can be arranged by sharing O-corner atoms in many different ways to build a crystalline lattice (Figure 5.30).

The SiO$_4$ tetrahedra can be arranged in into several silicate units, for example, squares, six- or eight-membered rings, called secondary building blocks. Zeolite structures are then built up by joining a selection of building blocks into periodic structures.

The same periodic structures can also be formed from alternating AlO$_4$ and PO$_4$ tetrahedra; the resulting aluminophosphates are not called zeolites but AlPOs. Zeolites are made in hydrothermal synthesis under pressure in auto-

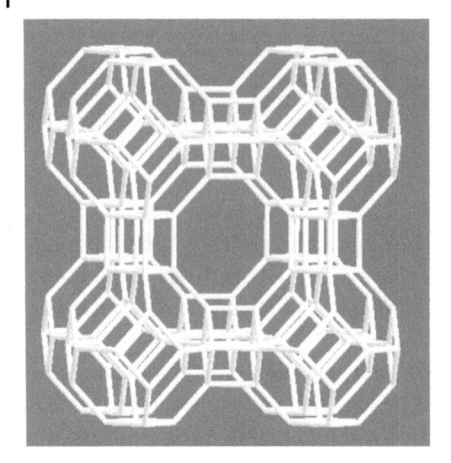

Figure 5.30 Example of the structure of a zeolite; every corner is a Si or Al ion, with O ions in between (halfway between the connecting lines). The structure is a combination of 4-, 6-, and 8-membered rings.

claves, in the presence of template molecules, such as tetramethylammonium, which act as structure directing agents.

Zeolites can have various pore systems. Zeolitel L (LTL) has parallel one-dimensional channels, Mordenite (MOR) has two different one-dimensional parallel channels. More complex is ZSM-5 (MFI), with straight and sinusoidal, hence two-dimensional, channels that are perpendicular to each other. Zeolite A (LTA) has cages connected by windows, ordered in a cubic array, while the Faujasites (FAU) of type X and Y possess a tetrahedrally ordered system of interconnected cages. Table 5.5 summarizes the properties of some of the most common zeolites.

5.7.2
Compensating Cations and Acidity

When Al^{3+} replaces Si^{4+} ions atoms in the tetrahedra, the units have a net charge of −1, and hence cations such as Na^+ are needed to neutralize the charge. The number of cations present within a zeolite structure equals the number of alumina tetrahedra in the framework. A zeolite in its sodium compensated form is indicated as Na-X, Na-ZSM-5, etc. If the sodium ions are replaced by protons (yielding H-X, H-ZSM-5, etc.) the zeolite becomes a gigantic polyacid. Figure 5.31 shows the structure of an acid site with a proton on a Si–O–Al bridge. Being a proton donor, the site is called a Brønsted acid. Its strength depends on the local environment of the proton, in particular on the number of other aluminum ions in the environment.

The AlPO family can also have acid sites, for example, if the Al^{3+} ions are replaced by divalent ions to yield MeAlPOs (Me = Co, Zn, Mg, Mn), or the P^{5+} ions are replaced by the tetravalent Si ion to give SAPOs. Of course, both types of substitution can occur, as in the CoAPSO. Substitution of AlO_4 and PO_4 by SiO_4

Table 5.5 Properties of some common zeolites.

Common name	3-Letter code	Channels	Window or channel diameter (nm)	Pore volume (mL g^{-1})	Si/Al
Zeolite A	LTA	3D, cages connected by windows, cubic	0.45	0.30	1
Zeolite X	FAU	3D, cages connected by windows, tetragonal	0.75	0.35	1–1.5
Zeolite Y	FAU	3D, cages connected by windows, tetragonal	0.75	0.35	≥ 2.5
Mordenite	MOR	1D, two straight, parallel channels	0.70; 0.65	0.20	≥ 5
Zeolite L	LTL	1D, one straight channel			
ZSM-5	MFI	2D, one straight, one sinusoidal, mutually perpendicular channels	0.55	0.15	≥ 10
Zeolite Beta	BEA	3D, two straight, one sinusoidal, perpendicular channels	0.76; 0.64	0.25	≥ 5

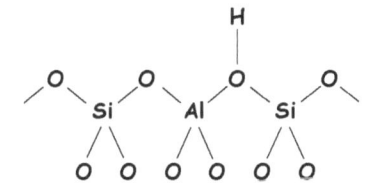

Figure 5.31 Principle of a solid acid: the proton compensates the deficient charge of the aluminum ion (3+) to match the valence of the silicon ion (4+).

tetrahedra is of course possible and yields the so-called SAPOs. Heteroatoms can also be introduced in the zeolite family, for example, trivalent iron instead of aluminum in ZSM-5 yields the Fe-ZSM-5 zeolite.

5.7.3
Applications of Zeolites

Zeolites have reactive surfaces, owing to the incorporation of Al^{3+} on sites where normally a Si^{4+} ion resides. This property, in addition to the crystalline system of micropores, enables several applications of zeolites, for example, in

- Adsorption: in drying, purification, and separation. Zeolites can absorb up to 25% of their weight in water.

- Ion exchange: zeolites are builders in washing powder, where they have gradually replaced phosphates to bind calcium. Calcium and, to a lesser extent, magnesium in water are exchanged for sodium in zeolite A. This is the largest application of zeolites today. Zeolites are essentially nontoxic, and pose no environmental risk whatsoever. Zeolite is also applied in toothpaste, again to bind calcium and prevent plaque.
- Catalysis: zeolites possess acid sites that are catalytically active in many hydrocarbon reactions, as we discuss in Chapter 9. The pore system only allows molecules that are small enough to enter; hence, it affects the selectivity of reactions by excluding the participation or the formation of molecules that are too large for the pores.

Unique properties of zeolites

- crystallinity
- microporosity
- uniform pore systems
- pore channels or cages
- high internal surface area
- high thermal stability
- ion exchange capabilities
- acidity
- nontoxic
- environmentally safe

Figure 5.32 illustrates how a zeolite can influence the selectivity of catalytic reactions. In the first case, one of the reactants is excluded because it cannot enter the zeolite. In the second, A reacts to give two produces, B and C, but molecule C is too large to leave the pore. In the third case, the onward reaction of B to C is prohibited, for example, because the transition state for this step does not fit.

The mechanistic aspects of zeolite-catalyzed reactions in the oil refinery are discussed briefly in Chapter 9.

5.8
Catalyst Testing

The only way to know if a material acts as a catalyst is to test it in a reaction [24–27]. Determining the activity of a catalyst is not as straightforward as it may seem. Particularly when working with single crystals and model systems, there are several pitfalls. For example, we prefer to measure the activity in the limit of zero conversion, to avoid results that are influenced by thermodynamic constraints, such as limitations due to equilibrium between reactants and products. We also want data under conditions of known gas composition and accurate temperature.

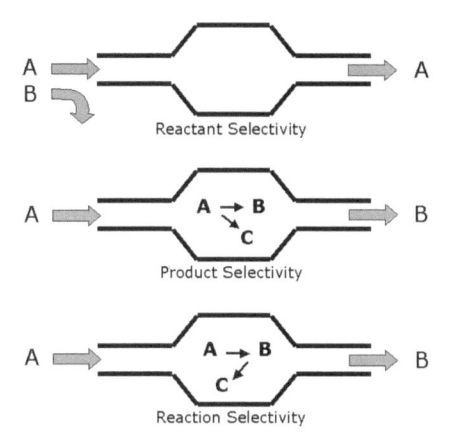

Figure 5.32 Examples of how zeolites can be used to increase selectivity for a reaction.

This may become problematic with fast and strongly endothermic or exothermic processes, since these will cool or heat the gas during reaction.

Often, a small sample (or even a diluted sample) of catalyst is placed in plug flow reactor (PFR) or in a continuously stirred tank reactor (CSTR) and the rate is determined. The latter is preferable since gradients are avoided, making the description of such reactions and kinetic modeling simpler. If everything is done properly, the rate will be given by the specific activity of each site multiplied by the number of sites present in the catalyst. In a metal catalyst, one often assumes that the number of sites is proportional to the metal area determined by chemisorption, as described above. Dividing the rate by the number of sites yields the turnover frequency. This number is important if we want to compare the intrinsic activity of catalysts. However, a turnover frequency is not necessarily the most relevant quantity for industrial applications where the activity per volume of catalyst is of greater interest. For industry many factors play a role – see the list at the beginning of this chapter.

5.8.1
Ten Commandments for Testing Catalysts

It is beyond the scope of this book to go through all the specifics of catalyst testing and to discuss all pitfalls that may arise. Instead we list the "Ten Commandments for the Testing of Catalysts," as have been formulated by Dautzenberg [24].

1. Specify objectives: What is the purpose of the test? Is it, for example, the specific reactivity and the reaction mechanism, or the long-term industrial use that is of interest?
2. Use an effective strategy for testing: what are the important parameters and how are these determined in the most efficient manner?
3. Choose the right type of reactor for testing: there are quite a number of different reactors. The above-mentioned plug flow reactor (PFR) and the continu-

ously stirred tank reactor (CSTR) are usually preferred for research laboratory use, but other set-ups may also be of interest for simulating real industrial conditions.

4. Establish ideal flow patterns: this is usually assumed to be the case for plug flow reactors (PFR) and continuously stirred tank reactor (CSTR), but are all conditions for ideal mixing fulfilled? For example, a rule of thumb is that the diameter d of the PFR should be at least $10\times$ the diameter d_p of the catalyst particles to eliminate the influence of the reactor wall. Also, the amount of catalyst should be sufficient to avoid axial gradients. Another rule is that the length of the bed L should be $50\times$ the particle diameter, that is, $L > 50\, d_p$. Higher values are preferable, but these may cause other problems such as temperature gradients and pressure drops.

5. Ensure isothermal conditions: even small deviations in the measured temperature may have a strong effect on the activity due to the exponential dependence of the rate on temperature as expressed by the Arrhenius equation. Temperature gradients may occur either in the catalyst bed or inside the catalyst particles due to exothermic or endothermic behavior of the reaction.

6. Minimize the effects of transport phenomena: if we are interested in the intrinsic kinetic performance of the catalyst, it is important to eliminate transport limitations, as these will lead to erroneous data. We will discuss later in this chapter how diffusion limitations in the pores of the catalyst influence the overall activation energy. Determining the turnover frequency for different gas flow velocities and several catalyst particle sizes is a way to establish whether transport limitations are present. A good starting point for testing catalysts is, therefore,
   - Small catalyst particles: eliminates transport limitation inside the particles.
   - Low conversion: eliminates temperature gradients and should always be chosen if it is the kinetics we are interested in; here, dilution with an inert material may be necessary.
   - Moderate temperature: to avoid gradients.

7. Obtain meaningful data on the catalyst: usually for kinetic purposes it is the turnover frequency per active site (TOF) that is of interest. But other parameters such as selectivity and yield are also of great importance for judging the potential of the catalyst. Instead of expressing the activity as a turnover frequency, it can also be given in terms of
   - Reaction rate per unit mass or unit volume of the catalyst.
   - The space velocity (the gas flow divided by the catalyst volume) at which a given conversion is obtained at a certain temperature.

   Expressing the activity of a catalyst as the temperature at which a certain conversion level is reached (not uncommon in applied catalysis!) is not recommended, as the overall rate of a catalytic reaction depends in a complicated way on the temperature.

8. Determine the stability of the catalyst: how fast does it lose its activity and what is the cause of the deactivation? How sensitive is the catalyst to various impurities that may be present in the feedstock under realistic conditions?

9. Follow good experimental practice: always make comparison with experiments performed under identical conditions. Check for reproducibility and that the experiments are not limited by such trivial factors as thermodynamics. Check for cleanliness and run blank experiments for the reactor and any inert filling material if such has been used.
10. Report the findings in an unambiguous manner: describe in great detail the conditions under which the catalyst was tested and give the relevant data such as activity, selectivity, and conversion in such a manner that they are uniquely defined.

5.8.2
Activity Measurements

The intrinsic activity of a catalyst can be measured by mounting it in a plug flow reactor and measuring its activity far away from equilibrium (preferable below 10% approach towards equilibrium), with well-defined gas mixtures and temperatures. The active metal area must be identified (not to be confused with the BET area which probes the entire surface area including the support). This can, for example, be done by adsorption of a molecular species that chemisorbs reversibly, like hydrogen or CO. Alternatively, it can be estimated if the size distribution of the nanoparticles is known from TEM measurements, together with the total metal loading and the degree of reduction. In this way, the intrinsic reactivity can be determined per unit of active surface area, or per active adsorption site. The rate of reaction expressed per site and per second is called the turnover frequency. It enables the comparison of catalytic activity of different materials on an objective basis.

For a series of catalysts, the activity can, for example, be plotted as a function of size of the nanoparticles, or their dispersion, D. The latter is defined as the ratio of the number of metal atoms at the surface, N_S, to those in the interior, N_V:

$$D \equiv \frac{N_S}{N_V} \tag{5.18}$$

One can estimate N_S and N_V as follows from the diameter of the particle and that of the atom:

$$N_S = \frac{\pi d^2}{A_x} \quad d = \text{particle diameter}; \quad A_x = \text{specific area of atom } x$$

$$N_V = \frac{\pi d^3}{6 V_x} \quad V_x = \text{specific volume of atom } x$$

$$\tag{5.19}$$

These equations obviously represent a continuum model of matter which fails for very small nanoparticles, but for practical use they work fairly well. The dispersion approaches zero for large nanoparticles, while it goes to 1 for very small nanoparticles where all the atoms are exposed in the surface, that is, particles with diameter smaller than 1 nm.

When reporting catalytic activity tests, another important piece of information is how much material was actually tested, what its volume was, and which amount of reactant gas passed over it per unit of time. At sufficiently low gas flow, one can get even the worst catalyst to make a conversion, without practical implication, since the amount of product would be insignificant. The quantity of interest is the gas-hourly space velocity, GSVH, which is the volume of gas passing through the catalyst bed per hour, divided by the volume of the catalyst. For highly active catalysts such number are large, for example, 30 000 h^{-1} (though lower numbers may well be acceptable depending of usage). Remember that in practice, the ideal is to have the exit gas mix from the reactor approach its equilibrium composition, to obtain as much product as possible with the highest possible space velocity, as this will maximize production. With respect to establishing turnover frequencies, a common problem may arise when the rate of reaction of a very active catalyst is too high. In such cases, operating close to the limit of zero conversion can always be achieved by diluting the catalyst with support or otherwise inert material.

Mass transport may constitute another problem [26, 28]. Since many catalysts are porous systems, diffusion of gases in and out of the pores may not be fast enough in comparison to the rate of reaction on the catalytic site. In such cases, diffusion limits the rate of the overall process.

Frequently, the catalyst to be tested has been pressed into a shaped body or pellet. For testing, it is crushed and a certain size fraction is mounted in a plug flow reactor, fulfilling the requirements described above. The dimensions of these particles are important. Figure 5.33 shows how gas flows around a catalyst particle. Note the stagnation layer around the particle: since the velocity of the gas is zero at the surface, there is a layer of gas through which the molecules have to be transported by diffusion. In addition, molecules need to diffuse in and out of pores, which is typically a slower process than movement through the stagnation layer. Transport limitations are likely to occur if the catalyst is very active.

5.8.2.1 Transport Limitations and the Thiele Diffusion Modulus

The effect of transport limitations can conveniently be evaluated by considering the spherical catalyst particle in Figure 5.33. We will introduce a dimensionless quantity called Thiele diffusion modulus (Φ_s) and an effectiveness factor to describe the loss of activity due to transport phenomena with respect to the ideal situation where such limitations are not present. We follow the treatment given by Satterfield, in whose book solutions for other geometries are also described [28].

We start with an ideal, porous, spherical catalyst particle of radius R. The catalyst is isothermal and we consider a reaction involving a single reactant. Diffusion is described macroscopically by the first and second laws of Fick, stating that

$$j(r, t) = -D\nabla C(r, t) \quad \text{and} \quad \frac{\delta C(r, t)}{\delta t} = D\nabla^2 C(r, t) \tag{5.20}$$

where D is the diffusion constant. (NB: the effective diffusion constant may be considerable smaller than the usual diffusion constant prevailing in the gas phase and in the stagnation layer. The reason is that if the pore in the catalyst becomes

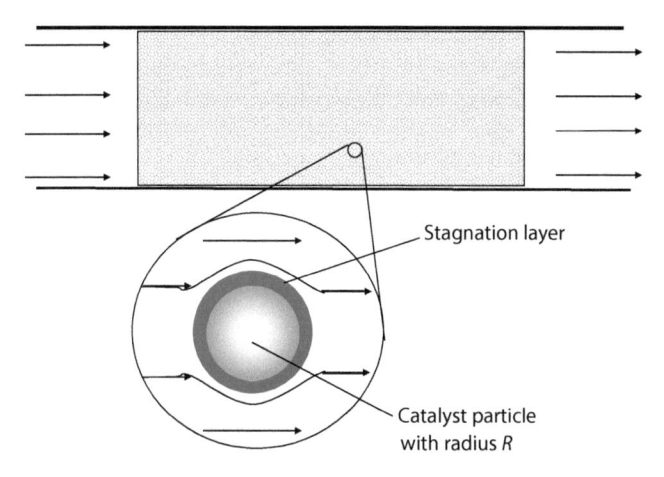

Figure 5.33 Catalyst particles in a plug-flow reactor surrounded by a gas flowing through the catalyst bed. Note the stagnation layer of gas around the particle that may cause transport limitations when diffusion is slow as compared to the reaction. However, limitation by pore diffusion is usually dominant.

too small, the gas molecules interact more with the walls than with each other. Here, they may be adsorbed and released again after some time. Diffusion in this regime is called Knudsen diffusion. Usually the effective diffusion constant describes an intermediate regime and is dependent on the pore size and distribution of the catalyst.)

The rate of the reaction has the form

$$\text{Rate}\left[\frac{\text{mol}}{\text{s}}\right] = V\left[\text{m}^3_{\text{cat}}\right] S \left[\frac{\text{m}^2_{\text{surf cat}}}{\text{m}^3_{\text{cat}}}\right] k \left[\frac{\text{m}^{3n}}{\text{m}^2_{\text{surf cat}} \text{ mol}^{n-1} \text{ s}}\right] \times C^n(r,t) \left[\frac{\text{mol}^n}{\text{m}^{3n}}\right] \quad (5.21)$$

where n is the order of reaction, V the volume of catalyst, S the surface area per volume catalyst, k the rate constant, and $C(r, t)$ the reactant concentration. We are only interested in the steady state solution. If we now consider a thin shell with inner radius r and outer radius $r + dr$ inside the porous particle, we can make a mass balance for the reactants in the shell of thickness dr (see Figure 5.34).

Under steady state conditions, the rate of diffusion inwards at the outside of the shell $(r + dr)$ minus the rate of diffusion inwards at the inside of the shell (r) corresponds to the rate of reaction in the shell. In other words, what came in and did not go out has reacted. The inward flux at position r is given by Fick's first law and hence the rate is the area of the shell $A(r) = 4\pi r^2$ times the flux:

$$\text{Rate}_{\text{in}}(r) = -A(r)j_{\text{in}}(r) = 4\pi r^2 D_{\text{eff}} \frac{dC(r)}{dr} \quad (5.22)$$

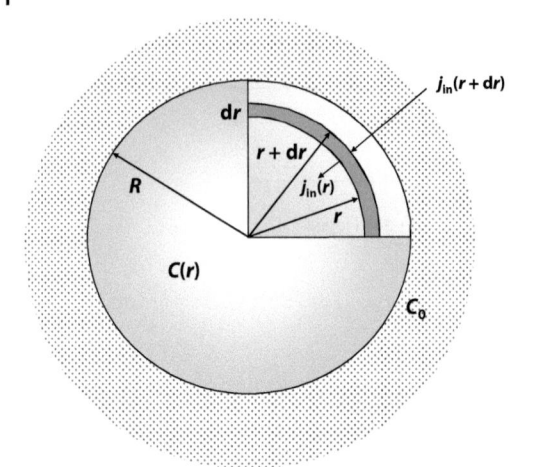

Figure 5.34 Schematic drawing of a spherical, porous catalyst particle of radius R. The reactant concentration is constant (C_0) outside the particle, but is lower inside the particle as a result of reaction. This makes the inward flux of reactants, $j_{in}(r)$, a function of the position inside the particle.

$$\text{Rate}_{in}(r+dr) = -A(r+dr)j_{in}(r+dr) = A(r+dr)D_{eff}\frac{dC(r+dr)}{dr}$$

$$= 4\pi(r+dr)^2 D_{eff}\left[\frac{dC(r)}{dr} + \frac{d^2C(r)}{dr^2}dr\right] \quad (5.23)$$

Here, we have used

$$\frac{C(r+dr) - C(r)}{dr} = \frac{dC(r)}{dr} \Rightarrow C(r+dr) = \left(\frac{dC(r)}{dr}\right)dr + C(r) \quad (5.24)$$

The rate of reaction is given by the volume of the shell $4\pi r^2\,dr$ [m$^3_{cat}$] multiplied by the pore area per volume catalyst S [m$^2_{surfcat}$/m$^3_{cat}$] multiplied by the rate constant k [m^{3n}/(m$^2_{surfcat} \times$ mol^{3n-1}s)] multiplied by the concentration of the reactant $C^n(r)$. Thus, the rate of reaction in the shell is

$$\text{Rate}_{react}(r) = 4\pi r^2\,dr S k C^n(r) \quad (5.25)$$

leading to the balance

$$\text{Rate}_{in}(r) - \text{Rate}_{in}(r+dr) = \text{Rate}_{react}(r) \quad (5.26)$$

$$4\pi r^2 D_{eff}\frac{dC(r)}{dr} - 4\pi(r+dr)^2 D_{eff}\left[\frac{dC(r)}{dr} + \frac{d^2C(r)}{dr^2}dr\right] \quad (5.27)$$

$$= 4\pi r^2\,dr S k C^n(r)$$

By eliminating higher orders in dr, this equation becomes

$$\frac{d^2C(r)}{dr^2} + \frac{2}{r}\frac{dC(r)}{dr} = \frac{SkC^n(r)}{D_{eff}} \equiv \frac{\Phi_s^2}{R^2}C(r) \quad (5.28)$$

5.8 Catalyst Testing

This defines the dimensionless Thiele diffusion modulus for a spherical particle

$$\Phi_s = R\sqrt{\frac{SkC^{n-1}}{D_{eff}}} \qquad (5.29)$$

The subscript s reminds us that Eq. (5.29) is only valid for a sphere. If we limit ourselves to a first-order reaction, the $n = 1$ and the differential equation transforms into

$$\frac{d^2 C(r)}{dr^2} + \frac{2}{r}\frac{dC(r)}{dr} = \frac{SkC(r)}{D_{eff}} = \frac{\Phi_s^2}{R^2}C(r) \equiv \omega^2 C(r) \qquad (5.30)$$

where

$$\Phi = R\sqrt{\frac{Sk}{D_{eff}}} \qquad \omega \equiv \frac{\Phi_s}{R} = \sqrt{\frac{Sk}{D_{eff}}} \qquad (5.31)$$

By introducing a new variable $\eta(r) = rC(r)$, Eq. (5.30) transforms into a homogeneous second-order differential equation

$$\frac{d^2\eta}{dr^2} = \omega^2 \eta \qquad (5.32)$$

for which the solution has the form

$$\eta = rC = B_1 e^{\omega r} + B_2 e^{-\omega r} \qquad (5.33)$$

We now have to find the function $C(r)$, using appropriate boundary conditions. These are that the concentrations outside the particle are constant, that is, $|C(r)|_{r=R} = C_0$ and that there is no concentration gradient at the center of the particle, that is, $\left|\frac{dC(r)}{dr}\right|_{r=0} = 0$. The latter condition leads to a solution for which $B_1 = -B_2$, and hence

$$C(r) = \frac{B_1}{r}(e^{\omega r} - e^{-\omega r}) = \frac{2B_1}{r}\sinh(\omega r) \qquad (5.34)$$

The first boundary condition allows us to determine B_1 as

$$B_1 = \frac{RC_0}{2\sinh(\omega r)} \qquad (5.35)$$

Consequently, the final solution for the concentration through the particle takes a relatively simple form

$$C(r) = \frac{RC_0 \sinh(\omega r)}{r \sinh(\omega R)} = \frac{C_0 \sinh\left(\Phi_s \frac{r}{R}\right)}{\frac{r}{R}\sinh(\Phi_s)} \qquad (5.36)$$

Hence, the concentration in the particle depends only on the Thiele diffusion modulus, Φ_s and a normalized radius $\frac{r}{R}$ as illustrated in Figure 5.35.

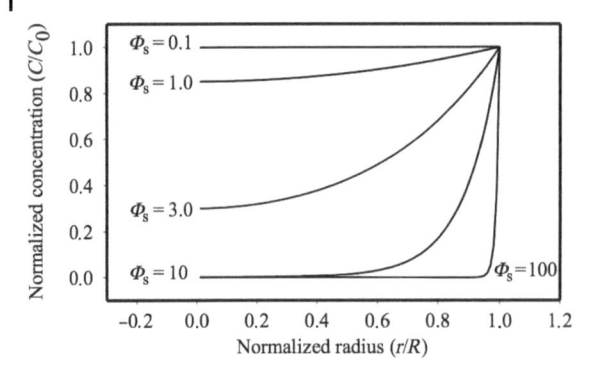

Figure 5.35 Normalized concentration profiles in a porous sphere for different values of the Thiele diffusion modulus. Note that if the latter is large, only a small part of the catalyst near the surface contributes to conversion.

Knowing the concentration profile, we find the overall rate of the particle, either by integration of the rate expression, or by realizing that all reactant diffusing into the particle will be converted. Thus, the total product rate is

$$\text{Rate}_{\text{diff}} = 4\pi R^2 D_{\text{eff}} \left| \frac{-dC(r)}{dr} \right|_{r=R} = 4\pi R D_{\text{eff}} \Phi_s C_0 \left[\frac{1}{\tanh(\Phi_s)} - \frac{1}{\Phi_s} \right] \quad (5.37)$$

where the subscript 'diff' refers to a process limited by diffusion. If there were no transport limitation, the entire particle would be active with a rate $\frac{4\pi}{3} R^3 S k C_0$ corresponding, as expected, to the situation when the Thiele diffusion modulus vanishes, $\Phi_s \to 0$ or

$$\Phi_s \left[\frac{1}{\tanh(\Phi_s)} - \frac{1}{\Phi_s} \right] \to \frac{\Phi_s^2}{3}$$

We define an effectiveness factor ε as the ratio of the diffusion-limited rate to the rate in the absence of diffusion limitation:

$$\varepsilon_s = \frac{3}{\Phi_s} \left[\frac{1}{\tanh(\Phi_s)} - \frac{1}{\Phi_s} \right] \quad (5.38)$$

The subscript refers to a spherical particle. One should also remember that we limited ourselves to a first-order irreversible reaction. Other expressions can be derived but are beyond the scope of this book. Nevertheless, Eq. (5.38) has important practical implications, since it is possible to discuss the effectiveness of the system by a single dimensionless parameter, Φ_s. Figure 5.36 shows the effectiveness factor as a function of Φ_s.

Note that the effectiveness approaches 1 as $\Phi_s \to 0$. Diffusion does not limit the rate if the diffusion constant is large, the particle is small, or the catalyst has a low activity. On the other, the effectiveness becomes small if Φ_s is large, that is, when D_{eff} is small, R is large, or the when the rate of reaction is high. The effectiveness

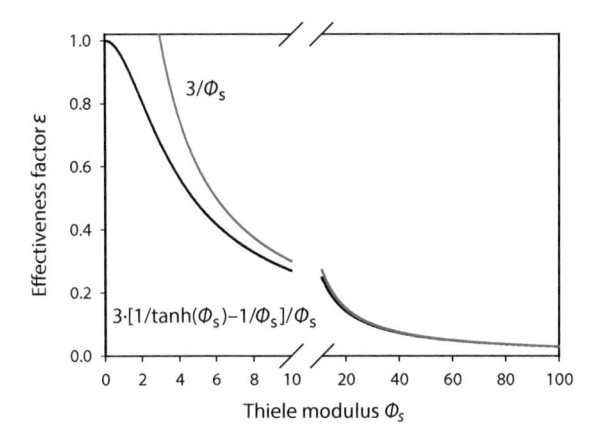

Figure 5.36 Effectiveness factor ε plotted as a function of the Thiele diffusion modulus Φ_s. The effective factor is well approximated by $\frac{3}{\Phi_s}$ for $\Phi_s > 10$.

approaches $\frac{3}{\Phi_s}$ for large Φ_s, as indicated in Figure 5.36. Thus, small particles are more efficiently used in a catalytic reaction than large particles.

However, there may be good reasons why a catalyst should *not* consist of particles that are too small, as we saw in the beginning of this chapter, for example, to avoid pressure gradients in the reactor. Based on an analysis such as above, one can decide whether it makes sense to use support particles that contain a homogeneous distribution of the catalytic phase. With expensive noble metals, one might perhaps decide to use an "egg-shell" type of arrangement, where the noble metal is only present on the outside of the particles.

The presence of diffusion limitations has a strong effect on the apparent activation energy one measures. We express both the rate constant, k, and the diffusion constant, D_{eff}, in Arrhenius form:

$$k = k_0 e^{\frac{-\Delta E_{act}}{RT}} \quad \text{and} \quad D_{eff} = D_0 e^{\frac{-\Delta E_{diff}}{RT}} \tag{5.39}$$

In the absence of diffusion limitation, the overall activation energy is

$$E^{app} = RT^2 \frac{\partial \ln(r)}{\partial T} = \Delta E_{act} \tag{5.40}$$

However, when diffusion is important, and Φ_s is large, the rate is

$$\text{Rate}_{diff} = 4\pi R D_{eff} \Phi_s C_0 \left[\frac{1}{\tanh(\Phi_s)} - \frac{1}{\Phi_s} \right]$$
$$\approx 4\pi R D_{eff} C_0 \Phi_s = 4\pi R^2 C_0 \sqrt{S k D_{eff}} \tag{5.41}$$

which results in the following apparent activation energy

$$E^{app} = RT^2 \frac{\partial \ln(\text{Rate}_{diff})}{\partial T} = \frac{\Delta E_{act}}{2} + \frac{\Delta E_{diff}}{2} \approx \frac{\Delta E_{act}}{2} \quad \text{when} \quad \frac{\Delta E_{diff}}{2} \to 0 \tag{5.42}$$

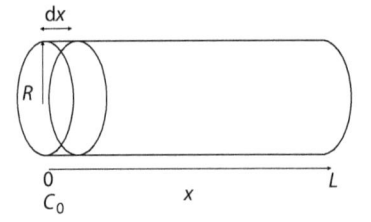

Figure 5.37 Schematic of a pore shaped as a cylinder.

Hence, the apparent activation energy is half the value we would obtain if there were no transport limitations. Obviously, it is important to be aware of these pitfalls when testing a catalyst. Indeed, apparent activation energies generally depend on the conditions employed (as discussed in Chapter 2), and diffusion limitation may cause them to change with temperature.

5.8.2.2 Pore Diffusion

Above we considered a porous catalyst particle, but we could similarly consider a single pore as shown in Figure 5.37. This leads to rather similar results. The transport of reactant and product is now determined by diffusion in and out of the pores, since there is no net flow in this region. We consider the situation in which a reaction takes place on a particle inside a pore. The latter is modeled by a cylinder with diameter R and length L (Figure 5.37). The gas concentration of the reactant is C_0 at the entrance of the pore and the rate is given by

$$r = kC(x) \tag{5.43}$$

If we take a thin slice of the cylinder of thickness dx, we can write an expression for the transport of mass through this slice at steady state. What goes in, either comes out or reacts, that is,

$$\begin{aligned}
Aj_{\text{in}} &= Aj_{\text{out}} - kC 2\pi R\, dx - \pi R^2 D_{\text{eff}} \frac{dC(x)}{dx} \\
&= -\pi R^2 D_{\text{eff}} \left(\frac{d\left(C(x) - \frac{dC(x)}{dx} dx\right)}{dx} \right) - kC(x) 2\pi R\, dx \frac{d^2 C(x)}{dx^2} \\
&= \frac{2k}{RD_{\text{eff}}} C(x) \equiv \frac{\Phi_p^2}{L^2} \equiv \omega^2 C(x)
\end{aligned} \tag{5.44}$$

$$\text{where} \quad \omega = \sqrt{\frac{2k}{RD_{\text{eff}}}} \quad \text{and} \quad \Phi_p = L\sqrt{\frac{2k}{D_{\text{eff}} R}}$$

In Eq. (5.44), the time has been omitted as it is a steady state solution, and we have introduced the Thiele diffusion modulus for a pore. We solve the differential equation with the following constraints:

$$|C(x)|_{x=0} = C_0 \quad \text{and} \quad \left|\frac{dC(x)}{dx}\right|_{x=L} = 0 \tag{5.45}$$

reflecting that the concentration of the reactants has reached a certain equilibrium condition at the end of the pore. The procedure to find a solution to this differential equation is the same as in the previous section, leading to

$$C(x) = C_0 \frac{\cosh(\omega(L-x))}{\cosh(\omega L)} = C_0 \frac{\cosh\left[\Phi_s \frac{L-x}{L}\right]}{\cosh(\Phi_p)} \qquad (5.46)$$

The reader may verify by insertion in the differential equation above that this is indeed a solution fulfilling the constraints.

We shall now introduce an efficiency factor, which is again defined as the ratio of the conversion in the pore with and without mass transport limitation:

$$\varepsilon \equiv \frac{\int_0^L 2\pi R k C(x)\, dx}{2\pi R L k C_0} = \frac{\int_0^L C(x)\, dx}{L C_0} = \frac{\int_0^L \cosh(\omega(L-x))\, dx}{L \cosh(\omega L)} \qquad (5.47)$$

$$= \frac{\tanh(\Phi_p)}{\Phi_p}$$

The rate in case of diffusion limitation is

$$r_{\text{diff}} = 2\pi R L k C_0 \frac{\tanh(\Phi_p)}{\Phi_p} \qquad (5.48)$$

Thus, we see again that the system can conveniently be described by the Thiele diffusion modulus Φ_p. It is not difficult to see that two limiting cases exist. There are no diffusion limitations when $D \to \infty$, or $L \to 0$, or $k' \to 0$, $R \to \infty$, since then

$$\sqrt{\frac{2kL^2}{DR}} \to 0 \quad \Rightarrow \quad \varepsilon \to 1 \quad \text{and} \quad r = kC_0$$

On the other hand, if the pore is large and/or the reaction fast (Lk large), or diffusion slow and/or the pore narrow (DR small), the efficiency is small and approaches

$$\varepsilon \longrightarrow \frac{1}{\Phi_p} = \frac{1}{L}\sqrt{\frac{R D_{\text{eff}}}{2k}}$$

while the rate becomes

$$r_{\text{diff}} \approx \frac{2\pi R L k C_0}{\Phi_p} = 2\pi C_0 \sqrt{k R^3 D_{\text{eff}}} \qquad (5.49)$$

When solving for the apparent activation energy in the diffusion limited region we obtain

$$E^{\text{app}} = RT^2 \frac{\partial \ln(r_{\text{diff}})}{\partial T} = \frac{\Delta E_{\text{act}}}{2} + \frac{\Delta E_{\text{diff}}}{2} \approx \frac{\Delta E_{\text{act}}}{2} \qquad (5.50)$$

when $\frac{\Delta E_{\text{diff}}}{2} \to 0$

Thus, considering diffusion in pores leads to very similar results to those we obtained when describing diffusion in catalyst particles.

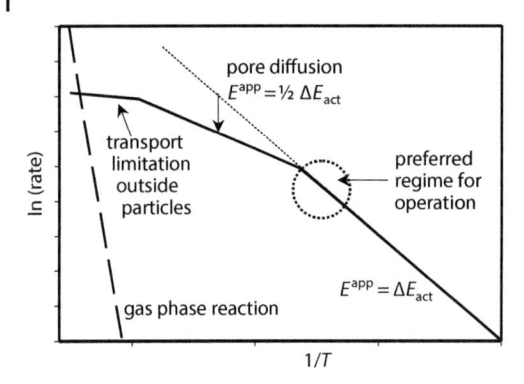

Figure 5.38 Arrhenius plot illustrating the effect on the apparent activation energy of pore diffusion and transport limitations through the stagnation layer surrounding a catalyst particle. Note that pore diffusion reduces the apparent activation energy to half its original value. At very high temperatures, gas phase reactions may play a role.

5.8.2.3 Consequences of Transport Limitations for Testing Catalysts

Normally, the activation energy for diffusion in the gas phase is much smaller than the activation energy for a catalyzed reaction and, hence, according to Eqs. (5.42) and (5.50), the overall or apparent activation energy for the diffusion-limited process is half of what it would be without transportation limitation. If we plot the rate as a function of reciprocal temperature, one observes a change in slope when transport limitations starts to set in.

As Figure 5.38 shows, at low temperatures the rate increases with increasing temperature according to the activation energy of the catalytic reaction (which may be an apparent one itself, see Chapter 2. At a certain temperature the intrinsic activity of the catalyst becomes so high that transport of the reactants to the sites cannot keep up with the rate of reaction at the sites. Hence, a concentration gradient starts to develop inside the particles. The effect is that the rate increases less steeply than in the absence of such limitations. At even higher temperatures yet another change in slope appears since the catalyst particles are now so effective that the diffusion through the stagnation layer around the particles becomes limiting. Finally, at very high temperature, gas phase reactions can take place, for which the activation energy is much higher than for the catalytic reaction. Thermodynamics determines whether conversion will be possible under such conditions. The ammonia synthesis represents a case where the equilibrium is entirely on the reactant side at temperatures where the activation energy for gas phase reaction can be surmounted, as we will see in Chapter 8.

For testing and optimizing catalysts, the temperature region just below that where pore diffusion starts to limit the intrinsic kinetics provides a desirable working point (unless equilibrium or selectivity considerations demand working at lower temperatures). In principle, we would like the rate to be as high as possible while also using the entire catalyst efficiently. For fast reactions such as oxidation, we may have to accept that only the outside of the particles is used. Consequently, we may decide to use a nonporous or monolithic catalyst, or particles with the catalytic material only on the outside.

Consideration of transport phenomena is of great importance for designing the shapes of catalysts, as discussed in the Section 5.5 (see also Figure 5.26). Although small particles may allow the efficient use all material, one cannot permit too large a pressure drop over the reactor. Mechanical strength is also an issue. Hence, the assembly of particles in larger bodies is necessary and the shapes will in general have been optimized with mass and heat transfer in mind. Designing catalytic processes that are efficient and economically feasible requires a coherent approach on all levels.

Rigorous, well-established kinetic data, obtained in the way as described above, enables comparison of tests from different laboratories [29]. Some of the top journals in the catalysis field require that data are presented as turn-over frequencies, or at least rates normalized by catalytic surface area, and not simply in terms of the conversion of a reactant. For comparing product selectivities, it is important that data are used at similar conversion levels, as selectivities often depend on these. All catalytic measurements need to be carried out in the kinetically controlled domain, while confirming tests need to be carried out and reported, especially for liquid phase reactions [30].

References

1 Busca, G. (1994) *Heterogeneous Catalytic Materials, Solids State Chemistry, Surface Chemistry and Catalytic Behavior*, Elsevier, Amsterdam.
2 Ertl, G. Knözinger, H., Schüth, F., Weitkamp, J. (2008) *Handbook of Heterogeneous Catalysis*, Wiley-VCH Verlag GmbH, Weinheim.
3 De Jong K.P. (2009) *Synthesis of Solid Catalysts*, Wiley-VCH Verlag GmbH, Weinheim.
4 Lloyd, L. (2011) *Handbook of Industrial Catalysis*, Plenum, New York.
5 Cejka, J., Corma, A. and Jones, S. (2010) *Zeolites and Catalysis: Synthesis, Reactions and Applications*, Wiley-VCH Verlag GmbH, Weinheim.
6 Regalbuto, J.E. (2006) *Catalyst Preparation, Science and Engineering*, CRC Press, Taylor and Francis Group, Boca Raton.
7 Elliot, S. (1998) *The Physics and Chemistry of Solids*, John Wiley & Sons, Inc., New York.
8 Kittel, C. (1976) *Introduction to Solid State Physics*, John Wiley & Sons, Inc., New York.
9 Ashcroft, N. and Mermin, N.D. (1976) *Solid State Physics*, Holt-Saunders, Philadelphia.
10 Somorjai, G.A. and Li, Y. (2010) *Introduction to Surface Chemistry and Catalysis*, John Wiley & Sons, Inc., New York.
11 Kolasinski, K. (2012) *Surface Science: Foundations of Catalysis and Nanoscience*, John Wiley & Sons, Inc., New York.
12 Klink, C., Olesen, L., Besenbacher, F., Stensgaard, I., Laegsgaard, E. and Lang, N.D. (1993) Interaction of C with Ni(100): Atom-resolved studies of the "clock" reconstruction. *Physical Review Letters*, **71**, 4350–4353.
13 Henrich, V.E. and Cox, P.A. (1994) *The Surface Science of Metal Oxides*, Cambridge University Press, Cambridge.
14 Göpel, W., Rocker, G. and Feierabend, R. (1983) Intrinsic defects of TiO_2(110): Interaction with chemisorbed O_2, H_2, CO, and CO_2, *Physical Review B*, **28**, 3427–3438.
15 Christensen, A., Ruban, A.V., Stoltze, P., Jacobsen, K.W., Skriver, H.L.,

Nørskov, J.K. and Besenbacher, F. (1997) Phase diagrams for surface alloys. *Physical Review B*, **56**, 5822–5834.

16 Van Helden, P., Ciobîcă, I.M., and Coetzer, R.L.J. (2016) The Size-Dependent Site Composition of FCC Cobalt Nanocrystals, *Catalysis Today*, **261**, 48–59.

17 Hansen, K.H., Worren, T., Stempel, S., Laegsgaard, E., Baumer, M., Freund, H.J., Besenbacher, F. and Stensgaard, I. (1999) Palladium nanocrystals on Al_2O_3: Structure and adhesion energy, *Physical Review Letters*, **83** 4120–4123.

18 Henry, C.R. (1998) Surface studies of supported model catalysts. *Surface Science Reports*, **31**, 235–325.

19 Thomas, J.M. and Thomas, W.J. (2014) *Principles and Practice of Heterogeneous Catalysis*, Wiley-VCH Verlag GmbH, Weinheim.

20 Jacobsen, H. and Kleinschmit, P. (1999) Flame Hydrolysis in *Preparation of Solid Catalysts* (eds G. Ertl, H. Knözinger and J. Weitkamp), Wiley-VCH Verlag GmbH, Weinheim.

21 Busca, G. (2014) Structural Surface, and Catalytic Properties of Aluminas, *Advances in Catalysis*, **57**, 319–404.

22 Regalbuto, J.E. (2006) Catalyst Preparation, *Science and Engineering*, CRC Press, Taylor and Francis Group, Boca Raton.

23 Meijers, A., De Jong, A.M., Van Gruijthuijsen, L.M.P. and Niemantsverdriet, J.W. (1991) Preparation of zirconium-oxide on silica and characterization by x-ray photoelectron-spectroscopy, secondary ion mass-spectrometry, temperature programmed oxidation and infrared-spectroscopy. *Applied Catalysis*, **70**, 53–71.

24 Dautzenberg, F.M. (1989) 10 Guidelines for catalyst testing. *ACS Symposium Series*, **411**, 99–119.

25 Davis, M.E. and Davis, R.J. (2003) *Fundamentals of Chemical Reaction Engineering*, McGraw-Hill, New York.

26 Wijngaarden, R.J. and Kronberg, A. Westerterp, K.R. (1998) *Industrial Catalysis: Optimizing Catalysts and Processes*, Wiley-VCH Verlag GmbH, Weinheim.

27 Bartholomew, C.H. and Farrauto, R.J. (2006) *Fundamentals of Industrial Catalytic Processes*, John Wiley & Sons, Inc., New York.

28 Satterfield, C.N. (1970) *Mass Transfer in Heterogeneous Catalysis*, MIT Press, Cambridge.

29 Ribeiro, F.H., Von Wittenau, A.E.S., Bartholomew, C.H. and Somorjai, G.A. (1997) Reproducibility of turnover rates in heterogeneous metal catalysis: Compilation of data and guidelines for data analysis. *Catalysis Reviews – Science and Engineering*, **39**, 49–76.

30 Guide for Authors (2015) *The Journal of Catalysis*.

Chapter 6
Surface Reactivity

6.1
Introduction

Gas-surface interactions and reactions on surfaces play a crucial role in many technologically important areas such as corrosion, adhesion, synthesis of new materials, electrochemistry, and heterogeneous catalysis. This chapter aims to describe the interaction of gases with metal surfaces in terms of chemical bonding. Molecular orbital and band structure theory are the basic tools for this. We limit ourselves to metals.

Computational chemistry has reached a level in which adsorption, dissociation, and formation of new bonds can be described with reasonable accuracy. Consequently, trends in reactivity patterns can be very well predicted nowadays. Such theoretical studies have had a strong impact in the field of heterogeneous catalysis, particularly because many experimental data are available for comparison from surface science studies (e.g., heats of adsorption, adsorption geometries, vibrational frequencies, activation energies of elementary reaction steps) to validate theoretical predictions. As a concise introduction into this growing subdiscipline of catalysis, we refer to recent text books [1–3].

As explained in the previous chapters, catalysis is a cycle, which starts with the adsorption of reactants on the surface of the catalyst. Often at least one of the reactants is dissociated, and it is often in the dissociation of a strong bond that the essence of catalytic action lies. Hence, we shall focus on the physics and chemistry involved when gases adsorb and dissociate on the surface, in particular metal surfaces.

When an atom or molecule approaches the surface, it feels the potential energy set up by the metal atoms in the solid. The interaction is usually divided into two regimes, namely physisorption and chemisorption, which we discuss separately.

Concepts of Modern Catalysis and Kinetics, Third Edition. I. Chorkendorff and J.W. Niemantsverdriet.
© 2017 WILEY-VCH Verlag GmbH & Co. KGaA. Published 2017 by WILEY-VCH Verlag GmbH & Co. KGaA.

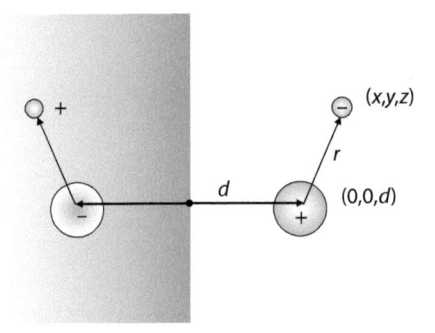

Figure 6.1 Schematic drawing of an atom outside a surface at distance d.

6.2 Physisorption

Physisorption is a weak interaction characterized by the lack of a true chemical bond between adsorbate and surface, that is, no electrons are shared. The physisorption interaction is conveniently divided into two parts: A strongly repulsive part at close distances and Van der Waals interactions at medium distances of a few Å.

6.2.1 The Van der Waals Interaction

When an atom or molecule approaches a surface, the electrons in the particle – due to quantum fluctuations – set up a dipole, which induces an image dipole in the polarizable solid. Since this image dipole has the opposite sign and is correlated with fluctuations in the particle, the resulting force is attractive. In the following, we construct a simple model to elucidate the phenomenon.

Consider a hydrogen atom with its nucleus at the origin located above the surface of a conducting metal at the position $\boldsymbol{d} = (0, 0, d)$ and an electron at $\boldsymbol{r} = (x, y, z)$ (Figure 6.1). The nucleus and the electron both induce image charges in the metal equal to

$$q = \frac{1-\varepsilon}{1+\varepsilon} e = -e; \quad \varepsilon \longrightarrow \infty \tag{6.1}$$

The potential due to the interactions between the charges is given by

$$V_W(\boldsymbol{d}, \boldsymbol{r}) \propto -\frac{e^2}{2d} - \frac{e^2}{2(d+z)} + 2\frac{e^2}{|2\boldsymbol{d} + \boldsymbol{r}|} \tag{6.2}$$

where the first two terms describe the interaction between the nucleus and its image and between the electron and its image. The third term originates from the two repulsive cross terms. Expanding Eq. (6.2) in a Taylor series in powers of $\frac{r}{d}$

Table 6.1 Typical heats of physisorption for a number of small molecules.

Molecule	Enthalpy (kJ mol^{-1})
CH$_4$	−21
CO	−25
CO$_2$	−25
N$_2$	−21
O$_2$	−21

yields

$$V_W(\mathbf{d}, \mathbf{r}) \propto -\frac{e^2}{d^3}\left(\frac{x^2}{2} + \frac{y^2}{2} + z^2\right) \tag{6.3}$$

Thus, the net result of all interactions is an attractive potential behaving as

$$V_W(d) \propto -\frac{C_{vdW}}{d^3} \tag{6.4}$$

where C_{vdW} is the so-called Van der Waals constant for the system, which depends on the polarizability of the atom and the response of the metal. Note that the interaction does not require a permanent dipole. Hence, rare gas atoms such as argon and xenon may also physisorb on the surface. Physisorption is the basis for the BET method described in detail in Chapter 5.

6.2.2
Including the Repulsive Part

The attraction between the atom and its image cannot continue at small distances as the electrons of the atom begin to interact strongly with the electrons of the surface. The kinetic energy of the electrons increases as they will have to orthogonalize to the localized electrons of the atom as a consequence of the Pauli principle. Some energy will be gained as the same electrons also become attracted to the positive nucleus. This may lead to chemisorption, but if the atom is a rare gas then repulsion dominates. A simple approximation of the potential in this regime is provided by the exponential function

$$V_R(d) \propto e^{-\frac{d}{a}} \tag{6.5}$$

which is a reasonable choice since the density of electrons away from the surface also decays in this manner. The resulting potential is then given by the sum

$$V(d) = V_R(d) + V_{vdW}(d) \cong C_R e^{-\frac{d}{a}} - \frac{C_{vdW}}{d^3} \tag{6.6}$$

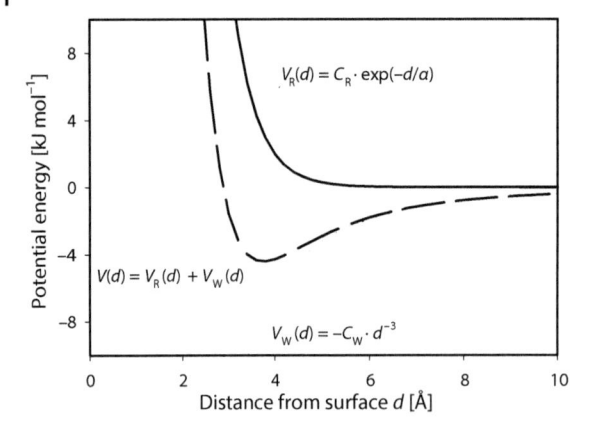

Figure 6.2 Potential energy diagram showing the attractive Van der Waals interaction and the repulsive interaction due to Pauli repulsion, leading to a physisorption well of 4 kJ mol^{-1} at 3.8 Å from the surface.

and is shown in Figure 6.2. The depth of the well is usually on the order of 20–25 kJ mol^{-1} or less (Table 6.1) and the minimum lies several Å outside the surface. Molecules physisorbed on the surface are not chemically altered by the interaction; they largely conserve the spectroscopic characteristics they exhibit in the gas phase.

Physisorption is very similar to the molecular van der Waals interaction, which makes gases condense in multilayers. The Van der Waals interaction between molecules is often described by the Lennard-Jones potential, which has the form

$$V(d) = \frac{C_n}{d^n} - \frac{C_6}{d^6} \tag{6.7}$$

Equation (6.7) is often referred to as the $(N,6)$ Lennard Jones potential. In particular, (12,6) is popular for mathematical reasons, despite the fact that an exponential form as in Eq. (6.6) usually describes the repulsive part of the potential better. The potentials shown in Figure 6.2 form a good description for the physisorbed molecule, but they break down for small distances, where the attractive term in Eq. (6.6) starts to dominate in an unrealistic way, because for $d \to 0$ the repulsive part becomes constant ($V_R \to C_R$), while the Van der Waals part continuously goes towards infinity ($V_{vdW} \to -\infty$).

6.3
Chemical Bonding

If molecules or atoms form a chemical bond with the surface upon adsorption, we call this chemisorption. To describe the chemisorption bond, we need to briefly review a simplified form of molecular orbital theory. This is also necessary to appreciate, at least qualitatively, how a catalyst works. As described in Chapter 1, the

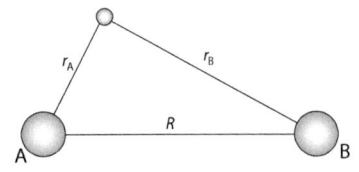

Figure 6.3 Diatomic molecule AB along with definitions of distances.

essence of catalytic action is often that it assists in breaking strong intramolecular bonds at low temperatures. We aim to explain how this happens in a simplified, qualitative electronic picture.

6.3.1
Bonding in Molecules

6.3.1.1 The Diatomic Molecule

Consider two well-separated atoms A and B with electron wave functions Ψ_a and Ψ_b, which are eigen functions of the atoms, with energies ε_a and ε_b. If we bring these atoms closer, the wave functions start to overlap and form combinations that describe the chemical bonding of the atoms to form a molecule. We will neglect the spin of the electrons. The procedure is to construct a new wave function as a linear combination of atomic orbitals (LCAO), which for one electron has the form

$$\Psi = c_1 \Psi_a + c_2 \Psi_b \tag{6.8}$$

The Hamiltonian is

$$H = \left(-\frac{\hbar^2}{2m_e}\right)\nabla^2 + \frac{e^2}{4\pi\varepsilon_0}\left(-\frac{1}{r_a} - \frac{1}{r_b} + \frac{1}{R}\right) \tag{6.9}$$

where r_a and r_b are the distances of the single electron from the nuclei A and B, respectively, and R is the distance between the nuclei (Figure 6.3).

By using the variation principle and solving the secular equations,

$$\sum_i c_i \left(H_{ik} - ES_{ik}\right) = 0 \tag{6.10}$$

in which H_{ik} are the matrix elements of the Hamiltonian and S_{ik} those for the overlap. Equations (6.10) have solutions only when the determinant vanishes:

$$\begin{vmatrix} \alpha_A - E & \beta - SE \\ \beta - SE & \alpha_B - E \end{vmatrix} = 0 \tag{6.11}$$

with

$$\alpha_A = H_{aa} = \int \psi_a H \psi_a \, dV; \quad \alpha_B = H_{bb} = \int \psi_b H \psi_b \, dV \tag{6.12}$$

$$\beta = H_{ab} = H_{ba} = \int \psi_a H \psi_b \, dV \tag{6.13}$$

$$S = S_{ab} = \int \psi_a \psi_b \, dV \tag{6.14}$$

while $S_{aa} = S_{bb} = 1$ since ψ_a and ψ_b are normalized. S is called the overlap integral. Solving Eq. (6.11) for the energy E leads to

$$E_{\pm} = \frac{\alpha_a + \alpha_b - 2S\beta \pm \sqrt{(\alpha_a - \alpha_b)^2 - 4(\alpha_a + \alpha_b)S\beta + 4\beta^2 + 4\alpha_a\alpha_b S^2}}{2(1 - S^2)} \tag{6.15}$$

Hence, we find two energy levels for the diatomic molecule where the electron can reside, one bonding, and the other antibonding. This simplified approach does not describe the situation quantitatively too well, but in a qualitative sense it captures all the important effects. In the following, we consider a few illustrative cases in the limit where the overlap S is small (this is the usual approximation for elementary work). In this limit Eq. (6.15) reduces to

$$E_{\pm} = \frac{\alpha_a + \alpha_b - 2S\beta \pm \sqrt{(\alpha_a - \alpha_b)^2 + 4\beta^2 - 4(\alpha_a + \alpha_b)S\beta}}{2} \tag{6.16}$$

6.3.1.2 Homonuclear Diatomic Molecules

The first case is that of a homonuclear molecule, such as H_2, N_2, and O_2, for which $\alpha_a = \alpha_b = \alpha$, and if we also use that $\alpha S/\beta \ll 1$, the energies reduce to

$$E_{\pm} \cong \alpha \pm \beta - S\beta \tag{6.17}$$

Since $\beta < 0$ and $S > 0$, this means that both levels are shifted upwards by the repulsion term $S|\beta|$. Next, they form a downward shifted bonding state E_+ and an upward shifted antibonding state E_- as shown in Figure 6.4.

If we only consider the one-electron energies, which naturally is not entirely correct, we can estimate the energy gained by making the chemical bond by summing over all occupied orbitals and take the differences:

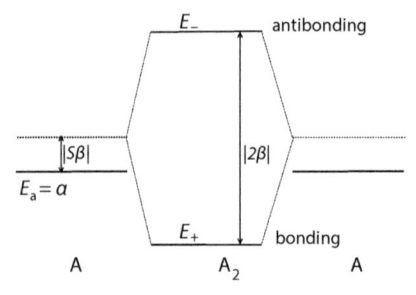

$$\Delta E = \sum_{i_{occ}} E_i^{ab} - \sum_{i_{occ}} E_i^a - \sum_{i_{occ}} E_i^b \tag{6.18}$$

Figure 6.4 Schematic energy diagram of a homonuclear diatomic molecule. Note that the splitting $|2\beta|$ is proportional to the overlap of the atomic orbitals.

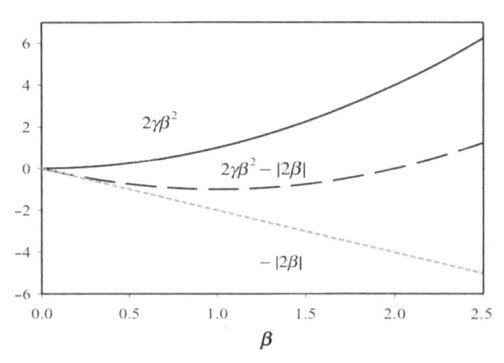

Figure 6.5 Change in energy as function of the overlap β.

If $\Delta E < 0$, the molecule is stable and the work required for dissociation, E_{diss} equals $-\Delta E$ with

$$\Delta E \cong 2E_+ - 2\alpha = -2|\beta| + 2S|\beta| \tag{6.19}$$

Since β is proportional to the density of the wave functions Ψ and the overlap S, we expect that a linear relation exists between S and β, that is, $S = -\gamma\beta$ resulting in

$$\Delta E \cong 2\gamma\beta^2 - 2|\beta| \tag{6.20}$$

The first term is a repulsion term and the second represents the hybridization leading to the chemical bonding, as indicated in Figure 6.5.

Figure 6.6 shows the molecular orbital energy diagrams for a few homonuclear diatomic molecules. The stability of the molecules can be estimated from the

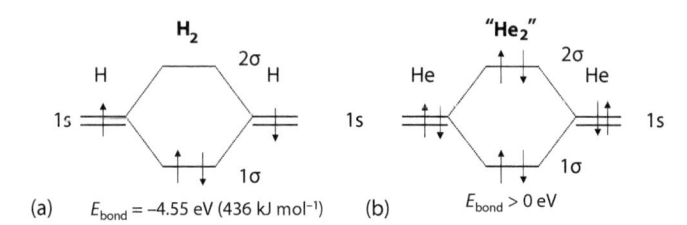

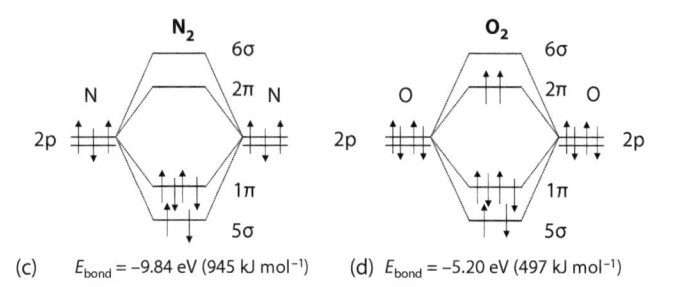

Figure 6.6 Energy levels of the frontier orbitals of selected molecules. In reality this picture is too simple. Due to interaction with the 2s orbitals the 5σ levels lie above the 1π level, but this does not change the overall stability of the molecules.

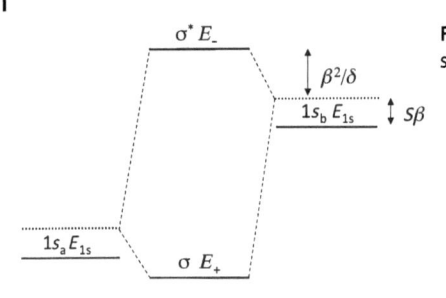

Figure 6.7 Schematic energy diagram of the splitting in a heteronuclear molecule.

number of electrons occupying bonding orbitals compared to the number of electrons in the antibonding orbitals. (Antibonding orbitals are sometimes denoted with the superscript ∗, as in $2\pi^*$.)

It is easily seen that He does not form a stable molecule because both the bonding and the antibonding orbital are occupied by two electrons, which gives a net repulsive interaction. The N_2 molecule is much more stable than the O_2 molecule, since in the latter the two additional electrons are located in antibonding orbitals. Note that in O_2, the outer electrons are parallel; hence, the molecule is in the triplet state.

6.3.1.3 The Heteronuclear System

In the heteronuclear diatomic molecule, we have a system where $\alpha_a \neq \alpha_b$. We assume that the difference between the two atomic levels $\delta = \alpha_b - \alpha_a > 0$ is much larger than the interaction β. If we again assume that the overlap S is small, we can expand the energies in a Taylor series

$$E_+ \cong \alpha_a - \frac{\beta^2}{\delta} + S|\beta|$$
$$E_- \cong \alpha_b + \frac{\beta^2}{\delta} + S|\beta|$$
(6.21)

showing that the splitting between antibonding and bonding orbital decreases with the energy difference between the interacting levels. Note that the upward and downward shifts are again not symmetrical (Figure 6.7) due to the repulsion term – the electrons have to orthogonalize (Pauli repulsion).

The wave function for the bonding and antibonding orbitals can be written as

$$\psi_+ = \psi_a + \frac{\beta^2}{\delta}\psi_b$$
$$\psi_- = \psi_b + \frac{\beta^2}{\delta}\psi_a$$
(6.22)

and the energy gain by forming the bond, calculated without using the Taylor expansion, is

$$\Delta E = 2E_+ - \alpha_a - \alpha_b$$
$$\Delta E \cong (\alpha_a + \alpha_b) - 2S\beta + \sqrt{(\alpha_a - \alpha_b)^2 + 4\beta^2} - \alpha_a - \alpha_b$$
$$\cong \sqrt{4\beta^2 + \delta^2} - 2S\beta$$
(6.23)

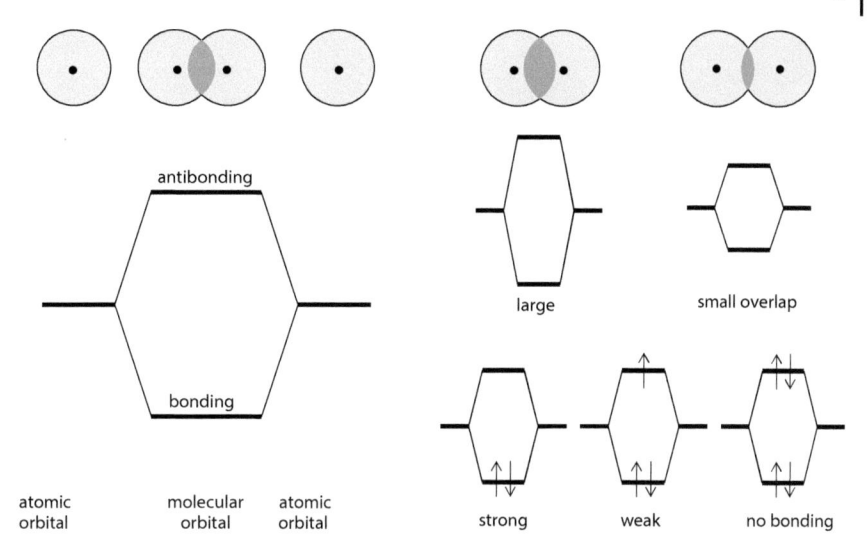

Figure 6.8 Summary of molecular orbital theory for homonuclear molecules. Note how the stability of a chemical bond depends both on the interaction strength and the filling of the orbitals.

Note that the homonuclear limit is obtained when $\delta \to 0$. By defining the hybridization term

$$W_{ab} = \sqrt{4\beta^2 + \delta^2} \tag{6.24}$$

we obtain

$$\Delta E = 2S|\beta| - W_{ab} \tag{6.25}$$

This is in principle all we need to understand chemical bonding on surfaces and trends in reactivity. For a more accurate description of molecular orbital theory, we refer to Atkins' famous text book [4]. The main results from molecular orbital theory are summarized in Figure 6.8.

6.3.2
The Solid Surface

Here, we briefly summarize some major results from solid state theory that are relevant for understanding the behavior of adsorption on surfaces. For a more detailed description please consult excellent books from Ascroft and Mermin, Elliot, and Kittel [5–7].

The metallic bond can be understood by extending the orbital picture we sketched above. The metals all have extended outer s or p orbitals, which ensure large overlap. The formation of the metallic bond from such s and p levels is illustrated in Figure 6.9.

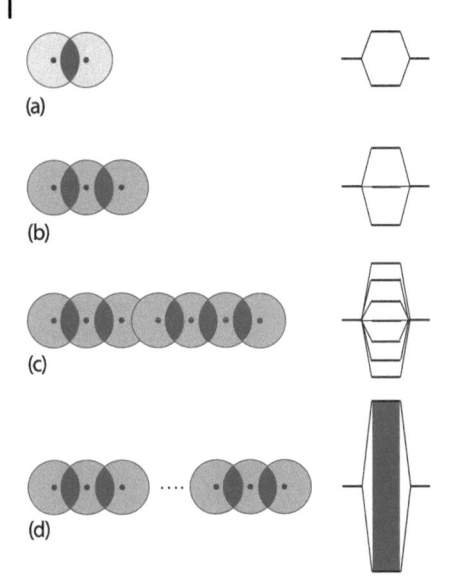

Figure 6.9 Formation of an electron band by addition of atoms and their orbitals. Note that the splitting between the bonding and antibonding levels increases by enhanced overlap. Eventually when a high number of orbitals are added, a continuum band is formed as illustrated by the shaded region in (d).

As this figure illustrates, the orbitals are very close in a metal and form an almost continuous "band" of levels. In fact, it is impossible to detect the separation between levels. The bands behavior is in many respects similar to the orbitals of the molecule in Figure 6.8: if there is little overlap between the electrons, the interaction is weak and the band is narrow. Such is the case for d orbitals (Figure 6.10), which have pronounced shapes and orientations that are largely retained in the metal. Hence, the overlap between the individual d orbitals of the atoms is much smaller than that of the outer s and p electrons. The latter are strongly delocal-

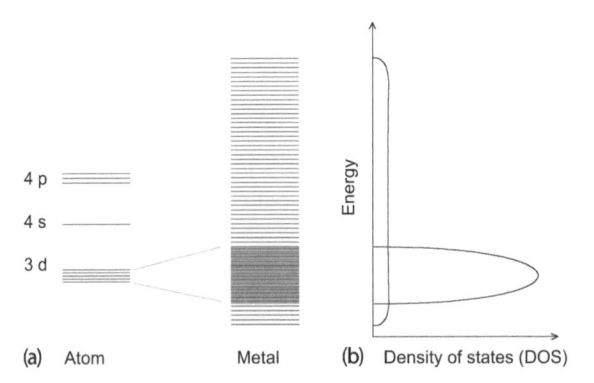

Figure 6.10 Schematic representation of the energy levels of a typical 3d transition metal. The extended s and p orbitals forms the broad sp band. The more localized d orbitals lead to a narrow d band.

ized, that is, not restricted to specific atoms. As a result they form an almost free-electron gas that spreads out over the entire metal. Hence, the atomic sp electron wave functions overlap to a great extent, and consequently the band they form is much broader (Figure 6.10).

Knowledge of the behavior of d orbitals is essential to understand differences and trends in reactivity of the transition metals. The width of the d band decreases as the band is filled when going to the right in the periodic table since the molecular orbitals become ever more localized and the overlap decreases. Eventually, as in copper, the d band is completely filled due to its location below the Fermi level. In zinc, however, it lowers even further in energy, thus, reaching a so-called core level, localized on the individual atoms. If we look through the transition metal series 3d, 4d, and 5d, we see that the d band broadens since the orbitals become ever larger and, therefore, the overlap increases.

Thus, a band is simply a collection of many molecular orbitals. The lower levels in the band are bonding, the upper ones antibonding, and the ones in the middle nonbonding. Metals with a half-filled band, as we shall see later, have the highest cohesive energy, the highest melting points, and the highest surface free energies.

Each energy level in the band is called a state. The important quantity to look at is the density of states (DOS), that is, the number of states at a given energy. The DOS of transition metals are often depicted as the smooth curves (Figure 6.10), but in reality DOS curves show complicated structure, due to crystal structure and symmetry. The bands are filled with valence electrons of the atoms up to the Fermi level. In a molecule, one would call this level the highest occupied molecular orbital or HOMO.

6.3.2.1 Work Function

The work function is the minimum energy needed to remove an electron from a solid and take it infinitely far away at zero potential energy. The weakest bound electrons in a solid are the electrons at the Fermi level. At this level, all sp and d bands are filled until all the valence electrons are used and this level is called the Fermi level. The vacuum level is the minimum energy that an electron has that is not bound, but free to move in any direction. Thus, the work function equals the energy difference between the vacuum level and the Fermi level (Figure 6.11). In chemical language, it is similar to the ionization potential – the energy needed to remove an electron from the highest occupied molecular orbital (HOMO).

6.3.2.2 The Free-Electron Gas and the Jellium Model

The simplest approach is to describe the valence electrons in the solid as a free noninteracting electron gas in a box with the volume V, as we did in Chapter 3. We have to find the ground state for the Schrödinger equation

$$[H - \varepsilon_k] \psi_k = 0 \tag{6.26}$$

where the wave functions ψ_k are plane waves

$$\psi_k = \frac{1}{V} e^{ikr} \tag{6.27}$$

6 Surface Reactivity

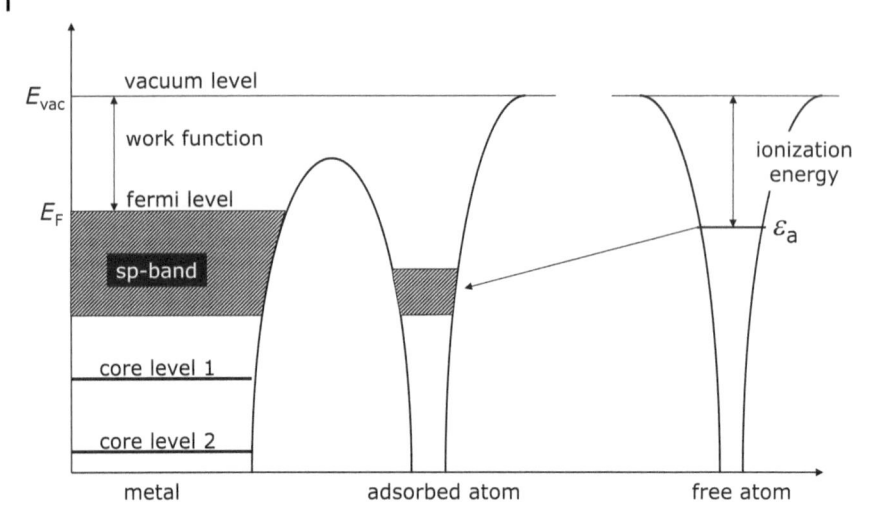

Figure 6.11 Schematic energy diagram of an atom approaching a free-electron metal with a sp band. Notice that the vacuum level is the common reference point and that the ionization energy of an atom is similar to the work function of a solid.

The electrons must fulfill the Pauli principle and the wave function must be zero at the walls of the box with length $L = V^{-\frac{1}{3}}$ resulting in the following values of the **k** vector:

$$\mathbf{k} = \frac{\pi}{L}(n_x, n_y, n_z) \tag{6.28}$$

The corresponding energy eigen values to this problem are

$$\varepsilon_k = \frac{\hbar^2 k^2}{2m} \tag{6.29}$$

where m is the mass of the electron. Since the number of electrons present, N, will typically be of the order 10^{23}, k is continuous, and the electron wave vectors occupy a sphere in k space with radius k_F and volume $V_F = \frac{4\pi}{3} k_F^3$. The number of quantum states is then the volume divided by the volume of each state, leading to

$$\frac{V_F}{\left(\frac{2\pi}{L}\right)^3} = \frac{k_F^3 V}{6\pi} = \frac{N}{2} \tag{6.30}$$

since there can be two electrons in each state (spin up and down). At $T = 0\,\text{K}$, all states below k_F are occupied and for higher temperatures the electrons are distributed according to the Fermi distribution:

$$f(\varepsilon)\,d\varepsilon = \frac{1}{e^{\frac{\varepsilon-\mu}{k_B T}} + 1} \tag{6.31}$$

where μ is the chemical potential of the electrons, which at $T = 0\,\text{K}$ equals $\mu_{T=0} = \varepsilon_F$.

Normally, when describing occupancies of states, we expect the Boltzman distribution. However, this only occurs for cases where the number of available states is much higher than the number of occupied states, that is, cases of high temperature or low density. When describing the occupancy of the electrons above, we have to take into account that the number of states is small compared to the number of electrons for normal temperatures. This implies that we need to consider the fundamental statistics for fermions and bosons. (In the case of fermions the N-body wave function will be antisymmetric upon interchange of two identical particles, whereas in the case of bosons it will be symmetrical.) Examples of fermions are electrons, protons, and their respective antiparticles, whereas photons, phonons, and plasmons are bosons. All known particles fall into these two classes. The important difference between these two sets of particles is that there is no restriction of the distribution of bosons on the available energy states, whereas no identical fermions can occupy the same single particle energy state (Pauli repulsion). For the interesting consequences of this fundamental difference, we refer to McQuarrie [8]. Note how the Fermi distribution for the electrons approaches the Boltzman distribution for high temperatures in Figure 6.14 below.

To find the density of states per energy $n(\varepsilon)$, we start from

$$\frac{N}{V} = 2\int_0^\infty \frac{f(\varepsilon(k))}{\left(\frac{8\pi^3}{V}\right)} dV_k = 2V\int_0^\infty \frac{f(\varepsilon(k))4\pi k^2}{8\pi^3} dk \tag{6.32}$$

and using

$$\frac{d\varepsilon}{dk} = \frac{2\hbar^2 k}{2m} = \sqrt{\frac{2\hbar^2 \varepsilon}{m}} \tag{6.33}$$

leads to

$$\frac{N}{V} = 2V\int_0^\infty f(\varepsilon)\frac{2m\varepsilon}{2\pi^2\hbar^2}\sqrt{\frac{m}{2\hbar^2\varepsilon}} d\varepsilon = V\int_0^\infty f(\varepsilon)\frac{m}{\pi^2\hbar^2}\sqrt{\frac{2m\varepsilon}{\hbar^2}} d\varepsilon \tag{6.34}$$

which can also be written in a form that contains the desired quantity:

$$\frac{N}{V} = \int_0^\infty f(\varepsilon(k))n(\varepsilon) d\varepsilon \tag{6.35}$$

finally resulting in

$$n(\varepsilon) = \frac{Vm}{\pi^2\hbar^2}\sqrt{\frac{2m\varepsilon}{\hbar^2}} = \frac{V}{\pi^2}\left(\frac{m}{\hbar^2}\right)^{\frac{3}{2}}\sqrt{2\varepsilon} \propto \sqrt{\varepsilon} \tag{6.36}$$

Thus, the resulting density of states (DOS) per energy is proportional to the square root of the energy. Although the free-electron gas represents a very simple model,

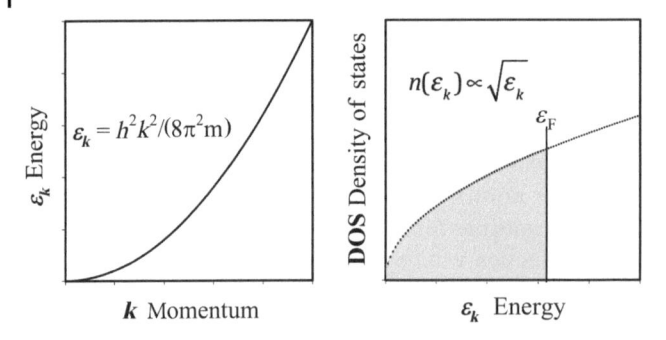

Figure 6.12 Energy as a function of the reciprocal wave vector and the density of states for a free-electron gas.

it nevertheless describes, to a good approximation, the situation for a number of the so-called free-electron metals like Na, K, Cs, Ca, Ba, Mg, and to some extent even Al. Here, the s and p valence electrons are smeared out on the positive background given by the nuclei and can be described as a free-electron gas filled to the Fermi level. The energy of the electrons as a function of the wave vector and the density of states as a function of energy are shown in Figure 6.12.

The hypothetical metal "jellium" consists of an ordered array of positively charged metal ions surrounded by a structureless sea of electrons that behaves as a free-electron gas (Figure 6.13).

The attractive potential due to the positively charged cores is not strong enough to keep the valence electrons inside the metal. As a result, the electrons spill out

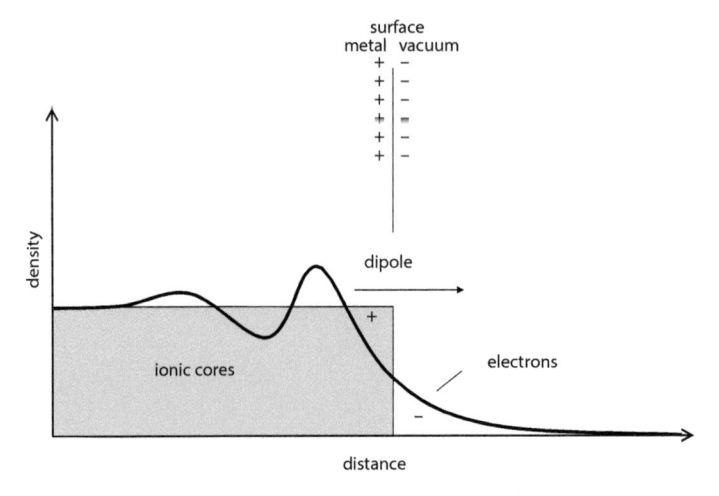

Figure 6.13 The electron distribution in the model metal jellium gives rise to an electric double layer at the surface, which forms the origin of the surface contribution to the work function. The electron wave function reaches out into the vacuum, but diminishes exponentially, as electrons are classically forbidden in this region. Nevertheless, these are the wave functions that are probed in scanning tunneling microscopy (STM).

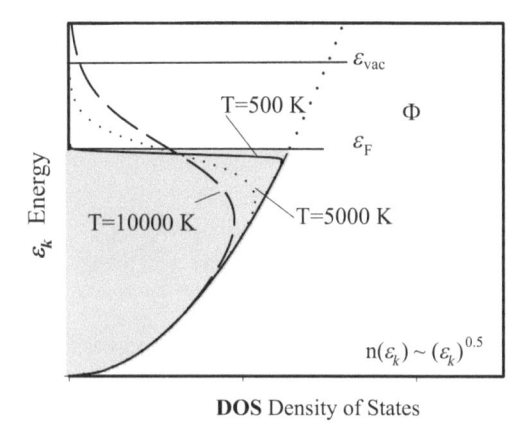

Figure 6.14 DOS as a function of temperature. Note how the electrons occupy the higher energy states as the temperature increases.

into the vacuum, that is, the electron density just outside the surface is not zero. Because the charge of these electrons is not compensated by positive ions, a dipole layer exists at the surface, with the negative end to the outside. An electron traveling from the solid to the vacuum must overcome this barrier of height Φ.

The energy needed to surmount the surface dipole layer is the surface contribution to the work function. It depends very much on the structure of the surface: for the fcc metal, the (111) surface is the most densely packed surface and has the largest work function because the dipole barrier is high. A more open surface such as fcc (110) has a smaller work function. In addition, when a surface contains many defects, the work function is lower than for the perfect surface. For a given surface structure, the work function increases from left to right in the periodic system.

The work function, Φ, is defined as the work required to remove an electron infinitely far away from the surface. Typical values are between 2.5 and 5.5 eV. The electron is free to travel through space when it has been excited to at least the vacuum level, ε_{vac}, as illustrated in Figure 6.14.

From Figure 6.14 it becomes clear why one must heat filaments to very high temperatures to see electron emission in electron guns. Only the part of the electron distribution that has obtained energies above Φ can be utilized. The occupation number at the vacuum level can be approximated by $e^{-\frac{\Phi}{k_B T}}$ leading to the well-known Richardson–Dushman formula, which describes the flux j of electrons evaporating from a surface with work function Φ at temperature T:

$$j = \frac{4\pi m e}{h^3} e^{-\frac{\Phi}{k_B T}} \qquad (6.37)$$

Filaments are usually refractory metals such as tungsten or iridium, which can sustain high temperatures for a long time ($T > 3000$ K). The lifetime of filaments for electron sources can be prolonged substantially if an adsorbate can be introduced that lowers the work function on the surface so it may be operated at lower

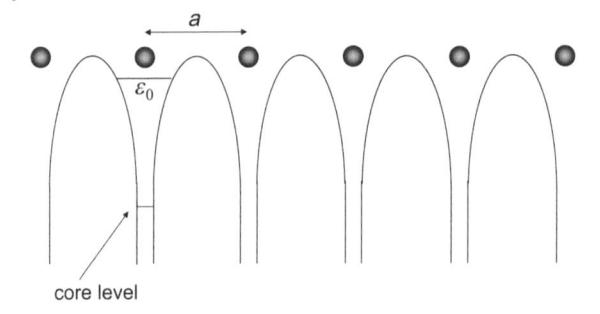

Figure 6.15 Potential energy of a one-dimensional solid with lattice distance a between the atoms.

temperature. Thorium fulfills this function by being partly ionized, donating electrons to the filament, which results in a dipole layer that reduces the work function of the tungsten. In catalysis, alkali metals are used to modify the effect of the work function of metals, as we will see later.

6.3.2.3 Tight Binding Model

To describe the band structure of metals, we use the same approach as employed above to describe the bonding in molecules. First, we consider a chain of two atoms. The result is the same as obtained for a homonuclear diatomic molecule; we find two energy levels, the lower one bonding and the upper one antibonding. Now let us add additional atoms. Upon adding additional atoms, we obtain an additional energy level per added electron, until a continuous band arises (Figure 6.9). To describe the electron band of a metal in a more quantitative manner, we have to start with the Hamiltonian of such a system. For simplicity, we will restrict ourselves to one dimension.

Each atom in the chain has an outer electron at an energy ε_0 and a number of deeper lying core levels, which we neglect in this approximation (Figure 6.15).

The total potential of the chain of atoms with mutual distances a can be described as the sum of all the individual potentials from the atom as

$$V(x) = \sum_n V(x - na) \tag{6.38}$$

resulting in a periodic potential that follows the periodicity of the lattice.

The energy eigen values of the isolated atom follow from

$$H_{atom}\psi = \left[-\frac{1}{2}\nabla^2 + V_{atom}\right]\psi = \varepsilon_0 \psi \tag{6.39}$$

The total wave function of the chain is now constructed by forming a Linear Combination of the wave functions of the individual Atomic Orbitals as

$$\Psi = \sum_n c_n \psi(x - na) \tag{6.40}$$

giving the name LCAO approximation to this approach. The Hamiltonian for the chain is approximated by the atomic Hamiltonian plus a potential accounting for the overlap between the orbitals of adjacent atoms:

$$H = H_{atom} + \Delta V(x) \tag{6.41}$$

and, hence, the wave function ψ from Eq. (6.39) is a solution when $\Delta V(x) = 0$. The total wave function Ψ is constructed to obey Blochs theorem, which states that the wave function for a periodic potential must be the product of a plane wave and a wave that has the same periodicity as the potential. The density of the electron states, $n(x)$, follows from the intensity of the wave function as

$$n(x) = |\Psi|^2 = \sum_n c_n^2 |\psi(x - na)|^2 \tag{6.42}$$

Neglecting edge effects at the end of the chain, we have $n(x) = n(x + \ell a)$, implying that $c_n^2 = c_{n+l}^2$ for any integer value of ℓ. By choosing $c_n = \frac{1}{N} e^{ikna}$ (with $i = \sqrt{-1}$), we find the desired Bloch wave:

$$\Psi = \frac{1}{N} \sum_n e^{ikna} \psi(x - na) \tag{6.43}$$

The secular determinant can now be set up and results in a tridiagonal determinant since we only have nonzero matrix elements in the diagonal and the two neighboring elements. Hence,

$$\begin{aligned} \int \psi_l \psi_m &= \delta_{l,m} \\ \int \psi_l H_{atom} \psi_m &= \varepsilon_0 \delta_{l,m} \\ \int \psi_l \Delta V(x) \psi_m &= -\alpha \delta_{l,m} \\ \int \psi_l \Delta V(x) \psi_m &= -\beta \delta_{l,m\pm 1} \end{aligned} \tag{6.44}$$

where $\delta_{l,m}$ is the Kronecker delta ($\delta_{l,m} = 1$ for $l = m$; $\delta_{l,m} = 0$ if $l \neq m$). This leads to the following solution for the kth level:

$$E_k = \varepsilon_0 - \alpha - \beta(e^{ika} + e^{-ika}) = \varepsilon_0 - \alpha - 2\beta \cos(ka) \tag{6.45}$$

representing a continuous band of states with a width of 4β, as shown in Figure 6.16.

The important lesson is that the valence electrons in our one-dimensional chain of atoms occupy a continuous band of states having the bonding states at the bottom and the antibonding states at the top and a bandwidth that is 4× the overlap integral to the neighbor sites. We have given the one-dimensional case, but it can easily be generalized into three dimensions. The width will then be proportional to the overlap integral multiplied by the number of nearest neighbors. This has the interesting consequence that a band becomes narrower at the surface of the metal where a number of atoms are missing neighbors. This band narrowing for

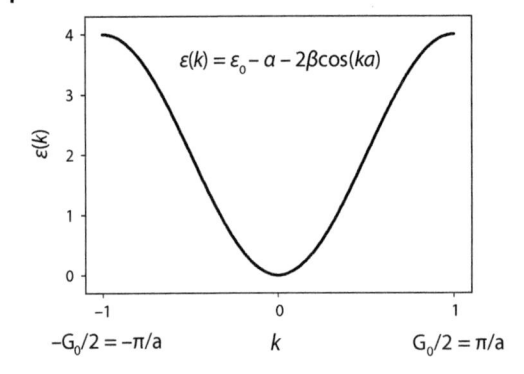

Figure 6.16 Band structure with bonding orbitals at the bottom and antibonding at the top. The band splitting (maximum difference between bonding and antibonding states) equals 4β. The band is only shown in the reduced zone, that is, corresponding to one reciprocal lattice constant.

surface atoms has indeed been observed experimentally and we will use this fact later to understand how reactivity changes with coordination and with strain or compression in surface overlayers.

The description derived above gives useful insight into the general characteristics of the band structure in solids. In reality, band structure is far more complex than suggested by Figure 6.16, as a result of the inclusion of three dimensions, and due to the presence of many types of orbitals that form bands. The detailed electronic structure determines the physical and chemical properties of the solids, in particular whether a solid is a conductor, semiconductor, or insulator (Figure 6.17).

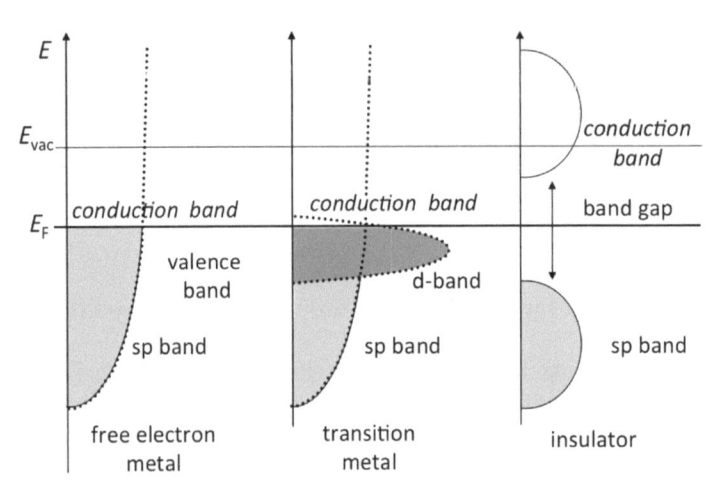

Figure 6.17 Figure 6.17. DOS diagrams schematically showing the electron density around the Fermi level for a free-electron metal, a transition metal, and an insulator.

For a material to be a good conductor, it must be possible to excite an electron from the valence band (the states below the Fermi level) to the conduction band (an empty state above the Fermi level) in which it can move freely through the solid. The Pauli principle forbids this in a state below the Fermi level, where all states are occupied. In the free-electron metal of Figure 6.14, there will be plenty of electrons in the conduction band at any nonzero temperature – just as there will be holes in the valence band – that can undertake the transport necessary for conduction. This is the case for metals such as sodium, potassium, calcium, magnesium, and aluminum.

The transition metals are also good conductors as they have a similar sp band as the free-electron metals, plus a partially filled d band. The Group IB metals, that is, copper, silver and gold, represent borderline cases, as the d band is filled and located a few eV below the Fermi level. Their sp band, however, ensures that these metals are good conductors.

However, in oxides (e.g., Al_2O_3), the sp band of the aluminum hybridizes with the p orbitals of the oxygen forming a new band below the Fermi level which leaves a gap of 7 eV to the antibonding part of the band. The lower part is the valence band, the upper part the conduction band, and the separation between them is the band gap. This material is an insulator, as it will be hard for an electron to become excited to the conduction band so that it can move through the oxides. The band gap of insulators is typically above 5 eV (e.g., 5.4 eV for diamond).

Semiconductors have a considerably smaller band gap (e.g., silicon: 1.17 eV). Their conductivity, which is zero at low temperature but increases to appreciable values at higher temperature, depends greatly on the presence of impurities or, if added advertently, dopants. This makes it possible to manipulate the band gap and tune the properties of semiconductors for applications in electronic devices [5, 6].

In particular, oxides such as MgO, Al_2O_3, and SiO_2 are very good insulators and since the electrons are taking part in chemical bonding and are moved down in energy, away from the Fermi level, they are also chemically inactive. Only if defects or impurities that furnish unpaired electrons are present will they tend to show some activity. In the following, we will concentrate on the more reactive metals.

6.3.2.4 Simple Model of a Transition Metal

Here, we try to gain insight in the trends in reactivity of the metals without getting lost in too much detail. Therefore, we invoke rather crude approximations. The electronic structure of many metals shows numerous similarities with respect to the sp band, with the metals behaving as a free-electron metals. Variations in properties are due to the extent of filling of the d band. We completely neglect the lanthanides and actinides where a localized f orbital is filled, as these metals hardly play a role in catalysis.

Our representation of a metal is shown in Figure 6.18. It possesses a block-shaped, partly filled sp band behaving as a free electron gas and a d band that is filled to a certain degree. The sp band is broad as it consists of highly delocalized electrons smeared out over the entire lattice. In contrast, the d band is much

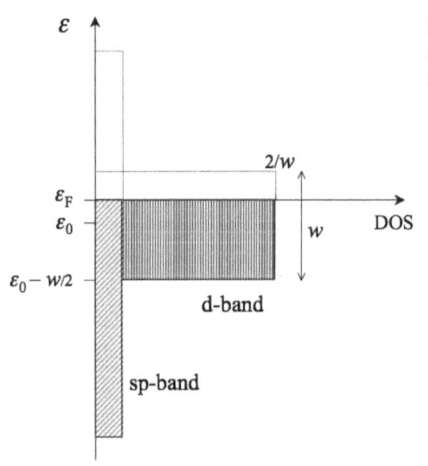

Figure 6.18 Density of states of a transition metal with a nearly filled d band on top of a partly filled sp band.

narrower because the overlap between d states, which are more localized on the atoms, is much smaller.

If we consider the energy gained by forming the metal from the individual atoms, the sp band gives a contribution of approximately 5 eV for all metals. The variation in bonding energy across the transition metals is due to the d band. We will look at its properties and contribution to bonding in more detail.

We approximate the d band by a narrow rectangular box, as shown in Figure 6.18, where

$$n(\varepsilon) = \frac{2}{w} \quad \text{for} \quad \varepsilon_0 - \frac{w}{2} < \varepsilon < \varepsilon_0 + \frac{w}{2} \vee \varepsilon_F$$
$$n(\varepsilon) = 0 \quad \text{for} \quad \varepsilon < \varepsilon_0 - \frac{w}{2} \vee \varepsilon > \varepsilon_0 + \frac{w}{2}$$
(6.46)

ε_0 is the center of the d band, $n(\varepsilon) = \frac{2}{w}$ is the density of states (DOS), and w is the width of the d band due to the overlap. A degree of filling f, a fraction between 0 and 1, is defined as

$$f \equiv \frac{1}{2} \int_{\varepsilon_0 - \frac{w}{2}}^{\varepsilon_F} n(\varepsilon) \, d\varepsilon = \frac{\varepsilon_F - \varepsilon_0 + \frac{w}{2}}{w}$$
(6.47)

The energy gained by the hybridization of the d orbitals is

$$\Delta E_{\text{hyp}} = \int_{\varepsilon_0 - \frac{w}{2}}^{\varepsilon_F} n(\varepsilon)(\varepsilon - \varepsilon_0) \, d\varepsilon = \frac{(\varepsilon_F - \varepsilon_0)^2}{w} - \frac{w}{4}$$
(6.48)

Introducing the definition of f, we find

$$\Delta E_{\text{hyp}} = -f(1 - f)w$$
(6.49)

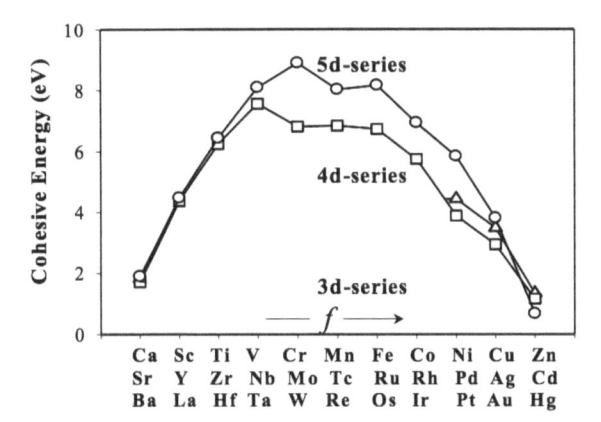

Figure 6.19 Cohesive energy for the three groups of transition metals. Note that the maximum cohesive energy of the 4d and 5d metals occurs when the d band is approximately half full, as predicted by Eq. (6.49). The deviations of the 3d series are largely caused by the magnetic behavior of these elements (adapted from Kittel [5]).

Remembering that there is always a repulsive term which is proportional to the overlap of the orbitals and the number of participating neighbors, we obtain

$$\Delta E = E_0 N_{nn} S |\beta| + \Delta E_{hyp} = E_0 N_{nn} S |\beta| - f(1-f)w \quad (6.50)$$

which shows that the cohesive energy of the d metals behaves parabolically in terms of the filling degree f.

The cohesive energy is the energy required to remove one atom from the solid. Alternatively, one can say that it is the energy gained per atom by forming the metal from the atoms. A high cohesive energy corresponds to a high melting point. Such metals are sometimes referred to as being refractory.

Despite neglecting the variations in the overlap term, while magnetic properties also cause deviations, Figure 6.19 shows that the approximation works reasonably well. The trends in Figure 6.19 are easily understood if one realizes that in d bands that are less than 50% filled only bonding orbitals are occupied, while antibonding orbitals become filled for $f > 0.5$. Thus, we expect to find the highest cohesive (bonding energy) energy for transition metals with a half filled d band.

Another interesting feature explained by Eqs. (6.49) or (6.50) is the increase in cohesive energy in going from 3d to 4d to 5d metals (Figure 6.19). As we go down in the periodic system, the orbitals become larger and the overlap increases. This implies that the band becomes broader, leading to larger value of w in Eqs. (6.49) and (6.50).

After these exercises in solid-state physics we continue to pursue our original goal, namely to describe the interaction between an adsorbate and surface.

6.4
Chemisorption

The previous sections have set the stage for describing the essentials of what happens when a molecule approaches the surface of a metal. The most important features of chemisorption are well captured by the Newns–Anderson model, which we describe in Section 6.4.1 [9, 10]. Readers who are not particularly fond of quantum mechanics and its somewhat involved use of mathematics, but merely want to learn the outcome of this model, may skip this section and go directly to Section 6.4.2, where we present a summary in qualitative terms. The same readers may also want to consult Roald Hoffmann's *Solids and Surfaces*, a book that we warmly recommend [11].

6.4.1
The Newns–Anderson Model

Initially, we consider a simple atom with one valence electron of energy ε_a and wave function Φ_a which adsorbs on a solid in which the electrons occupy a set of continuous states Ψ_k with energies ε_k. When the adsorbate approaches the surface, we need to describe the complete system by a Hamiltonian H, including both systems and their interaction. The latter comes into play through matrix elements of the form $V_{ak} = \int \Phi_a H \Psi_k$. We assume that the solutions Ψ_i to this eigen value problem can be described as linear combinations of the isolated wave functions:

$$\Psi_i = c_{ai}\Phi_a + \sum_k c_{ik}\Psi_k \tag{6.51}$$

In the following we will also assume that the basis set is orthogonal, that is, matrix elements of the type $\int \Phi_a \Psi_k$ vanish. The solution is found from the eigen value problem

$$H\Psi_i = \varepsilon_i \Psi_i \tag{6.52}$$

where the matrix elements have the form

$$\begin{aligned} \int \Phi_a H \Phi_a &= H_{aa} = \varepsilon_a \\ \int \Psi_k H \Psi_k &= H_{kk} = \varepsilon_k \\ \int \Phi_a H \Psi_k &= V_{ak} = V_{ka} \end{aligned} \tag{6.53}$$

6.4 Chemisorption

This results in a determinant of the form

$$\begin{vmatrix} H_{aa} - \varepsilon & V_{a1} & V_{a2} & V_{a3} & \cdots & \cdots & V_{an} \\ V_{1a} & H_{11} - \varepsilon & 0 & 0 & \cdots & \cdots & 0 \\ V_{2a} & 0 & H_{22} - \varepsilon & 0 & \cdots & \cdots & 0 \\ V_{3a} & 0 & 0 & H_{33} - \varepsilon & \cdots & \cdots & 0 \\ \cdots & \cdots & \cdots & \cdots & \cdots & \cdots & \cdots \\ \cdots & \cdots & \cdots & \cdots & \cdots & \cdots & \cdots \\ V_{na} & 0 & 0 & 0 & \cdots & \cdots & H_{nn} - \varepsilon \end{vmatrix} = 0 \quad (6.54)$$

As we already have seen there are infinitely many k states ($n \to \infty$) in the metal and it is rather problematic to keep track of all of them. It is better to consider the projection of the new states Ψ_i onto the original adsorbate state Φ_a. This maps out the development of the adsorbates state as the atom approaches the surface and is technically carried out by determining the quantity

$$n_a(\varepsilon) = \sum_i \left| \int \Psi_i \Phi_a \right|^2 \delta(\varepsilon_i - \varepsilon_a) \quad (6.55)$$

in which the summation runs over all eigen functions Ψ_i. In order to proceed, we need to introduce a Green's function, which allows us to reformulate Eq. (6.55) as

$$n_a(\varepsilon) = \frac{-1}{\pi}\bigg|_{\delta \to 0^+} \mathrm{Im}\left(\sum_i \frac{\int \Psi_i \Phi_a \int \Phi_a \Psi_i}{\varepsilon - \varepsilon_i + i\delta} \right) \quad (6.56)$$

by utilizing that a Lorentzian

$$f(\varepsilon) = \frac{\delta}{\pi\left((\varepsilon - \varepsilon_i)^2 + \delta^2\right)} \quad (6.57)$$

becomes a delta function for $\delta \to 0^+$, that is,

$$\int_{-\infty}^{+\infty} f(\varepsilon) \, d\varepsilon = \int_{-\infty}^{+\infty} \frac{\delta}{\pi\left((\varepsilon - \varepsilon_i)^2 + \delta^2\right)} \, d\varepsilon = 1 \quad (6.58)$$

$f(\varepsilon) \to \infty$ for $\varepsilon = \varepsilon_i$ and $f(\varepsilon) = 0$ for $\varepsilon \neq \varepsilon_i$

The conversion takes place as

$$\big|_{\delta \to 0^+} \mathrm{Im}\left(\frac{1}{\varepsilon - \varepsilon_i + i\delta} \right) = \big|_{\delta \to 0^+} \mathrm{Im}\left(\frac{\varepsilon - \varepsilon_i}{(\varepsilon - \varepsilon_i)^2 + \delta^2} - \frac{i\delta}{(\varepsilon - \varepsilon_i)^2 + \delta^2} \right)$$

$$= \big|_{\delta \to 0^+} \frac{\pi \delta}{\pi \left((\varepsilon - \varepsilon_i)^2 + \delta^2\right)} = \pi \delta(\varepsilon - \varepsilon_i)$$

$$(6.59)$$

6 Surface Reactivity

The so-called one-particle Green's function is now introduced as

$$n_a(\varepsilon) = \frac{-1}{\pi}\bigg|_{\delta \to 0^+} \mathrm{Im}\left(\sum_i \frac{\int \Psi_i \Phi_a \int \Phi_a \Psi_i}{\varepsilon - \varepsilon_i + i\delta}\right) \equiv \frac{-1}{\pi}\mathrm{Im}\left(G_{aa}(\varepsilon)\right) \quad (6.60)$$

which can then be written as

$$G(\varepsilon)^{-1} = \sum_i \frac{\int \Psi_i \int \Phi_a}{\varepsilon - \varepsilon_i + i\delta} \quad (6.61)$$

where $G(\varepsilon)^{-1}$ is the inverse matrix to $\varepsilon - H + i\delta$, which is actually the formal definition of $G(\varepsilon)^{-1}$, that is,

$$(\varepsilon - \hat{H} + i\delta)G(\varepsilon)^{-1} = I \quad (6.62)$$

The latter can easily be proven:

$$(\varepsilon - \hat{H} + i\delta)G(\varepsilon)^{-1} = \sum_i (\varepsilon - H + i\delta)\left(\frac{\int \Psi_i \int \Psi_i}{\varepsilon - \varepsilon_i + i\delta}\right)$$

$$= \sum_i \left(\frac{(\varepsilon - \varepsilon_i + i\delta)\int \Psi_i \int \Psi_i}{\varepsilon - \varepsilon_i + i\delta}\right) = \sum_i \int \Psi_i \Psi_i = I \quad (6.63)$$

And, hence, the Green's function is in principle defined as an inverse matrix

$$G(\varepsilon)^{-1} = (\varepsilon - \hat{H} + i\delta)^{-1} \quad (6.64)$$

To proceed, we write the product in the left-hand of Eq. (6.62) in matrix form and set it equal to the unit matrix

$$\begin{pmatrix} \varepsilon' - \varepsilon_a & -V_{a1} & \cdots & -V_{an} \\ -V_{1a} & \varepsilon' - H_{11} & \cdots & 0 \\ \cdots & \cdots & \cdots & \cdots \\ -V_{na} & 0 & \cdots & \varepsilon' - H_{nn} \end{pmatrix} \begin{pmatrix} G_{aa} & G_{a1} & \cdots & G_{an} \\ G_{1a} & G_{11} & \cdots & G_{1n} \\ \cdots & \cdots & \cdots & \cdots \\ G_{na} & G_{n1} & \cdots & G_{nn} \end{pmatrix}$$

$$= \begin{pmatrix} 1 & 0 & \cdots & 0 \\ 0 & 1 & \cdots & 0 \\ \cdots & \cdots & \cdots & \cdots \\ 0 & 0 & 0 & 1 \end{pmatrix} = I \quad (6.65)$$

where $\varepsilon' = \varepsilon + i\delta$. This immediately gives us the following relations involving G_{aa}:

$$G_{aa}(\varepsilon - \varepsilon_a + i\delta) - \sum_k V_{ak}G_{ka} = 1$$

$$-V_{ka}G_{aa} + (\varepsilon - \varepsilon_k + i\delta)G_{ka} = 0 \quad (6.66)$$

allowing us to solve for G_{ka}:

$$G_{ka} = \frac{-V_{ka} G_{aa}}{\varepsilon - \varepsilon_k + i\delta} \tag{6.67}$$

By substituting this into Eq. (6.66), G_{ka} can be eliminated, leading to

$$G_{aa} = \frac{1}{\varepsilon - \varepsilon_a + i\delta - \sum_k \frac{V_{ak}^2}{\varepsilon - \varepsilon_k + i\delta}} \tag{6.68}$$

Letting $\delta \to 0^+$ and introducing a complex function

$$q(\varepsilon) \equiv \Lambda(\varepsilon) - i\Delta(\varepsilon) \equiv \sum_k \frac{V_{ak}^2}{\varepsilon - \varepsilon_k + i\delta} \tag{6.69}$$

and using the fact that G_{aa} can be written as

$$G_{aa} = \frac{1}{\varepsilon - \varepsilon_a - q(\varepsilon)}, \quad \delta \to 0^+ \tag{6.70}$$

we find an expression for $\Delta(\varepsilon)$:

$$\Delta(\varepsilon) = -\mathrm{Im}\left(\sum_k \frac{V_{ak}^2}{\varepsilon - \varepsilon_k + i\delta}\right)$$
$$\Delta(\varepsilon) = \pi \sum_k V_{ak}^2 \delta(\varepsilon - \varepsilon_k), \quad \delta \to 0^+ \tag{6.71}$$

This is basically a hopping matrix element between the adsorbate state labeled "a" and the metal state "k". Thus, $\Delta(\varepsilon)$ is the local projection of metal states on the state of the adsorbate. The real part $\Lambda(\varepsilon)$ is found by using the Kronig–Kramer transformation. By having the imaginary part of a complex function, the real part can be found by evaluating the principal value of the following integral

$$\Lambda(\varepsilon) = \frac{1}{\pi} \int_{-\infty}^{\infty} \frac{\Delta(x)}{x - \varepsilon} \, dx \tag{6.72}$$

Remember that the quantity of interest is $n_a(\varepsilon)$. Now it becomes evident why it was worth introducing the Greens function as

$$n_a(\varepsilon) = \frac{-1}{\pi} \mathrm{Im}\left(G_{aa}\right) = \frac{-1}{\pi} \mathrm{Im}\left(\frac{1}{\varepsilon - \varepsilon_a - \Lambda(\varepsilon) + i\Delta(\varepsilon)}\right)$$
$$= \frac{-1}{\pi} \mathrm{Im}\left(\frac{\varepsilon - \varepsilon_a - \Lambda(\varepsilon) - i\Delta(\varepsilon)}{(\varepsilon - \varepsilon_a - \Lambda(\varepsilon))^2 + \Delta(\varepsilon)^2}\right) \tag{6.73}$$
$$= \frac{1}{\pi} \frac{\Delta(\varepsilon)}{(\varepsilon - \varepsilon_a - \Lambda(\varepsilon))^2 + \Delta(\varepsilon)^2}$$

In other words, we have expressed the interaction between the adsorbate and the metal in terms of $\Delta(\varepsilon)$ and $\Lambda(\varepsilon)$, which essentially represent the overlap between the states of the metal and the adsorbate multiplied by a hopping matrix element; $\Lambda(\varepsilon)$ is the Kronig–Kramer transform of $\Delta(\varepsilon)$. Let us consider a few simple cases in which the results can be easily interpreted.

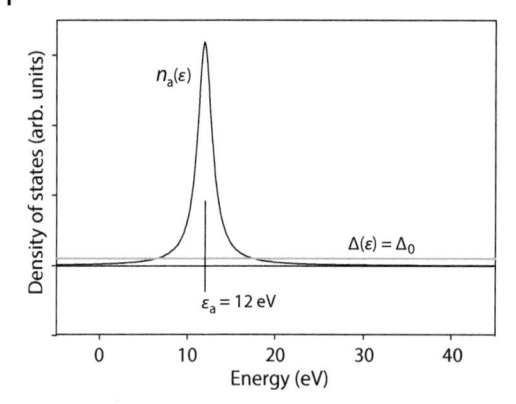

Figure 6.20 Solution to $n_a(\varepsilon)$ for an adsorbate level located arbitrarily at $\varepsilon_a = 12\,\text{eV}$ approaching a surface with a constant sp band $\Delta(\varepsilon) = \Delta_0$. The solution $n_a(\varepsilon)$ is a band of Lorentzian shape centered at the original energy of the adsorbate. The choice of the energy zero is the same as in the subsequent two figures, but is irrelevant in this case.

6.4.1.1 Case 1: Atom on a Metal of Constant Electron Density

Let us describe the solid as having a constant electron density for all energies. Of course, such metals do not exist, but the situation gives us the simplest case. The matrix element V_{ak} has also been assumed constant, meaning that $\Delta(\varepsilon)$ is just proportional to the electron density of the metal,

$$\Delta(\varepsilon) = \Delta_0 \tag{6.74}$$

From the Kronig–Kramer relation, it immediately follows that $\Lambda(\varepsilon) = 0$, as the function to be integrated is odd, and hence the resulting projected density of states becomes

$$n_a(\varepsilon) = \frac{1}{\pi} \frac{\Delta_0}{(\varepsilon - \varepsilon_a)^2 + \Delta_0^2} \tag{6.75}$$

reflecting that the adsorbate level is broadened into a band of Lorentzian line shape with width Δ_0. This means that the electrons hop back and forth between the metal and the adsorbate, leading to a broadening in energy as a result of the Heisenberg uncertainty in time and energy, that is, $\Delta \varepsilon \Delta t \geq \frac{\hbar}{2}$. The solution is shown in Figure 6.20.

6.4.1.2 Case 2: Atom on an sp Metal

We now consider a more realistic metal with a density of states as given by Eq. (6.36) and Figure 6.14. In Figure 6.21 we have placed the zero of the energy axis at the bottom of the valence band and plotted the density of states of the sp band for comparison. The weak variation of the metal density of state at the energies where the overlap with the atomic state occurs makes the interaction nearly constant, and hence $\Delta(\varepsilon)$ in this energy region has approximately the same shape as the sp band, that is, it is proportional to $\sqrt{\varepsilon}$. At higher energies, however, the interaction becomes weaker. This is understood if we consider the wave functions

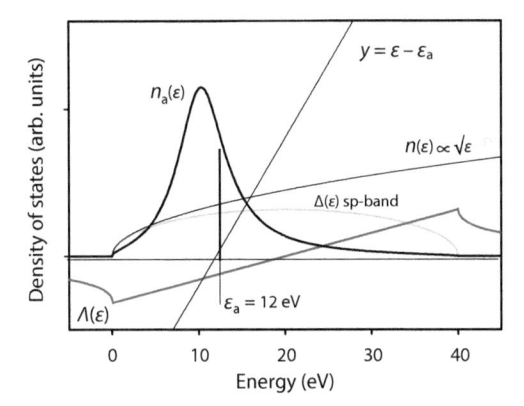

Figure 6.21 Projected density of states $n_a(\varepsilon)$ when an adsorbate level located at $\varepsilon_a = 12.0$ eV approaches a surface with an sp band. The function $\Delta(\varepsilon)$ follows the shape of an sp band a low energies, but decreases at higher energies due to a vanishing overlap. See text for further explanation.

for high-energy electrons. These have many nodes and, therefore, the overlap with the localized adsorption states eventually averages out to zero. The effect is taken into account by letting $\Delta(\varepsilon)$ go to zero at higher energies, as illustrated in Figure 6.21. To account for this behavior, $\Delta(\varepsilon)$ was arbitrarily approximated by an elliptic function.

We consider the same atom as in Case 1, with a valence electron at an orbital energy of $\varepsilon_a = 12.0$ eV above the bottom the sp band, when the atom is far away from the surface. This level is narrow, like a delta function. When approaching the surface, the adsorbate level broadens into a Lorentzian shape for the same reasons as described above, and falls in energy to a new position at 10.3 eV. From Eq. (6.73) for $n_a(\varepsilon)$, we see that the maximum occurs for $\varepsilon = \varepsilon_a + \Lambda(\varepsilon)$, that is, when the line described by $y = \varepsilon - \varepsilon_a$ crosses $\Lambda(\varepsilon)$, as shown in Figure 6.21. Thus, in this case the interaction between the metal and the adsorbate level results in a shift of the adsorbate level to lower energy (implying stronger bonding) and a broadening.

6.4.1.3 Case 3: Atom on a Transition Metal

Transition metals have both a broad sp band, leading to the interaction described for Case 2, and also a narrow d band, which interacts strongly with an adsorbate. The latter interaction is illustrated by including a strong contribution with the shape of the d band DOS to $\Delta(\varepsilon)$. There are now three solutions where $\varepsilon = \varepsilon_a + \Lambda(\varepsilon)$, that is, where the line $y = \varepsilon - \varepsilon_a$ crosses $\Lambda(\varepsilon)$ (Figure 6.22). At the lower energy, there is a down-shifted bonding state, in the middle a weak, nonbonding state, and at higher energy an upshifted antibonding state. Hence, the narrow d band almost functions as a molecular orbital itself towards the adsorbate: we find a bonding and an antibonding level for the adsorbate–metal complex. The filling of these orbitals depends, among other factors, on the extent to which the d band

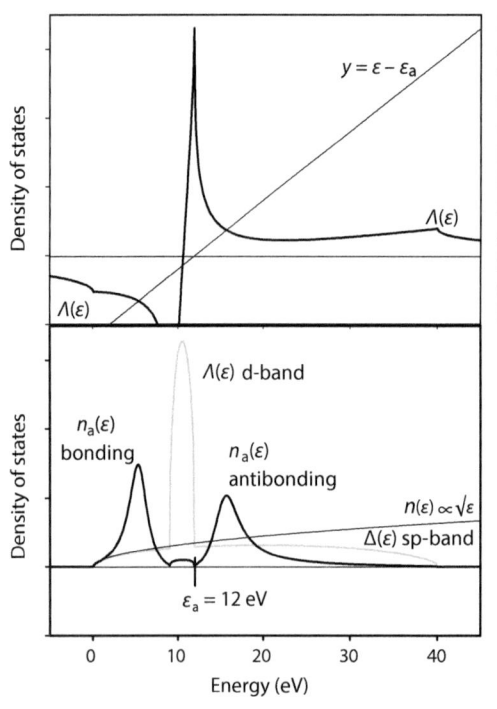

Figure 6.22 Adsorption of an atom on a d metal. The valence electron of the adsorbate, initially at 12 eV above the bottom of the metal band, interacts both weakly with a broad sp band and strongly with a narrow d band located between 9 and 12 eV. Note the significant splitting of the adsorbate density of states into bonding and antibonding orbitals of $n_a(\varepsilon)$ due to the interaction with the d band.

is filled. In the example of Figure 6.22, the Fermi level of the metal would fall somewhere between 9 and 12 eV.

6.4.2
Summary of the Newns–Anderson Approximation in Qualitative Terms

The Newns–Anderson approximation successfully accounts for the main features of bonding when an adsorbate approaches the surface of a metal and its wave functions interact with those of the metal. It can also be used to describe features of the dynamics in the scattering of ions, atoms, and molecules on surfaces. In particular, the neutralization of ions at surfaces is well understood in this framework. The subject is beyond the scope of this book and the reader is referred to the literature [1].

In this section we summarize the main results in simple and idealized schemes. We consider adsorption on a free-electron metal, and on a transition metal. In particularly the adsorption of a molecule on a metal with d states is of great interest for catalysis.

6.4.2.1 Adsorption on a Free-Electron Metal
A free-electron metal only possesses a broad sp band. Upon approach, the electron levels of the adsorbate broaden and shift down in energy, implying that the adsorbate becomes more stable when adsorbed on the metal. The interaction re-

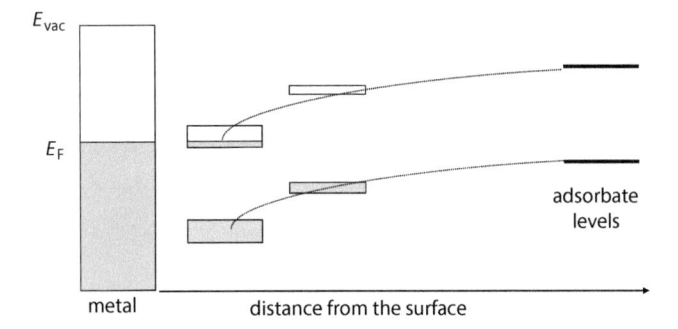

Figure 6.23 The energy levels of an adsorbate broaden and are lowered in energy when it approaches a free-electron metal with a broad sp band. Note that the initially empty upper electron level of the adsorbate becomes filled when the bonding interaction shifts it below the Fermi level of the metal.

sults in a bonding energy of typically 5 eV for atomic adsorbates on metals. The situation is illustrated in Figure 6.23.

6.4.2.2 Atomic Adsorption on a Transition or d Metal

The density of states of a transition or d metal consists of a broad sp band and a narrow but intense d band, both of which are partly filled (see, for example, Figure 6.17). In addition to the broadening and shifting of the adsorbate level, as in the case above, there is the strong interaction with the d band. In fact, the d band can be considered as a broad orbital itself, and its interaction with the adsorbate level leads to a pair of bonding and antibonding chemisorption orbitals, as indicated in Figure 6.24. The antibonding component will be occupied as far as it falls below the Fermi level of the metal, which weakens the adsorption bond somewhat.

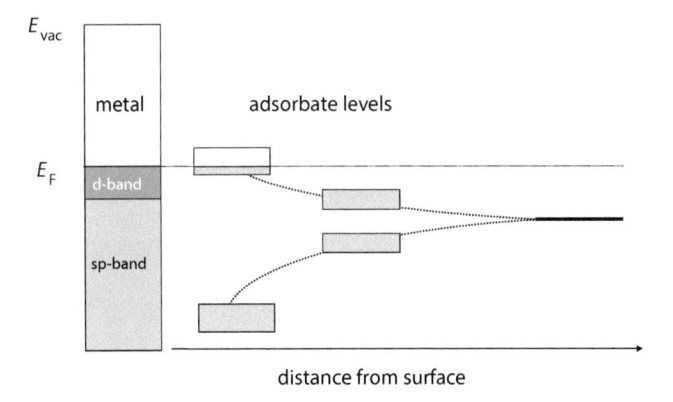

Figure 6.24 Interaction between an atomic adsorbate with one valence level and a transition metal, which possesses a broad sp band and a narrow d band located at the Fermi level. The strong interaction with the d band causes splitting of the adsorbate level into a bonding and an antibonding level. The part of the adsorbate levels below the Fermi level is occupied by electron density.

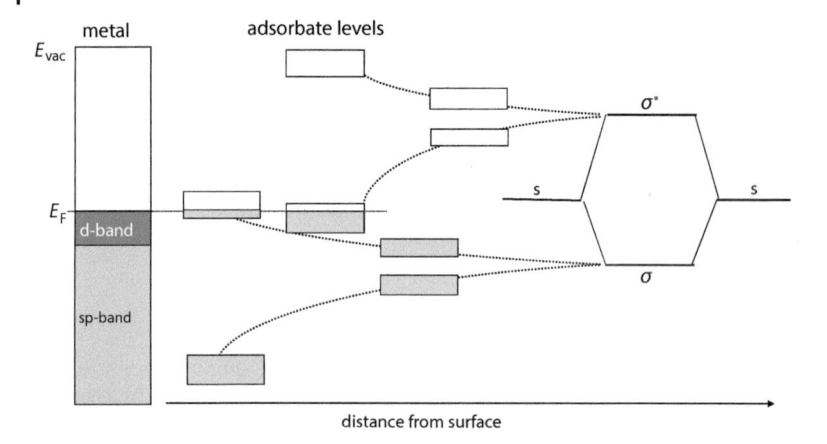

Figure 6.25 A molecule with a bonding σ and antibonding orbitals σ* interacts with both the sp band and the narrow d band of the transition metal. The former leads to the lowering and broadening of the bands, while the latter results in splitting into bonding and antibonding orbitals. Note that if electrons fill the antibonding orbital of the molecule then the internal bonding in the molecule becomes weaker, which may lead to dissociation of the molecule.

6.4.2.3 Adsorption of a Molecule on a Transition Metal

Finally, we look at the chemisorption of a molecule with a pair of bonding and antibonding orbitals on a transition metal (Figure 6.25). This situation can be simply visualized with H_2, for which the bonding orbital contains two electrons and the antibonding orbital is empty; other molecules can be examined in a similar way. In principle, we simply apply Section 6.4.2.2 twice, once to the bonding orbital and once to the antibonding orbital of the molecule. This has been done in Figure 6.25.

Interestingly, both the bonding and the antibonding orbitals of the molecule contribute to the chemisorption bond in Figure 6.25. It is very important to realize that the filling of the originally antibonding orbital of the molecule strengthens the interaction with the surface, but weakens the intramolecular bond of the adsorbed molecule! This is the key to understand how a surface dissociates molecules. One should note, however, that sufficient filling of the antibonding orbital for the molecule to dissociate may occur in a transition state, for example, when a molecule bends towards the surface to increase its overlap with the d states. Finally, such filling of the antibonding orbital of the metal by electron density from the metal is often called "back donation."

6.4.3
Electrostatic Effects in Atomic Adsorbates on Jellium

It is worth considering what sort of charge transfer adsorption may cause, since this may strongly influence the work function.

Consider an atom approaching the surface in Figure 6.23. If the upper level of the atom contained originally an electron, then upon adsorption it will transfer

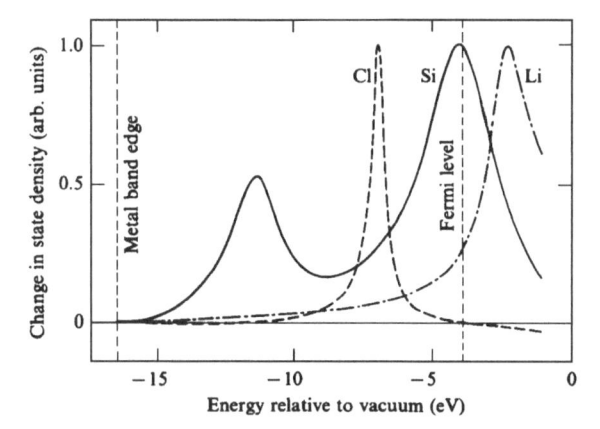

Figure 6.26 Density functional calculations show the change in the density of states induced by adsorption of Cl, Si, and Li on jellium. Lithium charges positively and chlorine negatively (reprinted with permission from Lang and Williams [12]; copyright 1978 by the American Physical Society).

part of this electron density to the metal and become positively charged. This is the case with alkali atoms. The atom forms a dipole with the positive end towards the outside, which counteracts the double layer that constitutes the surface contribution to the work function of the metal (Figure 6.13). Thus, alkali atoms reduce the work function of a metal surface simply because they all have a high lying s electron state that tends to donate charge to the metal surface.

Conversely, an atom in Figure 6.23 with an affinity level that initially is empty becomes partly occupied upon adsorption. Hence, charge is transferred from the metal to the atom. This sets up a dipole that increases the surface contribution to the work function. This is the case for adsorbed halides, which will be negatively charged at the surface. We will later see that such dipole fields can explain promotion and inhibition effects caused by various adsorbates in catalysis.

More detailed information can be obtained by calculating the charge contours by a procedure called density functional method. We will not explain how it works but merely give the results for the adsorption of Cl and Li on jellium, reported by Lang and Williams [12]. Figure 6.26 shows the change in density of states due to chemisorption of silicon, chlorine, and lithium atoms.

Note that the zero of energy has been chosen at the vacuum level and that all levels below the Fermi level are filled. For lithium, we are looking at the broadened 2s level with an ionization potential in the free atom of 5.4 eV. The density functional calculation tells that chemisorption has shifted this level above the Fermi level (this is in contradiction to the general picture presented above, but it does not change the overall conclusions) so that it is largely empty. Thus, lithium atoms on jellium are present as Li$^{\delta+}$, with δ almost equal to 1. Chemisorption of chlorine involves the initially unoccupied 3p level, which has a high electron affinity of 3.8 eV. This level has shifted down in energy upon adsorption and ended up below

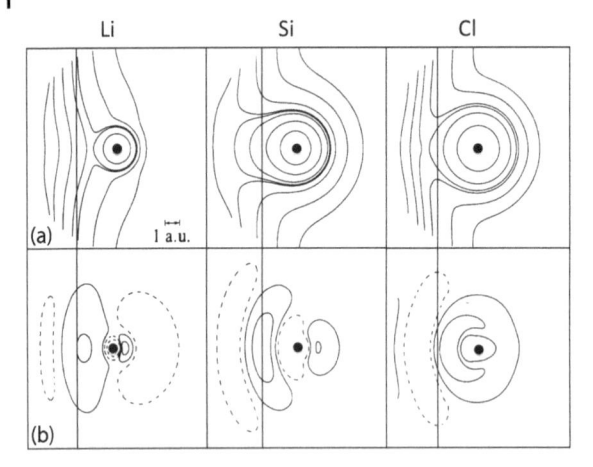

Figure 6.27 Charge density contours for the adsorption of Cl, Si, and Li on jellium. (a) Total charge and (b) induced charge; solid lines indicate an increase in electron density, dashed a decrease. Note the formation of strong dipoles for Li and Cl adsorption (reprinted with permission from Lang and Williams [12]; copyright 1978 by the American Physical Society).

the Fermi level, where it has become occupied. Hence, the charge on the chlorine atom is about 1. Silicon, however, remains largely neutral upon adsorption.

A map of the electron density distribution around these atoms provides important information. It tells us to what distance from the adatom the surface is perturbed or, in catalytic terms, how many adsorption sites are promoted or poisoned by the adatom. The charge density contours in Figure 6.27 are lines of constant electron density. Note that these contours follow the shape of the adsorbed atom closely and that the electrons are very much confined to the adsorbed atom and the adsorption site.

Even more interesting are the difference plots, indicating the difference between the charge contours of Figure 6.27a and the situation in which there would be no interaction between the adatom and the metal surface. Solid lines stand for an increase in electron density (excess negative charge), dashed lines for a decreased electron density (depletion of negative charge). These plots clearly illustrate the above-mentioned dipole effects and their influence on the work function.

6.5
Important Trends in Surface Reactivity

By combining the results of the Newns–Andersons model and the considerations from the tight binding model, it is now possible to explain a number of trends in surface reactivity. This has been done extensively by Nørskov and coworkers [1, 13]. We will discuss adsorption of atoms and molecules in separate sections.

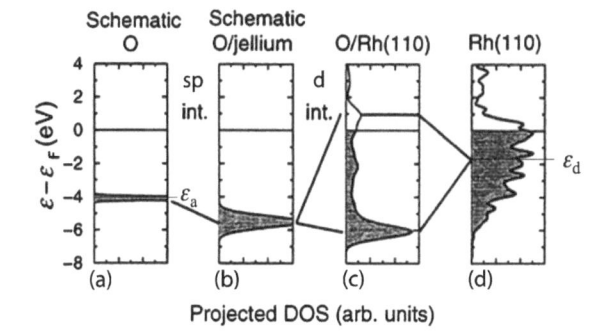

Figure 6.28 Schematic illustration of the change in local electronic structure of an oxygen atom adsorbing on the late transition metal rhodium, the DOS of which is shown in (d). (b) The interaction of the oxygen 2p orbital with the sp band of the transition metal is illustrated through the interaction with the idealized free-electron metal jellium. The result is a downward shift and a broadening of the 2p level. (c) Interaction of this level with the d band of the rhodium results in a density of states for the chemisorbed atom in which bonding and antibonding parts are recognized (adapted from Hammer and Nørskov [13]).

6.5.1
Trend in Atomic Chemisorption Energies

When an atom with a filled level at energy ε_a approaches a metal surface, it will first of all chemisorb due to the interaction with the sp electrons of the metal. Consider, for example, an oxygen atom. The 2p level contains four electrons when the atom is isolated, but as it approaches the metal the 2p levels broaden and shift down in energy through the interaction with the sp band of the metal. Figure 6.28a,b shows this for adsorption on jellium, the ideal free-electron metal.

All levels below the Fermi level are filled with additional electrons leading to a strong sp-induced bonding, which amounts to about −5 eV. The O-sp interaction cannot explain the trends in bonding through the transition metals, since the sp bands of these metals are very similar. Instead, we need to look at the d band.

We consider the interaction between the O 2p and the relatively narrow d band located around energy ε_d as a perturbation on top of the one caused by the sp electrons. This enables us to treat the system essentially as a two level problem, as before. The result is shown in Figure 6.28c where the downshifted oxygen level interacts with the d band of rhodium so that the adsorbate orbitals split into bonding and antibonding components, allowing an estimate of the additional gain in bonding energy due to the hybridization energy $E_\text{d-hyb}$:

$$E_\text{d-hyb} = 2\varepsilon_+ + 2f\varepsilon_- - 2f\varepsilon_d - 2\varepsilon_a \tag{6.76}$$

Here, f is the filling degree of the d band, and ε_+ and ε_- are the energies of the bonding and antibonding adsorbate orbitals, respectively. These energies can be estimated from Eq. (6.16):

$$2\varepsilon_+ = 2\left(\frac{(\varepsilon_a + \varepsilon_d) - 2S\beta - \sqrt{(\varepsilon_a - \varepsilon_d)^2 + 4\beta^2}}{2}\right)$$

$$= \varepsilon_a + \varepsilon_d - 2S\beta - \sqrt{\delta^2 + 4\beta^2}$$

$$2f\varepsilon_- = 2f\left(\frac{(\varepsilon_a + \varepsilon_d) - 2S\beta + \sqrt{(\varepsilon_a - \varepsilon_d)^2 + 4\beta^2}}{2}\right) \tag{6.77}$$

$$= f\left(\varepsilon_a + \varepsilon_d - 2S\beta + \sqrt{\delta^2 + 4\beta^2}\right)$$

$$\delta = \varepsilon_a - \varepsilon_d > 0$$

By introducing the hybridization term

$$W_{ad} = \sqrt{\delta^2 + 4\beta^2} \tag{6.78}$$

and by utilizing the earlier argument that the overlap is proportional to the interaction matrix element $S = -\gamma\beta$, we obtain

$$E_{d\text{-hyb}} = -(1 - f)(W_{ad} - \delta) + 2(1 + f)\gamma\beta^2 \tag{6.79}$$

If $\delta \gg 4\beta^2$, then $W_{ad} \cong \delta + \frac{2\beta^2}{\delta}$, which leads us to the final expression for the hybridization energy

$$E_{d\text{-hyb}} = -(1 - f)\frac{2\beta^2}{\delta} + 2(1 + f)\gamma\beta^2$$

$$= -(1 - f)\frac{2\beta^2}{\varepsilon_a - \varepsilon_d} + 2(1 + f)\gamma\beta^2 \tag{6.80}$$

The first term is attractive (it increases the bonding energy), while the second is repulsive (decreases the bonding energy). Hence, three parameters play a role in determining the bond strength between the metal d band and the atomic adsorbate:

1. The degree of filling of the d band.
2. The interaction matrix element β between the wave functions of the electron on the atom and those for the d states of the metal.
3. The energy difference between the original electron level of the adsorbate and the center of the d band.

The interaction matrix element β varies with the metal for a fixed adsorbate and adsorption site on the surface. The variation between different adsorbates only needs to be included by a proportionally constant, $V_{xad}^2 = \lambda_{ax}\beta^2$, where x labels the different adsorbates. This means that we can easily obtain an overview of the effect

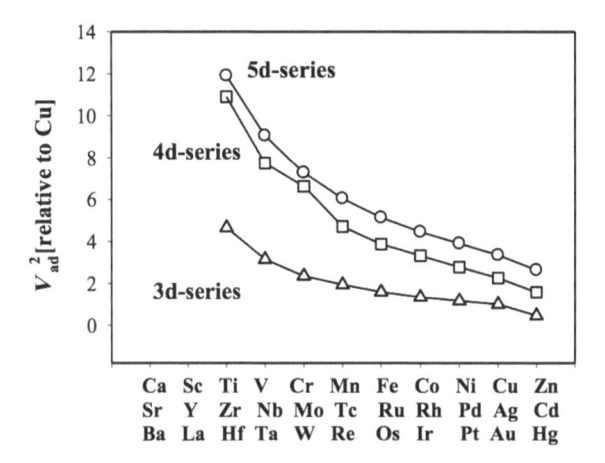

Figure 6.29 The matrix element V_{ad}^2 expresses how the d band of the metal couples with the s or p level of the atomic adsorbate, for the three transition metal series. Note how the matrix element increases when moving to the left. It also increases when moving down a column, due to the larger geometrical extent of the d orbitals (adapted from [13, 14]).

of the important parameters β and $\delta = \varepsilon_d - \varepsilon_a$ by calculating them for all transition metals. These data have been supplemented by Nørskov and Hammer [13] and published by Anderson *et al.* [14] and are shown in Figures 6.29 and 6.30.

Figure 6.29 shows that β^2 decreases with the filling degree f and increases when going from the 3d series to the 5d series, since the geometrical extent of the d orbitals decreases with filling degree (they are becoming more localized on the atom) and the geometrical extent increases when going up through the series. Since β is correlated with the band width, this means that with increased filling

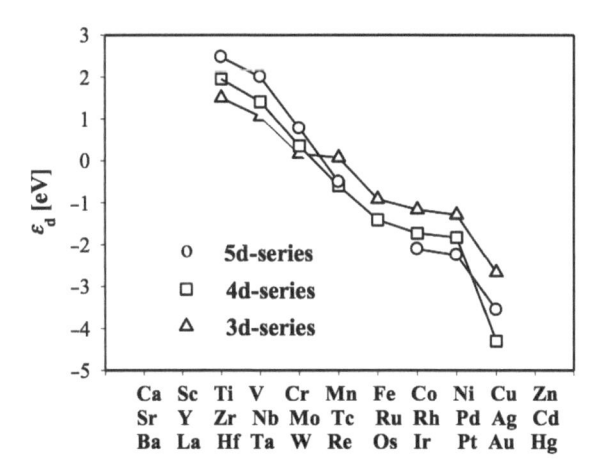

Figure 6.30 Position of the center of the d band for the three series of transition metals. Note that the d band center shifts down towards the right of the periodic table. When the d band is completely filled, it shifts further down and becomes, effectively, a core level with little influence on the chemical behavior of the metal (adapted from [13, 14]).

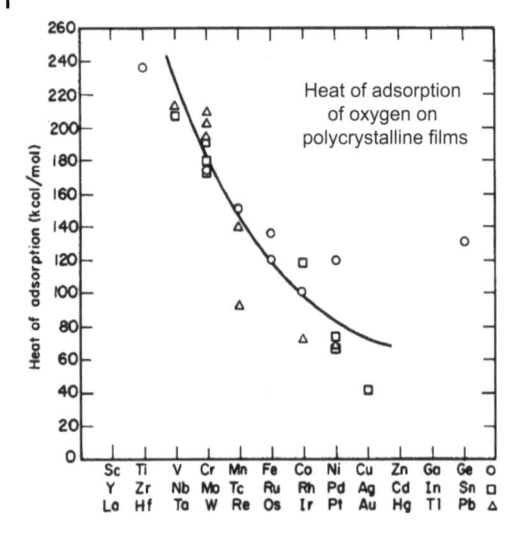

Figure 6.31 Experimental heats of adsorption of O_2 on polycrystalline films (adapted from Toyoshima and Somorjai [15]).

the d band levels also become narrower (less overlap due to smaller geometric extent) and they shift down until they behave like core levels when the d band is filled, as in Cu, Ag, Au, Zn, Cd, and Hg.

Overall, when realistic numbers are substituted in Eq. (6.80), we conclude that the repulsion term always dominates the contribution of the d band to bonding. Because the attractive term in Eq. (6.80) weakens with increased filling f and since the repulsion increases with filling, we expect to find a decreasing bonding energy for atomic adsorption when moving to the right through the transition metals. Remember that the total bonding energy is given by two terms, the contribution due to the sp interaction E_{sp} and the one form the d band, $\Delta E_{\text{d-hyb}}$, where the first term is large and negative at all times. Thus, if we consider the bonding energy of oxygen to different transition metal surfaces, the heat of adsorption decreases from left to right. In essence, the bonding at the left of the periodic table is so large because there are no d electrons that cause repulsion (Figure 6.31).

We can also understand why gold assumes such a special place among the noble metals with respect to reactivity. If we apply Eq. (6.80) to O adsorption on Cu, Ag, and Au, the d band is full, and consequently $f = 1$. As a result, the attractive term in Eq. (6.80) vanishes and only the repulsive term remains, leading to

$$E_{\text{bond}} = E_{sp} + 3 \times 2(1 + f)\gamma\beta^2 \tag{6.81}$$

The factor of 3 stems from the fact that the p orbitals are three times degenerated in the oxygen case. As we discussed above, β^2 increases as we move down through the periodic table implying that the repulsive term is the largest for gold, making the Au–O bond the least stable among the d metals. A similar argumentation holds for the adsorption of other atomic adsorbates.

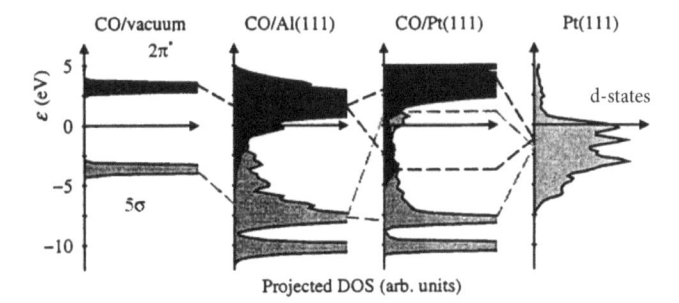

Figure 6.32 Self-consistent calculation of the electronic structure of CO adsorbed on Al and Pt. The sharp 5σ and 2π shift down and broaden upon interaction with the sp band of Al. This band undergoes splitting into bonding and antibonding orbitals upon interaction with the d band of Pt. The diagram for CO on Pt(111) reveals that the contribution from the 5σ orbital to the chemisorption bond is small, whereas the 2π–d interaction clearly strengthens the bond, as only the bonding region of this orbital is occupied (reproduced with permission from Hammer et al. [16]; copyright 1996 by the American Physical Society).

6.5.2
Trends in Molecular Chemisorption

In diatomic molecules such as N_2, O_2, and CO, the valence electrons are located on the 5σ, 1π, and 2π orbitals, as shown by Figure 6.6. (Note that the 5σ level is below the 1π level due to the interaction with the 4σ level, which was not included in the figure.) In general, the 1π level is filled and sufficiently low in energy that the interaction with a metal surface is primarily though the 5σ and 2π orbitals. Note that the former is bonding and the latter antibonding for the molecule. We discuss the adsorption of CO on d metals. CO is the favorite test molecule of surface scientists, as it is stable and shows a rich chemistry upon adsorption that is conveniently tracked by vibrational spectroscopy.

Upon adsorption, there is again a strong interaction of the 5σ and 2π orbitals and the metal sp electrons, resulting, as above, in a downward shift and broadening of these two levels. Also, in this case the variation of the adsorption energy is accounted for by the interaction with the d band of the metal, which will cause the levels to split in bonding and antibonding parts. The result is shown in Figure 6.32, which should be seen as a realistic alternative to the more qualitative representation of Figure 6.25.

In Figure 6.32, the 5σ and 2π levels of CO are first allowed to interact with the sp band of the free-electron metal aluminum. Both levels are seen to shift down and broaden. The additional structure in the downshifted 5σ orbital is due to interaction with the 4σ orbital of CO, making things more complicated without changing the overall effect. A very similar result is obtained if one lets the CO orbitals interact with the sp band of platinum. The interaction with the d electrons leads to the expected splitting into bonding and antibonding orbitals, where it should be remembered that what we see in Figure 6.32 is the projection of these orbitals on the adsorbate levels. Thus, it appears that the 5σ–5d interaction largely yields a combination of a filled bonding and an almost filled antibonding orbital for the chemisorption (of which only the part on the adsorbate is shown in Figure 6.32), whereas the 2π–5d interaction gives orbitals of bonding nature.

With respect to the adsorption energy, the interaction of the 5σ and 2π orbitals with the sp band again gives a large and negative (that is, stabilizing) contribution, E_{sp}, to the bond. The hybridization $\Delta E_{d\text{-hyb}}$ is estimated in a form similar to the case of atomic adsorbates.

The contribution from the 5σ–d interaction is very similar to that of a filled state ε_a of an atom:

$$\Delta E^{5\sigma}_{d\text{-hyb}} = \frac{-2(1-f)\beta^2_{5\sigma}}{|\varepsilon_d - \varepsilon_{5\sigma}|} + 2(1+f)\gamma_{5\sigma}\beta^2_{5\sigma} \tag{6.82}$$

The term due to the 2π interaction looks somewhat different. Initially there are no electrons in this orbital and, hence, the hybridization energy becomes

$$\Delta E^{2\pi}_{d\text{-hyb}} = 2f\varepsilon_+ - 2f\varepsilon_d$$

$$= 2f\left(\frac{(\varepsilon_{2\pi} + \varepsilon_d)}{2} - \frac{\sqrt{4\beta^2_{2\pi} + \Delta^2_{2\pi}}}{2} + \gamma_{2\pi}\beta^2_{2\pi} - \varepsilon_d\right) \tag{6.83}$$

If $\Delta^2_{2\pi} \equiv (\varepsilon_{2\pi} - \varepsilon_d)^2 \gg \beta^2_{2\pi}$ (which is actually not fulfilled for the transition metals to the left-hand side of the periodic table), we obtain

$$\Delta E^{2\pi}_{d\text{-hyb}} = 2f\left(\frac{-\beta^2_{2\pi}}{\Delta_{2\pi}} + \gamma_{2\pi}\beta^2_{2\pi}\right) \tag{6.84}$$

leading to the final expression

$$\Delta E_{d\text{-hyb}} = \Delta E^{5\sigma}_{d\text{-hyp}} + \Delta E^{2\pi}_{d\text{-hyp}}$$

$$= \frac{-2(1-f)\beta^2_{5\sigma}}{|\varepsilon_d - \varepsilon_{5\sigma}|} - 2f\frac{\beta^2_{2\pi}}{\varepsilon_{2\pi} - \varepsilon_d} + 2(1+f)\gamma_{5\sigma}\beta^2_{5\sigma} + 2f\gamma_{2\pi}\beta^2_{2\pi} \tag{6.85}$$

where the first two terms are attractive, while the latter two are repulsive. If we restrict ourselves to trends in chemisorption for the late transition metals (from Fe, Ru, Os, and to the right), where the d band is rather narrow and $\varepsilon_{2\pi} - \varepsilon_d$ is large, the above approximation holds, but it is not as easily interpreted as the corresponding expression for atomic adsorption. Nevertheless, it is easily seen why CO adsorbs so much better on Ni, Pd, and Pt than it does on Cu, Ag, and Au. For all these metals towards the right of the transition metal series, the d band couples more strongly to 2π than to 5σ orbitals, that is, $\beta^2_{2\pi} \gg \beta^2_{5\sigma}$. Hence, if we only consider the terms involving 2π, then we find

$$\Delta E_{d\text{-hyb}} \approx -2f\frac{\beta^2_{2\pi}}{\varepsilon_{2\pi} - \varepsilon_d} + 2f\gamma_{2\pi}\beta^2_{2\pi} \tag{6.86}$$

The first term of the interaction is attractive, as before adsorption the 2π orbital is empty, and upon adsorption only the bonding region of the orbital becomes occupied. The attractive interaction is even present for the noble metals on the right-hand side of the transition metals. As we move to the left in the periodic table, the bond strength increases, mainly because the center of the d band moves up in energy. Again we see the essence of how a catalyst weakens an internal bond of an adsorbate molecule by ensuring that the originally antibonding orbital of the latter becomes occupied.

6.5.2.1 The Effects of Stress and Strain on Chemisorption

Stress and strain in the surface of a metal affects the overlap of the electron orbitals between the atoms and, therefore, affects the electronic properties of the surface and hence its reactivity. Although the degree of d band filling remains the same, the width of the d band and the energy of its center, ε_d, change when the overlap changes. If one pulls the atoms apart, then the average coordination diminishes, leading to decreased overlap of the d orbitals, and consequently the band narrows. To maintain the filling degree, the center of the band moves upwards in energy. Hence, applying stress or strain to a surface provides a way of influencing its reactivity. We will, therefore, investigate how the hybridization energy varies with the position of the d band for a constant filling degree.

By differentiating expression Eq. (6.86) with respect to the center energy of the d band, we obtain

$$\delta \Delta E_{\text{d-hyb}} \approx -2f \frac{\beta_{2\pi}^2}{\left(\varepsilon_{2\pi} - \varepsilon_d\right)^2} \delta \varepsilon_d \equiv \kappa \, \delta \varepsilon_d \tag{6.87}$$

This expression indicates that the change in hybridization energy is opposite and proportional to the shift of the d band center. Thus, if the d band shifts upwards, the hybridization energy increases and vice versa. Strain and the associated shift of the d band can be brought about by growing the desired metal pseudomorfically on another material with a different lattice constant. Pseudomorfic means that the overlayer grows with the same lattice constant as the substrate. The overlayer may hereby be strained or compressed depending on the lattice constants of the two materials.

The strain or stress will either lead to narrower or broader d bands that are shifted up or down in energy, respectively. An upward shift leads to a stronger interaction with the orbital of adsorbed CO and, thus, to a stronger chemisorption bond. Stress has the opposite effect.

Figure 6.33 gives an overview of the changes in d band position for monolayers of one metal on top of another [17]. Such calculations have also been performed for isolated impurities of late transition metals alloyed into the surface other transition metals, resulting in the same trends. The accuracy of the numbers in Figure 6.33 is limited since many approximations had to be made in order to calculate them. Nevertheless, they reflect trends very well and give useful insight in

Overlayer Substrate	Fe	Co	Ni	Cu	Ru	Rh	Pd	Ag	Ir	Pt	Au
Fe	-0.92	0.24	-0.04	-0.05	-0.73	-0.72	-1.32	-1.25	-0.95	-1.48	-2.19
Co	-0.01	-1.17	-0.20	-0.06	-0.70	-0.95	-1.65	-1.36	-1.09	-1.89	-2.39
Ni	0.96	0.11	-1.29	0.12	-0.63	-0.74	-1.32	-1.14	-0.86	-1.53	-2.10
Cu	0.25	0.38	0.18	-2.67	-0.22	-0.27	-1.04	-1.21	-0.32	-1.15	-1.96
Ru	0.30	0.37	0.29	0.30	-1.41	-0.12	-0.47	-0.40	-0.13	-0.61	-0.86
Rh	0.31	0.41	0.34	0.22	0.03	-1.73	-0.39	-0.08	0.03	-0.45	-0.57
Pd	0.36	0.54	0.54	0.80	-0.11	0.25	-1.83	0.15	0.31	0.04	-0.14
Ag	0.55	0.74	0.68	0.62	0.50	0.67	0.27	-4.30	0.80	0.37	-0.21
Ir	0.33	0.40	0.33	0.56	-0.01	-0.03	-0.42	-0.09	-2.11	-0.49	-0.59
Pt	0.35	0.53	0.54	0.78	0.12	0.24	0.02	0.19	0.29	-2.25	-0.08
Au	0.53	0.74	0.71	0.70	0.47	0.67	0.35	0.12	0.79	0.43	-3.56

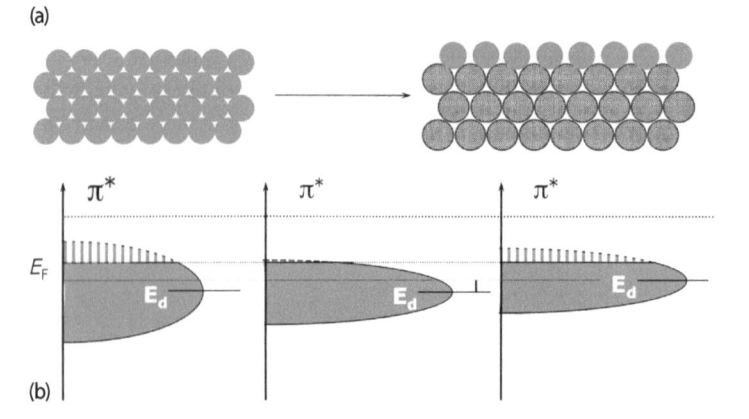

(a)

(b)

Figure 6.33 Trends in reactivity for an overlayer deposited pseudomorfically on a substrate. The diagonal gives the position of the center of the d band for the pure metals. The other numbers indicate the shift of the d band by formation of a pseudomorfic overlayer, irrespective of whether it can be realized. Notice that in the lower left-hand corner the d bands shift upwards, leading to higher reactivity, while the opposite trend is seen in the upper right-hand corner. (b) The diagram shows schematically the change in band shape and position when a metal with a smaller lattice constant is deposited on a substrate with a larger lattice constant (adapted from Ruban et al. [17]).

reactivity trends that have actually been measured for a number of pseudomorfic overlayers [18].

The same theory, that is, Eqs. (6.86) and (6.87), allows us to understand why CO and similar molecules adsorb so much more strongly on undercoordinated sites, such as steps and defects on surfaces. Since the surface atoms on these sites are missing neighbors they have less overlap with and their d band will be narrower. Consequently, the d band shift upwards, leading to a stronger bonding.

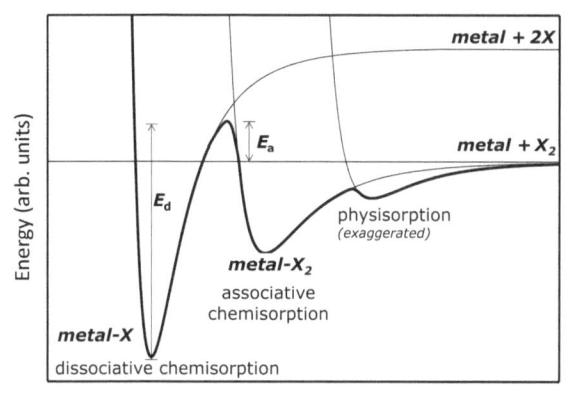

Figure 6.34 Schematic, strongly simplified potential energy diagram along the reaction coordinate of a X$_2$ molecule approaching a metal surface. First, the molecule feels the weak Van der Waals interaction, leading to physisorption. The next stage is associative chemisorption, in which the molecule interacts chemically with the surface. If the molecule overcomes the barrier E_a, it may dissociate into two chemisorbed atoms. The energy required for desorbing these atoms again is indicated as E_d.

Based on insight gained by Figure 6.33, we can start to design surfaces that are fine tuned towards desired chemisorption bond strength and reactivity, as we shall see in the following.

6.5.3
Trends in Surface Reactivity

In the previous two sections, we have described trends in chemisorption energies of atoms and molecules on metallic surfaces. These express the final situation of the adsorption process. Here, we consider what happens when a molecule approaches a surface.

6.5.3.1 Physisorption, Chemisorption, and Dissociation

Figure 6.34 shows potential energy curves for a hypothetical diatomic molecule X$_2$, which approaches a surface, coming from the right-hand side of the diagram. First the molecule encounters the weak Van der Waals interaction. If the molecule loses energy upon interaction with the surface, it may be trapped in the weak attractive potential and become physisorbed on the surface. Physisorption at low temperature is useful for determining the surface area of a catalyst, as we described in Chapter 5.

If the molecule can rearrange its electronic configuration, for instance, by interacting with the sp electrons and the d electrons of the metal, it may become chemisorbed, as we discussed extensively in the previous sections. Depending on where, in terms of distance from the surface, and on how strong the chemisorption bond is, the molecule may have to overcome an energy barrier, as indicated

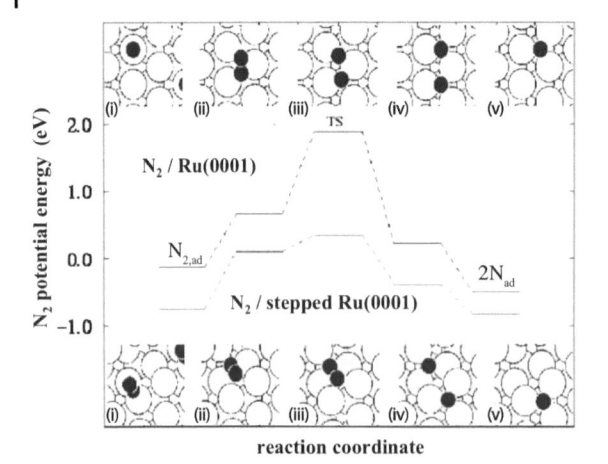

Figure 6.35 Potential energy diagrams for adsorption and dissociation of N_2 on a Ru(0001) surface and on the same surface with a monoatomic step, as calculated with a density functional theory procedure (reproduced with permission from Dahl *et al.* [19]; copyright 1999 by the American Physical Society).

in Figure 6.34. Usually this barrier between physisorption and chemisorption is small, but it depends on the system under investigation. Since the chemisorption energy is also dependent on the orientation of the molecule relative to the surface, the actual potential is much more complicated than the simple one-dimensional representation in the figure suggests.

If we move the chemisorbed molecule closer to the surface, it will feel a strong repulsion and the energy rises. However, if the molecule can respond by changing its electron structure in the interaction with the surface, it may dissociate into two chemisorbed atoms. Again the potential is much more complicated than drawn in Figure 6.34, since it depends very much on the orientation of the molecule with respect to the atoms in the surface. For a diatomic molecule, we expect the molecule in the transition state for dissociation involves the molecule to bind parallel to the surface. The barriers between the physisorption, associative and dissociative chemisorption are activation barriers for the reaction from gas phase molecule to dissociated atoms and all subsequent reactions. It is important to be able to determine and predict the behavior of these barriers since they have a key impact on if and how and at what rate the reaction proceeds.

6.5.3.2 Dissociative Adsorption: N_2 on Ruthenium Surfaces

Figure 6.35 shows a surface potential energy diagram for N_2 on two ruthenium surfaces – one perfectly flat, the other with a monoatomic step – as calculated by density functional theory (DFT). The total energy of the system is minimized at all stages as the molecule approaches the surface and reacts, implying that the configuration of N_2 and the surface is at all times oriented such that the potential energy barriers are minimized. It is important to identify this path since this will be the one with the highest probability for reaction. In the calculation of Figure 6.35, physisorption has been neglected, and hence the first state considered is the chemisorbed nitrogen molecule, which stands vertically on top of a ruthenium atom (Figure 6.35 (i)). There is no barrier for the chemisorption. For this molecule to dissociate, it must lie with the N–N bond parallel to the surface (stage (ii)) and the bond must stretch until it breaks. This is the transition state (stage (iii)) with a barrier of almost 2 eV (200 kJ mol^{-1}) as measured from the adsorbed molecule. Thus, this system is highly activated and requires a substantial amount of energy to overcome this transition state.

The barrier for dissociation at the surface is much lower than that for dissociation in the gas phase, which for N_2 would require about 9.8 eV (945 kJ mol^{-1}) and would lead to two free N radicals. The surface reaction offers an energetically much more favorable end situation, in which the two N atoms are bound to metal atoms. Initially, the N atoms arrive in adjacent threefold adsorption sites and are in contact with the same metal atom (stage (iv)). Such a situation is repulsive, and the minimum energy configuration is found when the two N atoms diffuse away from each other, for which they have to overcome a small activation energy of diffusion (not indicated).

Note that the dissociation proceeds with a much lower barrier on the stepped surface. As the structure diagrams show, at all stages in the dissociation the species are more strongly bound on the stepped surface, for reasons discussed in connection with Eq. (6.87). However, the transition state is most affected because two N atoms are bound to four metal atoms in the transition state on a perfect surface, whereas that on the stepped surface consists of five metal atoms. As noted above, geometries in which atoms bind to different metal atoms are always more stable than when the two adsorbate atoms share one metal atom. Hence, dissociation is favored over step sites, and if a surface contains such defects, they may easily dominate the kinetics.

For dissociation, the antibonding orbitals of the N_2 molecule become filled with electron density from the metal. We have seen that the bonding of the molecule to the surface occurs primarily through the sp band, but for dissociation the interaction of the metal d band with the antibonding 2π orbital is particularly important because this leads to the filling of these orbitals. This is in reality the essence of catalysis: breaking bonds and allowing new ones to form. The essential action is in the cleavage of the strong internal bond of the N_2 molecule. Through interaction with the surface, the barrier for dissociation of the molecule is lowered and the subsequent fragments become available to form new compounds.

6.5.3.3 Trends in Dissociative Adsorption

Identifying the transition state and the associated energy barrier is essential for understanding the course of a reaction. Of course, details of the shape of the potential matter, for example, steric hindrance and entropic effects, may impede the system from crossing the barrier. The barrier energy (which is not very different from the activation energy, as we explained in Chapter 3) is again determined by the interaction of the molecule's electrons with the substrate in the geometry of the transition state, and hence we can in principle use the same type of formu-

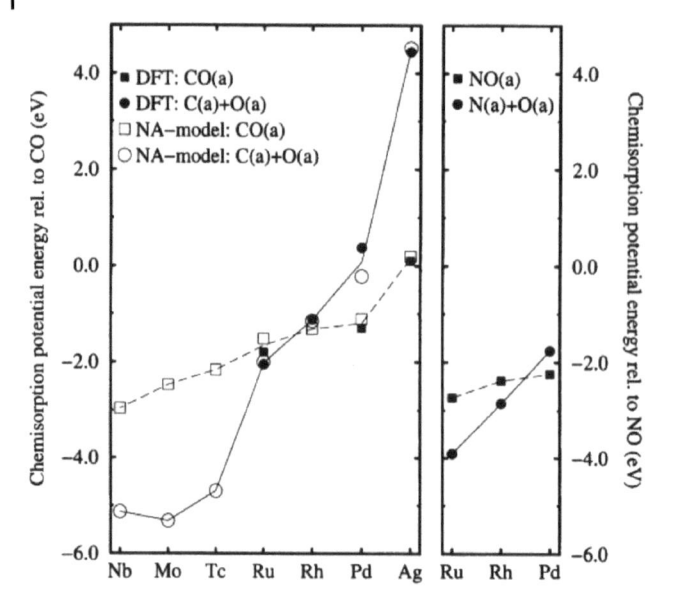

Figure 6.36 Calculated variation in the heats of adsorption of molecular CO and NO compared with the heats of adsorption of the dissociation products. Open symbols follow from the Newns–Anderson model, closed symbols from density functional theory (reproduced from Hammer & Nørskov [13] with permission of Elsevier).

lae as for chemisorption to see how activation barriers change from one metal to another.

If we restrict ourselves to the late transition metals, the trends will, as for the CO chemisorption energy, be dominated by the interaction of the antibonding orbital with the d band and the leading term is

$$\Delta E_{\text{d-hyb}} \approx -2f \frac{\beta_{2\pi}^{2}}{\varepsilon_{2\pi} - \varepsilon_{d}} + 2f\gamma_{2\pi}\beta_{2\pi}^{2} \tag{6.88}$$

Hence, a high-lying d band ($\varepsilon_{2\pi} - \varepsilon_{d}$ is small) is favorable for a stronger interaction and consequently a lower barrier for chemisorption. This explains why CO cannot be dissociated on Cu and why the reactivity increases when going to the left in the transition series. However, there is more to it.

The energy of the final state also has a significant effect on the barrier to dissociation. If we consider CO dissociation on group 10 metals (Ni, Pd, and Pt), Eq. (6.86) suggests that the barrier varies only slightly. Although the d band lies lower in platinum, the interaction matrix element is larger and compensates for this. However, CO does dissociate on nickel, but certainly not on platinum. This is because the end products, C and O atoms, bind much more strongly to nickel than to platinum.

Looking at the trends in dissociation probability across the transition metal series, dissociation is favored towards the left and associative chemisorption towards the right. This is nicely illustrated for CO on the 4d transition metals in Figure 6.36, which shows how, for Pd and Ag, molecular adsorption of CO is more

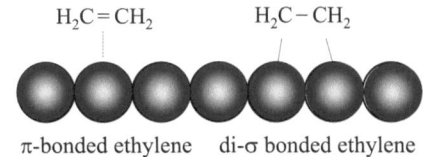

H₂C=CH₂   H₂C–CH₂   **Figure 6.37** Modes of ethylene adsorption on noble metals.

π-bonded ethylene   di-σ bonded ethylene

stable than adsorption of the dissociation products. Rhodium is a borderline case, and to the left of rhodium, dissociation is favored. Note that the heat of adsorption of the C and O atoms changes much more steeply across the periodic table than that for the CO molecule. A similar situation occurs with NO, which, however, is more reactive than CO, and hence, barriers for dissociation are considerably lower for this molecule.

6.5.3.4 Transition States and the Effect of Coverage: Ethylene Hydrogenation

As noted in the previous section, identifying the transition step is essential to obtain a fundamental understanding of a catalytic reaction. Unfortunately, transition states are difficult to access by experiments, as their lifetime is too short, but computational chemistry is a great tool to explore the structure of transition states, as the reader will by now agree. Here, we discuss a slightly more complicated reaction than the dissociation of a diatomic molecule, namely the catalytic hydrogenation of ethylene (C_2H_4) to ethane (C_2H_6) on a noble metal. This has long been a popular test reaction for experimenters in catalysis and surface science [20].

Ethylene (C_2H_4) can adsorb in two modes: (1) the weaker π-bonded ethylene, in which the C=C double bond is above a single metal atom; and/or (2) the stronger di-σ bonded ethylene in which the two C atoms of the ethylene molecule bind to two metal atoms (Figure 6.37). We consider the (111) surface. Hydrogen adsorbs dissociatively and is believed to reside in the threefold hollow sites of the metal.

The reaction scheme is

$$H_2 + 2* \longrightarrow 2H*$$
$$C_2H_4 + * \longrightarrow \pi\text{-}C_2H_4* \quad (\pi\text{-bonded ethylene})$$
$$\pi\text{-}C_2H_4* + * \longrightarrow *C_2H_4* \quad (\text{di-}\sigma \text{ bonded ethylene})$$
$$*C_2H_4* + H* \longrightarrow C_2H_5* + 2* \quad (\text{ethyl; rate-determining step})$$
$$C_2H_5* + H* \longrightarrow C_2H_6 + 2*$$

(6.89)

Neurock and coworkers [21] performed density functional calculations for this reaction scheme up to the formation of the ethyl fragment, for a palladium (111) surface. Figure 6.38a shows the potential energy diagram, starting from the point at which H atoms are already at the surface. As the diagram shows, ethylene adsorbs in the π-bonded mode with a heat of adsorption of 30 kJ mol⁻¹ and conversion of the latter into the di-σ bonded mode stabilizes the molecule by another 32 kJ mol⁻¹.

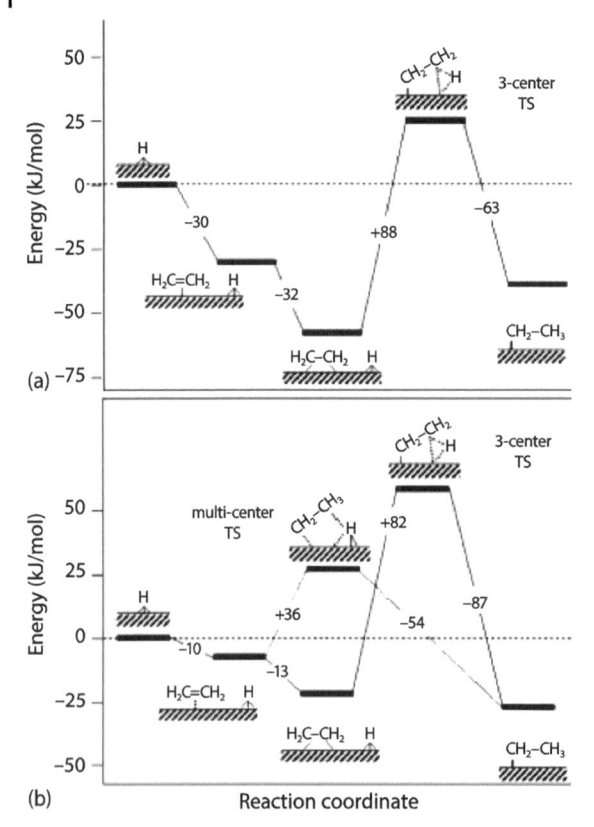

Figure 6.38 Potential energy diagram for the hydrogenation of ethylene to the ethyl (C_2H_5) intermediate on a palladium(111) surface (reaction scheme (6.89)). The zero of energy has been set at that of an adsorbed H atom. (a) Situation at low coverage: ethylene adsorbed in the relatively stable di-σ bonded mode, in which the two carbon atoms bind to two metal atoms. In the three-centered transition state, hydrogen and carbon bind to the same metal atom, which leads to a considerable increase in the energy and, hence, a high barrier for the hydrogenation to the ethyl fragment. (b) At high coverage an additional route opens up, in which the less strongly bound π-C_2H_4 forms a multicentered transition state with a nearby H atom that is bound to adjacent metal atoms. This is the predominant route to hydrogenation (adapted from Neurock et al. [21]).

In the rate-determining step, the reacting fragments have to overcome a significant barrier energy of 88 kJ mol^{-1}. The most likely transition state (or the least unfavorable in this case) is shown in Figure 6.39a. It is a three-centered Pd–C–H complex, in which the H atom binds to the same metal atom as the carbon atom from the ethylene. In fact, as we noted previously, situations in which adsorbate atoms bind to the same metal atom are always much less stable than when they bind to separate metal atoms, and as a consequence the three-centered transition state of Figure 6.39a lies 88 kJ mol^{-1} higher in energy than the adsorbates in their most stable state. The transition state involves breaking of the Pd–C and the Pd–H bonds along with the addition of the H atom to the CH_2

(a)

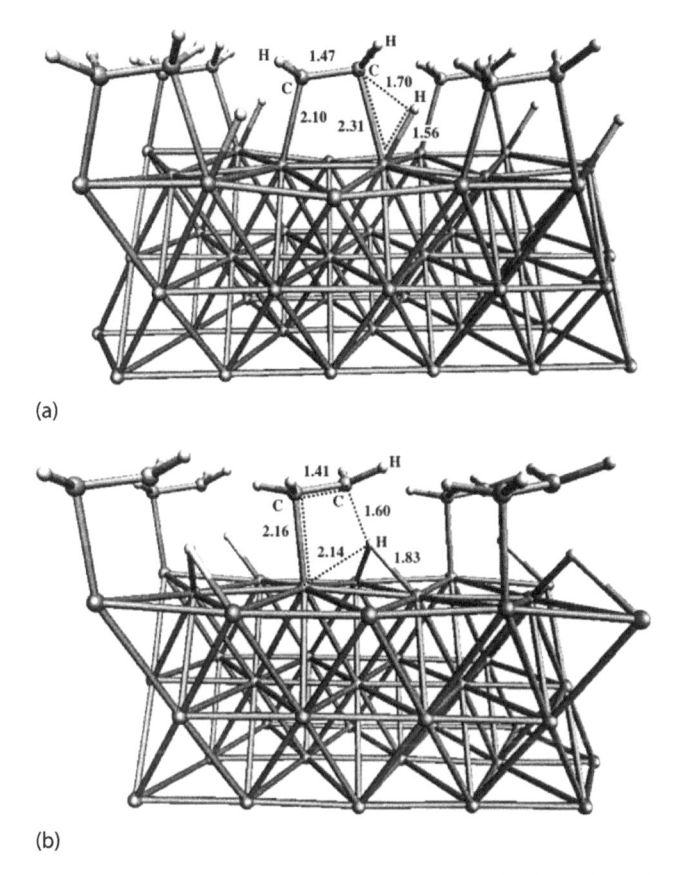

(b)

Figure 6.39 Structure of the transition states for ethylene hydrogenation, corresponding to Figure 6.38. (a) Three-centered transition state; (b) multicentered transition state. See text for details (reproduced from Neurock *et al.* [21] with permission of the American Chemical Society).

group of the adsorbed ethylene. Once the transition state has been reached, the ethyl fragment, C_2H_5* forms readily at an adsorption energy of 37 kJ mol^{-1} below that of the adsorbed H atom, which is the reference energy of Figure 6.38. Note that the energy scheme also gives the barrier energy for the reverse reaction, that is, the dehydrogenation of ethyl fragments.

However, our discussion so far applies to low coverages, as is usually the case in kinetic modeling. With highly covered surfaces another mechanism prevails, which offers an alternative for the energetically unfavorable three-centered transition state of Figure 6.39a. At higher coverages, ethylene and hydrogen are forced closer and another reaction path opens up between the weakly π-bonded ethylene and a nearby H atom. The π-C_2H_4, bound to a single metal atom, slides towards and partly over the adjacent H atom, such that the ethylene adsorbs via one σ bond to the metal, while the other C atom engages in a multicentered transition state complex as drawn in Figure 6.39b. Analysis of the bond distances clearly shows that the H atom is predominantly bound to metal atoms other than those associated with the ethylene molecule, which avoids the repulsion that gave rise to the high energy barrier of the three-centered transition state in Figure 6.39a. The corresponding energy scheme of Figure 6.38b reveals that this route to ethyl has a barrier energy of only 36 kJ mol^{-1} for the elementary step. Of course, the original route via the di-σ-C$_2$H$_4$ remains available; its barrier energy is actually decreased somewhat as a result of the higher coverage. However, it can never compete with the route via the slide mechanism and the multicentered transition state.

Note that the entire discussion has been given on the level of isolated elementary steps. In a steady state reaction situation, one would have to compare the two routes on the basis of the apparent activation energies. We leave this as an exercise for the reader.

Spectroscopic evidence that ethylene preferably hydrogenates directly via the π-bonded adsorbate and not via the more stable di-σ bonded state has been provided by Somorjai and coworkers by *in situ* sum-frequency generation measurements [22].

This example illustrates several points: first, reaction mechanisms, adsorbed intermediates, and transition states can nowadays be investigated very well by computational chemistry, not only for simple diatomics but also for more complicated configurations. The agreement between calculated and measured adsorption energies and barrier energies is sufficiently good that computational predictions can be quite reliable. Such studies are providing a lot of detailed insight into reaction mechanisms.

Second, catalytic reactions do not necessarily proceed via the most stable adsorbates. In the ethylene case, hydrogenation of the weakly bound π-C$_2$H$_4$ proceeds much faster than that of the more stable di-σ bonded C$_2$H$_4$. In fact, on many metals, ethylene dehydrogenates to the highly stable ethylidyne species, $\equiv$C–CH$_3$, bound to three metal atoms. This species dominates at low coverage, but is not reactive in hydrogenation. Therefore, it is sometimes referred to as a spectator species. Hence, weakly bound adsorbates may dominate in catalytic reactions, and to observe them experimentally *in situ* spectroscopy is necessary.

Third, the energy diagrams of Figure 6.38 illustrate that hydrogenation is favored if the hydrocarbon adsorbs less strongly (compare, for example, the barrier for the reaction between di-σ ethylene at low and high coverage). This can be brought about by the choice of the metal, the addition of promoters, by alloying, or by the effect of coverage. Note that the situation, thus, differs from dissociation of diatomic molecules discussed in Sections 6.5.3.2 and 6.5.3.3, where dissociation is favored if the interaction with the surface is stronger.

6.5.3.5 Sabatier's Principle

The results of the previous sections show that catalytic reactions proceed best if the interaction between the adsorbates and the surface is not too strong and not too weak. Sabatier realized that there must be an optimum of the rate of a catalytic reaction as a function of the heat of adsorption. If the adsorption is too weak, the catalyst has little effect and will, for example, be unable to dissociate a bond. If

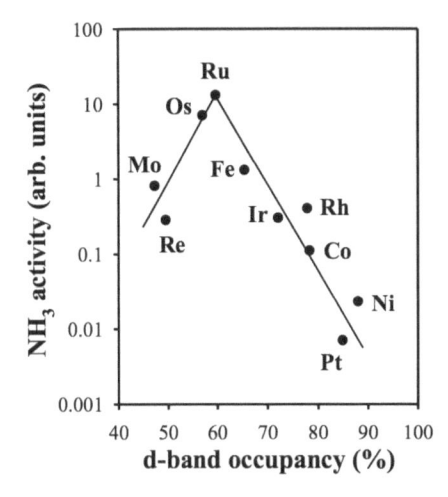

Figure 6.40 Catalytic activity of various supported metals for the synthesis of ammonia (adapted from Ozaki and Aika [23]).

the interaction is too strong, the adsorbates will not be able to desorb from the surface. Both extremes result in small rates of reaction.

Sabatier's Principle is illustrated in Figure 6.40 where the ammonia rate is plotted for similar conditions versus type of transition metals supported on graphite. The theory outlined so far readily explains the observed trends: metals to the left of the periodic table are perfectly capable of dissociating N_2 but the resulting N atoms will be bound very strongly and therefore less reactive. The metals to the right are unable to dissociate the N_2 molecule. This leads to an optimum for metals such as Fe, Ru, and Os. This type of plot is common in catalysis and is usually referred to as a volcano plot.

6.5.3.6 Opportunities for Tuning Surface Reactivity

Given a certain metal, what can we do to alter its reactivity? First, there is the structure of the surface. More open surfaces expose atoms of lower coordination. This narrows the d band and shifts its position (up if it is more than half filled, down if the d band is less than half filled). To illustrate the point, Table 6.2 shows experimentally determined activation energies of NO dissociation on the (111) and (100) surfaces of rhodium.

Table 6.2 Activation energies for NO dissociation and N_2 desorption from two rhodium surfaces.

	Rh(111) (kJ mol^{-1}) [24]	Rh(100) (kJ mol^{-1}) [25]
NO∗ + ∗ ⟶ N∗ + O∗	65 ± 6	37 ± 5
2N∗ ⟶ N_2 + 2∗	118 ± 5	225 ± 5

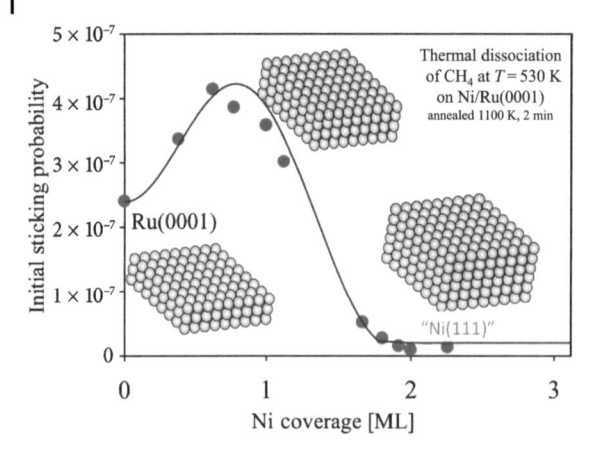

Figure 6.41 Reactivity of a pseudomorfic overlayer of Ni deposited on Ru(0001) for the dissociative adsorption of methane. At zero coverage, the measurements reveals the sticking of methane on pure Ru. When nickel atoms are deposited on the surface, the dissociation probability reaches a maximum when the ruthenium surface is covered by a complete monolayer of nickel. The formation of multilayers of nickel reduces the reactivity to that of Ni(111) (adapted from Egeberg & Chorkendorff [26]).

Secondly, as we discussed in Section 6.5.2.1, inducing stress or strain in the metal by supporting a monolayer pseudomorfically on another metal is a powerful way to change its reactivity. Figure 6.33 gives guidance on what to expect.

An example is provided by the dissociation of methane on nickel. Nickel itself is not particularly active for this reaction. To enhance its reactivity, Eq. (6.87) suggests that we should find ways to shift its d band upward. This can be achieved by growing a pseudomorphic overlayer of nickel on a suitable substrate, that is, one with a larger lattice constant than nickel. Figure 6.33 indicates that this should be the case for nickel atoms deposited on ruthenium since the d band of Ni should be shifted upwards by 0.29 eV. Moreover, nickel atoms have a strong tendency to segregate to the surface of ruthenium, and hence such a pseudomorfic overlayer is stable. Figure 6.41 shows the reactivity of such surfaces for dissociative chemisorption, expressed in the form of a reactive sticking probability.

Methane is a stable molecule and is, therefore, difficult to activate. As a result, the sticking probability for dissociative chemisorption is small, on the order of 10^{-7} only, and ruthenium is more reactive than nickel. However, a stretched overlayer of nickel is significantly more active than nickel in its common form, in agreement with expectation.

6.5.4
Universality in Heterogeneous Catalysis

Calculations based on density functional theory (DFT) have provided further understanding of the reasons underlying Sabatier's principle in reactions where dissociation is the essential step in the reaction mechanism. In essence these reac-

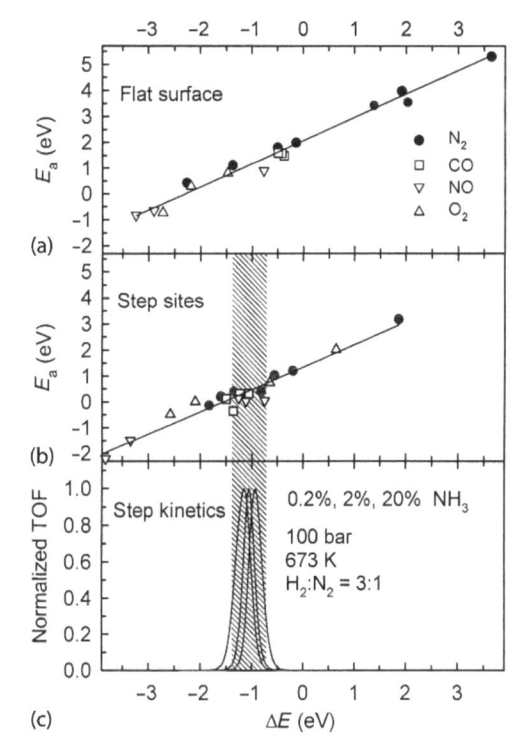

Figure 6.42 (a) Brønsted–Evans–Polanyi relationship between the activation energy for dissociation of diatomic molecules and the heat of adsorption of the dissociation products on a number of atomically flat and close-packed surfaces such as fcc(111), bcc(110), and hcp(0001). Similar results are shown in (b) for stepped surfaces where the active site is a so-called B5 site. Note that the slope is unchanged but that the activation energies are lower for the reaction on step sites. (c) Normalized reactivity for ammonia synthesis at different ammonia concentrations (reproduced from Nørskov et al. [27] with permission of Elsevier).

tions have two main parts: (1) the dissociation of a molecule with a strong bond (e.g., N_2), and (2) the onward reactions of the dissociated atoms (e.g., $N + 3H \rightarrow NH_3$). Nørskov and coworkers found that plots of the activation energy for dissociation, E_a, versus the adsorption energy of the dissociated atoms, indicated by ΔE, give a straight line (Figure 6.42) [27]. The result is remarkable, as the plot includes data for four different diatomic gases and a large number of d metals. Such linear relationships also exist for close-packed surfaces and for stepped surfaces; the latter relationship runs parallel to the former but is shifted to lower activation energies. The plots suggest that the linear relation between activation energy and heat of reaction of the dissociation products is universal, albeit that there is probably a different line for every different surface structure.

Linear relations between the activation energies and heats of adsorption or heats of reaction have been assumed to be valid. Such relations are called Brønsted–Evans–Polanyi relations [1]. In catalysis, such relations have recently been found to hold for the dissociation reactions summarized in Figure 6.42, and also for a number of reactions involving small hydrocarbon fragments such as hydrogenation of ethylene, and dehydrogenation of ethylene and ethyl fragments, by Neurock and coworkers [28].

When the results for N_2 dissociation on stepped surfaces are combined with a kinetic model for ammonia synthesis (Figure 6.42c), the rate of reaction is seen to be a function of the heat of adsorption of the N atoms. One immediately recognizes the curves as idealized volcano plots of the type shown in Figure 6.40. Hence, an optimum catalyst for ammonia synthesis should be one with a stepped surface that binds N atoms with a adsorption energy around 100 kJ mol^{-1}. Stronger bonding would aid in the dissociation but would imply that the N atoms do not easily engage in subsequent reactions. Weaker bonding would imply that the activation energy for dissociation becomes too high. Hence, the optimum catalyst is one in which activation energy and heats of adsorption form a carefully balanced compromise, a statement that is in full agreement with Sabatier's Principle but is in fact more detailed.

The fact that universal Brønsted–Evans–Polanyi relations appear to exist for these dissociation reactions raises the following questions. Why is the relationship between the activation energy and the adsorption energy of the dissociation products linear? Why does it depend on structure? Why is it independent of the adsorbates?

All these questions can be answered if we consider the transition states for the dissociation reactions, which are all very similar. The transition state structure for a given substrate geometry is essentially independent of the type of molecule and substrate. Thus, the close packed surfaces as well as the stepped surfaces considered in Figure 6.42 each form a group. Furthermore, dissociation is characterized by a late transition state, in which the two atoms have already separate to a large extent and have lost their molecular identity. This means that variations of the transition state closely follow the variations in the final state of dissociated atoms, as expressed by the heats of adsorption of the latter. As the structure of the transition state depends on surface structure (see, for example, Figure 6.35c for flat and stepped ruthenium), there is a Brønsted–Evans–Polanyi relation for every different surface structure.

6.5.5
Scaling Relations

In recent years, the universality principle has been further developed in order to simplify the search for new active and selective catalysts. Imagine one investigates a new catalytic reaction. First, we want to identify the reaction pathway by computational chemistry and construct a microkinetic model which would enable us to predict a rate of reaction (see Chapter 7). Then we want to get a feeling for what happens with another material as the catalyst, or to get some insight in how the catalyst needs to be modified, for example, by alloying or adding promoters. It would be very time-consuming to calculate all the intermediates on all these dif-

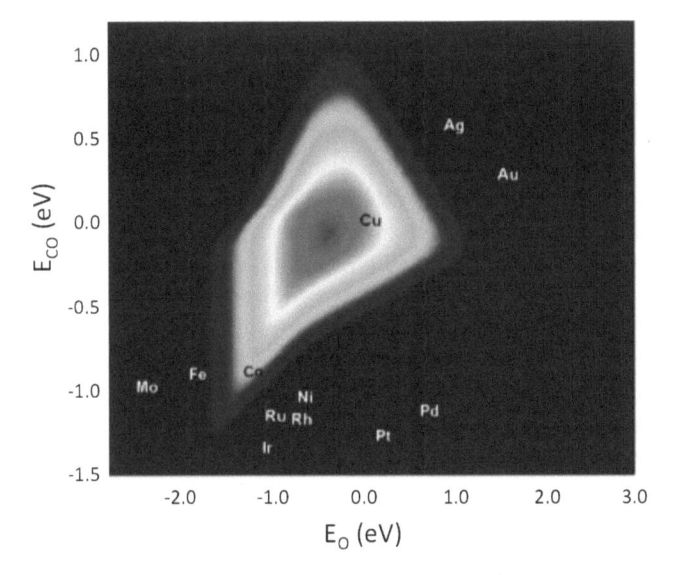

Figure 6.43 Contour plot of the calculated turnover frequency of the water–gas shift reaction as a function of CO and O bonding energy relative to copper. Note that the theoretical maximum falls at a system (metal or alloy) that is slightly more reactive than copper.

ferent surfaces. The universality principle, however, suggests that the adsorption energies of the various intermediates are interrelated, by so-called scaling relations. For example, intermediates bound to the surface through a metal–oxygen bond often tend to respond similarly when the metal changes. These similarities can go quite far, such that even the adsorption energies of O- and C-bonded intermediates change in similar ways when they bind to different metals in the same geometry. In such cases, all we need to do is find suitable descriptors to describe the trends in activity when we change the bond energy of the characteristic intermediate to the surface.

An early example of a reaction where this approach was successful is the water–gas shift reaction

$$CO + H_2O \rightleftharpoons CO_2 + H_2 \qquad (6.90)$$

where the activity can be described as a function of CO (carbon bonding) and oxygen bonding energy relative to pure copper [29]. The contour plot in Figure 6.43 shows that copper is close to optimal for the reaction. Ideally one would like CO and O to bind a bit stronger, which would mean moving slightly to the left and downward in Figure 6.43. According to what we have learned in this chapter, this could be achieved by depositing copper on a platinum surface where it forms a strained overlayer. The elongation of the Cu–Cu bond would render it more reactive towards both CO and O chemisorption. Unfortunately, in practice such a strained Cu/Pt overlayer system is not stable under CO, and it reconstructs to a Pt-rich outer layer (because CO binds stronger to Pt than to Cu). This illustrates that reality may prove more complex than the idealized world of computational modeling.

Similar approaches have been taken in search for a good catalyst for converting syngas (CO + H_2) into alcohols such as ethanol. This reaction also presents a challenge in selectivities, as side reactions to methane and higher hydrocarbons as well as to methanol need to be avoided [30]. The outcome of the analysis is that none of the single metals are predicted to produce ethanol at high selectivities, Metals such as nickel, rhodium, ruthenium, and iridium are close, but the probability of producing methane (or methanol) is significantly higher. We refer to Medford *et al.* for more examples [30–32].

Of course, one should be aware that schemes based on scaling relations of two descriptors in combination with two-dimensional surfaces will inevitably have limitations. Reaction mechanisms may also be more complex than the idealized schemes of elementary steps to which we necessarily have to limit ourselves. As the example of unstable alloy surfaces under reaction conditions illustrates, it will not always be possible to realize the predicted ideal in reality.

Another direction to explore is whether scaling relations can be broken by incorporating the third dimension in considering catalyst structures for a particular reaction. This way of thinking is very common in homogeneous and enzyme catalysis, where ligands and protein configurations around the active site are critical for optimizing the reaction pathway for a specific molecule. Traditionally, heterogeneous catalysis tends to focus on two-dimensional surface structures and sites thereon, perhaps with the exception of zeolite applications, and the use of enantioselective surface modifiers. The advent of nanostructured materials, such as carbon nanotubes and other well-defined structures may offer new venues in three-dimensional site engineering approaches.

Notwithstanding these considerations, the progress in predictive computational catalysis has been impressive, and predictions on the basis of scaling relations can certainly give the experimentalist valuable guidance in exploring new, promising catalyst compositions without having to resort on trial and error only.

6.5.6
Appendix: Density Functional Theory (DFT)

In this chapter we have largely relied on computational chemistry, in particular on density functional theory. Quantum mechanical calculations of a macroscopic piece of metal with various adsorbed species adsorbed on it are as yet impossible, but it is possible to obtain realistic results on simplified systems. One approach is to simulate the metal by a cluster of 3–30 atoms on which the molecule adsorbs and then describe all the involved orbitals. Many calculations have been performed on this basis with many useful results. Obviously, the cluster must be sufficiently large that the results do not represent an artifact of the particular cluster size chosen, which can be verified by varying the cluster size.

The other approach which originates from solid state physics relies on describing the many electrons in an average manner, using the symmetry in a metal and also making use of the fact that valence electrons screen the core levels of the atoms effectively. Symmetry allows us to describe the metal as a unit cell that is repeated infinitely in all three dimensions. Hence, description of the electrons in the unit cell would in principle be sufficient. Core electrons do not participate in the chemistry but screen the nuclear charge and give rise to a periodic potential in which the valence electrons are moving.

The three-dimensional symmetry is broken at the surface, but if one describes the system by a slab of 3–5 layers of atoms separated by 3–5 layers of vacuum, the periodicity has been reestablished. Adsorbed species are placed in the unit cell, which can exist of 3×3 or 4×4 metal atoms. The entire construction is repeated in three dimensions. By this trick, one can again use the computational methods of solid state physics. The slab must be thick enough that the energies calculated converge and the vertical distance between the slabs must be large enough to prevent interaction.

Instead of treating all electrons in the metal plus adsorbate system individually, one considers the electron density of the system. Hohenberg and Kohn (Kohn received the 1999 Noble Prize in Chemistry for his work in this field) showed that the ground state E_0 of a system is a unique functional of the electron density in its ground state n_0. Neglecting electron spin, the energy functional can be written as

$$E(n) = T(n) + \int n(r) v^{\text{nucl}}(r) \, dr + \iint \frac{n(r)n(r')}{|r-r'|} \, dr \, dr' \qquad (6.91)$$

where $T(n)$ is the kinetic energy of the interacting electrons. The kinetic energy is a sum of the noninteracting electrons plus the exchange and correlation energy.

$$T(n) = T_{\text{ks}}(n) + E_{\text{xc}}(n) \qquad (6.92)$$

where by all the problematic terms have been isolated in the expression for the exchange and correlation energy. By solving iteratively the so-called Kohn–Sham equation, which is a Schrödinger-like one-electron equation

$$\left(-\frac{\hbar^2}{2m} \frac{\delta^2}{\delta^2 r} + v \right) \Psi_i = \varepsilon_i \Psi_i \qquad (6.93)$$

where

$$\sum_{i \text{ occ}} |\Psi_i|^2 = n \qquad (6.94)$$

it would be possible to solve the problem if the form of the exchange and correlation were known. This, however, is only the case for a homogeneous electron gas, where

$$E_{\text{xc}}(n) = \int n \varepsilon_{\text{xc}}(n) \, dr \qquad (6.95)$$

and $\varepsilon_{\text{xc}}(n)$ is the exchange and correlation per electron. The approach where the results of the homogeneous gas are utilized is referred to as the LDA or the local density approximation. Since this approximation is derived from first principles, it is a so-called *ab inito* method. Unfortunately, Eq. (6.93) is not a very appropriate description of the strongly varying electron density, and so the method often overestimates bonding energies by several eV. A substantial improvement is obtained by adding a term that depends on the gradient of the electron density, which is referred to as the Density Functional Theory-Generalized Gradient Approximation (DFT-GGA).

DFT-GGA calculations are very useful for investigating plausible reaction pathways of various molecules on surfaces. The method provides detailed information on the bonding geometry, on bond energies as well on activation barriers, and transition states that are otherwise not accessible. Typical accuracies in such numbers amount to a few tenths of an eV, making the method particularly useful to investigate trends as demonstrated in Figures 6.42 and 6.43.

References

1 Nørskov, J.K., Studt, F., Abild-Pedersen, F., and Bligaard, T. (2014) *Fundamental Concepts in Heterogeneous Catalysis*, John Wiley & Sons, Inc., Hoboken.

2 R.A. Van Santen and Neurock, M. (2006) *Molecular Heterogeneous Catalysis: A Conceptual and Computational Approach*, Wiley-VCH Verlag GmbH, Weinheim.

3 Gross, A. (2009) *Theoretical Surface Science: A Microscopic Perspective*, Springer, Berlin.

4 Atkins, P.W. (1998) *Physical Chemistry*, Oxford University Press, Oxford.

5 Kittel, C. (1976) *Introduction to Solid State Physics*, John Wiley & Sons, Inc., New York.

6 Ashcroft, N. and Mermin, N.D. (1976) Solid State Physics, Holt-Saunders, Philadelphia.

7 Elliot, S. (1998) *The Physics and Chemistry of Solids*, John Wiley & Sons, Inc., New York.

8 McQuarrie, D.A. (1976) Statistical Mechanics, Harper & Row, New York.

9 Newns, D.M. (1969) Self-consistent model of hydrogen chemisorption. *Physical Review*, **178**, 1123–1135.

10 Anderson, P.W. (1961) Localized magnetic states in metals. *Physical Review*, **124**, 41–53.

11 Hoffmann, R. (1988) *Solids and Surfaces: A Chemist's View of Bonding in Extended Structures*, Wiley-VCH Verlag GmbH, Weinheim.

12 Lang, N.D. and Williams, A.R. (1978) Theory of atomic chemisorption on simple metals. *Physical Review B*, **18**, 616–636.

13 Hammer, B. and Nørskov, J.K. (2000) Theoretical surface science and catalysis – Calculations and concepts, in *Advances in Catalysis* (eds B.C. Gates, H. Knözinger), Vol 45: Impact of Surface Science on Catalysis, pp. 71–129.

14 Andersen, O.K., Jepson, O., and Glötzel, D. (1985) Canonical description of the band structure of metals, in *Highlights of Condensed Matter Theory LXXXIX*, Corso Soc. Italiana di Fisica, Bologna, pp. 59–176.

15 Toyoshima, I. and Somorjai, G.A. (1979) Heats of chemisorption of O_2, H_2, CO, CO_2 and N_2 on polycrystalline and single-crystal transition-metal surfaces. *Catalysis Reviews – Science and Engineering*, **19** 105–159.

16 Hammer, B., Morikawa, Y., and Nørskov, J.K. (1996) CO chemisorption at metal surfaces and overlayers. *Physical Review Letters*, **76** 2141–2144.

17 Ruban, A., Hammer, B., Stoltze, P., Skriver, H.L., and Nørskov, J.K. (1997) Surface electronic structure and reactivity of transition and noble metals. *Journal of Molecular Catalysis A: Chemical*, **115** 421–429.

18 Rodriguez, J.A. and Goodman, D.W. (1992) The nature of the metal-metal bond in bimetallic surfaces. *Science*, **257** 897–903.
19 Dahl, S., Logadottir, A., Egeberg, R.C., Larsen, J.H., Chorkendorff, I., Tornqvist, E., and Nørskov, J.K. (1999) Role of steps in N_2 activation on Ru(0001). *Physical Review Letters*, **83** 1814–1817.
20 Somorjai, G.A. and Li, Y. (2010) *Introduction to Surface Chemistry and Catalysis*, John Wiley & Sons, Inc., Hoboken.
21 Neurock, M., Pallassana, V., and van Santen R.A. (2000) The importance of transient states at higher coverages in catalytic reactions. *Journal of the American Chemical Society*, **122**, 1150–1153.
22 Cremer, P.S., Su, X.C., Shen, Y.R., and Somorjai, G.A. (1996) Ethylene hydrogenation on Pt(111) monitored in situ at high pressures using sum frequency generation. *Journal of the American Chemical Society*, **118**, 2942–2949.
23 Ozaki, A. and Aika, K. (1981) Catalytic activation of dinitrogen, in *Catalysis Vol 1* (eds. J. Anderson and M. Boudart), Springer, Berlin, pp. 87–158.
24 Borg, H.J., Reijerse, J., Van Santen, R.A., and Niemantsverdriet, J.W. (1994) The dissociation kinetics of NO on Rh(111) as studied by temperature-programmed static secondary-ion mass-spectrometry and desorption. *Journal of Chemical Physics*, **101**, 10052–10063.
25 Hopstaken, M.J.P. and Niemantsverdriet, J.W. (2000) Lateral interactions in the dissociation kinetics of NO on Rh(100). *Journal of Physical Chemistry B*, **104**, 3058–3066.
26 Egeberg, R.C. and Chorkendorff, I. (2001) Improved properties of the catalytic model system Ni/Ru(0001). *Catalysis Letters*, **77**, 207–213.
27 Nørskov, J.K., Bligaard, T., Logadottir, A., Bahn, S., Hansen, L.B., Bollinger, M., Bengaard, H., Hammer, B., Sljivancanin, Z., Mavrikakis, M., Xu, Y., Dahl, S., and Jacobsen, C.J.H. (2002) Universality in heterogeneous catalysis. *Journal of Catalysis*, **209**, 275–278.
28 Pallassana, V. and Neurock, M. (2000) Electronic factors governing ethylene hydrogenation and dehydrogenation activity of pseudomorphic Pd-ML/Re(0001), Pd-ML/Ru(0001), Pd(111), and Pd-ML/Au(111) surfaces. *Journal of Catalysis*, **191**, 301–317.
29 Schumacher, N., Boisen, A., Dahl, S., Gokhale, A.A., Kandoi, S., Grabow, L.C., Dumesic, J.A., Mavrikakis, M., and Chorkendorff, I. (2005) Trends in low-temperature water-gas shift reactivity on transition metals. *Journal of Catalysis*, **229**, 265–275.
30 Medford, A.J., Lausche, A.C., Abild-Pedersen, F., Temel, B., Schjodt, N.C., Nørskov, J.K., and Studt, F. (2014) Activity and Selectivity Trends in Synthesis Gas Conversion to Higher Alcohols. *Topics in Catalysis*, **57**, 135–142.
31 Medford, A.J., Wellendorff, J., Vojvodic, A., Studt, F., Abild-Pedersen, F., Jacobsen, K.W., Bligaard, T., and Nørskov, J.K. (2014) Assessing the reliability of calculated catalytic ammonia synthesis rates. *Science*, **345**, 197–200.
32 Medford, A.J., Vojvodic, A., Hummelshoj, J.S., Voss, J., Abild-Pedersen, F., Studt, F., Bligaard, T., Nilsson, A., and Nørskov, J.K. (2015) From the Sabatier principle to a predictive theory of transition-metal heterogeneous catalysis. *Journal of Catalysis*, **328**, 36–42.

Chapter 7
Kinetics of Reactions on Surfaces

7.1
Elementary Surface Reactions

Unraveling catalytic mechanisms in terms of elementary reactions and determining the kinetic parameters of such steps is at the heart of understanding catalytic reactions at the molecular level. As explained in Chapters 1 and 2, catalysis is a cyclic event that consists of elementary reaction steps. Hence, to determine the kinetics of a catalytic reaction mechanism, we need the kinetic parameters of these individual reaction steps. Unfortunately, these are rarely available. Here, we discuss how sticking coefficients, activation energies, and pre-exponential factors can be determined for elementary steps as adsorption, desorption, dissociation, and recombination.

Once the kinetic parameters of elementary steps, as well as thermodynamic quantities such as heats of adsorption (Chapter 6), are available one can construct a microkinetic model to describe the overall reaction. Otherwise, one has to rely on fitting a rate expression that is based on an assumed reaction mechanism. Examples of both cases are discussed in this chapter.

7.1.1
Adsorption and Sticking

The rate of adsorption of a gas on a surface is determined by the rate of collision between the gas and the surface and by the sticking coefficient:

$$r_{ads} = \frac{p_A \theta_*}{N_0 \sqrt{2\pi m k_B T}} S_0(T) = k^+ p_A \theta_* \tag{7.1}$$

In equilibrium, the rate of adsorption equals the rate of desorption,

$$r_{des} = k^- \theta_A \tag{7.2}$$

and equating r_{ads} and r_{des} leads to one of the Langmuir isotherms that we derived in Chapter 2:

$$\theta_A = \frac{K_A p_A}{1 + K_A p_A} \tag{7.3}$$

Concepts of Modern Catalysis and Kinetics, Third Edition. I. Chorkendorff and J.W. Niemantsverdriet.
© 2017 WILEY-VCH Verlag GmbH & Co. KGaA. Published 2017 by WILEY-VCH Verlag GmbH & Co. KGaA.

To describe the adsorption, we need to know the sticking coefficient. As discussed in Chapter 3, it can conveniently be expressed in the Arrhenius form:

$$S = S_0^0 e^{\frac{-\Delta E_{act}}{k_B T}} \tag{7.4}$$

in which we note that the activation energy may be zero. Strongly activated sticking is often associated with the breaking of a bond, such as in the adsorption of methane, or in the dissociative adsorption of N_2.

7.1.1.1 Determination of Sticking Coefficients

Measuring the uptake of a gas by a surface as a function of the dose to which the surface is exposed is the most straightforward way to determine a sticking coefficient. In such experiments, great care should be taken to ensure that gas and surface are in thermal equilibrium. In addition, we need to determine the coverage, either by surface sensitive methods (XPS, AES, IR) or by thermal desorption and ensure that adsorption is not accompanied by desorption.

For example, consider the dissociative adsorption of methane on a Ni(100) surface. If the experiment is performed above 350 K, methane dissociates into carbon atoms and hydrogen that desorbs instantaneously. Consequently, one determines the uptake by measuring (e.g., with Auger electron spectroscopy) how much carbon is deposited after exposure of the surface to a certain amount of methane. A plot of the resulting carbon coverage against the methane exposure represents the uptake curve.

In general, if an adsorbing molecule A occupies a single adsorption site, the adsorption process follows first-order kinetics:

$$\frac{d\theta'}{dt} = \frac{p_A \theta_*}{N_0 \sqrt{2\pi m k_B T}} S_0(T) \equiv F(\theta_{sat} - \theta') S_0(T) \tag{7.5}$$

where θ' is the actual coverage of A on the surface and θ_{sat} its saturation coverage, F is the flux of gas molecules, and $S_0(T)$ the sticking coefficient at zero coverage, which depends on temperature. For methane decomposition, the saturation coverage of carbon atoms is 0.5 with respect to the number nickel atoms in the surface. In the following, we normalize the coverage with respect to its saturation coverage by using

$$\theta = \frac{\theta'}{\theta_{sat}} \tag{7.6}$$

Hence, Eq. (7.3) reduces to the simple relation:

$$\frac{d\theta}{dt} = \frac{d\theta'}{\theta_{sat} dt} = F(1 - \theta) S_0(T) \tag{7.7}$$

This differential equation is readily solved by separation of variables, leading to

$$\int_0^\theta \frac{d\theta}{1 - \theta} = \int_0^t F S_0(T) dt \quad \Rightarrow \quad \theta(t) = 1 - e^{-FS_0(T)t} \tag{7.8}$$

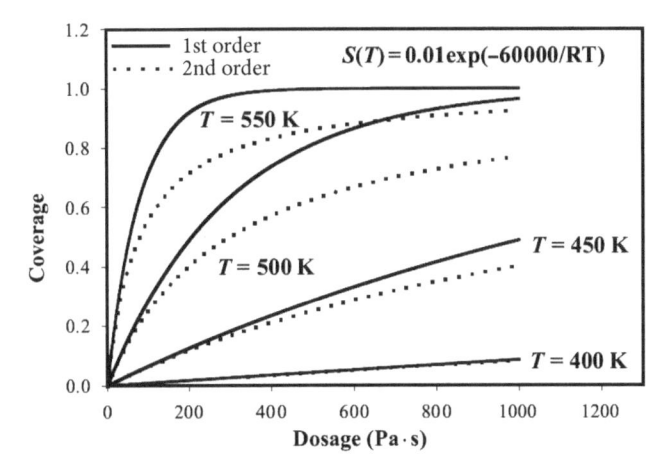

Figure 7.1 Uptake curves for first and second order adsorption as a function of gas dose given in Pascal × second. In this example, the sticking coefficient is strongly activated, with an activation energy of 60 kJ mol^{-1}.

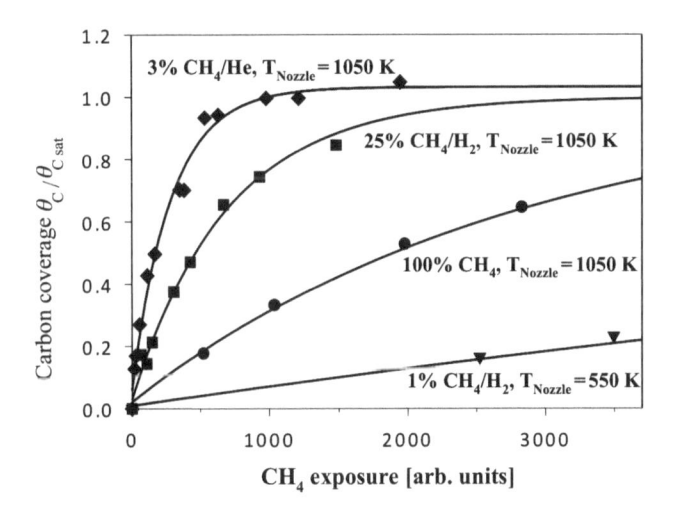

Figure 7.2 The uptake of carbon by dissociative adsorption of methane on Ni(111) follows first-order kinetics. The experiment involved dosing the surface with a supersonic beam of molecular methane at the indicated gas temperatures. The surface was kept at 500 K to enable hydrogen atoms to desorb instantaneously upon decomposition of the methane (reproduced from Larsen and Chorkendorff [1] with permission from Elsevier).

Figure 7.1 shows a calculated set of uptake curves obtained at different temperatures, while experimental results for methane on nickel are given in Figure 7.2.

The gas dose or exposure is normally expressed in the units of time multiplied by the pressure. Of course, it is more meaningful to give the exposure D in terms of a number of molecules hitting a site during the experiment:

$$D = Ft = \frac{pt}{N_0\sqrt{2\pi m k_B T}} \tag{7.9}$$

The reader is left to make this trivial conversion. Please note that the slope of the uptake curve at zero coverage equals $S_0(T)$ and that the above derivation implicitly assumes that the adsorbates do not interact, which is seldom the case. Hence, sticking coefficients in the limit of zero coverage are the most meaningful quantity.

Frequently, adsorption proceeds via a mobile precursor, in which the adsorbate diffuses over the surface in a physisorbed state before finding a free site. In such cases the rate of adsorption and the sticking coefficient are constant until a relatively high coverage is reached, after which the sticking probability declines rapidly. If the precursor resides only on empty surface sites, it is called an intrinsic precursor, while if it exists on already occupied sites, it is called extrinsic. Here, we simply note such effects, without further discussion.

While first-order kinetics are observed with most diatomic gases that adsorb in molecular form, dissociative adsorption of gases such as H_2 and N_2 follows second-order kinetics. In the limit of the empty surface, the rate of adsorption is

$$\frac{d\theta}{dt} = \frac{2p\theta_*^2}{N_0\sqrt{2\pi m k_B T}} S_0(T) = 2FS_0(T)(1-\theta)^2 \tag{7.10}$$

which can be solved by separation of variables, leading to

$$\theta(t) = \frac{2FS_0(T)t}{1 + 2FS_0(T)t} \tag{7.11}$$

for the uptake curve (Figure 7.1). The initial slope of the curve ($t \approx 0$) is proportional to $S_0(T)$. An Arrhenius plot of $\ln[S_0(T)]$ versus $1/T$ yields the activation energy for the dissociative adsorption. Figure 7.3 shows an example for dissociative adsorption of hydrogen and deuterium on Cu(100) at different temperatures. The Arrhenius plots (insert Figure 7.3) reveal activation energies of 48 and 56 kJ mol^{-1} for H_2 and D_2, respectively. Similar considerations as discussed above for first-order adsorption apply for second-order processes.

The pressures necessary for performing experiments as in Figures 7.2 and 7.3 depend on the magnitude of the sticking coefficient. As a rule of thumb, an exposure of 1×10^{-6} Torr during 1 s corresponds approximately to a monolayer of adsorbate if the sticking probability is 100%. In other words, at this dose, a gas molecule has on average hit every atom in the surface once. This unit of 10^{-6} Torr s is called a Langmuir (denoted simply as L) – a convenient unit that is widely used in surface science. However, the correspondence between a Langmuir unit and a monolayer is not entirely accurate and for cases such as H_2 adsorption on Cu(100) (Figure 7.3) one should calculate the dose more precisely to derive the correct sticking coefficients.

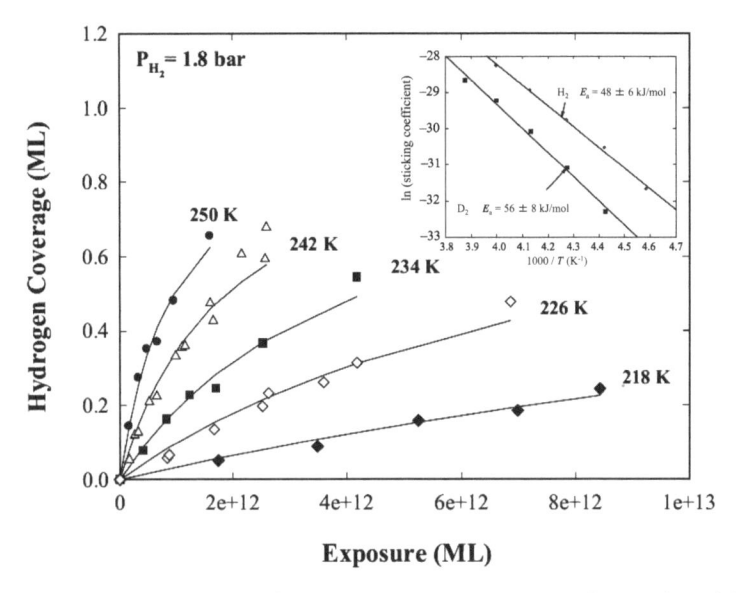

Figure 7.3 Uptake curves of hydrogen on Cu(100). Here, the dosage has been converted into the equivalent number of monolayers (ML). Note that the sticking coefficient is very low and that 1.8 bar of H_2 was required. The insert shows Arrhenius plots of the extracted sticking coefficients for both H_2 and D_2 experiments; the latter having been determined in a similar manner (reproduced from Rasmussen et al. [3] with permission from Elsevier).

In general, one requires that gas and surface be equilibrated, such that they are at the same temperature. This may be a problem at low pressures, where the gas molecules collide more often with the walls of the vacuum vessel than with the surface under study. Reducing the volume and increasing the pressure to the millibar regime by adding an inert gas helps to establish a region around the crystal where the gas is in thermal equilibrium with the surface. Such measurements are commonly referred to as bulb experiments.

Contamination is another pitfall that can influence the measurement of a sticking coefficient, particularly when its value is small. For example, if the sticking coefficient of the gas of interest is of the order of 10^{-6}, an impurity gas present at the ppm level that adsorbs with a high sticking coefficient (e.g., CO) competes efficiently with the gas under study. Special filters need to be used in such cases. Surface inhomogeneities may also be important, as, for example, when N_2 adsorbs on a clean, well-prepared Ru(0001) surface (Figure 7.4). Here, one finds a sticking coefficient of the order of 10^{-10} with an activation energy for dissociation of 42 kJ mol^{-1}. However, all surfaces have defects, such as the steps between terraces. On ruthenium, such defects can be decorated by gold atoms, which are inactive for N_2 dissociation. Repeating the N_2 dissociation experiment on a surface where all defect sites are blocked by gold atoms leads to a much lower sticking coefficient of 10^{-15}–10^{-14} and a considerably higher activation energy, in excess of 130 kJ mol^{-1}. Hence, the reactivity of a surface can be entirely dominated by the

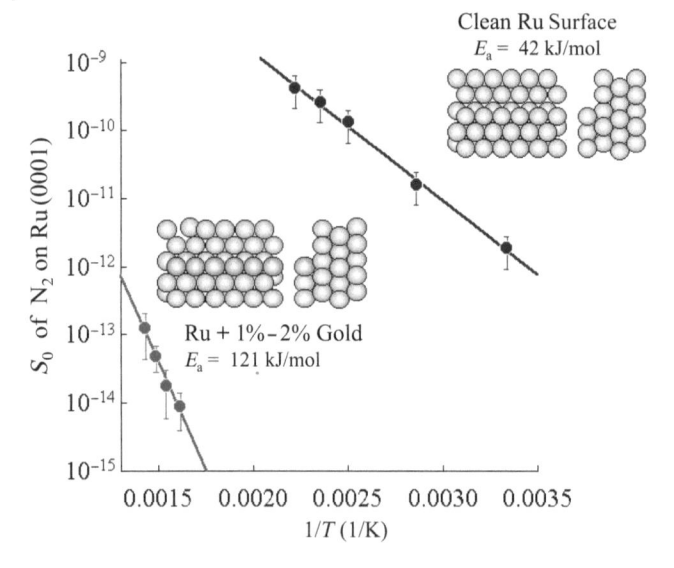

Figure 7.4 Dissociative sticking of N_2 on both clean and gold-modified Ru(0001). The clean surface contains a small fraction of defects that are responsible for the dissociation of N_2. When the defects are blocked by gold atoms, which are known to decorate the steps, the sticking coefficient decreases by several orders of magnitude (adapted from Dahl et al. [4]).

Table 7.1 Sticking coefficients for dissociative adsorption of selected gases.

Molecule	Surface	T (K)	S_0	E_{act} (kJ mol^{-1})	Reference
H_2	Cu(100)	250	5×10^{-13}	48	[3]
D_2	Cu(100)	250	2×10^{-13}	56	[3]
CH_4	Ni(111)	500	2×10^{-8}	74	[1]
CH_4	Ni(100)	500	7×10^{-8}	59	[1]
CH_4	Cu(100)	1000	8.6×10^{-9}	201	[1]
N_2	Ru(001)	400	1×10^{-10}	38	[4]
N_2	Au/Ru(001)	670	5×10^{-15}	> 130	[4]

small number of ever-present defects! Table 7.1 lists several sticking coefficients for dissociative adsorption, along with the corresponding activation energies.

Sticking coefficients for the molecular adsorption of gases such as CO, NO, C_2H_4, etc. are usually between 0.1 and 1. Those for ethylene (C_2H_4) and acetylene (C_2H_2) on a rhodium surface are shown in Figure 7.5. For both gases, the initial sticking coefficients, that is, of the gas on the empty surface, are high, but they decrease as the surface gradually becomes more occupied. Although this is generally observed with many molecules, there are also cases in which the sticking coefficient remains constant up to a substantial filling of the surface. In such

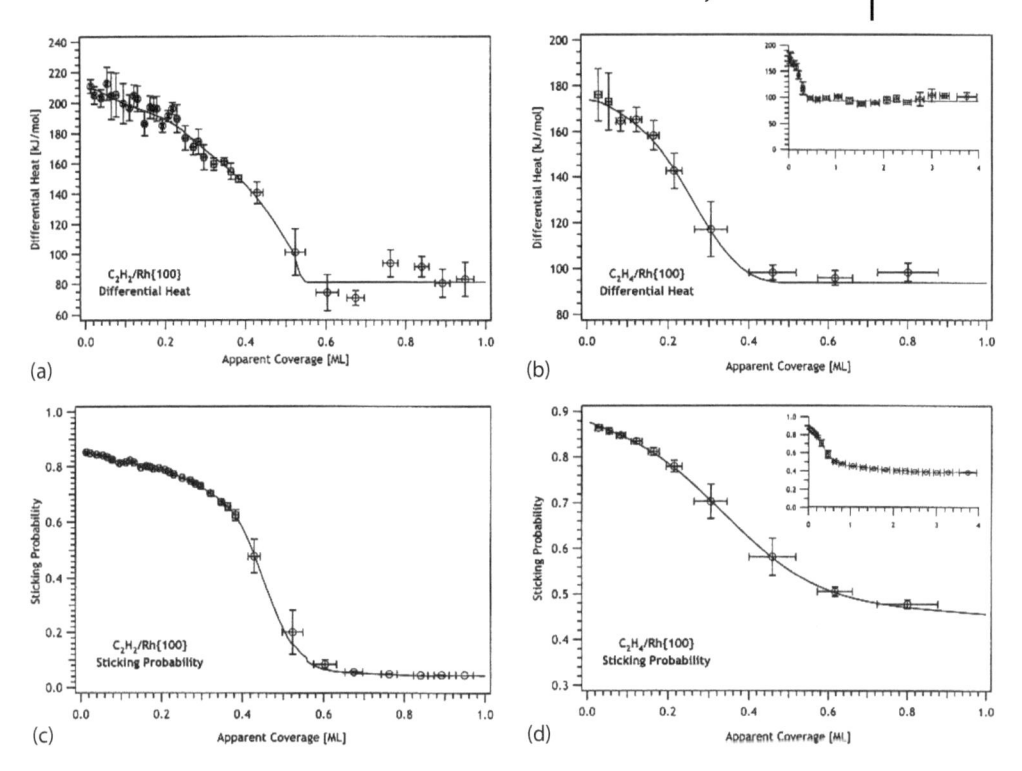

Figure 7.5 Sticking coefficients along with differential heats of adsorption as measured by microcalorimetry for ethylene and acetylene on Rh(100) (reproduced from Kose, Brown, and King [2] with permission from Elsevier).

cases, the incoming molecule adsorbs first in a precursor state in which it moves freely across the surface, whether sites are empty or occupied.

Figure 7.5 also shows the differential heat of adsorption of ethylene and acetylene on rhodium, as measured by microcalorimetry [2]. Note that the differential heat of adsorption is the energy released when a small amount of adsorbate is added to the partly covered surface, whereas the integral heat of adsorption is the energy released when all molecules adsorb at the same time on a previously empty surface. As Figure 7.5 shows, both the sticking coefficients and the heat of adsorption decrease with increasing surface coverage, illustrating that adsorption on partly covered surfaces proceeds, in general, less efficiently than on empty ones.

7.1.2
Desorption

Desorption is important both because it represents the last step in a catalytic cycle and because it is also the basis of temperature-programmed desorption (TPD), a powerful tool used to investigate the adsorption, decomposition, and reaction of

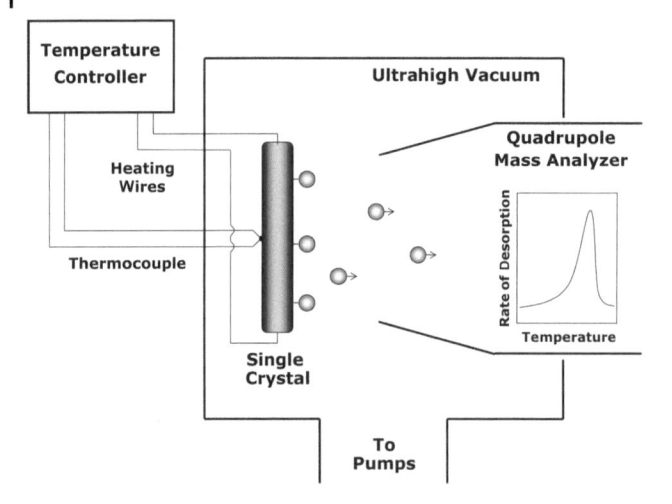

Figure 7.6 Experimental set up for temperature-programmed desorption in ultrahigh vacuum. The heat dissipated in the tantalum wires resistively heats the crystal; the temperature is measured by a thermocouple spot-welded to the back of the crystal. A temperature programmer heats the crystal at a rate of typically 1–5 K s^{-1}. Desorption of gases is followed with a mass spectrometer. Provided the pumping speed is sufficiently high, the mass spectrometer signal is proportional to the rate of desorption (reproduced from [5] with permission of Wiley-VCH).

species on surfaces. This method is also called thermal desorption spectroscopy (TDS), or sometimes temperature-programmed reaction spectroscopy (TPRS; although strictly speaking the method has nothing to do with spectroscopy) [5].

Figure 7.6 shows a schematic set up for TPD. The crystal, mounted on a manipulator in an ultrahigh vacuum chamber, is heated such that the temperature increases linearly in time. The concentration of desorbing species is monitored with a mass spectrometer or with a simple pressure gauge.

If the pumping speed is infinitely high, readsorption may be ignored, and the relative rate of desorption, defined as the change in adsorbate coverage per unit of time, is given by

$$r = -\frac{d\theta}{dt} = k_{des}\theta^n = \nu(\theta)\theta^n \exp\left[-\frac{E_{des}(\theta)}{RT}\right] \tag{7.12}$$

$$T = T_0 + \beta t$$

in which r is the rate of desorption, θ the coverage in monolayers, t the time, n the order of desorption, ν the pre-exponential factor, and E_{des} the activation energy of desorption. T_0 is the temperature at which the experiment starts, and β is the heating rate, equal to dT/dt. If the adsorption is not activated, E_{des} equals the heat of adsorption of the desorbing gas. Attractive or repulsive interactions between the adsorbate molecules render the desorption parameters E_{des} and ν dependent on coverage. Before deriving kinetic parameters from the measurements, we shall look at a few example cases.

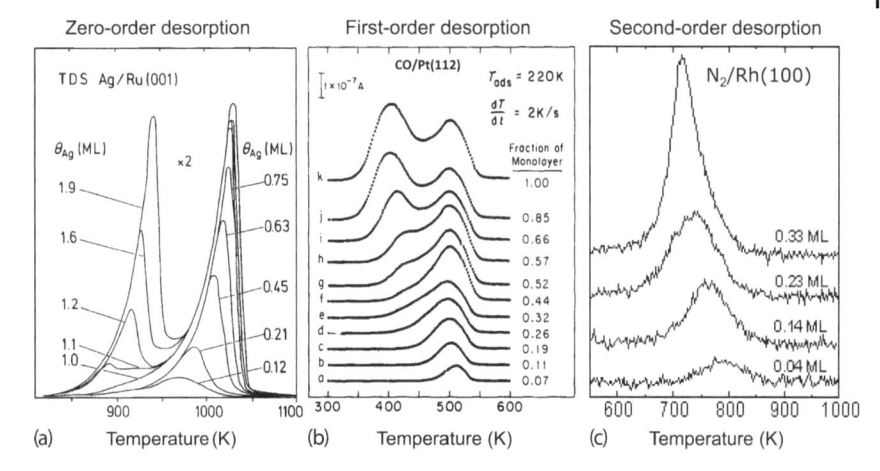

Figure 7.7 Temperature-programmed desorption measurements corresponding to (a) zero-, (b) first-, and (c) second-order kinetics of silver from ruthenium, CO from a stepped platinum surface, and N_2 from rhodium, respectively. The initial coverages indicated reflect the adsorbate coverage at the start of the experiment (data adapted from [6–8]).

Figure 7.7 shows three different sets of TPD measurements, corresponding to zero-, first-, and second-order desorption processes.

Zero-order desorption occurs if the rate of desorption does not depend on the adsorption coverage, as seen with relatively large silver islands on a ruthenium surface (Figure 7.7a), where the Ag atoms desorb from the edges of the island [7]. As the θ^n term in Eq. (7.12) vanishes, the curves exhibit a clearly recognizable exponential shape on the leading side. Such situations are rare.

Figure 7.7b corresponds to first-order desorption of CO from a stepped Pt(112) surface. This surface consists of (111) terraces and (100) steps. At coverages below one-third of a monolayer, CO only occupies the step sites, while at higher coverage the terraces are also populated, resulting in two clearly distinguishable peaks in the TPD pattern. As CO binds more strongly to the steps, the corresponding TPD peak occurs at the higher temperature of the two. Note that the temperature of the maximum peak of CO peaks does not shift with varying coverage – a characteristic signature of first-order desorption kinetics. The example illustrates how TPD reveals the existence of two different adsorption sites on the surface [6].

Figure 7.7c corresponds to the second-order desorption of nitrogen atoms from a rhodium surface. As the desorption reaction corresponds to $N* + N* \rightarrow N_2 + 2*$, the rate is indeed expected to vary with θ_N^2 [8]. A characteristic feature of second-order desorption kinetics is that the peaks shift to lower temperature with increasing coverage because of the strong dependence of the rate on coverage.

7.1.2.1 Quantitive Interpretation of TPD Data

The area under a TPD curve is proportional to the initial coverage of the adsorbate before desorption. If these areas can be calibrated, for example, against ordered patterns in LEED or against a known saturation coverage, TPD can be used to determine the surface coverage. A set of TPD curves contains highly valuable information on the concentration or surface coverage of species, and determining these, such that they can be combined with structures, vibrations or reactivity patterns, is one of the most useful applications of the technique.

However, the temperature at which a molecule desorbs also reflects how strongly it is bound to the surface (Eq. (7.12)). The activation energy in Eq. (7.12) equals the heat of adsorption provided the adsorption of the molecule occurred without an activation barrier. This condition is usually fulfilled.

Equation (7.12) also contains a pre-exponential factor. In Section 3.8.4, we treated desorption kinetics in terms of transition state theory (Figure 3.14 summarizes the situations we may encounter). If the transition state of a desorbing molecule resembles the chemisorbed state, we expect pre-exponential factors on the order of $ek_B T/h \cong 10^{13}$ s^{-1}. However, if the molecule is adsorbed in an immobilized state but desorbs via a mobile precursor, the pre-exponential factors may be two to three orders of magnitude higher than the standard value of 10^{13} s^{-1}.

How do we derive the activation energy of desorption from TPD data? Unfortunately, the differential equation in Eq. (7.12) can not be solved analytically. Hence, analyzing TPD curves can be a cumbersome task, in particular, because the kinetic parameters usually depend on surface coverage [9, 10].

A simple, often used approach is to analyze the curves in terms of easily accessible parameters such as the temperature T_m at which the peak reaches its maximum, and the width of the peak at half maximum.

For a first-order desorption, a useful relation between E_{des} and ν arises if we consider the peak maximum, which occurs when the derivative of the rate becomes zero:

$$\frac{d^2\theta}{dt^2} = \frac{E_{des}\beta}{k_B T_m^2}\frac{d\theta}{dt} - \nu e^{\frac{-E_{des}}{k_B T_m}}\frac{d\theta}{dt} = 0 \qquad (7.13)$$

resulting in

$$E_{des} = k_B T_m \ln\left(\frac{k_B T_m^2 \nu}{E_{des}\beta}\right) \qquad (7.14)$$

which can be solved iteratively, provided a suitable choice for ν (typically 10^{13} s^{-1}) is made. The procedure is to read T_m from the measurement, insert an estimated value for E_{des} in the right-hand side of Eq. (7.14) and calculate the resulting E_{des}. This value is then fed back into Eq. (7.14) to yield an improved value. One continues until the difference between two subsequent iterations becomes negligible, typically after no more than 3–6 iterations.

A similar though more complicated expression exists for second-order desorption kinetics:

$$\frac{d^2\theta}{dt^2} = \frac{E_{des}\beta}{k_B T_m^2}\frac{d\theta}{dt} - \nu e^{\frac{-E_{des}}{k_B T_m}} 2\frac{d\theta}{dt}\theta = 0 \tag{7.15}$$

which leads to

$$E_{des} = k_B T_m \ln\left(\frac{k_B T_m^2 \nu}{E_{des}\beta 2\theta}\right) \tag{7.16}$$

Hence, θ needs to be known (or estimated, e.g., at half the initial coverage) at the point where T_m is reached. Again, a couple of iteration steps is usually sufficient to arrive at an estimate for the desorption energy.

Slightly more elaborate schemes include the peak width in the analysis. The Chan–Aris–Weinberg procedure offers a reliable way to determine both the desorption energy and the pre-exponential factor, provided one performs the analysis for a series of coverages and extrapolates the kinetic parameters to zero coverage. Values at higher coverages are usually meaningless, but the extrapolated values for desorption of an isolated molecule on an empty surface are reliable [11].

If the desorption process follows straightforward first-order kinetics, one may divide the rate at any temperature by the actual coverage, and plot this logarithmically against the reciprocal temperature, to construct an Arrhenius plot. This procedure usually works well in cases where the initial coverage is sufficiently low that lateral interactions play no role. For example, it would work well for CO desorption with an initial coverage below 0.3 ML (Figure 7.7).

If lateral interactions play a role, and E_{des} and ν depend on coverage, Arrhenius plots deviate from straight lines. An alternative analysis is based on the realization that at the onset of the curve there is a small temperature range where the coverage changes only minimally. Provided the measurements in this range are sufficiently accurate, one can convert them straightforwardly into an Arrhenius plot. The slope and intercept then yield desorption parameters valid for the initial coverage. This procedure is often called "leading-edge analysis" [12].

Finally, one can also perform a complete analysis by integrating the TPD curves, and finding sets of rates and temperatures corresponding to the same coverage. Such data can be plotted in the form of Arrhenius plots. However, this is both tedious and time-consuming, and has is rarely performed [13].

7.1.2.2 Compensation Effect in Temperature Programmed Desorption

TPD data that have been analyzed to reveal the coverage dependence of the kinetic parameters often show a compensating effect between desorption energy and pre-exponential factor in the way that $E(\theta)$ and $\log[\nu(\theta)]$ are proportional (Figure 7.8). Although the compensation effect has intrigued many, it is largely an artifact of a mathematical derivation, as we point out in the following [9].

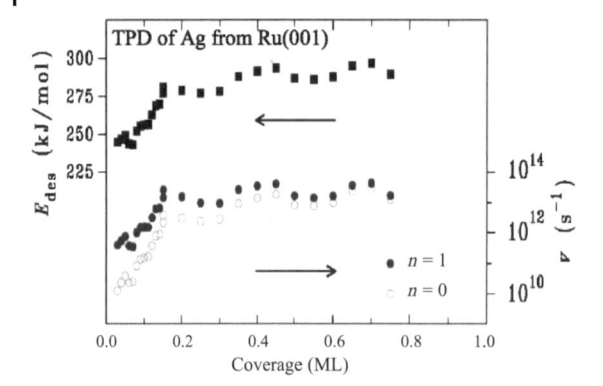

Figure 7.8 The compensation effect in the desorption of Ag from a ruthenium surface: activation energy and pre-exponential factor depend in the same way on coverage. The effect is often caused by a usually overlooked flaw in the kinetic analysis (reproduced from [7] copyright 1987 by the American Vacuum Society).

The usual derivation of an activation energy from a set of temperature dependent rates as the slope of an Arrhenius plot gives

$$E_{app} \equiv RT^2 \frac{\partial \ln\left(\frac{1}{\theta^n}\frac{d\theta}{dt}\right)}{\partial T} = E_{meas} = RT^2 \frac{\partial \ln\left[\nu(\theta)e^{\frac{-E_{des}(\theta)}{RT}}\right]}{\partial T} \quad (7.17)$$

However, if the prefactor and the desorption energy depend on coverage, the derivative will take the form

$$E_{meas} = E_{des}(\theta) + RT^2 \frac{\partial \theta}{\partial T}\left\{\frac{\partial \ln \nu(\theta)}{\partial \theta} - \frac{1}{RT}\left[\frac{\partial E_{des}(\theta)}{\partial \theta}\right]\right\} \quad (7.18)$$

Hence, the activation energy contains coverage-dependent, second-order terms, which are usually ignored. This is only allowed in three cases. The trivial cases are when the kinetic parameters are constant, or when the coverage does not change with temperature, which is approximately correct for a narrow temperature window at the onset of the desorption. The third case arises when the term in brackets on the right-hand side of Eq. (7.18) becomes zero, that is, when

$$\frac{\partial \ln \nu_d(\theta)}{\partial \theta} = \frac{1}{RT}\left[\frac{\partial E_d(\theta)}{\partial \theta}\right] \Rightarrow \ln \nu(\theta) \propto E(\theta) \quad (7.19)$$

Such behavior is known as the "compensation effect." The important point is that if we ignore the additional term in Eq. (7.18), we essentially force the kinetic parameters to satisfy Eq. (7.19) resulting in a correlation between the prefactor and the desorption energy according to the compensation effect!

Finally, although both temperature-programmed desorption and reaction are indispensable techniques in catalysis and surface chemistry, they do have limitations. First, TPD experiments are not performed at equilibrium, since the temperature increases constantly. Second, the kinetic parameters change during TPD, due to changes in both temperature and coverage. Third, temperature-dependent surface processes such as diffusion or surface reconstruction may accompany desorption and exert an influence. Hence, the technique should be used judiciously and the derived kinetic data should be treated with care!

7.1.3
Lateral Interactions in Surface Reactions

Kinetics expressions such as Eq. (7.12), where the rate depends on the coverage, representing the average concentration of adsorbates on the surface, often break down when molecules and/or atoms start to feel each other's presence. When an adsorbed atom or molecule has neighbors, its bonding to the surface is affected and its reactivity changes. As lateral interactions between adsorbates are predominantly repulsive, their effect becomes notable at highly covered surfaces, which is very often the situation for catalytic reactions at normal pressures. The issue of lateral interactions and how these affect kinetics is of great relevance in so-called "Bridging the Gap" strategies that attempt to translate results obtained from surface science experiments to the situation in an industrial reactor.

A proper analysis of how the reactivity of an adsorbate changes when neighbors surround it requires that the local environment of each reacting molecule be taken into account separately. This is possible in Monte Carlo simulations [14, 15]. Figures 7.9 and 7.10 illustrate how interactions affect the distribution of adsorbates over the surface. The mean-field approximation and the associated averaged coverage are no longer a useful descriptor for the multitude of different configurations present. Instead, one has to assign to each molecule an adsorption energy that depends on the number of nearest and next-nearest neighbors:

$$E^i_{ads} = E^i_0 - \sum_j n_j \omega_{ij} \tag{7.20}$$

Here, j numbers the neighbors and ω^{ij} is the interaction energy between adsorbate i and its neighbor j. Positive values of ω^{ij} correspond to repulsion, negative values to attraction. The assumption in Eq. (7.20) is that all pair-wise interaction energies can be added, which is not necessarily correct. For reactions, one can either use Eq. (7.20) to describe the activation energy, or by assuming that the change in activation energy due to neighbors is a constant fraction, that is, 50%, of the change in adsorption energies of the reacting species, according to the Brønsted–Evans–Polanyi relations. Lateral interaction energies are difficult to derive experimentally but can be estimated by computational chemistry.

The simulations of Figure 7.9 concern the distribution of a single species over the surface for different values of the interaction energies for nearest neighbors and next-nearest neighbors. The maps clearly show that repulsion between nearest neighbor species leads to spreading over the surface, and at higher coverages to ordering in regular patterns. The combination of repulsion between nearest neighbors and attraction between atoms on next-nearest neighbor positions induces clear $c(2 \times 2)$ structures at high coverages.

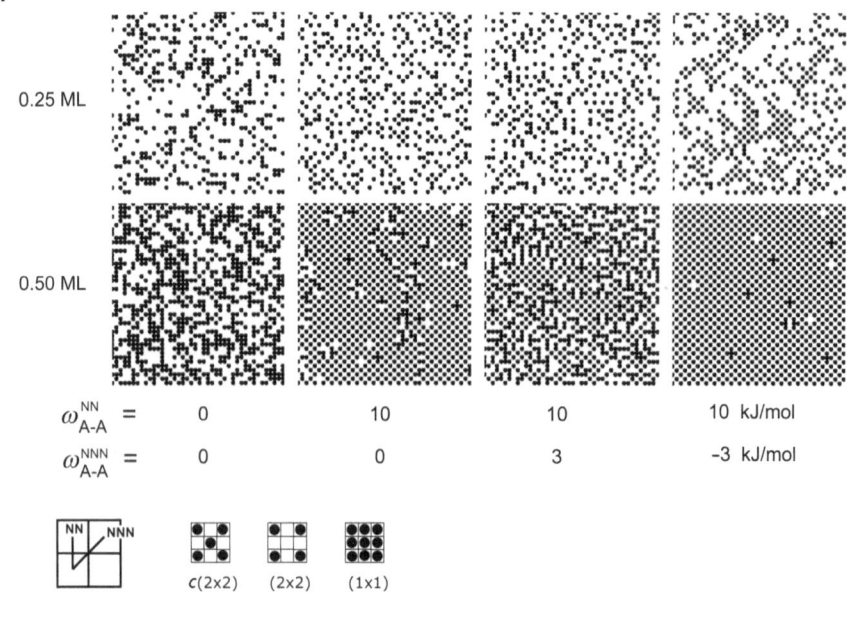

Figure 7.9 Effect of lateral interactions on the distribution of a single adsorbate species A on the surface. The adsorption energy of each atom A is calculated using Eq. (7.20) and the interaction energies indicated under the maps. Negative energies correspond to attraction, positive to repulsion. Nearest neighbor (NN) and next-nearest neighbor (NNN) interaction energies are included. Also indicated are some common ordering patterns, see also Chapter 5 (courtesy of A.P. van Bavel, Eindhoven)

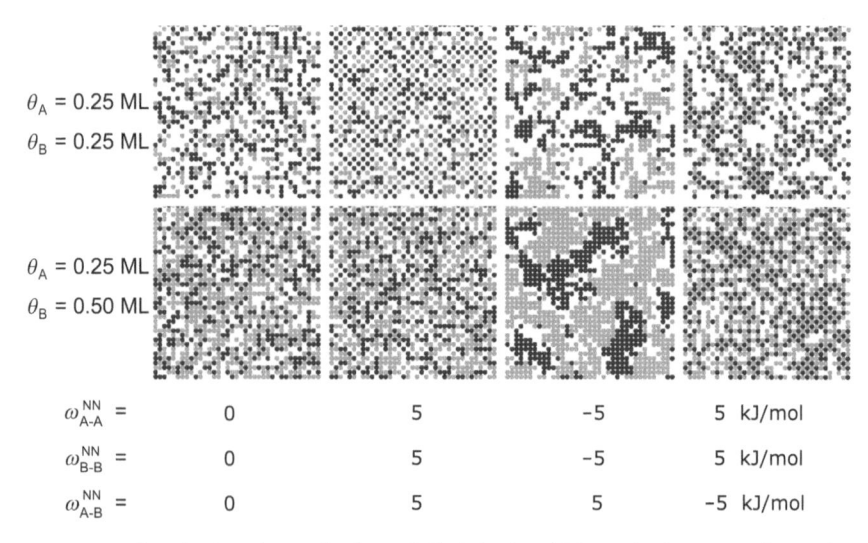

Figure 7.10 Distribution of two adsorbates A (dark dots) and B (grey dots) over a surface with different combinations of attractive and repulsive interactions, as predicted by a Monte Carlo simulation (courtesy A.P. van Bavel, Eindhoven)

Figure 7.10 shows the case of two adsorbate species, A and B, on a square grid of equivalent sites. For simplicity, next-nearest neighbor interactions are ignored. Let us consider a surface reaction between A and B to give AB and determine whether the rate can be described in terms of $r = k\theta_A \theta_B$. Two relevant cases need to be considered. The first arises when like species attract and unlike species repel each other. This leads to segregation of A and B into islands with very little mixing (see Figure 7.10). The rate will certainly not be proportional to the coverages of A and B, but to the fractions of A and B at the edge of the segregated regions. Conversely, attraction between A and B, but repulsion between like species, leads to almost perfect ordering between A and B and is certainly much more favorable for a surface reaction between the two. Note that the "all repulsive case" also leads to a considerable mixing of A and B across the surface.

Figures 7.9 and 7.10 are simply simulations based on assumed interaction energies. The lateral interactions for specific cases can only be derived from measurement in favorable cases, such as the following example on the co-adsorption of CO and N atoms.

Figure 7.11 shows the desorption of CO from a Rh(100) surface covered by an ordered array of N atoms in a $c(2 \times 2)$ structure [16]. The heat of adsorption is 75 kJ mol^{-1} less than that of CO on the clean surface (135 ± 5 kJ mol^{-1}). As CO has four N neighbors, one can assign an interaction energy of $75/4 = 19$ kJ mol^{-1} to repulsion between CO and each surrounding N neighbor, assuming that pairwise lateral interactions may be added. Quantification is possible in this example because we know the number of neighbors in ordered structures. Alternatively, one may estimate interaction energies on the basis of density functional calculations, or by fitting Monte Carlo simulations to the experiment, using the interaction energies as fit parameters.

7.1.4
Dissociation Reactions on Surfaces

Elementary steps in which a bond is broken form a particularly important class of reactions in catalysis. The essence of catalytic action is often that the catalyst activates a strong bond that cannot be broken in a direct reaction, but which is effectively weakened in the interaction with the surface, as we explained in Chapter 6. To monitor a dissociation reaction, we need special techniques. Temperature-programmed desorption is an excellent tool for monitoring reactions in which products desorb. However, when the reaction products remain on the surface, one needs to employ different methods such as infrared spectroscopy or secondary-ion mass spectrometry (SIMS).

Taking the dissociation of NO on a rhodium(100) surface (Figure 7.12) as an example, the chemisorbed NO molecule can be monitored by SIMS of the characteristic secondary ions RhNO$^+$ or Rh$_2$NO$^+$, while the dissociation products are visible via ions of the type Rh$_2$N$^+$ [8]. Figure 7.12a shows that at low coverage, dissociation of NO on Rh(100) sets in at about 175 K and is completed around 250 K, corresponding to an activation energy of 37 kJ mol^{-1}. The N atoms recom-

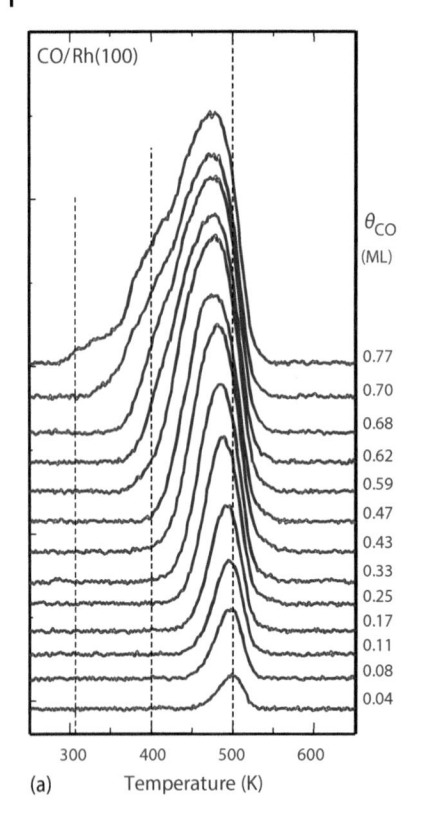

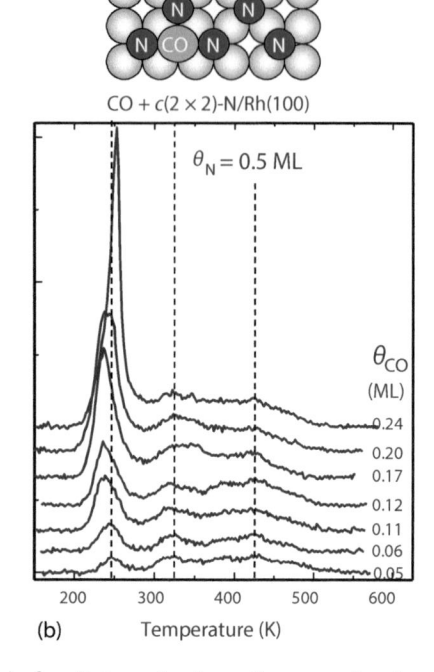

Figure 7.11 Effect of repulsive interactions between N atoms and CO molecules. (a) The TPD of CO from a clean Rh(100) surface is characterized by a desorption energy of 135 kJ mol^{-1}. (b) When CO desorbs out of a structure where it is, on average, surrounded by four N atoms, the desorption energy is only about 60 kJ mol^{-1}. Hence, the lateral repulsion energy ω_{N-CO} is approximately 19 kJ mol^{-1} (reproduced from Van Bavel et al. [16] with permission of AIP Publishing).

bine and desorb as N_2 from the surface at much higher temperatures, between 600 and 800 K, with an activation energy of 225 kJ mol^{-1}. Oxygen atoms remain on the surface in the temperature range studied. Because the coverage is low and all interactions between species on the surface are repulsive when adsorbates reside on nearest neighbor sites, the NO molecules avoid each other as well as the dissociated atoms. Hence, the overall rate of dissociation as measured in the experiment follows undisturbed first-order kinetics and the rate constant satisfies Arrhenius behavior. The situation changes drastically at higher coverage.

Figure 7.12b shows the case when the surface is initially filled by 0.37 ML of NO molecules. If all molecules dissociated, the total coverage would become 0.75 ML, which is very high. This does not happen because repulsive interactions between NO and the already formed atoms slows the dissociation reaction down, such that at about 400 K desorption becomes preferred over dissociation and part of the NO

7.1 Elementary Surface Reactions | 299

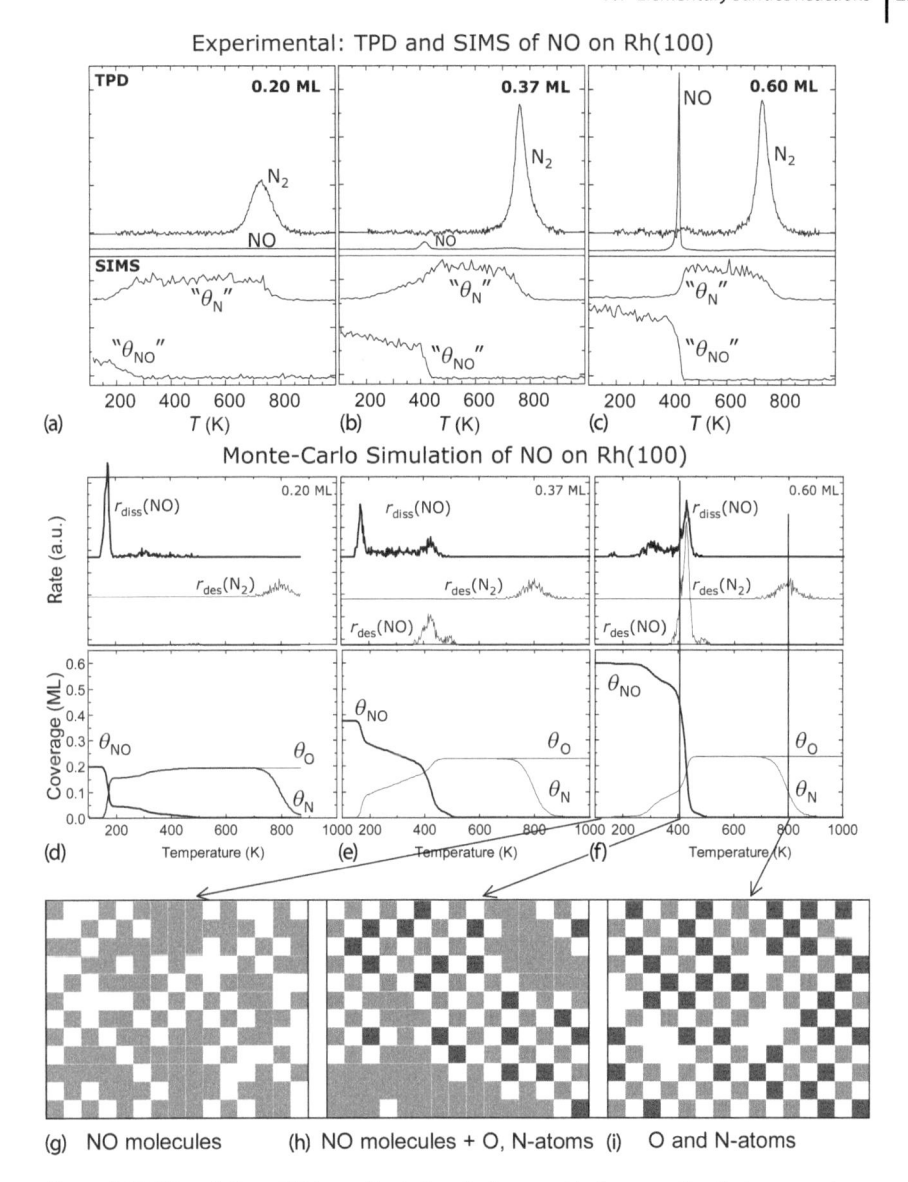

Figure 7.12 Dissociation of NO∗ → N∗ + O∗ in a temperature-programmed desorption and static SIMS experiment, along with Monte Carlo simulations, showing the effect of lateral interactions (see text for explanation). (g–i) Representative arrangements of NO molecules (grey), and N or O atoms (speckled) on a grid of square sites that represent the Rh(100) surface are shown ((a) reproduced with permission from Hopstaken & Niemantsverdriet [8], copyright 2000 American Chemical Society; simulations (d–i) courtesy of A.P. van Bavel, J. Lukkien and P.A.J. Hilbers, Eindhoven).

molecules leave the surface. The desorbed molecules leave space for the remaining molecules, which then dissociate instantaneously, as on the empty surface.

If the surface is fully occupied by NO, all dissociation is blocked until temperatures are reached at which NO desorbs from the surface, after which dissociation follows instantaneously (Figure 7.12c). Here, dissociation is initially suppressed by site blocking rather than by lateral interactions.

To describe the kinetics represented by these experiments, one must include the local environments of the reacting species. The summation of all local events then can be compared with experiment. The Monte Carlo simulations in Figure 7.12d–f, showing rates and coverages of all species on the surface, reproduce most of the experimental phenomena. The Rh(100) surface has been modeled by a square lattice consisting of one type of site [16]. Hence, subtleties such as whether NO adsorbs on different sites to the N and O atoms do not occur. Kinetic parameters for dissociation or desorption in the limit of zero coverage are taken from experiments. However, these are modified by the effect of neighbors according to Eq. (7.20), with a small repulsion between NO molecules ($\omega_{NO-NO} = 6$ kJ mol^{-1}), a larger repulsion between NO and the atoms ($\omega = 17-20$ kJ mol^{-1}), and an even larger repulsion between the atoms themselves ($\omega = 30$ kJ mol^{-1}). Repulsion occurs only when adsorbates are on adjacent (near neighbor) sites, implying that atoms will preferably order in $c(2 \times 2)$ or more open patterns, where occupation of adjacent sites is avoided.

Each simulation starts by randomly placing a number of NO molecules on the surface corresponding to the desired coverage. The molecules are allowed to diffuse over the surface, such that energetically unfavorable distributions (e.g., with several molecules on adjacent sites) are largely avoided, at least at low coverage. For crowded surfaces, NO molecules are forced to occupy the less favorable neighbor positions as well (Figure 7.12c,f). The effect of the stronger repulsion between NO and the N and O atoms is clear in Figure 7.12h: Because NO–NO repulsion is weaker than between NO–N or NO–O, the NO molecules are compressed into islands, while the atoms optimize their arrangement in $c(2 \times 2)$ patterns because of the strong repulsion between the atoms on adjacent sites. In these islands, all chemistry is blocked until the temperature is high enough to break the Rh–NO and molecules start to desorb. The created free sites are immediately used for dissociation of NO, and the surface fills with atoms in an almost perfect $c(2 \times 2)$ pattern (Figure 7.12i).

Figure 7.12 neatly illustrates how dissociation reactions on surface are greatly affected by interactions with neighboring adsorbate species. This is because dissociations often require ensembles of sites in a specific arrangement. From our discussion of surface reactivity in Chapter 6, a molecule such as NO dissociates when there is sufficient overlap between its $2\pi^*$ orbitals and electron density from the metal. This usually requires the molecule to bend over a metal atom, starting from a hollow position [17]. This automatically implies that an ensemble of several metal atoms is involved.

This section also illustrates how complicated the kinetics of surface reactions may become if ensemble requirements and lateral interactions come into play. Compare this to the standard Langmuir–Hinshelwood approach, which limits itself in essence to the zero-coverage limit. Monte Carlo simulation, however, incorporates local events into the global picture given by the experimentally observed kinetics. As of 2003, such simulations can successfully be carried out on a personal computer. However, these simulations are still based on several simplifying assumptions about the types of adsorption sites, the validity of pair-wise additive interactions, and the Brønsted–Evans–Polanyi relationships to convert adsorption energies into activation energies. Also, experimentally determined lateral interaction energies are rarely available, and often one has to rely on fitting them, immediately making such energies model dependent. Nevertheless, such simulations provide useful insight into the kinetics of reactions on covered surfaces, a situation that is prevalent in many catalytic reactions.

7.1.5
Intermediates in Surface Reactions

If an adsorbed species, for example, an intermediate in a catalytic reaction cycle, decomposes into products that desorb instantaneously, TPD can be used to monitor the reaction step.

This is illustrated by the TPD spectra of formate adsorbed on Cu(100). To prove that formate is a reaction intermediate in the synthesis of methanol from CO_2 and H_2, a Cu(100) surface was subjected to methanol synthesis conditions and the TPD spectra recorded (Figure 7.13 (ii)). For comparison, Figure 7.13 (i) represents the decomposition of formate obtained by dosing formic acid on the surface. As both CO_2 and H_2 desorb at significantly lower temperatures than those of the peaks in Figure 7.13, the measurements represent decomposition-limited desorptions. Hence, the fact that both decomposition profiles are identical is strong evidence that formate is present under methanol synthesis conditions.

In principle, TPD can also be applied to high-surface area catalysts in plug flow reactors. Often, however, the curves are seriously broadened by mass-transport phenomena. Hence, the use of single crystals or particles on planar supports offers great advantages for these investigations.

7.1.6
Association Reactions

Temperature-programmed reaction spectroscopy offers a straightforward way to monitor the kinetics of elementary surface reactions, provided that the desorption itself is not rate limiting. Figure 7.14 shows the example of the reaction between adsorbed CO and O atoms to CO_2 [19]. Here, small amounts of CO and O atoms were adsorbed at relatively low temperature, after which the surface was heated

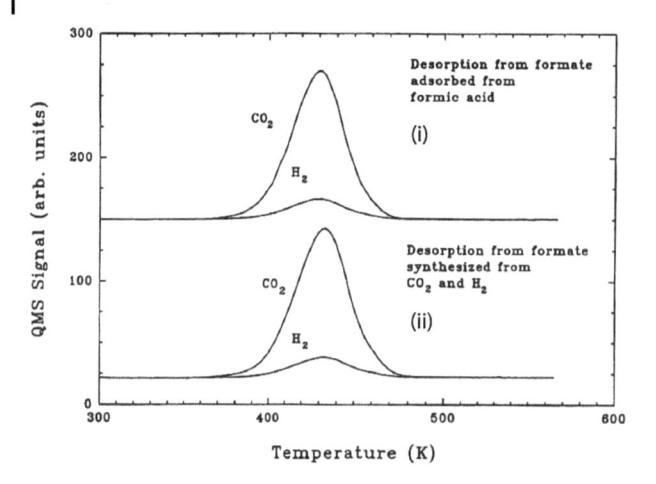

Figure 7.13 TPD curves of CO_2 and H_2 originating from formate on a Cu(100) surface. (i) Formate deposited by dosing formic acid. (ii) Formate synthesized in a high-pressure cell (2 bar) from a mixture of CO_2 and H_2 (reproduced from Taylor et al. [18] with permission from Elsevier).

linearly in time, and the CO_2 formation monitored by mass spectrometry. The reaction sequence for this process is

$$CO* + O* \longrightarrow CO_2* + *$$
$$CO_2* \longrightarrow CO_2 + * \tag{7.21}$$

As long as the desorption of CO_2 is faster than the surface reaction between $CO*$ and $O*$, the rate of desorption equals that of the preceding reaction:

$$r = k\theta_{CO}\theta_O = \nu e^{-E_{act}/RT} \theta_{CO}\theta_O \tag{7.22}$$

As the initial coverages of $CO*$ and $O*$ are known, and the surface is free of CO at the end of the temperature-programmed experiment, the actual coverages of $CO*$ and $O*$ can be calculated for any point of the TPD curves in Figure 7.14. Hence, an Arrhenius plot of the rate of desorption divided by the coverages, against the reciprocal temperature yields the activation energy and the pre-exponential factor:

$$\ln\left(r/(\theta_{CO}\theta_O)\right) = \ln \nu - \frac{E_{act}}{RT} \tag{7.23}$$

Because the Arrhenius plots of both TPD experiments are straight lines over a large portion of the data points, the reaction between $CO*$ and $O*$ is, most likely, an elementary step, with an activation energy of 103 ± 5 kJ mol^{-1} and a pre-exponential factor of $10^{12.7\pm0.2}$ s^{-1}. This analysis is again only valid if coverage dependencies play no role. Here, the straight line of the Arrhenius plot in Figure 7.14 indicates that this requirement is met.

Elementary surface reactions in which the product remains on the surface can sometimes be monitored by secondary-ion mass spectrometry. Figure 7.15 shows the example in which adsorbed carbon and nitrogen atoms react to form a CN species on the surface [21]. The measurements in Figure 7.15 represent the ratio of the mass spectrometric signals of Rh_2CN^+ and Rh_2^+ in the SIMS spectrum. Both mass signals were measured simultaneously during temperature programming. The figure shows the formation of CN at 500–600 K ($E_{act} = 110 \pm 10\,\text{kJ}\,\text{mol}^{-1}$; $\nu = 10^{11\pm1}\,\text{s}^{-1}$) and the subsequent decomposition at 700–800 K ($E_{act} = 210 \pm 15\,\text{kJ}\,\text{mol}^{-1}$; $\nu = 10^{13\pm1}\,\text{s}^{-1}$), which is instantaneously followed by desorption of N_2. Whenever a technique is capable to determining the concentration of an adsorbed species in real time, temperature-programmed reaction offers the possibility of studying reaction steps on surfaces.

Based on surface science and methods such as TPD, most of the kinetic parameters of the elementary steps that constitute a catalytic process can be obtained. However, short-lived intermediates cannot be studied spectroscopically, and then one has to rely on either computational chemistry or estimated parameters. Alternatively, one can try to derive kinetic parameters by fitting kinetic models to overall rates, as demonstrated below.

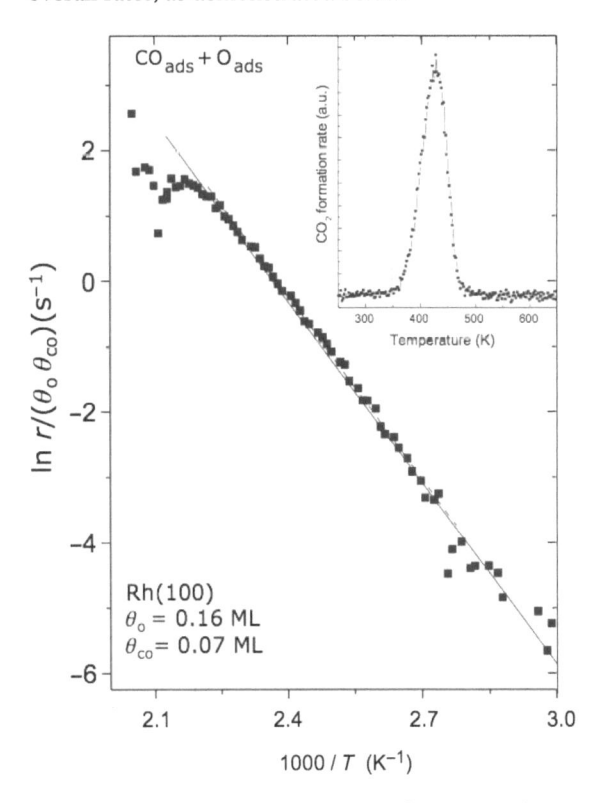

Figure 7.14 The temperature-programmed reaction and corresponding Arrhenius plot based on rate expression (7.22) enables the calculation of kinetic parameters for the elementary surface reaction between CO and O atoms on a Rh(100) surface (adapted from [20]).

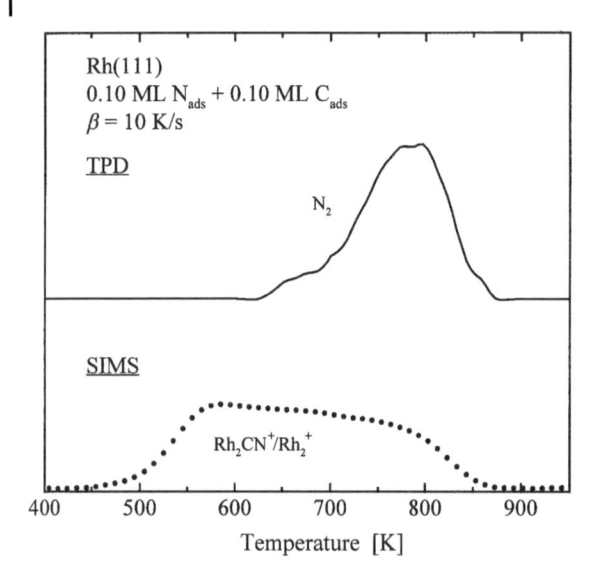

Figure 7.15 Temperature-programmed SIMS experiment showing the surface reaction between adsorbed C and N atoms to give a surface cyanide species at 475 and 600 K; decomposition of CN into C + N, followed by instantaneous desorption of N_2, occurs at higher temperatures. In the absence of C atoms and a route to adsorbed CN, the N atoms would already form N_2 at 500 and 600 K (reprinted with permission from Van Hardeveld et al. [21], copyright 1997 American Chemical Society).

7.2
Kinetic Parameters from Fitting Langmuir–Hinshelwood Models

Here, we illustrate how to use kinetic data to establish a power rate law, and how to derive rate constants, equilibrium constants of adsorption and even heats of adsorption when a kinetic model is available. We use the catalytic hydrodesulfurization of thiophene over a sulfidic nickel-promoted MoS_2 catalyst as an example:

$$C_4H_4S + 2H_2 \rightleftharpoons C_4H_6 + H_2S \qquad (7.24)$$

and

$$C_4H_6 + H_2 \rightleftharpoons C_4H_8 \qquad (7.25)$$

The subsequent hydrogenation of butadiene to but-1-ene and but-2-ene is kinetically insignificant, and these hydrocarbons have no influence on the rate of the first step. H_2S, however, does influence the rate. Briefly, the reaction proceeds over a site where a sulfur atom in the catalyst is missing (see Chapter 9 for details). A high pressure of H_2S simply reduces the number of these vacancies and, therefore, adversely affects the rate.

Establishing the power rate law implies that we need to measure the rate of the reaction for different pressures of the gases thiophene, hydrogen, and hydrogen sulfide. Figure 7.16 shows data measured at 400 °C along with power law fits [22].

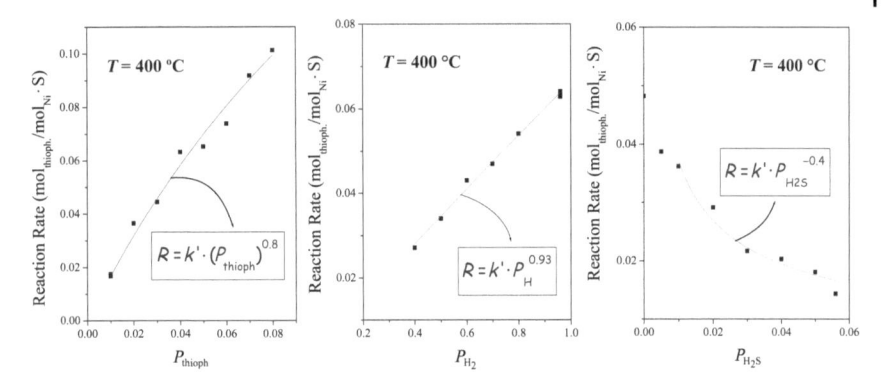

Figure 7.16 Dependence of the rate of thiophene hydrodesulfurization on the partial pressures of the reactants thiophene and hydrogen and of the product hydrogen sulfide, along with fitted power laws (data adapted from Borgna et al. [22]).

Alternatively, one could make a doubly logarithmic plot of rate versus partial pressure and determine the order from the slope. As expected, we find positive orders in thiophene and hydrogen, and a moderately negative order in H_2S (at 400 °C):

$$r = -\frac{dp_T}{dt} = kp_T^{0.8} p_{H_2}^{0.93} p_{H_2S}^{-0.4} \qquad (7.26)$$

where p_T is the partial pressure of thiophene.

Next we will adopt a kinetic scheme and see if it describes the data of Figure 7.16. Several treatments of HDS kinetics are available in the literature. Here we use a simplified scheme in which thiophene (T) exclusively adsorbs on sulfur vacancies, denoted by Δ, and H_2 adsorbs dissociatively on all the sites (indicated by $*$) to form butadiene (B) and H_2S in a rate-determining surface reaction (we ignore the kinetically insignificant hydrogenation steps of butadiene):

$$T + \Delta \underset{}{\overset{K_1}{\rightleftharpoons}} T\Delta \qquad (7.27)$$

$$H_2 + 2* \underset{}{\overset{K_2}{\rightleftharpoons}} 2H* \qquad (7.28)$$

$$T\Delta + 2H* \overset{k_3}{\rightarrow} B + S\Delta + 2* \qquad (7.29)$$

$$S\Delta + 2H* \underset{}{\overset{K_4}{\rightleftharpoons}} H_2S + 2* + \Delta \qquad (7.30)$$

Although the precise mechanism of the HDS reaction is still under debate, we deliberately chose this scheme because it illustrates the kinetics of processes involving two kinds of sites. Consequently, two site balances exist:

$$\theta_T + \theta_S + \theta_\Delta = 1; \quad \theta_H + \theta_* = 1 \qquad (7.31)$$

The reader may verify that the scheme leads to the following Langmuir–Hinshelwood rate equation:

$$r = N_\Delta k_3 \theta_T \theta_H^2 = \frac{N_\Delta k_3 K_1 K_2 p_T p_{H_2}}{(1 + K_1 p_T + K_4^{-1} p_{H_2S})(1 + K_2^{1/2} p_{H_2}^{1/2})^2} \qquad (7.32)$$

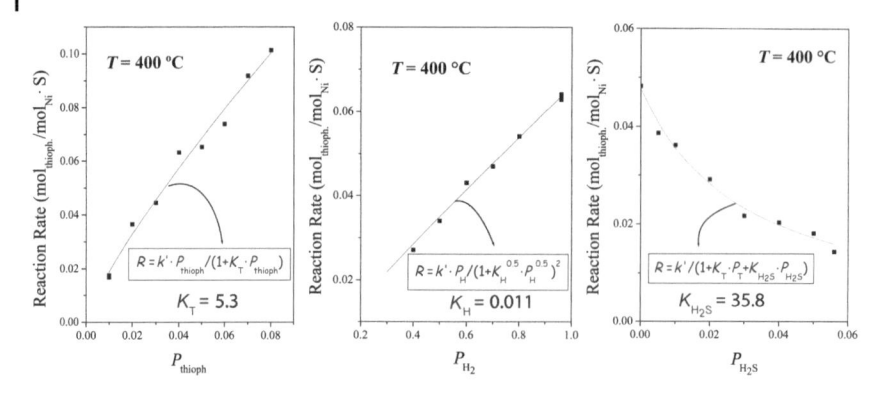

Figure 7.17 Dependence of the rate of thiophene hydrodesulfurization on the partial pressures of the reactants thiophene and hydrogen and of the product hydrogen sulfide, along with fits according to the Langmuir–Hinshelwood model, and the corresponding values of the equilibrium constants (data adapted from Borgna et al. [22]).

Figure 7.17 shows fits of this equation along with optimized values for the equilibrium constants at 400 °C.

The fits indicate that the Langmuir–Hinshelwood model describes the measurements very well. The equilibrium constants point to a relatively strong adsorption of thiophene and, in particular, H_2S, while adsorption of hydrogen is weak. Hence, the term $K_2^{1/2} p_{H_2}^{1/2}$ may safely be ignored in Eq. (7.32). The order in H_2 is 0.93, that is, close to one, which is another indication that hydrogen adsorbs only weakly.

The data in Figures 7.16 and 7.17 are for a single reaction temperature of 400 °C. If one varies the temperature, the equilibrium constants as well as the rate constant will change. Figure 7.18 shows the dependence of the rate on the thiophene partial pressure at different temperatures, along with fits based on the Langmuir–Hinshelwood model. The variation of the equilibrium constants with temperature corresponds to a heat of adsorption for thiophene of 58 ± 5 kJ mol^{-1}, whereas the activation energy for the rate-determining step amounts to 84 ± 5 kJ mol^{-1}.

This example illustrates how the parameters of interest are derived from kinetic measurements. Of course, one should have ensured that the data are free from diffusion limitations and represent the intrinsic reaction kinetics. The data, reported by Borgna, that we used here satisfy these requirements, as the catalyst was actually a nonporous surface science model in a batch reactor.

7.3
Microkinetic Modeling

Microkinetic modeling represents the state of the art in describing the kinetics of catalytic reactions. It was pioneered by Stoltze and Nørskov [23, 24] in the mid-1980s and was further explored by Dumesic and coworkers in the early 1990s [25, 26]. Ideally, as many parameters as can be determined by surface science studies of

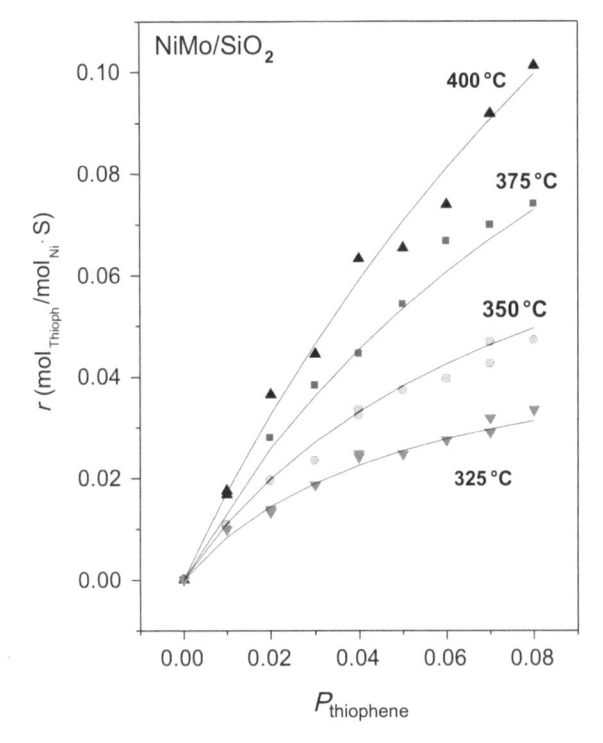

Figure 7.18 Dependence of the rate of thiophene hydrodesulfurization on the partial pressures of thiophene at different temperatures, along with fits according to the Langmuir–Hinshelwood model, Eq. (7.32) (reproduced from Borgna et al. [22] with permission from Elsevier).

adsorption and of elementary steps, as well as results from computational studies, are used as the input in a kinetic model, so that fitting of parameters, as employed in Section 7.2, can be avoided. We shall use the synthesis of ammonia as a worked example.

7.3.1
Reaction Scheme and Rate Expressions

The first step in constructing a microkinetic model is to identify all the elementary reaction steps that may be involved in the catalytic process we want to describe, in this case the synthesis of ammonia. The overall reaction is

$$N_2 + 3H_2 \rightleftharpoons 2NH_3 \tag{7.33}$$

We will list the elementary steps and decide which is rate-limiting and which are in quasi-equilibrium. For ammonia synthesis, a consensus exists that the dissociation of N_2 is the rate-limiting step, and we shall make this assumption here. With quasi-equilibrium steps the differential equation, together with equilibrium condition, leads to an expression for the coverage of species involved in terms of the partial pressures of reactants, equilibrium constants, and the coverage of other intermediates.

The elementary steps in the kinetic model for ammonia synthesis are

$$N_2 + * \rightleftharpoons N_2* \qquad \frac{d\theta_{N_2}}{dt} = P_{N_2}k_1^+\theta_* - k_1^-\theta_{N_2} = 0 \Rightarrow \theta_{N_2} = K_1 P_{N_2}\theta_* \quad (7.34)$$

$$N_2* + * \xrightarrow{RDS} 2N* \qquad r = r^+ - r^- = k_2^+ \theta_{N_2}\theta_* - k_2^- \theta_N^2 \quad (7.35)$$

here RDS stands for Rate Determining Step,

$$N* + H* \rightleftharpoons NH* + *$$

$$\frac{d\theta_{NH}}{dt} = k_3^+ \theta_N \theta_H - k_3^- \theta_{NH}\theta_* = 0 \Rightarrow \theta_N = \frac{\theta_{NH}\theta_*}{K_3 \theta_H} \quad (7.36)$$

$$NH* + H* \rightleftharpoons NH_2* + *$$

$$\frac{d\theta_{NH_2}}{dt} = k_4^+ \theta_{NH}\theta_H - k_4^- \theta_{NH_2}\theta_* = 0 \Rightarrow \theta_{NH} = \frac{\theta_{NH_2}\theta_*}{K_4 \theta_H} \quad (7.37)$$

$$NH_2* + H* \rightleftharpoons NH_3* + *$$

$$\frac{d\theta_{NH_3}}{dt} = k_5^+ \theta_{NH_2}\theta_H - k_5^- \theta_{NH_3}\theta_* = 0 \Rightarrow \theta_{NH_2} = \frac{\theta_{NH_3}\theta_*}{K_5 \theta_H} \quad (7.38)$$

$$NH_3* \rightleftharpoons NH_3 + *$$

$$\frac{d\theta_{NH_3}}{dt} = -k_6^+ \theta_{NH_3} + k_6^- P_{NH_3}\theta_* = 0 \Rightarrow \theta_{NH_3} = \frac{1}{K_6} P_{NH_3}\theta_* \quad (7.39)$$

$$H_2 + 2* \rightleftharpoons 2H*$$

$$\frac{d\theta_H}{dt} = k_7^+ P_{H_2}\theta_*^2 - k_7^- \theta_{H_2} = 0 \Rightarrow \theta_H = \sqrt{K_7 P_{H_2}}\theta_* \quad (7.40)$$

All coverages of adsorbed species (which, of course, can not be measured experimentally) can now be expressed as equilibrium constants and partial pressures. The reader may verify that this leads to

$$\theta_{N_2} = K_1 P_{N_2}\theta_* \equiv a_1 \theta_* \quad (7.41)$$

$$\theta_N = \frac{P_{NH_3}}{K_3 K_4 K_5 K_6 (K_7 P_{H_2})^{3/2}}\theta_* \equiv a_3 \theta_* \quad (7.42)$$

$$\theta_{NH} = \frac{P_{NH_3}}{K_4 K_5 K_6 K_7 P_{H_2}}\theta_* \equiv a_4 \theta_* \quad (7.43)$$

$$\theta_{NH_2} = \frac{P_{NH_3}}{K_5 K_6 \sqrt{K_7 P_{H_2}}}\theta_* \equiv a_5 \theta_* \quad (7.44)$$

$$\theta_{NH_3} = \frac{1}{K_6} P_{NH_3} \theta_* \equiv a_6 \theta_* \tag{7.45}$$

$$\theta_H = \sqrt{K_7 P_{H_2}} \theta_* \equiv a_7 \theta_* \tag{7.46}$$

The fraction of free sites follows from the condition that the sum of all the coverages is unity, and hence

$$\theta_* = 1 - \sum_i a_i \theta_* \quad \Rightarrow \quad \theta_* = \frac{1}{1 + \sum_i a_i} \tag{7.47}$$

where the constants a_i contain equilibrium constants and partial pressures only. The number of free sites can now be expressed in terms of pressures and equilibrium constants as

$$\theta_* = \frac{1}{\left(\begin{array}{c}1 + K_1 P_{N_2} + \dfrac{P_{NH_3}}{K_3 K_4 K_5 K_6 \sqrt{(K_7 P_{H_2})^3}} + \dfrac{P_{NH_3}}{K_4 K_5 K_6 K_7 P_{H_2}} \\ + \dfrac{P_{NH_3}}{K_5 K_6 \sqrt{K_7 P_{H_2}}} + \dfrac{1}{K_6} P_{NH_3} + \sqrt{K_7 P_{H_2}}\end{array}\right)} \tag{7.48}$$

Returning to the overall reaction rate expression, Eq. (7.35), we find

$$r = r^+ - r^- = k_2^+ \theta_{N_2} \theta_* - k_2^- \theta_N^2$$
$$r = k_2^+ K_1 P_{N_2} \theta_*^2 - k_2^- \left[\frac{P_{NH_3}}{K_3 K_4 K_5 K_6 (K_7 P_{H_2})^{3/2}}\right]^2 \theta_*^2 \tag{7.49}$$

A more user friendly form, which expresses how far the reaction is removed from equilibrium, is obtained by utilizing $K_2 = k_2^+/k_2^-$ to give

$$r = k_2^+ K_1 P_{N_2} \left(1 - \frac{P_{NH_3}^2}{K_1 K_2 K_3^2 K_4^2 K_5^2 K_6^2 K_7^3 P_{H_2}^3 P_{N_2}}\right) \theta_*^2 \tag{7.50}$$

Upon equilibrium, the rate must equal zero, that is,

$$\frac{P_{NH_3}^2}{P_{H_2}^3 P_{N_2}} = K_1 K_2 K_3^2 K_4^2 K_5^2 K_6^2 K_7^3 = K_G \tag{7.51}$$

showing that the product of the equilibrium constants in the denominator of Eq. (7.50) equals the equilibrium constant for the ammonia synthesis reaction in Eq. (7.33). Thus, the rate can now conveniently be written as

$$r = k_2^+ K_1 P_{N_2} \left(1 - \frac{P_{NH_3}^2}{K_G P_{H_2}^3 P_{N_2}}\right) \theta_*^2 \tag{7.52}$$

where the term in brackets expresses how far the reaction is removed from equilibrium and the term in front of the brackets describes the rate at the limit of zero conversion. Note that the rate is also proportional to the square of the coverage of free sites.

7.3.2
Activation Energy and Reaction Orders

Important observable parameters such as the apparent activation energy and the reaction order can be derived using our knowledge gained in Chapter 2:

$$n_i \equiv P_i \frac{\partial \ln(r_+)}{\partial P_i} \quad \text{and} \quad E^{\text{app}} \equiv RT^2 \frac{\partial \ln(r_+)}{\partial T} \tag{7.53}$$

Note that we are interested only in the forward rate. In kinetics studies, we prefer to carry out measurements far from equilibrium. Performing the necessary differentiations we obtain the orders:

$$n_{N_2} = 1 - 2\theta_{N_2} \tag{7.54}$$

$$n_{H_2} = 3\theta_N + 2\theta_{NH} + \theta_{NH_2} - \theta_H \tag{7.55}$$

$$n_{NH_3} = -2(\theta_N + \theta_{NH} + \theta_{NH_2} + \theta_{NH_3}) \tag{7.56}$$

and the activation energy:

$$\begin{aligned} E^{\text{app}} = & E_2^{\text{act}} + \Delta H_1 - 2\Delta H_1 \theta_{N_2} - 2(\Delta H_3 + \Delta H_4 + \Delta H_5 + \Delta H_6) \\ & + 3/2\Delta H_7)\theta_N - 2(\Delta H_4 + \Delta H_5 + \Delta H_6 + \Delta H_7)\theta_{NH} \\ & - 2(\Delta H_5 + \Delta H_6 + 1/2\Delta H_7)\theta_{NH_2} - 2\Delta H_6 \theta_{NH_3} - \Delta H_7 \theta_H \end{aligned} \tag{7.57}$$

We see again that, through the coverages in Eqs. (7.54)–(7.57), the overall kinetic parameters are very much dependent on the actual experimental conditions.

The rate of the ammonia production can now be predicted if we can estimate all of the participating equilibrium constants and k_2^+. Where possible, one should take experimental values for the different constants. For instance, it is possible to measure the uptake of atomic nitrogen on the Fe or Ru surface and thereby determine experimentally the factor $k_2^+ K_1$. One can also compare this to calculated values if a potential energy diagram is available that provides sufficient detail about the transition state. It is noteworthy that although the N_2 molecules in the model are in equilibrium with an adsorbed phase, the state of the latter, and in particular its partition function, plays no role, as

$$\theta_{N_2} = K_1 P_{N_2} \theta_* = \left(\frac{q_{N_2} V}{q_{\text{gas}} k_B T}\right) P_{N_2} \theta_* \tag{7.58}$$

and

$$\begin{aligned} r_+ &= \frac{2\theta_\# \nu M}{N_0 A} = \frac{2q'_\# k_B T}{q_{N_2} h} \theta_{N_2} \theta_* = \frac{2q'_\#}{q_{N_2}} \frac{q_{N_2} V}{q_{\text{gas}} h} P_{N_2} \theta_*^2 = \frac{2q'_\# V}{q_{\text{gas}} h} P_{N_2} \theta_*^2 \\ &= k_2^+ K_1 P_{N_2} \theta_*^2 \end{aligned} \tag{7.59}$$

Hence, the properties of the molecularly adsorbed N_2 cancel as soon as we take $k_2^+ K_1$ together, which is the relevant term in the formation of atomic nitrogen. Similarly, but on a much larger scale, partition functions cancel in the term K_G in Eqs. (7.51) and (7.52). Returning to Eq. (7.59), the factor of two arises because the rate describes the number of nitrogen atoms, whereas the transition state refers to the molecule, which dissociates into two atoms.

The forward rate in Eq. (7.59) can be written in the usual manner, introducing the sticking coefficient $S_0(T)$:

$$\begin{aligned} r_+ &= k_2^+ K_1 P_{N_2} \theta_*^2 = \frac{2q'_\#}{q_{gas}} \frac{V}{h} P_{N_2} \theta_*^2 = \frac{2q'_{\#0} e^{\frac{-\Delta E_{act}}{k_B T}}}{q_{3D\text{-}trans} q_{rot} q_{vib}} \frac{V}{h} P_{N_2} \theta_*^2 \\ &= \frac{2q'_{\#0} e^{\frac{-\Delta E_{act}}{k_B T}} V h^3}{V(2\pi m k_B T)^{3/2} q_{rot} q_{vib} h} P_{N_2} \theta_*^2 \\ &= \frac{2q'_{\#0} e^{\frac{-\Delta E_{act}}{k_B T}} h^2}{N_0 \sqrt{2\pi m k_B T} \frac{1}{N_0}(2\pi m k_B T) q_{rot} q_{vib}} P_{N_2} \theta_*^2 \\ &= \frac{2 P_{N_2} S_0(T)}{N_0 \sqrt{2\pi m k_B T}} \theta_*^2 \end{aligned} \quad (7.60)$$

The sticking coefficient is calculated directly from transition state theory, as in Chapter 3. The vibrational partition function is

$$q_{vib} = \frac{e^{\frac{-h\nu}{2k_B T}}}{1 - e^{\frac{-h\nu}{k_B T}}} \cong e^{\frac{-h\nu}{2k_B T}} \quad (7.61)$$

since $h\nu \gg k_B T$. The rotational partition function is given by $q_{rot} = k_B T/(2\varepsilon_{rot})$ ($\varepsilon_{rot} = 0.248$ meV). The density of sites on a Fe(111) surface is $N_0 = 1/\sqrt{3}a^2 = 1/14.3$ Å^2. The transition state consists of five frustrated modes (three translational and two rotational modes), while the sixth mode is the reaction coordinate for dissociation. All these modes have relatively low frequencies, but since they are restricted in the transition state we shall assume that $h\nu_i \gg k_B T$ and that the modes are degenerate, that is,

$$q'_{\#0} = \prod_{i=1}^{5} \frac{e^{\frac{-h\nu_i}{2k_B T}}}{1 - e^{\frac{-h\nu_i}{k_B T}}} \cong e^{\frac{-h\sum_i \nu_i}{2k_B T}} \quad (7.62)$$

All this combines to give

$$S_0(T) = \frac{e^{\frac{-\Delta E_{act} + \frac{1}{2}h\nu - \frac{1}{2}h\sum_i \nu_i}{k_B T}} h^2 \varepsilon_{rot}}{\sqrt{3}a^2 \pi m (k_B T)^2} = S_0^0 e^{\frac{-\Delta E_{act} + \frac{1}{2}h\nu - \frac{1}{2}h\sum_i \nu_i}{k_B T}} \quad (7.63)$$

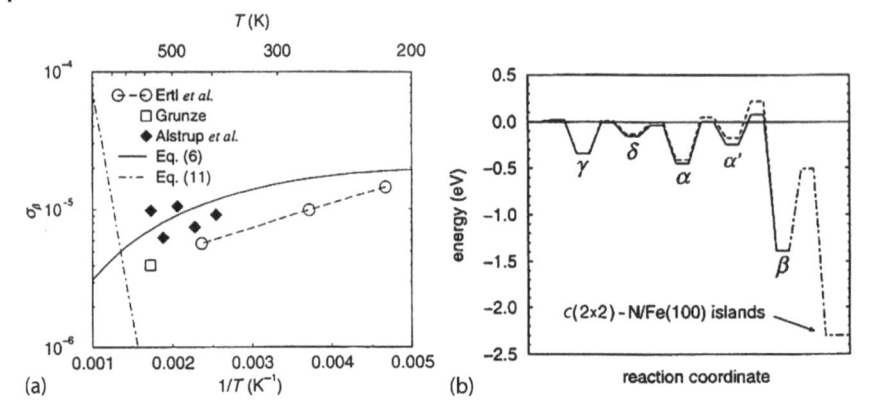

Figure 7.19 (a) Comparison between experimental sticking coefficients of N$_2$ on Fe(111) and the prediction on the basis of Eq. (7.57) with an activation energy of 0.03 eV. (b) Potential energy diagram for molecular nitrogen dissociating on Fe(111). Note that the model predicts an activation energy of 0.1 eV, in good agreement with the 0.03 eV used in the plot on the left (reproduced from Mortensen et al. [27] with permission from Elsevier).

The prefactor at 500 K is $S_0^0 = 1.49 \times 10^{-5}$, illustrating the reduction in entropy when a molecule from the gas phase becomes restricted in motion on an adsorption site on the surface. The activation energy has been measured to be slightly negative in this case, that is, $\Delta E_{\text{measured}} = -0.034$ eV, which should be compared with

$$\Delta E^{\text{app}} = k_B T^2 \frac{\partial \ln[S_0(T)]}{\partial T} = \Delta E_{\text{act}} - \frac{1}{2}h\nu + \frac{h}{2}\sum_i \nu_i - 2k_B T \qquad (7.64)$$

If we assume that

$$\Delta E_{\text{act}} - \frac{1}{2}h\nu + \frac{h}{2}\sum_i \nu_i = 0.03 \text{ eV} \qquad (7.65)$$

we obtain a sticking coefficient with not only the correct magnitude as compared to the experimental measurements but also with the right apparent activation energy for sticking (e.g., Figure 7.19a). The figure shows the experimentally determined sticking coefficients together with the theoretical prediction (solid line). Figure 7.19b describes the results of theoretical investigation, displaying the reaction coordinate a N$_2$ molecule undergoes when approaching a surface. The theoretically predicted curve (solid curve) suggests an activation energy of 0.1 eV. Thus, nitrogen dissociation is rate-limiting for the ammonia synthesis on the Fe(111) surface because of the reduction in entropy between gas phase N$_2$ and the transition state for dissociation on the surface, and not the energy barrier!

Usually, activation energies for dissociation are much higher than $k_B T$ and situations like this, where the apparent activation energy becomes negative for the rate-limiting process, are rare. Nevertheless, the present case illustrates nicely that entropy changes may play an important role and that it is not always the activation energy that dominates the process.

Example

Having estimated the sticking coefficient of nitrogen on the Fe(111) surface above, we now consider the desorption of nitrogen, for which the kinetic parameters are readily derived from a TPD experiment. Combining adsorption and desorption enables us to calculate the equilibrium constant of dissociative nitrogen adsorption from

$$\frac{d\theta_N}{dt} = 0 = \frac{2P_{N_2}S_0(T)}{N_0\sqrt{2\pi m k_B T}}(1-\theta_N)^2 - \nu_0 e^{\frac{-E_{des}}{k_B T}}\theta_N^2 \qquad (7.66)$$

Solving for the coverage of atomic nitrogen yields

$$\theta_N = \frac{\sqrt{K_{N_2}P_{N_2}}}{1+\sqrt{K_{N_2}P_{N_2}}} \qquad (7.67)$$

where

$$K_{N_2} = \frac{2S_0^0 e^{-\left(\Delta E_{act} - \frac{h}{2}\nu + \frac{h}{2}\Sigma_i \nu_i - E_{des}\right)/k_B T}}{N_0 \nu_0 \sqrt{2\pi m k_B T}} \qquad (7.68)$$

$$= \frac{2h^2 \varepsilon_{rot}}{\pi m (k_B T)^2} \frac{e^{-\left(\Delta E_{act} - \frac{h}{2}\nu + \frac{h}{2}\Sigma_i \nu_i - E_{des}\right)/k_B T}}{\nu_0 \sqrt{2\pi m k_B T}}$$

The N_2 desorbs as a symmetric second-order TPD peak around 740 K, with $E_{des} = 190$ kJ mol^{-1} and $\nu = 1 \times 10^{13}$ s^{-1}. Using these values, the nitrogen coverage can be calculated as a function of temperature and pressure according to Eqs. (7.67) and (7.68), as shown in Figure 7.20. The conclusion is that an iron surface under N_2 gas will be largely covered by N atoms up to fairly high temperatures. Is this also true under ammonia synthesis conditions?

7.3.3
Ammonia Synthesis Catalyst under Working Conditions

For ammonia synthesis, we still need to determine the coverages of the intermediates and the fraction of unoccupied sites. This requires a detailed knowledge of the individual equilibrium constants. Again, some of these may be accessible via experiments, while the others will have to be determined from their respective partition functions. In doing so, several partition functions will again cancel in the expressions for the coverage of intermediates.

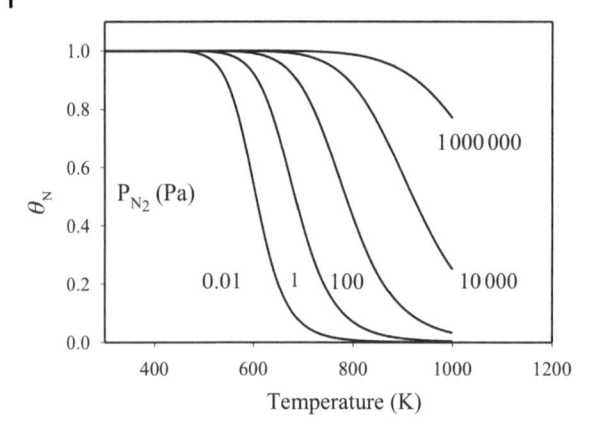

Figure 7.20 Coverage of atomic nitrogen on Fe(111) on the basis of Eqs. (7.67) and (7.68). Desorption in ultrahigh vacuum is observed as a symmetric TPD peak around 740 K. Note that at reaction conditions employing 0.1 bar (10^4 Pa) of nitrogen the surface will be largely covered by nitrogen atoms.

Figure 7.21 shows the results for the ammonia synthesis on real catalysts in a reactor. The surface is predominantly covered by atomic nitrogen and by NH intermediates. This actually limits the rate of the reaction as soon as an appreciable partial pressure of ammonia has built up. In fact, ammonia poisons the reaction.

What happens in a plug flow reactor will give us a feeling for the problems encountered in the industry. Consider a small volume moving through the reactor. At the beginning of the reactor, the ammonia concentration will initially be zero (in practice this is not true because ammonia is recycled); hence, we can treat the volume as a batch reactor approaching equilibrium. The initial reaction is fast since there is no site blocking, but as soon as conversion increases the catalyst

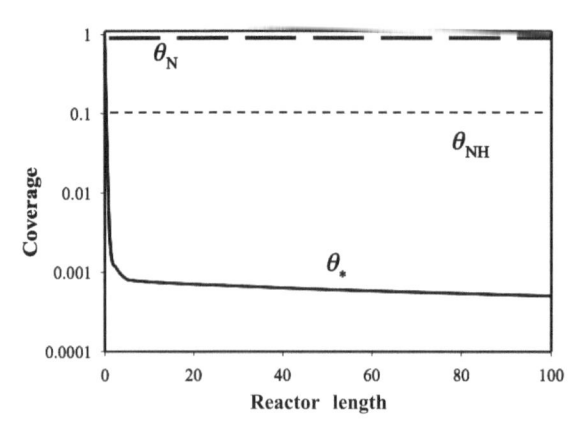

Figure 7.21 Fraction of unoccupied sites and of sites occupied by atomic nitrogen and NH, as a function of reactor length on a potassium-promoted iron ammonia catalyst at 673 K, 100 bar, and approaching 68% of the equilibrium ammonia concentration (calculation based on Stoltze [28]).

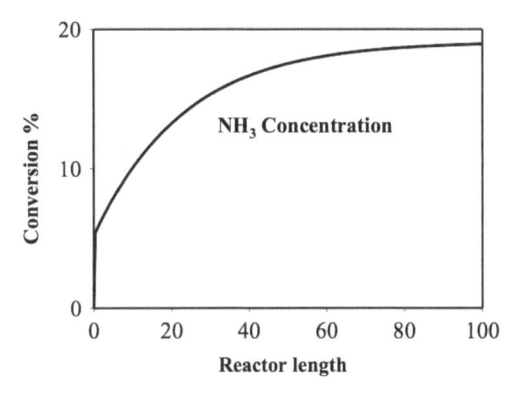

Figure 7.22 NH$_3$ concentration as a function of reactor length in the synthesis of ammonia with a potassium-promoted iron catalyst. The exit concentration is 19% and corresponds to 75% of the obtainable equilibrium conversion at 100 bar and 673 K (calculation based on Stoltze [28]).

surface becomes blocked by nitrogen and the rate decreases rapidly. Figure 7.22 displays the concentration of ammonia as a function of reactor length. At the reactor exit, 19% conversion is obtained, corresponding to 75% of the equilibrium conversion. This maximum is only possible if the reactor is very long, which is not desirable, and so clever engineering is needed to circumvent this effect and optimize the conversion. A further complication is that the process is exothermic, causing the temperature to increase along the bed, and implying that the maximum obtainable conversion into ammonia also decreases along the bed. Cooling of the reactor is a solution. We shall return to these problems in Chapter 8, in which we treat the ammonia synthesis process in detail. A microkinetic model greatly enhances our insight into the process. For example, intuitively one might have anticipated that a more reactive surface would help in dissociating the nitrogen more efficiently. However, this would increase the concentration of N atoms on the surface and make the conversion into ammonia more difficult. Rather, one should try to maintain the same sticking coefficient, but destabilize the nitrogen on the surface.

Finally, the constructed microkinetic model must of course be tested against measurements performed with real catalysts. Figure 7.23 shows a plot of the calculated output from the reactor against experimental values. Apparently, the microkinetic model describes the situation very well. This does not prove that the model is correct since models based on another series of elementary steps might also work.

Overall, catalytic processes in industry are more commonly described by simple power rate law kinetics, as discussed in Chapter 2. However, power rate laws are simply a parameterization of experimental data and provide little insight into the underlying processes. A microkinetic model may be less accurate as a description, but it enables the researcher to focus on those steps in the reaction that are critical for process optimization.

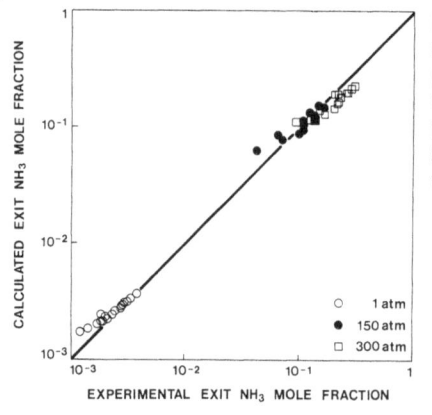

Figure 7.23 Comparison of predicted and experimentally measured rates for various pressures. Note the good correspondence, indicating that the model describes the experiment well (reproduced with permission from Stoltze and Nørskov [23], copyright 1985 by the American Physical Society).

References

1. Larsen, J.H. and Chorkendorff, I. (1999) From fundamental studies of reactivity on single crystals to the design of catalysts. *Surface Science Reports*, **35**, 165–222.
2. Kose, R., Brown, W.A., and King, D.A. (1999) Determination of the Rh–C bond energy for C_2H_2 and C_2H_4 reactive adsorption on Rh(100). *Chemical Physics Letters*, **311**, 109–116.
3. Rasmussen, P.B., Holmblad, P.M., Christoffersen, H., Taylor, P.A., and Chorkendorff, I. (1993) Dissociative adsorption of hydrogen on Cu(100) at low temperatures. *Surface Science*, **287**, 79–83.
4. Dahl, S., Logadottir, A., Egeberg, R.C., Larsen, J.H., Chorkendorff, I., Tornqvist, E., and Nørskov, J.K. (1999) Role of steps in N_2 activation on Ru(0001). *Physical Review Letters*, **83**, 1814–1817.
5. Niemantsverdriet, J.W. (2007) *Spectroscopy in Catalysis, an Introduction*, Wiley-VCH Verlag GmbH, Weinheim.
6. Siddiqui, H.R., Guo, X., Chorkendorff, I., and Yates, J.T. (1987) CO adsorption site exchange between step and terrace sites on Pt(112). *Surface Science*, **191**, L813–L818.
7. Niemantsverdriet, J.W., Dolle, P., Markert, K. and Wandelt, K. (1987) Thermal-desorption of strained monoatomic ag and au layers from Ru(001). *Journal of Vacuum Science & Technology A – Vacuum Surfaces and Films*, **5**, 875–878.
8. Hopstaken, M.J.P. and Niemantsverdriet, J.W. (2000) Lateral interactions in the dissociation kinetics of NO on Rh(100). *Journal of Physical Chemistry B*, **104**, 3058–3066.
9. Miller, J.B., Siddiqui, H.R., Gates, S.M., Russell, J.N., Yates, J.T., Tully, J.C., and Cardillo, M.J. (1987) Extraction of kinetic-parameters in temperature programmed desorption – A comparison of methods. *Journal of Chemical Physics*, **87**, 6725–6732.
10. De Jong, A.M. and Niemantsverdriet, J.W. (1990) Thermal-desorption analysis – Comparative test of 10 commonly applied procedures. *Surface Science*, **233**, 355–365.
11. Chan, C.M., Aris, R., and Weinberg, W.H. (1978) Analysis of thermal desorption mass spectra 1. *Applied Surface Science*, **1**, 360–376.
12. Habenschaden, E. and Küppers, J. (1984) Evaluation of flash desorption spectra. *Surface Science*, **138**, L147–L150.
13. King, D.A. (1975) Thermal desorption from metal-surfaces. *Surface Science*, **47**, 384–402.
14. Jansen, A.P.J. (2012) *An Introduction to Kinetic Monte Carlo Simulations of Surface Reactions*, Springer, Berlin.
15. Nieminen, R.M. and Jansen, A.P.J. (1997) Monte Carlo simulations of surface reactions. *Applied Catalysis A – General*, **160**, 99–123.

16 A.P. van Bavel, Hopstaken, M.J.P., Curulla, D., Niemantsverdriet, J.W., Lukkien, J.J., and Hilbers, P.A.J. (2003) Quantification of lateral repulsion between coadsorbed CO and N on Rh(100) using temperature-programmed desorption, low-energy electron diffraction, and Monte Carlo simulations. *Journal of Chemical Physics*, **119** 524–532.

17 De Koster, A. and Van Santen, R.A. (1990) Dissociation of CO on Rh surfaces. *Surface Science*, **233**, 366–380.

18 Taylor, P.A., Rasmussen, P.B., Ovesen, C.V., Stoltze, P., and Chorkendorff, I. (1992) Formate synthesis on Cu(100). *Surface Science*, **261**, 191–206.

19 Hopstaken, M.J.P., van Gennip W.J.H., and Niemantsverdriet, J.W. (1999) Reactions between NO and CO on rhodium(111): An elementary step approach. *Surface Science*, **433**, 69–73.

20 Hopstaken, M.J.P. and Niemantsverdriet, J.W. (2000) Structure sensitivity in the CO oxidation on rhodium: Effect of adsorbate coverages on oxidation kinetics on Rh(100) and Rh(111). *Journal of Chemical Physics*, **113**, 5457–5465.

21 van Hardeveld, R.M., van Santen, R.A., and Niemantsverdriet, J.W. (1997) C–N coupling in reactions between atomic nitrogen and ethylene on Rh(III), *Journal of Physical Chemistry B*, **101**, 7901–7907.

22 Borgna, A., Hensen, E.J.M., van Veen, J.A.R., and Niemantsverdriet, J.W. (2004) Intrinsic kinetics of thiophene hydrodesulfurization on a sulfided NiMo/SiO$_2$ planar model catalyst. *Journal of Catalysis*, **221**, 541–548.

23 Stoltze, P. and Nørskov, J.K. (1985) Bridging the pressure gap between ultrahigh-vacuum surface physics and high-pressure catalysis. *Physical Review Letters*, **55**, 2502–2505.

24 Stoltze, P. and Nørskov, J.K. (1988) An interpretation of the high-pressure kinetics of ammonia-synthesis based on a microscopic model. *Journal of Catalysis*, **110**, 1–10.

25 Dumesic, J.A., Rudd, D.A., Apuvicio, L.A., Bekoske, J.E., and Trevino, A.A. (1993) *The Microkinetics of Heterogeneous Catalysis*, American Chemical Society, Washington DC.

26 Cortright, R.D. and Dumesic, J.A. (2001) Kinetics of heterogeneous catalytic reactions: Analysis of reaction schemes, in *Advances in Catalysis* (eds B.C. Gates and H. Knözinger), **46**, 161–264.

27 Mortensen, J.J., Hansen, L.B., Hammer, B., and Nørskov, J.K. (1999) Nitrogen adsorption and dissociation on Fe(111). *Journal of Catalysis*, **182**, 479–488.

28 Stoltze, P. (1987) Surface science as the basis for the understanding of the catalytic synthesis of ammonia. *Physica Scripta*, **36**, 824–864.

Chapter 8
Catalysis in Practice: Synthesis Gas and Hydrogen

8.1
Introduction

Catalysis involves phenomena on a range of length and time scales (see also Chapter 1). While previous chapters were mainly concerned with catalysis on the molecular level, the following three chapters will also include the dimensions of applied catalysis. These range from the immense capacity of an oil refinery (which processes some 5 bathtubs of crude oil per second!) to the small-scale application of a monolith in an automotive exhaust converter. In this chapter, we discuss processes in which the large-scale production and use of hydrogen are central: steam reforming, water-gas shift, and the synthesis of ammonia, methanol, and synthetic fuels. Finally, we discuss hydrogen as a potential fuel for the future. Chapter 9 deals with the oil refinery and petrochemical industry, and Chapter 10 with important environmental applications of catalysis. All these processes are essential for sustaining our modern lifestyle in a responsible way.

8.2
Synthesis Gas and Hydrogen

Hydrogen plays a central role in many large-scale processes, such as the synthesis of ammonia, the hydrotreating of crude oil fractions, the hydrogenation of edible fats, and many others. On a large scale, hydrogen is produced from the intermediate synthesis gas or syngas, a mixture of CO and H_2. Therefore, we treat these two subjects in one chapter.

Syngas can be produced from many sources, including natural gas, coal, biomass, oil residue, or virtually any hydrocarbon feedstock, by reaction with steam or oxygen, see Figure 8.1. Its formation is strongly endothermic and requires high temperatures. Steam reforming of natural gas (or shale gas) proceeds in tubular reactors that are heated externally. The process uses nickel catalyst on a special support that is resistant against the harsh process conditions. Waste heat from the oven section is used to preheat gases and to produce steam. This plant gen-

Concepts of Modern Catalysis and Kinetics, Third Edition. I. Chorkendorff and J.W. Niemantsverdriet.
© 2017 WILEY-VCH Verlag GmbH & Co. KGaA. Published 2017 by WILEY-VCH Verlag GmbH & Co. KGaA.

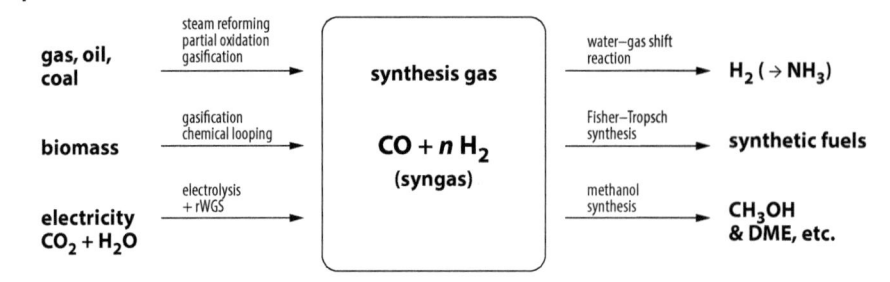

Figure 8.1 Syngas production and primary use.

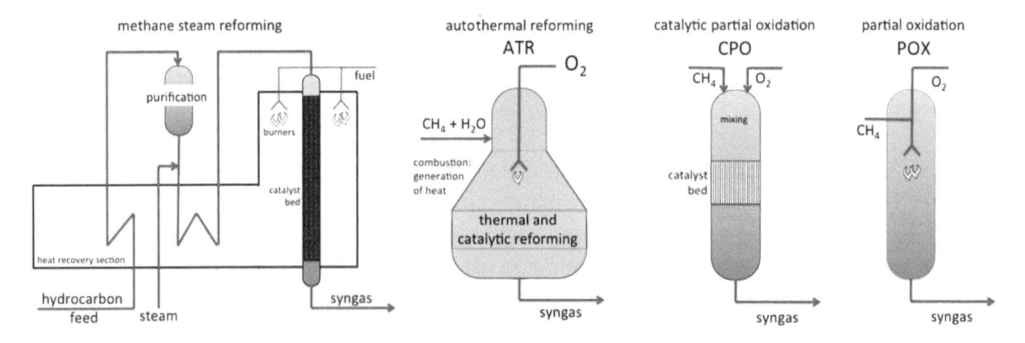

Figure 8.2 Reactors and process layout for syngas production from natural gas and shale gas.

erates syngas with H_2/CO ratios in the range of 3–4 and is suitable for hydrogen production. We discuss it in more detail below.

Partial oxidation of methane (or hydrocarbons) is a noncatalytic, large-scale process and yields syngas with a H_2/CO ratio of about 2. This is an optimal ratio for producing methanol, or synthetic fuels in gas-to-liquids plants. A catalytic version of partial oxidation (CPO), based on short-contact time conversion of methane, hydrocarbons, or biomass on, for example, rhodium catalysts, is suitable for small-scale applications.

Autothermal reforming (ATR) is a hybrid which combines methane steam reforming and oxidation in one process. The heat needed for reforming is generated inside the reactor by oxidation of the feed gas. As partial oxidation, ATR is also suitable for large-scale production of syngas for gas-to-liquids or large-scale methanol synthesis processes.

Alternative routes to syngas, such as reduction of CO_2 from flue gas with H_2 from electrolytic splitting of water may become interesting from the viewpoint of storage of wind or solar energy in the near future.

Figure 8.2 summarizes the processes for making syngas from natural gas. We refer to Rostrup-Nielsen and Christiansen [1] for further details.

Table 8.1 Typical composition of natural gas from the North Sea.

Component	CH$_4$	C$_2$H$_6$	C$_3$H$_6$	C$_{4+}$	CO$_2$	N$_2$	S
Vol.%	94.9	3.8	0.2	0.1	0.2	0.8	4 ppm

8.2.1
Steam Reforming: Basic Concepts of the Process

Steam reforming largely owes its importance to the extensive and relatively inexpensive amounts of methane that are available world-wide. It is the first step in several very important large-scale chemical processes that use hydrogen. Steam reforming produces synthesis gas (also called syngas), a mixture of H$_2$, CO, and CO$_2$ that can either be used or be converted to hydrogen by using the water-gas shift reaction, where the reducing capacity of CO is employed to convert more steam into hydrogen.

Steam reforming proceeds according to the following reactions [1]:

$$CH_4 + H_2O \rightleftharpoons CO + 3H_2 \quad \Delta H^\circ_{298} = +206 \text{ kJ mol}^{-1} \quad (8.1)$$

$$C_nH_m + nH_2O \rightleftharpoons nCO + \left(n + \frac{m}{2}\right)H_2 \quad \Delta H^\circ_{298} > 0 \text{ kJ mol}^{-1} \quad (8.2)$$

Simultaneously, the so-called water–gas shift reaction also produces CO$_2$:

$$CO + H_2O \rightleftharpoons CO_2 + H_2 \quad \Delta H^\circ_{298} = -41 \text{ kJ mol}^{-1} \quad (8.3)$$

Steam reforming was developed in Germany at the beginning of the twentieth century to produce hydrogen for ammonia synthesis and was further developed in the 1930s when natural gas and other hydrocarbon feedstocks such as naphtha became available on a large scale.

Natural gas consists mainly of methane together with some higher hydrocarbons (Table 8.1). Sulfur, if present, must be removed to a level of about 0.2 ppm prior to the steam reforming process as it poisons the catalyst. This is typically done by catalytically converting the sulfur present as thiols, thiophenes, or COS into H$_2$S, which is then adsorbed stochiometrically by ZnO, at 400 °C, upstream of the reactor.

Many late transition metals such as Pd, Pt, Ru, Rh, and Ir can be used as catalysts for steam reforming, but nickel-based catalysts are, economically, the most feasible. More reactive metals such as iron and cobalt are in principle active but they oxidize easily under process conditions. Ruthenium, rhodium, and other noble metals are more active than nickel, but are less attractive due to their costs. A typical catalyst consists of relatively large Ni particles dispersed on an Al$_2$O$_3$ or an AlMgO$_4$ spinel. The active metal area is relatively low, on the order of only a few m^2/g.

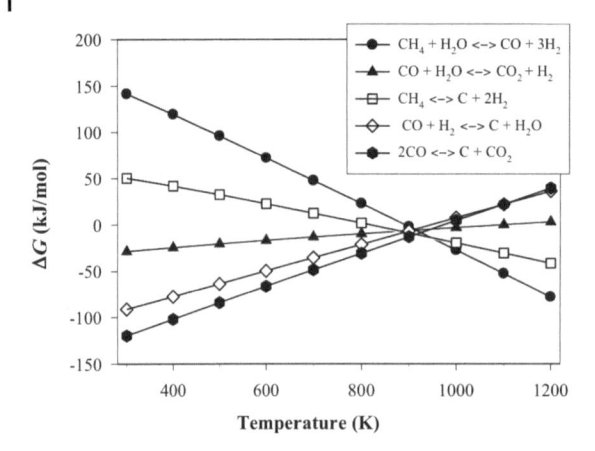

Figure 8.3 Free energy change for steam reforming and related reactions, including those leading to deposition of carbon on the catalyst. The plot clearly illustrates why steam reforming needs to be carried out at high temperatures.

Figure 8.4 The heart of a steam reforming plant consists basically of a six-story oven in which the reactor tubes stand vertically (courtesy of Haldor Topsøe A/S).

The process is operated at high temperatures (up to 1000 °C) and moderate pressures (25–35 bar), for reasons that are easily understood by examining both Eqs. (8.1)–(8.3) and the plots of ΔG versus temperature shown in Figure 8.3. Reactions that lead to carbon deposition are also included because minimizing these is a critical issue in steam reforming.

Traditionally, the steam reforming reactor has a tubular design in which vertical tubes, loaded with catalyst, are surrounded by furnaces to supply the heat required for the strongly endothermic process, see Figure 8.4. Combustion of natural gas supplies the heat to the tubes.

Recently, more sophisticated reactors, such as the autothermal reformer, use a burner at the reactor entrance to heat the catalyst bed directly through combustion of some of the methane.

8.2.2
Mechanistic Detail of Steam Reforming

Methane is a stable and highly symmetrical molecule. It is activated in a direct dissociative adsorption step, which is associated with substantial energy and entropy barriers. Given, also, the extremely low sticking coefficients for methane adsorption (see beginning of Chapter 7, it is no surprise that dissociative adsorption is the rate-limiting step in the steam reforming process. This strongly activated elementary step has been studied extensively, including dynamic effects. For example, Holmblad *et al.* [2] found that vibrational excitation of the molecule substantially enhances its sticking. Complete studies of the steam reforming on well-defined single crystals are not available as yet, but the reverse reaction – methanation – has been studied extensively by Goodman and coworkers [3]. Here, we discuss recent results from computational chemistry. The overall reaction scheme is

$$CH_4 + 2* \longrightarrow CH_3* + H* \tag{8.4}$$

$$CH_3* + * \rightleftharpoons CH_2* + H* \tag{8.5}$$

$$CH_2* + * \rightleftharpoons CH* + H* \tag{8.6}$$

$$CH* + * \rightleftharpoons C* + H* \tag{8.7}$$

$$H_2O + 2* \rightleftharpoons HO* + H* \tag{8.8}$$

$$HO* + * \rightleftharpoons O* + H* \tag{8.9}$$

$$C* + O* \rightleftharpoons CO* + * \tag{8.10}$$

$$CO* \rightleftharpoons CO + * \tag{8.11}$$

$$2H* \rightleftharpoons H_2 + 2* \tag{8.12}$$

The energy diagram given in Figure 8.5 indicates that C or CH species are the most stable intermediates, depending on the surface. For the reverse reaction, methanation, CO dissociation has a large barrier on the perfect (111) surface but is favored on the stepped (211). Hence, steps on the (111) surface are predicted to be the sites where CO dissociates, which is in good agreement with surface science studies as shown below. Despite their higher reactivity, one cannot automatically assume that they will indeed dominate the reaction, since they may be blocked by the stable C or CH intermediates. Research is ongoing in this area, particularly because carbon deposition causes severe problems in the steam reforming process.

That steps are essential and much more active for the dissociation of both methane and CO has been demonstrated in surface science experiments, by using single crystals cut in certain directions giving rise to steps of the desired geometry. For example, one can cut a nickel crystal such that it displays a B5 type

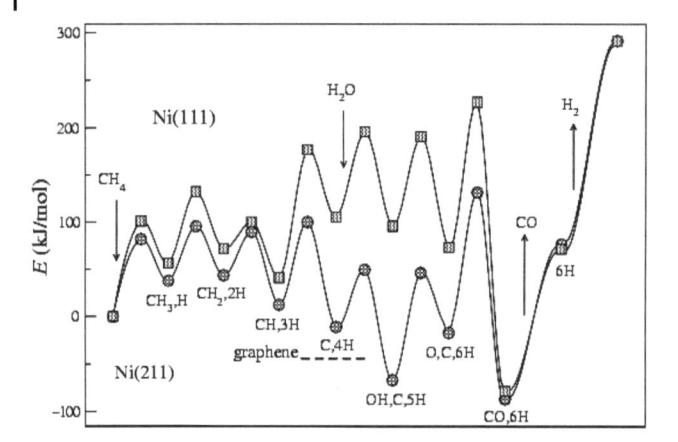

Figure 8.5 Potential energy diagram based on DFT calculations. Notice how the reaction pathway is strongly modified by the presence of atomic steps on the Ni(112) surface. First of all, steps lower the barrier for the initial methane dissociation. Although this barrier is not the largest of the series, the large loss in entropy gives a very low pre-exponential factor, causing this step to be rate limiting (reproduced from Bengaard et al. [4] with permission from Elsevier).

step for every 25 terrace atoms, as shown in Figure 8.6. Also shown is the STM picture of this surface after exposure to CO at 500 K. Notice that the steps are not as regular as indicated in the model, but on average they will have the right density. The carbon causes reconstruction of the (111) terraces seen as protrusions growing out from the steps [5].

When measuring the amount of carbon on the surface after a certain CO dose, it is possible to make an uptake curve as shown in Figure 8.7. In these experiments, the amount of carbon is very low, < 0.05 ML, and it is therefore a challenge to measure it by conventional methods like AES or XPS. Instead, the carbon coverage was determined by exposing the crystal to O_2 at 500 K and measuring how much CO could be reacted off again. This method proved to be at least an order of magnitude more sensitive that the spectroscopic methods.

In the upper curve the surface was clean when exposed to the CO, that is, steps sites were available. Since the CO can dissociate here and since the oxygen is subsequently reacted of with the CO, the procedure leaves carbon on the surface. If on the contrary, the steps are blocked by 0.05 ML sulfur, no CO dissociation and, thus, no carbon is observed, indicating that the terrace atoms are much less reactive than the B5 sites. This is in very good agreement with the potential energy diagram shown in Figure 8.5 where the barrier for dissociation is prohibitive for the (111) surface, while it is smaller for the (112) surface, which contains step sites. The example of Figure 8.7 is analogous to the dissociation of N_2 on ruthenium steps discussed in Chapter 7 (Figure 7.4) where the enhancement of reactivity is due both to an electronic effect (undercoordinated atoms at the steps) and an ensemble effect (since the reactive site consist of five surface atoms in the B5 sites instead of only four at the terrace).

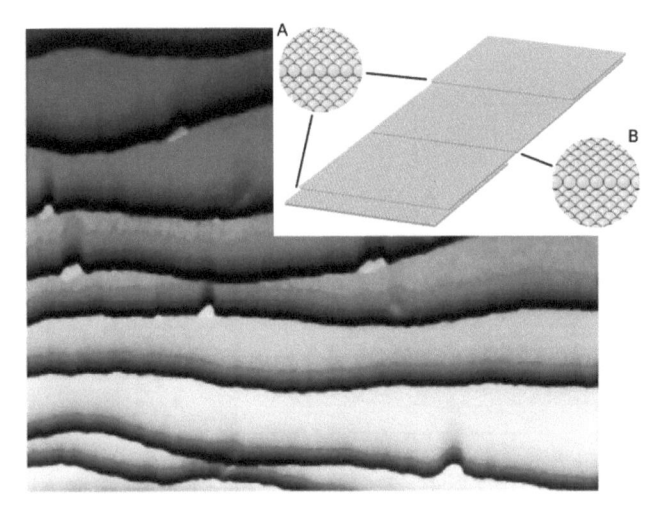

Figure 8.6 Shown is an STM of a Ni(14 13 13) where roughly 0.5% carbon has been deposited by dissociation of CO at 500 K. The carbon reacts with the surface and causes a new structure growing out from the steps. The hard ball model of a surface is shown as an insert. The Ni(14 13 13) exhibits exclusively B5 type steps and (111) terraces (adapted from Andersson et al. [5]).

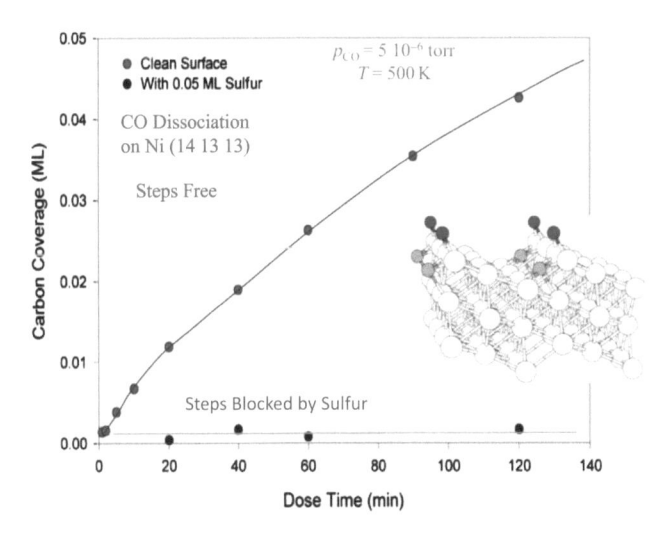

Figure 8.7 Carbon deposition as a function of CO dose for the stepped Ni(14 13 13) surface when the B5 steps are available and when they are blocked by 0.05 ML sulfur, demonstrating that unoccupied steps are crucial for CO dissociation. The insert shows the transition state for CO dissociation on the B5 steps. Note that the dissociation is enabled by both electronic and ensemble effects, similarly as observed in Figure 7.4 for the dissociation of the N_2 molecule on a stepped ruthenium surface (adapted from Andersson et al. [5]).

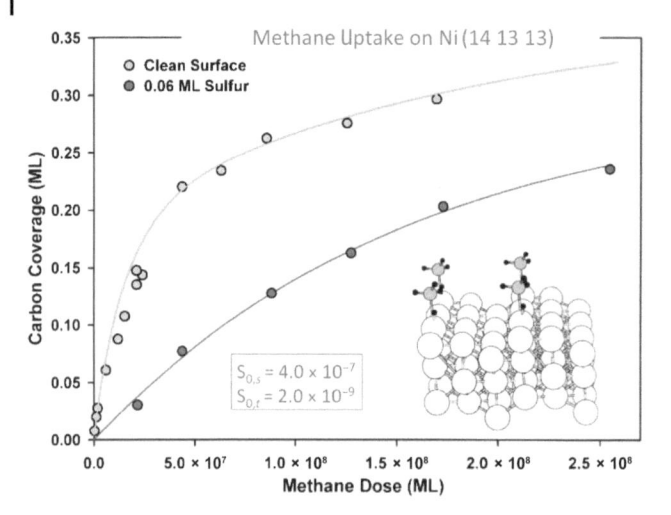

8 Catalysis in Practice: Synthesis Gas and Hydrogen

Figure 8.8 The carbon uptake is shown as a function of methane dose on a Ni(14 13 13) surface that is either clean (upper curve) or where the steps are blocked by sulfur (lower curve). The initial sticking coefficient of the methane is a factor of 200 higher for the stepped surface. The insert shows the transition state for methane activation; it is worth noticing that in this case there is only an electronic effect since the dissociation takes place on one atom only (reproduced from Abild–Pedersen et al. [6] with permission from Elsevier).

The influence of the steps on the opposite reaction – steam reforming – is somewhat different [6]. Figure 8.8 shows a similar experiment where the Ni(14 13 13) surface was exposed to doses of methane. Its dissociation is considered to be the rate-limiting step and the dissociation rate is again monitored by measuring the amount of carbon deposited. It is clearly seen that the clean stepped surface is more reactive since the carbon uptake is faster here. However, the difference is less as in the example of Figure 8.6, since here the terrace sites are also active for methane activation. The sticking coefficients differ by a factor of 200, which again is in good agreement with the potential energy diagram of Figure 8.5: the difference in activation barrier for methane dissociation on steps or terraces is about 20 kJ/mol. The enhancement is due to an electronic effect, related to the degree of coordination of Ni atoms at the step and in the terrace. This type of experiment confirms the importance of considering the influence of the detailed structure of nanoparticles in catalysts, as was also shown in Fig. 6.35 of Chapter 6.

8.2.3
Challenges in the Steam Reforming Process

The nickel steam-reforming catalyst is very robust but is threatened by carbon deposition [1]. As indicated in Figure 8.3, several reactions may lead to carbon (graphite), which accumulates on the catalyst. In general, the probability of carbon formation increases with decreasing oxidation potential, that is, lower steam

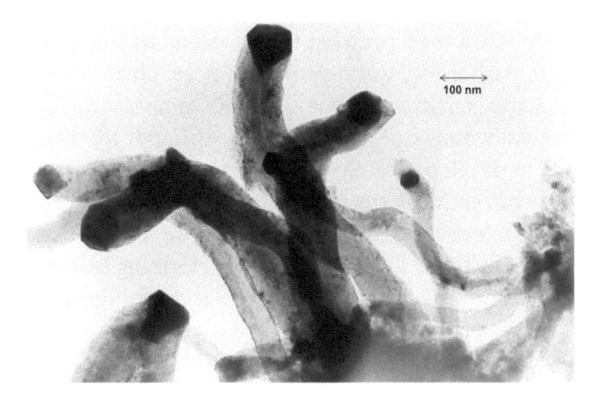

Figure 8.9 Transmission electron microscope picture of carbon formation and filament growth on a SiO$_2$-supported Ni catalyst after exposure to a CH$_4$ + H$_2$ gas mixture at 1 bar and 763 K. The dark pear-shaped areas are the Ni particles, which are approximately 1000 Å in diameter. (Courtesy of Haldor Topsøe A/S).

content (which may be desirable for economic reasons). The electron micrograph in Fig. 8.9 dramatically illustrates how carbon formation may disintegrate a catalyst and cause plugging of a reactor bed.

Carbon deposits come in several forms, but the typical morphology of carbon formed at high temperatures is that of whiskers, which are long filaments of graphitic carbon. Such filaments usually consist of more or less disordered, curved graphite planes formed into stacked cones or concentric tubes. The diameter of the filament is determined by the diameter of the catalytic particle from which it grows. The metal is commonly transformed into pear-shaped particles (Figure 8.9). By using *in situ* electron microscopes, real-time movies can be recorded of the filaments growing out from the metal particle. Filament formation is closely linked to the crystallographic orientation of the Ni particles and by, for example, modifying the Ni with Cu, multiple carbon filaments have been observed from each particle.

This process of filament growth is closely related to the synthesis of single-walled carbon nanotubes. Here, the aim is to selectively produce a single layer only of carbon that is as long as possible. Owing to their extreme mechanical strength and interesting electronic behavior, these materials have recently attracted substantial interest in materials science.

Carbon filament growth in the reforming process can lead to catastrophic blocking of the reactor and, because of the high strengths of the filaments, to destruction of the catalyst. This again may lead to hot spots in the tubular reactor since the catalyst no longer catalyzes the strongly endothermic process that adsorbs the supplied heat. The reactor may become overheated, enhancing the carbon formation process even further. Thus, although there might be economic gain by reducing the water content, fatal carbon formation may be the penalty, which is better avoided by keeping a safe margin in operating conditions.

Several approaches have been taken to circumvent this problem. Avoiding the thermodynamic regions where carbon deposition prevails is an obvious one. This severely restricts the available parameter space since the temperature and gas composition vary both radially and especially vertically through the reactor. Therefore, it is important to develop a catalyst with a much higher carbon formation resistance than the conventional steam reforming catalysts. Carbon formation can be considered as a selectivity problem, which may be solved by developing a catalyst with high selectivity for the desired steam reforming reaction and little propensity to form carbon. We shall discuss two successful developments in this respect.

8.2.4
The SPARG Process: Selective Poisoning by Sulfur

The operational carbon limit has been pushed substantially towards conditions corresponding to higher carbon formation potential by using sulfur passivation of the catalyst in the so-called SPARG (Sulfur PAssivated ReforminG) process, developed by scientists at Haldor Topsøe A/S [1]. Sulfur passivation is based on surface science experiments which indicate that a much larger ensemble of active surface-nickel atoms is required for carbon filament formation than for the reforming reaction. The idea is therefore that by controlling the number and size of unperturbed nickel sites on the surface, by adsorbing sulfur, could exert ensemble control. Figure 8.10 compares the deposition rate of carbon, or "coking rate," with the rate of reforming as a function of the number of sites not blocked by sulfur $(1 - \theta_S)$ on a real catalyst. Using a simple mean-field approximation, the rate is expressed in terms of Langmuir kinetics:

$$r = r_0(1 - \theta_S)^\alpha \tag{8.13}$$

where r_0 is the rate on a sulfur-free surface and α reflects the number of surface atoms required for the reaction. Fitting this expression to the data of Figure 8.10, the exponents α can be extracted, revealing that a much larger ensemble is required for the carbon formation process ($\alpha = 6.3$) than for steam reforming ($\alpha = 2.7$). For both reactions, the rate decreases rapidly with increasing sulfur coverage, confirming that sulfur is a strong poison. This explanation was founded on the concept that the active surface only consists of (111) surface atoms, but now we know that this is not the case. The important effect is that, as we have seen above, the adsorbed sulfur atoms suppress carbon formation by blocking the steps. Since the carbon grows out of the steps, its deposition could be even more suppressed.

It is important to note that the selectivity of sulfur-passivated catalysts towards steam reforming is greatly enhanced because carbon formation is effectively suppressed. The decrease in activity can largely be compensated for by selecting inherently more active catalysts and by operating at higher temperatures. Unfortunately, in the SPARG process, the sulfur coverages have to be maintained by continuously adding ppm levels of H_2S to the feed gas because adsorbed sulfur

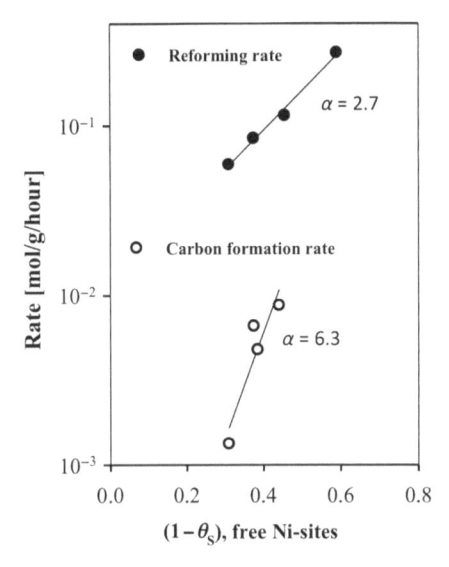

Figure 8.10 Rates of reforming and carbon deposition as a function of free Ni sites on partly sulfur-covered nickel surfaces (adapted from Rostrup-Nielsen and Christiansen [1]).

is hydrogenated and removed from the catalyst. As a result, the product gas contains H_2S, which calls for an additional purification step, since catalysts in downstream processes are very sensitive to sulfur poisoning. Hence, catalyst modifications providing a more permanent way to suppress carbon formation are highly desirable. Alloying offers another way to affect the nickel ensembles, as shown in the next section.

8.2.5
Gold–Nickel Alloy Catalyst for Steam Reforming

Alloying is a powerful tool to affect the selectivity of metal catalysts [7]. As explained in Chapter 5, two-dimensional alloys may form at the surface of metals that otherwise are immiscible in the bulk. Using scanning tunneling microscopy (STM), Besenbacher and coworkers [8] found that gold on nickel forms a random two-dimensional surface alloy in the submonolayer regime (Figure 8.11). Gold atoms, which have no activity for the steam reforming reaction, break up the large ensembles that favor carbon deposition. In addition, density functional theory calculations indicated that carbon is less strongly bound on nickel atoms in contact with gold atoms. This evidence suggests that Ni-Au might be a successful catalyst with high selectivity for steam reforming and low selectivity towards carbon deposition.

Figure 8.12 confirms that this is correct: a single nickel catalyst used for steam reforming of n-butane deactivates steadily and gains weight due to the accumulation of carbon, but a Ni-Au catalyst maintains its activity at a constant level [9].

Figure 8.11 Atomically resolved STM pictures (51 × 49 Å^2) of 0.3 ML Au deposited on Ni(111) at elevated temperatures. The Au atoms are most often depicted as depressions and the Ni atoms as protrusions, owing to differing electronic structures.

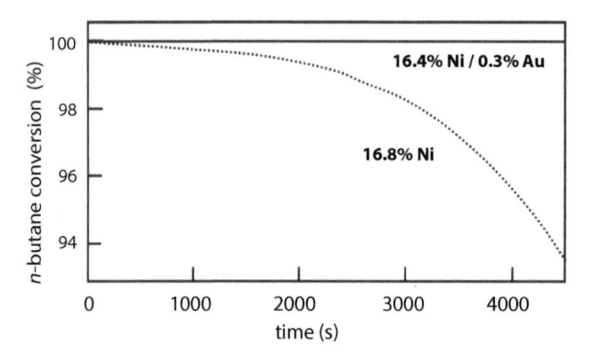

Figure 8.12 Steam reforming of *n*-butane as a function of time for a conventional Ni catalyst and a novel Ni-Au alloy catalyst, showing the superior stability of the latter (adapted from Besenbacher *et al.* [9]).

Please note that the approach taken to find a solution for carbon deposition started with leads from surface science and computational chemistry and led to the straightforward development of a new catalyst – a good example of rational catalyst development based on fundamental knowledge!

8.2.6
Direct Uses of Methane

Although not directly related to steam reforming, it is well worth considering how methane might be converted directly into useful chemicals. In industrialized or densely populated countries, methane can be readily transported and distributed through pipelines and, thus, it is a convenient fuel for heat and power generation. It is also quite clean and produces less carbon dioxide (greenhouse gas) than does the combustion of oil and coal. However, methane also accompanies the oil recov-

Figure 8.13 Natural gas (predominantly methane) is concomitant with oil production and is commonly flared, particularly at production sites in remote locations (copyright Digital Vision Ltd. Industry and Technology).

ered in remote locations, where transportation is economically unfeasible. Here, it is simply burned (flared, see Figure 8.13) to CO_2. Flaring is a better solution than letting it escape in the air because methane is, by about a factor of 25, a stronger greenhouse gas than CO_2. Building a complete steam reformer along with a plant for converting synthesis gas into methanol or liquid hydrocarbons is only feasible in special cases. Ideally, such methane would be activated directly into a condensable product (with a density roughly 1000 × higher than that of methane) that can be transported by ship or pipelines to the consumer.

8.2.6.1 Direct Methanol Formation

Direct oxidation of methane to methanol is an obvious dream reaction:

$$CH_4 + 1/2 O_2 \rightleftharpoons CH_3OH \quad \Delta H° = -126.4 \, \text{kJ mol}^{-1} \tag{8.14}$$

It has to compete with more exothermic oxidation to CO or CO_2:

$$CH_4 + 3/2 O_2 \rightleftharpoons CO + 2 H_2O \quad \Delta H° = -519.5 \, \text{kJ mol}^{-1} \tag{8.15}$$

$$CH_4 + 2 O_2 \rightleftharpoons CO_2 + 2 H_2O \quad \Delta H° = -792.5 \, \text{kJ mol}^{-1} \tag{8.16}$$

Obtaining a high selectivity towards methanol is strongly hampered by the fact that methane is a stable molecule. Abstracting the first hydrogen needs a relatively high temperature at which the route to total oxidation is open. Oxide catalysts to yield methanol have been reported, but always with low selectivity, or with a combination of high selectivity and low activity. Hence, the direct oxidation of methane to methanol awaits a breakthrough, perhaps one inspired by nature. Bacteria exist that use enzymes, called monooxygenases, that have active sites based on iron or copper and perform the desired reaction at low temperatures [10].

8.2.6.2 Catalytic Partial Oxidation of Methane

The catalytic partial oxidation of methane into CO and H_2 according to

$$CH_4 + 1/2 O_2 \rightleftharpoons CO + 2 H_2 \quad \Delta H° = -35.9 \, \text{kJ mol}^{-1} \tag{8.17}$$

is an alternative to steam reforming, particularly for producing smaller amounts of hydrogen, for example, for use in a fuel cell in automobiles. Here, the partial oxidation is ensured by only using a thin, white hot (1000 °C) monolith to keep the contact time of the methane-oxygen mixture with the catalyst in the millisecond range. This field was pioneered by Schmidt and coworkers and is currently being considered by oil companies and car manufacturers for the on-board generation of hydrogen in cars [11].

8.3
Reaction of Synthesis Gas

The term synthesis gas or syngas refers mostly to mixtures of CO and H_2 although sometimes the mixture of N_2 and H_2 used for the synthesis of ammonia is also called synthesis gas. Depending on origin or application, the various combinations of CO and H_2 can be called water–gas (CO + H_2 from steam and coal), crack gas (CO + $3H_2$ from steam reforming of natural gas), or oxogas (CO + H_2 for hydroformylation). The three most important applications of syngas are in the methanol synthesis, the hydroformylation of alkenes to aldehydes and alcohols, and the synthesis of larger hydrocarbons, the so-called Fischer–Tropsch synthesis of synthetic fuels. Hydroformylation is carried out with homogeneous catalysts and falls outside the scope of our treatment. We shall discuss the two other processes in the next sections.

8.3.1
Methanol Synthesis

8.3.1.1 Basic Concepts of the Process

Methanol was originally produced by the distillation of wood. In 1923, however, BASF developed the first catalyst that allowed large amounts of methanol to be synthesized. The process operated at high pressure (300 bar) and high temperatures (300–400 °C) over a Zn/Cr_2O_3 catalyst, which was replaced in 1966 by a substantially more active $Cu/Zn/Al_2O_3$ catalyst, developed by ICI, that allowed operation under much milder conditions. Today methanol is synthesized industrially from a nonstoichiometric mixture of hydrogen, carbon dioxide, and carbon monoxide (90 : 5 : 5) at 50–100 bar and 500–550 K over $Cu/ZnO/Al_2O_3$ catalysts. Furthermore, modern plant design has integrated methanol and steam reforming such that the exothermicity of methanol synthesis is utilized for the endothermic steam reforming [12, 13].

Methanol is used in industry as a raw material for formaldehyde (40–50%), as a solvent, and for producing, for example, acetic acid by hydroformylation. The latter is used, for example, to make precursors for polymers.

The reaction mechanism of methanol synthesis is complex since two processes are involved and coupled. Formally, the reaction can be written as the hydrogenation of CO by the overall reaction:

$$CO + 2H_2 \rightleftharpoons CH_3OH \quad \Delta H^\circ_{298} = -91 \text{ kJ mol}^{-1} \tag{8.18}$$

but actually this reaction is a combination of

$$CO_2 + 3H_2 \rightleftharpoons CH_3OH + H_2O \quad \Delta H^\circ_{298} = -47 \text{ kJ mol}^{-1} \tag{8.19}$$

and the water–gas shift reaction, Eq. (8.3). Both reactions are exothermic; hence, a relatively low temperature is preferred. The reduction in the number of molecules on going from reactants to products makes operation at elevated pressures desirable. Selectivity is important for the methanol catalyst since several side reactions are thermodynamically more favorable than the reaction towards methanol under the prevailing synthesis conditions. For example,

$$CO + 3H_2 \rightleftharpoons CH_4 + H_2O \tag{8.20}$$

$$2CO + 4H_2 \rightleftharpoons CH_3OCH_3 + H_2O \tag{8.21}$$

$$2CO + 4H_2 \rightleftharpoons CH_3CH_2OH + H_2O \tag{8.22}$$

The literature contains ample evidence that methanol is synthesized from CO_2 according to Eq. (8.19) and not from CO as in Eq. (8.18). Isotopic labeling experiments have demonstrated that CO_2 is the main source of carbon in methanol formed from synthesis gas [12]. In addition, it has been shown that a Cu(100) single crystal readily produces methanol from a mixture of CO_2 and H_2 at 2 bar (see Figure 8.14) [14].

A microkinetic model based on 13 elementary steps, of which the first 8 relate to the water–gas shift reaction, describes the process well:

$$H_2O(g) + * \rightleftharpoons H_2O* \tag{8.23}$$

$$H_2O* + * \rightleftharpoons HO* + H* \tag{8.24}$$

$$2HO* + * \rightleftharpoons H_2O* + O* \tag{8.25}$$

$$HO* + * \rightleftharpoons H* + O* \tag{8.26}$$

$$H_2(g) + 2* \rightleftharpoons 2H* \tag{8.27}$$

$$CO(g) + * \rightleftharpoons CO* \tag{8.28}$$

$$CO* + O* \rightleftharpoons CO_2* + * \tag{8.29}$$

$$CO_2(g) + * \rightleftharpoons CO_2* \tag{8.30}$$

$$CO_2* + H* \rightleftharpoons HCOO* + * \tag{8.31}$$

$$HCOO* + H* \rightleftharpoons H_2COO* + * \tag{8.32}$$

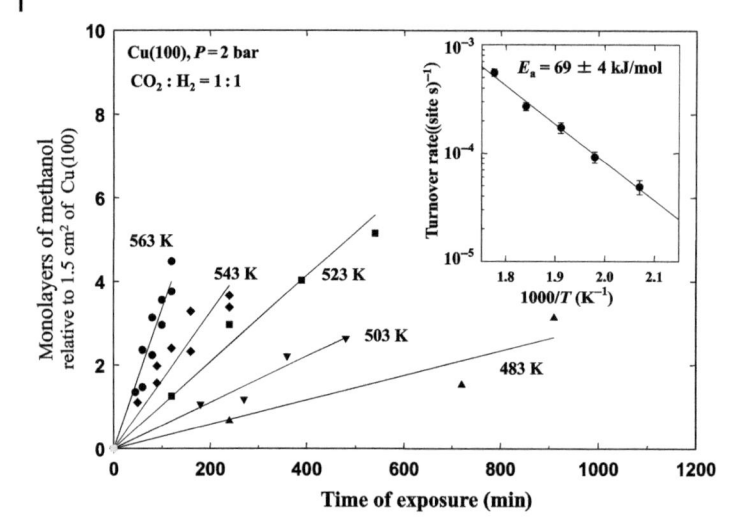

Figure 8.14 Methanol synthesis rate over a Cu(100) single crystal in a 1 : 1 gas mixture of H_2 and CO_2 at $p_{tot} = 2$ bar. The turnover frequency is very low and the crystal produces no more than 1 ppm of methanol per hour making it difficult to perform quantitative measurements. Nevertheless, the experiments confirm that methanol is formed directly from CO_2 (reproduced from Rasmussen et al. [14] with permission from Elsevier).

$$H_2COO* + H* \rightleftharpoons H_3CO* + O* \qquad (8.33)$$

$$H_3CO* + H* \rightleftharpoons CH_3OH* + * \qquad (8.34)$$

$$CH_3OH* \rightleftharpoons CH_3OH(g) + * \qquad (8.35)$$

The hydrogenation of dioxomethylene, step (8.33) is, most likely, the rate-limiting step, although the hydrogenation of formate in (8.32) is also a candidate. By assuming that Eqs. (8.24), (8.23), and (8.29) are slow for the water–gas shift reaction and that (33) is slow for methanol synthesis, we arrive at the following set of equations, in which one site is assumed to consist of two copper atoms:

$$K_1 p_{H_2O} \theta_* = \theta_{H_2O} \qquad (8.36)$$

$$r_2 = k_2 \theta_{H_2O} \theta_* - \frac{k_2}{K_2} \theta_{OH} \theta_H \qquad (8.37)$$

$$K_3 \theta_{OH}^2 = \theta_{H_2O} \theta_O \qquad (8.38)$$

$$r_4 = k_4 \theta_{HO} \theta_* - \frac{k_4}{K_4} \theta_O \theta_H \qquad (8.39)$$

$$\theta_H = \sqrt{K_5 p_{H_2}} \theta_* \qquad (8.40)$$

$$K_6 p_{CO} \theta_* = \theta_{CO} \qquad (8.41)$$

$$r_7 = k_7 \theta_{CO} \theta_* - \frac{k_7}{K_7} \theta_{CO_2} \theta_* \quad (8.42)$$

$$\theta_{CO_2} = K_8 p_{CO_2} \theta_* \quad (8.43)$$

$$K_9 \theta_{CO_2} \theta_H = \theta_{HCOO} \theta_* \quad (8.44)$$

$$K_{10} \theta_{HCOO} \theta_H = \theta_{H_2COO} \theta_* \quad (8.45)$$

$$r_{11} = k_{11} \theta_{H_2COO} \theta_H - \frac{k_{11}}{K_{11}} \theta_{CH_3O} \theta_H \quad (8.46)$$

$$K_{12} \theta_{CH_3O} \theta_H = \theta_{CH_3OH} \theta_* \quad (8.47)$$

$$K_{13} \theta_{CH_3OH} = p_{CH_3OH} \theta_* \quad (8.48)$$

By utilizing the mass balance for O∗ and OH∗:

$$r_7 - r_{11} = 1/2(r_2 + r_4) \quad (8.49)$$

and

$$\sum_i \theta_i = 1 \quad (8.50)$$

a closed form of the various coverages inclusive θ_* is found. For details, we refer the reader to the original paper [15].

After a rather lengthy calculation, the rate of methanol formation is found as

$$r = k_{11} K_5^{3/2} K_8 K_9 K_{10} p_{H_2}^{3/2} p_{CO_2} \left(1 - \frac{1}{K_G} \frac{p_{CH_3OH} p_{H_2O}}{p_{H_2}^3 p_{CO_2}} \right) \theta_*^2 \quad (8.51)$$

According to Eq. (8.51), the reaction orders in H_2 and CO_2 are 1.5 and 1, respectively – in the limit where there coverages of the different intermediates can be ignored – in good agreement with experiment where a total pressure dependence of 2.4 was found. Similarly, we expect the rate to be a maximum for a syngas mixture of 60% H_2 and 40% CO_2. If the rate-determining step were the hydrogenation of formate (8.32), the maximum would be expected with a 50/50 mixture, as the reader can easily verify. Figure 8.15 compares these predictions with experimental values. The data clearly support hydrogenation of dioxomethylene as the rate-determining step.

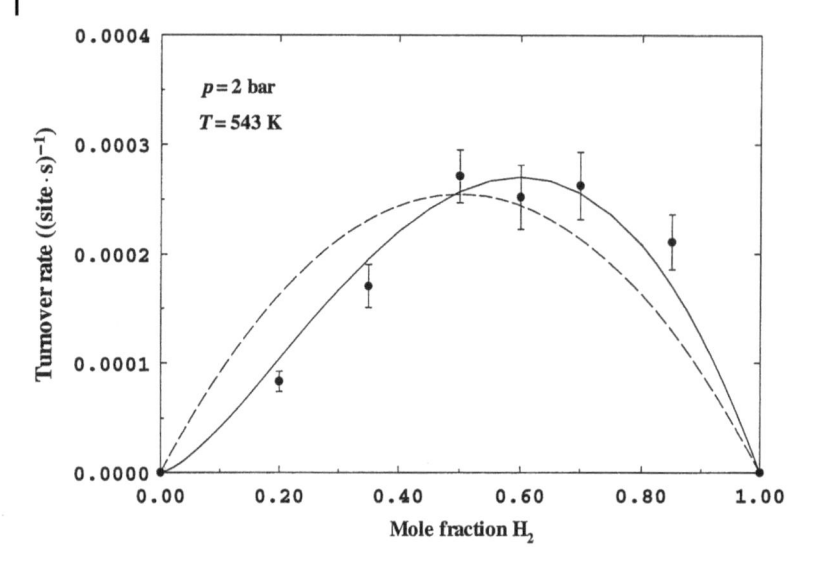

Figure 8.15 Methanol synthesis rate over a Cu(100) single crystal in the zero conversion limit as a function of the H_2 mole fraction. The solid line corresponds to the kinetic model in Eqs. (8.23)–(8.35) with reaction (8.33), dioxomethylene hydrogenation, as the rate-limiting step. The dashed line is that predicted if the hydrogenation of formate was rate limiting (reproduced from Rasmussen et al. [14] with permission from Elsevier).

A full analysis of the rate expression reveals that all data on the Cu(100) single crystal are modeled very well, as shown in Figure 8.15. Even more important is that the model also describes data obtained on a real catalyst measured under considerably different conditions reasonably well, indicating that the microkinetic model captures the most important features of the methanol synthesis (Figure 8.16).

The microkinetic model also predicts the coverages of the various intermediates on the surface. As shown in Table 8.2, the approximation of the surface being clean is quite reasonable. The highest coverages are observed for hydrogen and formate, but the majority of sites are free, even at 50 bar.

Other vicinal planes of copper single crystals and polycrystalline copper have been investigated and are also found to synthesize methanol in similar amounts as the Cu(100) surface [16, 17]. This strongly suggests that the synthesis takes place on the copper particles in the catalyst and not on the ZnO, which serves merely as the support.

The Cu/ZnO system is very dynamic. The morphology of the Cu particles responds immediately to a change in reduction potential of the gas mixture above it. EXAFS studies suggest that the change in morphology is associated with the extent that the metal particles wet the underlying support [18]. As the surface of ZnO becomes more reduced, the copper may bind more strongly to it. A model calculation based on the Wulff construction (see Chapter 5 taking the Cu-ZnO interface energy into account illustrates how the morphology may change (Fig-

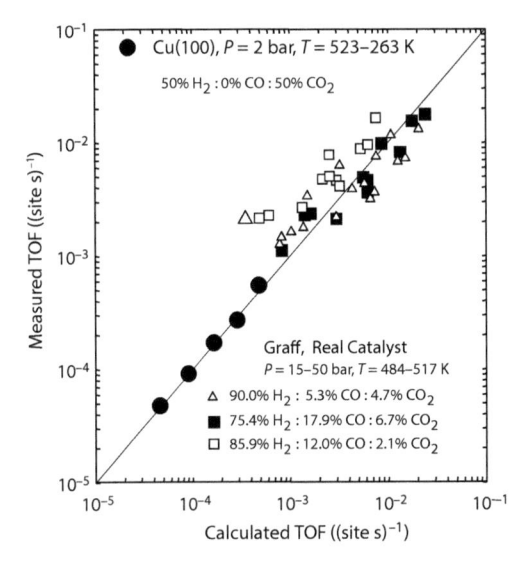

Figure 8.16 Comparison between the predictions of a microkinetic model and measurements on a Cu(100) model catalyst with a real methanol synthesis catalyst. The solid line represents the ideal match between model and experiment (reproduced from Rasmussen *et al.* [15] with permission of Springer).

Table 8.2 Coverages of the various intermediates in the methanol synthesis for a stoichiometric gas mixture at 500 K at 85% equilibrium; note that the surface is almost empty at low pressures, while H atoms and formate coverages become significant at high pressure.

Species	P_{tot} 2 bar	P_{tot} 50 bar
θ_*	8.8×10^{-1}	5.2×10^{-1}
θ_H	1.1×10^{-1}	3.3×10^{-1}
θ_{HCOO}	1.2×10^{-3}	7.0×10^{-2}
θ_O	2.4×10^{-6}	3.9×10^{-6}
θ_{H_2O}	4.5×10^{-5}	1.7×10^{-3}
θ_{OH}	2.2×10^{-3}	1.7×10^{-2}
θ_{CO}	3.3×10^{-3}	1.6×10^{-2}
θ_{CO_2}	3.3×10^{-3}	2.8×10^{-3}
θ_{CH_3OH}	1.8×10^{-5}	4.2×10^{-2}
θ_{CH_3O}	4.6×10^{-11}	2.2×10^{-8}
θ_{H_2COO}	1.7×10^{-8}	4.9×10^{-6}

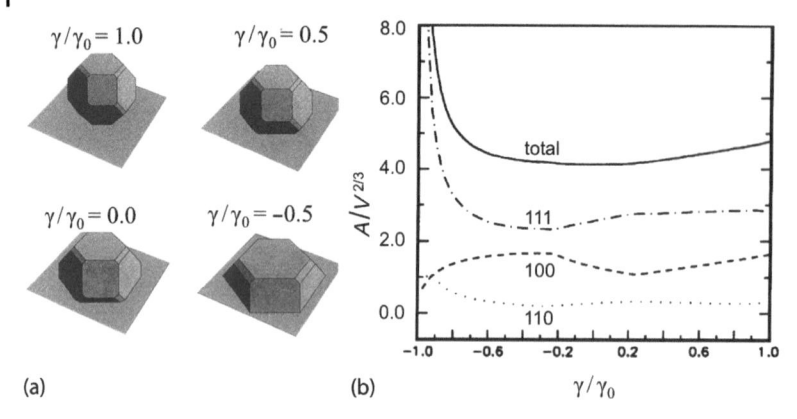

Figure 8.17 Shape of fcc metal particles for different values of the metal-support interface energy (γ/γ_0). For $\gamma/\gamma_0 = 1$, there is no energy gain and the particle will behave as a free metal particle, while for $\gamma/\gamma_0 = -1$, the particle will try to maximize contact, degenerating into a metal overlayer. (b) Surface area as a function of γ/γ_0 for the different vinicial planes of copper [19]. The dynamics of particle morphology can be used to an advantage, to counteract the effect of sintering of the copper particles. As Figure 8.18 shows, a Cu/ZnO catalyst slowly loses activity, which is attributed to sintering. Exposing the catalyst for a short time to a highly reducing mixture of CO_2-free synthesis gas restores the activity because the sintered particles spread to some extent over the support (similarly to the case in the model calculations of Figure 8.17). The procedure gives a considerable net gain in methanol production (data from Ovesen et al. [19]).

ure 8.17). The copper particles expose a (111) surface towards the ZnO surface and the surface energy here is called γ_0. By introducing a quantity

$$\gamma = (\gamma_{\text{interface}} - \gamma_{\text{substrate}}) \tag{8.52}$$

the range of no-wetting to complete wetting can be described by the ratio γ/γ_0 in the interval 1 to −1.

This can easily be realized by the following arguments: let us assume that there is no interaction between the two surfaces. This means that there is no energy gain β, which can be written as

$$\beta \equiv \gamma_{\text{substrate}} + \gamma_0 - \gamma_{\text{interface}} = 0 \tag{8.53}$$

From this, we find that the interface energy $\gamma_{\text{interface}}$ is the same as that of the two separate surfaces

$$\gamma = \gamma_{\text{interface}} - \gamma_{\text{substrate}} = \gamma_0 \quad \Rightarrow \quad \frac{\gamma}{\gamma_0} = 1 \tag{8.54}$$

and the cryallite will not be perturbed as also indicated in both Fig. 5.14 when $\beta = 0$ and Fig. 8.17 for $\gamma/\gamma_0 = 1$.

However, if $\beta \geq 2\gamma_0$, then there will be complete wetting corresponding to

$$\gamma = \gamma_{\text{interface}} - \gamma_{\text{substrate}} = \gamma_{\text{substrate}} - \gamma_0 - \gamma_{\text{substrate}} = -\gamma_0$$
$$\Rightarrow \frac{\gamma}{\gamma_0} = -1 \tag{8.55}$$

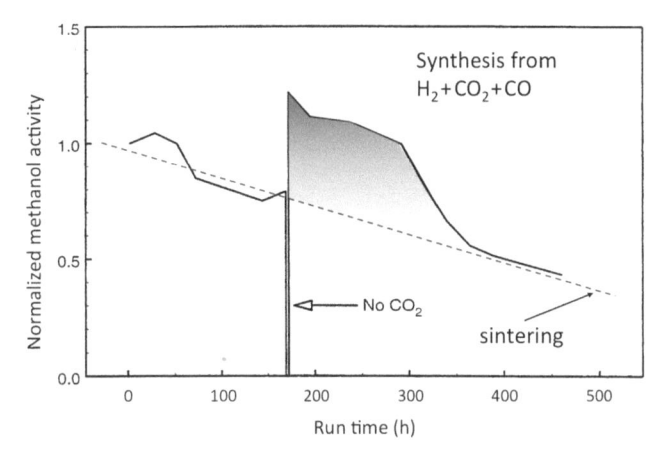

Figure 8.18 Rate of methanol synthesis of a Cu/ZnO/Al$_2$O$_3$ catalyst in a plug flow reactor as a function of time on stream. The catalyst was operated at 494 K and 63 bar in a gas steam of 5% CO, 5% CO$_2$, 88% H$_2$, and 2% N$_2$. Note the steady decrease in reactivity, which is ascribed to sintering of the copper particles. The CO$_2$ was removed from the reactants for 4 h after 168 h. After reintroduction the catalyst displays a restored activity which is associated with morphology changes of the copper particles resulting from increased wetting of the ZnO support (adapted from [20]).

corresponding to $\Delta h_0 = \beta h_0/\gamma_0$ or $\gamma/\gamma_0 = -1$ in Figures 5.16 and 8.17, respectively.

The contribution of different crystal planes to the overall surface area of the particle can, thus, be calculated and is shown in Figure 8.17b. The results have been included in a dynamical microkinetic model of the methanol synthesis, yielding a better description of kinetic measurements on working catalysts [19].

ZnO is, apparently, a very suitable support for the copper particles. Evidence exists, however, that its role does not have to be limited to that of a support only. Nakamura *et al.* [21] studied the influence of Zn on methanol synthesis on copper crystals by depositing Zn on the surface. They found that the rate was enhanced by a factor of six (see Figure 8.19), suggesting that Zn atoms also act as a chemical promoter. Whether some of the ZnO in the real catalyst is actually reduced to such a degree that it can alloy into the copper particles and segregate to the surface, as suggested by Nakamura *et al.*, is still a controversial topic. Recently, it was suggested that the active sites are Cu(211) B5 steps in the copper decorated by metallic Zn and stabilized by stacking faults in the copper nanoparticles [22].

That a metallic phase of Zn is present on the Cu surface of the commercial methanol synthesis catalysts has recently been confirmed spectroscopically using AES [21] and also by Kuld *et al.* [23], see Figure 8.20. Another indication is that the measured activity scales with the presence of metallic Zn. Thus, it appears that catalytic activity for methanol synthesis depends both on the dynamic area of Cu metal and metallic Zn, reduced out of the initially present ZnO.

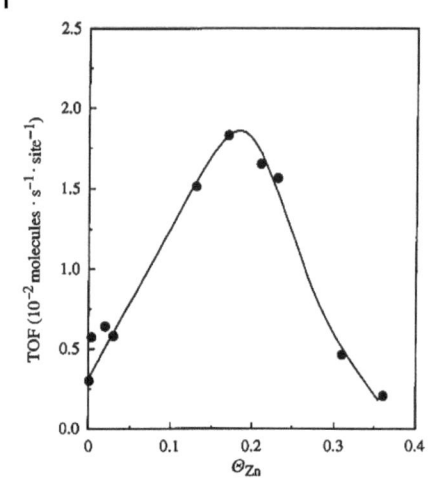

Figure 8.19 Turnover frequency for methanol synthesis from H_2 and CO_2 at 18 bar and 523 K as a function of Zn coverage on polycrystalline copper (reproduced from Nakamura et al. [21] with permission from Elsevier).

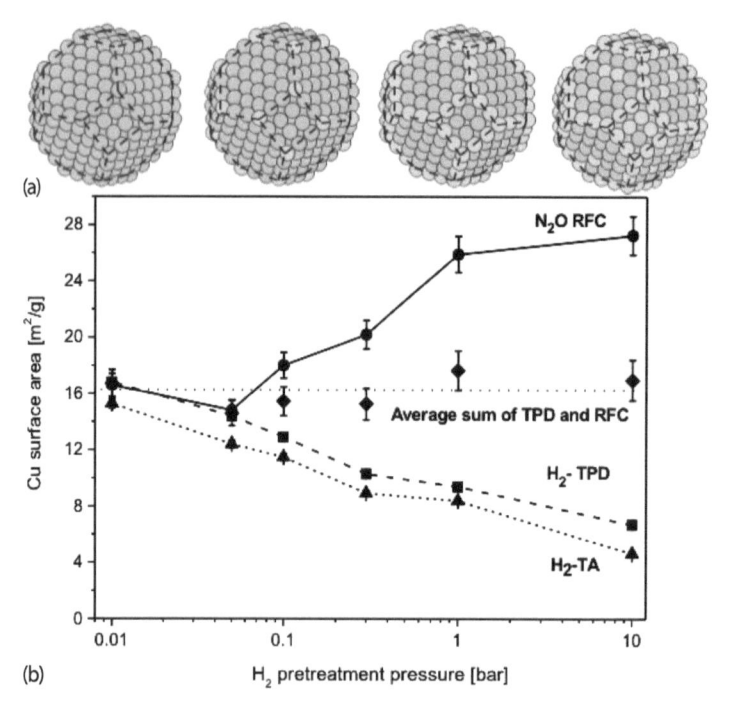

Figure 8.20 (a) Copper nanoparticles as a function of reduction potential. (b) Metallic Cu area measured by hydrogen temperature programmed desorption (TPD) and transient adsorption (TA) of hydrogen and by N_2O reactive frontal chromatography (RFC). Note that the hydrogen area decreases as Zn is alloyed into the Cu surface, while the area increases as measured by N_2O oxidation, since both the Zn and the Cu contribute to oxidation ((b) reproduced with permission from Kuld et al. [23], copyright 2014 by Wiley).

Table 8.3 Mole fractions of CO, H$_2$, and CH$_3$OH in the direct synthesis of methanol.

	CO	+ 2H$_2$	⇌ CH$_3$OH
Before reaction	$1 - x$	x	0
After reaction	$1 - x - t$	$x - 2t$	t
Normalized	$\dfrac{1-x-t}{1-2t}$	$\dfrac{x-2t}{1-2t}$	$\dfrac{t}{1-2t}$

8.3.1.2 Methanol Direcly Synthesized from CO and H$_2$

CO in the synthesis gas mixture for the methanol synthesis does not seem to take part directly in the reaction, but it does influence the process through two effects: first the water–gas shift reaction and, second, through its effect on the surface morphology (and possibly also composition). For thermodynamic reasons, however, it would be desirable if CO could be hydrogenated directly via Eq. (8.18) instead of going through two coupled Eqs. (8.3) and (8.19), since it would yield a higher equilibrium concentration of methanol at the reactor exit.

In the following, we demonstrate how one can analyze the influence of the gas composition on the rate of reaction. We mention, however, that today there are computer routines that allow to do this with substantially less effort, for example, the commercially available program HSC [http://www.outotec.com/HSC].

First, consider the direct formation of methanol from CO and H$_2$ If x is the mole fraction of H$_2$ and t is the mole fraction of methanol (Table 8.3 gives the other mole fractions), then summation of all concentrations gives $1 - 2t$ as the total after reaction, which is used to normalize the mole fractions after reaction.

The equilibrium constant is

$$K = \frac{p_{CH_3OH} p_0^2}{p_{H_2}^2 p_{CO}} \Rightarrow \tag{8.56}$$

$$K = \frac{\left(\frac{t}{1-2t}\right) p_0^2}{\left(\frac{x-2t}{1-2t}\right)^2 \left(\frac{1-x-t}{1-2t}\right)} = \frac{t(1-2t)^2 p_0^2}{(x-2t)^2(1-x-t)} \tag{8.57}$$

p_0 is the standard pressure and p_x is the partial pressure of component X. Rearrangement of Eq. (8.57) leads to a third-order equation in t, which can be solved iteratively. Notice that the pressures used in Eq. (8.56) should actually be replaced by activities, implying that they should be corrected by their respective fugacity coefficients, which are of importance when dealing with methanol and water. We leave it as an exercise for the reader to judge the influence of such effects, utilizing the relation between pressure and activity given in Eq. (2.39).

When methanol is produced from a mixture of CO$_2$, CO, and H$_2$, the reverse water–gas shift reaction complicates the system, since it competes with the methanol synthesis.

Table 8.4 Mole fractions of CO_2, H_2, CH_3OH, and H_2O in the direct synthesis of methanol.

	CO_2	+	$3H_2$	⇌	CH_3OH	+	H_2O
Before reaction	$1-x-y$		x		0		0
After reaction	$1-x-y-t$		$x-3t$		t		t
Normalized	$\dfrac{1-x-y-t}{1-2t}$		$\dfrac{x-3t}{1-2t}$		$\dfrac{t}{1-2t}$		$\dfrac{t}{1-2t}$

Table 8.5 Mole fractions in the water–gas shift reaction.

	CO_2	+	H_2	⇌	CO	+	H_2O
Before reaction	$1-x-y$		x		y		0
After reaction	$1-x-y-w$		$x-w$		$y+w$		w
Normalized	$1-x-y-w$		$x-w$		$y+w$		w

We use the same procedure as shown above where x again denotes the mole fraction of H_2, y is the mole fraction of CO, and t is the mole fraction of methanol and of water produced in this reaction. The total mole fraction after reaction is again $(1-2t)$ and the normalized mole fractions after reaction are given in Table 8.4.

The equilibrium constant for this reaction is

$$K_{CH_3OH} = \frac{p_{H_2O}\, p_{CH_3OH}\, p_0^2}{p_{H_2}^3\, p_{CO_2}} \tag{8.58}$$

The mole fractions of the reverse water–gas shift reaction are given in Table 8.5.

Combining all of the normalized mole fraction expressions, to account for all reactions:

$$p_{CO_2} = \frac{1-x-y-w-t}{1-2t} \tag{8.59}$$

$$p_{H_2} = \frac{x-w-3t}{1-2t} \tag{8.60}$$

$$p_{CO} = \frac{y+w}{1-2t} \tag{8.61}$$

$$p_{H_2O} = \frac{w+t}{1-2t} \tag{8.62}$$

Inserting these in the equilibrium expression for the methanol reaction yields:

$$K_{CH_3OH} = \frac{p_{H_2O}\, p_{CH_3OH}\, p_0^2}{p_{H_2}^3\, p_{CO_2}} \Rightarrow \tag{8.63}$$

$$K_{CH_3OH} = \frac{\left(\frac{w+t}{1-2t}\right)\left(\frac{t}{1-2t}\right)p_0^2}{\left(\frac{x-w-3t}{1-2t}\right)^3\left(\frac{1-x-y-w-t}{1-2t}\right)} \Rightarrow \tag{8.64}$$

$$K_{CH_3OH} = \frac{t(w+t)(1-2t)^2 p_0^2}{(x-w-3t)^3(1-x-y-w-t)} \tag{8.65}$$

For the equilibrium constant for the reverse water–gas shift reaction:

$$K_{RWGS} = \frac{p_{H_2O}p_{CO}}{p_{H_2}p_{CO_2}} \Rightarrow \tag{8.66}$$

$$K_{RWGS} = \frac{\left(\frac{x+t}{1-2t}\right)\left(\frac{y+w}{1-2t}\right)}{\left(\frac{x-w-3t}{1-2t}\right)\left(\frac{1-x-y-w-t}{1-2t}\right)} \Rightarrow \tag{8.67}$$

$$K_{RWGS} = \frac{(w+t)(y+w)}{(x-w-3t)(1-x-y-w-t)} \tag{8.68}$$

We now have two equations, Eqs. (8.65) and (8.68), with two unknowns (t and w), which can then be solved.

The solution is illustrated in Figure 8.21, which shows the equilibrium concentration of methanol for different initial gas mixtures. Note that the maximum methanol concentration occurs for the pure $CO + H_2$ mixture. Hence, in principle, a mixture of just CO and H_2 could be used, with minor amounts of CO_2, to produce the maximum amount of methanol. However, it is not the equilibrium constant that matters but the rate of methanol formation, and one must remember that methanol forms from CO_2 not CO. Hence, the rate is proportional to the CO_2 pressure and this is why the methanol synthesis is not performed with the simple stoichiometric 3 : 1 mixture of H_2 and CO_2 that Eq. (8.19) suggests.

The discussion above indicates that there are good reasons to develop a catalyst that can produce methanol directly from the hydrogenation of CO with high selectivity. Several studies indicate that this is possible with noble metals such as palladium, platinum, potassium-promoted copper, or several combinations of group 8 metals [24–26]. However, no such catalysts have been commercialized.

8.3.2
Fischer–Tropsch Process

In the Fischer–Tropsch (FT) process, synthesis gas ($CO + H_2$) is converted into hydrocarbons over iron or cobalt catalysts [27, 28]. It offers a way to convert coal, natural gas, oil residues, or biomass into gasoline, diesel fuel, and other useful chemicals. The process was developed by German scientists Franz Fischer and Hans Tropsch at the Kaiser-Wilhelm-Institut für Kohlenforschung in Mülheim an der Ruhr in 1923 and supplied Germany with fuels during the Second World War.

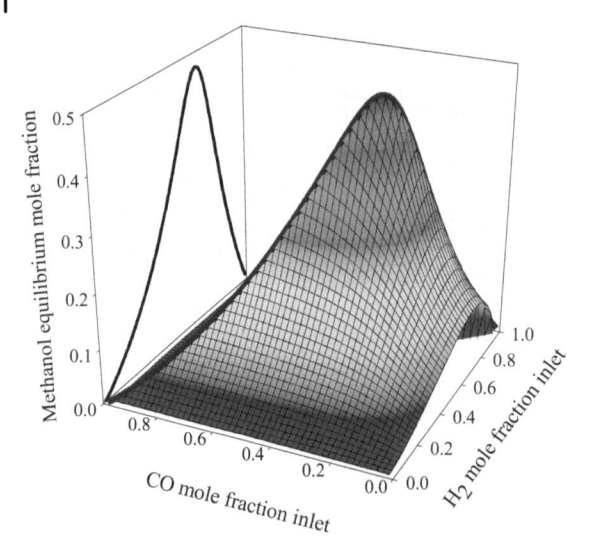

Figure 8.21 Outlet equilibrium methanol concentration as function of the inlet mole fraction of H_2, CO, and CO_2. Notice that the highest methanol concentration is for a mixture of only H_2 and CO at a ratio of 2 : 1 – the stoichiometric ratio. The solid curve is the methanol equilibrium without CO_2 in the gas mixture, which is also shown as a projection to the left.

Catalysts were based on cobalt or on iron. In 1944, nine plants produced 600 000 t of transportation fuel per year via gasification of coal and the Fischer–Tropsch process. South Africa has used the Fischer–Tropsch synthesis for its supply of fuels and base chemicals since 1955 in the SASOL plants in order to make the country less dependent of imported oil [28, 29].

In response to the oil crises of the 1970s, the Fischer–Tropsch process attracted renewed interest as a way to produce high quality, sulfur-free diesel fuel from natural gas and, possibly, an opportunity to utilize natural gas at remote oilfields. Large gas-to-liquids (GTL) plants were built in Qatar in the first decade of the twenty-first century, while others are being constructed in Nigeria and Uzbekistan. China developed a new version of the process for converting coal to liquids [30]. After 2014, the falling oil prices set the development of new projects under serious pressure, and plans for GTL facilities in the USA were put on hold. The Fischer–Tropsch process represents proven technology and is regarded as an alternative for when oil may no longer be widely available, and one may have to resort to natural gas and coal. For the more distant future, one may even contemplate the use of CO and H_2 produced by electro- or photocatalytic dissociation of CO_2 and water.

Catalysts for the Fischer–Tropsch synthesis are those materials that readily dissociate CO, and do not form oxides or carbides that are stable under reaction conditions. Hence, cobalt [31], ruthenium, and iron [32] are successful catalysts, although ruthenium is too expensive for practical use. Iron does form carbides in the synthesis, but its surface is dynamic under H_2, such that carbon atoms at the surface can participate in the reaction [33, 34].

The fact that Fischer–Tropsch fuels hardly contain sulfur or aromatics is a strong selling point for the process. Less sulfur in the fuel has, of course, a direct effect on the sulfur oxides in the emissions, and the newly developed exhaust purification systems for lean burning engines that can be introduced means that all emissions, including CO_2 and NO_x, will diminish. Aromatics promote particulate formation in the combustion of diesel fuels and are, therefore, undesirable. We discuss this further in Chapter 10.

8.3.2.1 Reactions and Mechanism

The Fischer–Tropsch process produces mostly linear alkanes and alkenes, but also small amounts of branched products, aromatics, alcohols and other oxygenates, and water, as well as smaller amounts of CO_2 [35]. The total number of different products can be well over a thousand. Not all of these are directly formed, but are the result of secondary reactions [36]. Limiting ourselves to the dominant products, we can write the overall reaction equations as

$$n\text{CO} + (2n+1)\text{H}_2 \longrightarrow \text{C}_n\text{H}_{2n+2} + n\text{H}_2\text{O} \tag{8.69}$$

$$n\text{CO} + 2n\text{H}_2 \longrightarrow \text{C}_n\text{H}_{2n} + n\text{H}_2\text{O} \tag{8.70}$$

To these overall reactions, we can add the water–gas shift reaction

$$\text{CO} + \text{H}_2\text{O} \rightleftharpoons \text{CO}_2 + \text{H}_2 \tag{8.71}$$

which plays a role when iron catalysts are used. In theory, it inherently adjusts for insufficient hydrogen in the feed, but in practice one rather adds more than the stoichiometric amount of H_2, to suppress CO_2 formation, and to prevent catalyst oxidation by limiting the H_2O/H_2 ratio in the reactor. Reactions (8.69) and (8.70) are highly exothermic, for example, one mole of $-CH_2-$ units generates 165 kJ of heat, and removal of heat is, therefore, a key issue in Fischer–Tropsch synthesis reactor technology.

The Fischer–Tropsch synthesis reaction mechanism has many stages, as summarized in Figure 8.22.

After adsorption of CO and H_2, both molecules have to dissociate, which for H_2 happens instantaneously upon contact with the surface. The close-packed surfaces of iron and cobalt, on the other hand, are virtually inactive for breaking the CO bond, and either more open facets or defects are needed to accomplish this. In addition, iron metal is unstable under Fischer–Tropsch synthesis conditions and it converts to iron carbides, which have even less surface reactivity than the metal. Also here, defects in the structure play a prominent role. The question whether

$\text{CO} + n\,\text{H}_2 \rightarrow$ [CO dissociation (direct or H-assisted)] $\rightarrow$ [O-removal as H_2O] $\rightarrow$ [Monomer formation] $\rightarrow$ [Initiation of a chain] $\rightarrow$ [Chain growth] $\rightarrow$ [Termination as alkene or alkane]

Figure 8.22 Steps in the Fischer–Tropsch reaction mechanism.

Table 8.6 Three chain-growth mechanisms for Fischer–Tropsch synthesis.

	Alkyl mechanism	Alkylidene mechanism	Alkylidyne mechanism
Inserting 'monomer'	CH_2	CH	CH
Initiation step(s)	$CH_2 + CH_3 \rightarrow CH_2CH_3$ (ethyl)	$CH + CH_2 \rightarrow CH=CH_2$ $CH=CH_2 + H \rightarrow CH-CH_3$ (vinyl and ethylidene)	$CH + CH \rightarrow CH\equiv CH$ $CH\equiv CH + H \rightarrow C-CH_3$ (acetylene and ethylidyne)
Growth	$CH_2 + C_nH_{2n+1}$ $\rightarrow C_{n+1}H_{2n+3}$	$CH + CH-C_nH_{2n+1}$ $\rightarrow CH=CH-C_nH_{2n+1}$ $CH=CH-C_nH_{2n+1} + H$ $\rightarrow CH-C_{n+1}H_{2n+3}$	$CH + C-C_nH_{2n+1}$ $\rightarrow CH\equiv C-C_nH_{2n+1}$ $CH\equiv C-C_nH_{2n+1} + H$ $\rightarrow C-C_{n+1}H_{2n+3}$
Termination			
Olefin	$C_nH_{2n+1} - H \rightarrow C_nH_{2n}$	$CH=CH-C_nH_{2n+1} + H$ $\rightarrow C_{n+2}H_{2n+4}$	$CH\equiv C-C_nH_{2n+1} + 2H$ $\rightarrow C_{n+2}H_{2n+4}$
Paraffin	$C_nH_{2n+1} + H \rightarrow C_nH_{2n+2}$	$CH-C_nH_{2n+1} + 2H$ $\rightarrow C_{n+1}H_{2n+4}$ or $CH=CH-C_nH_{2n+1} + 3H$ $\rightarrow C_{n+2}H_{2n+6}$	$C-C_{n+1}H_{2n+3} + 3H$ $\rightarrow C_{n+2}H_{2n+6}$ or $CH\equiv C-C_nH_{2n+1} + 4H$ $\rightarrow C_{n+2}H_{2n+6}$
Main comments	Adsorbed alkyl groups are largely unstable and either decompose or hydrogenate to alkanes	Adsorbed alkylidene groups are relatively unstable on metal surfaces	Adsorbed alkynes and alkylidynes are stable surface species on metals and iron carbides

dissociation of CO is assisted by coadsorbed H-atoms, by making a CHO complex first, or direct, for example on the B5 types of sites discussed in Chapters 6 and 7, is still debated in the literature.

The next step is the removal of O atoms, in the form of H_2O. This is an essential step in the mechanism, which has not had much attention in the literature so far. It can occur via stepwise hydrogenation of an O atom to OH and H_2O, or via the disproportionation of two OH groups, which is energetically somewhat more favorable than the former route.

The formation of hydrocarbons is a polymerization process, and needs a monomer, an initiation step, propagation (chain growth), and termination reactions. Several variations have been proposed in the literature, see Table 8.6.

With the advent of computational chemistry, elementary steps can be evaluated in detail, and the likeliness of their occurrence evaluated. It appears that chain-growth mechanisms based on less-hydrogenated species such as CH, acetylene, and alkylidynes are energetically more feasible, also as these are based on surface intermediates of greater stability. Figure 8.23 shows an example based on the alkylidyne mechanism of Table 8.6.

Termination reactions, being essentially hydrogenations of the stable surface intermediates, lead to olefins and paraffins. As the former are more strongly bound

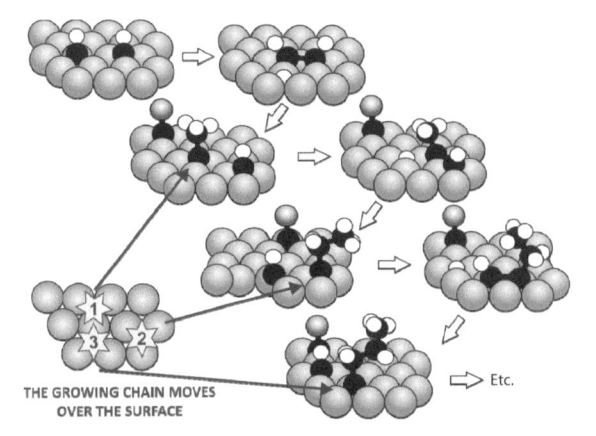

Figure 8.23 The alkylidyne chain-growth mechanism is one among several mechanisms used to describe the Fischer–Tropsch process. The version shown here is particularly attractive as it involves stable intermediates with known occurrence on group VIII metals, and it is energetically feasible on a close-packed cobalt surface. The polymerization initiates by a reaction between CH + CH to acetylene, followed by hydrogenation to ethylidyne (CCH$_3$), which with respect to surface coordination is chemically similar as the CH monomer, making all further coupling steps chemically similar as the initiation step. Note also that the growing chain changes adsorption site during a chain growth event and, thus, moves over the surface (adapted from Weststrate et al. [37]).

as the molecules grow longer, product olefins are mostly limited to the smaller hydrocarbons, while the longer hydrocarbons are almost exclusively paraffinic. Olefins and paraffins together form the vast majority of the product distribution, but small amounts of branched products, aromatics, alcohols, and other oxygenates are present as well, particularly on iron-based catalysts.

The hydrocarbon distribution ranges from methane up to heavy waxes, depending on the nature of the catalyst and the reaction conditions. The chain lengths of the hydrocarbons, where we do not distinguish between alkanes and alkenes, obey a statistical distribution named after Anderson, Schulz, and Flory. We derive it hereafter.

The competition between chain growth (yielding a surface intermediate with one higher carbon number) and chain termination (yielding a desorbed final product) is determined by the probability for growth, α. A high α-value results in longer hydrocarbons and, thus, a heavier product spectrum. The chance that a chain terminates as alkane or alkene equals $(1 - \alpha)$. If we assume that α is the same for all hydrocarbons of different length, the weight of products with n carbon atoms (C_i) can be formulated on a relative basis:

$$C_1 = 1(1 - \alpha) \tag{8.72}$$

$$C_2 = 2(1 - \alpha)\alpha \tag{8.73}$$

$$C_3 = 3(1 - \alpha)\alpha^2 \tag{8.74}$$

$$C_n = n(1 - \alpha)\alpha^{n-1} \tag{8.75}$$

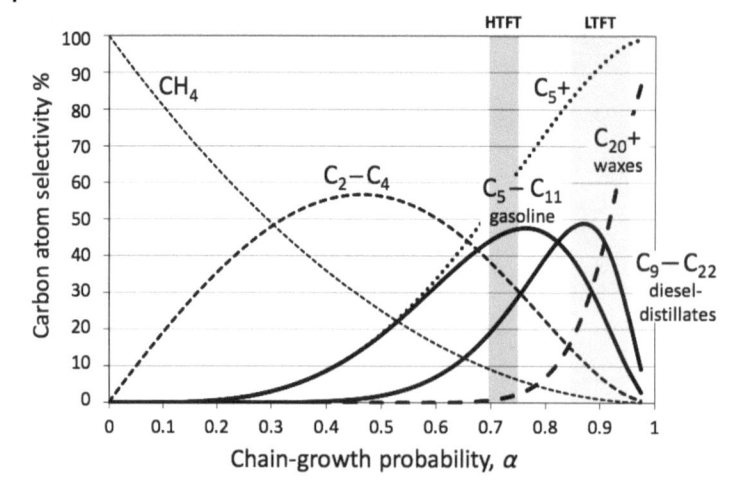

8 Catalysis in Practice: Synthesis Gas and Hydrogen

Figure 8.24 Weight fraction of hydrocarbon products in the Anderson–Schulz–Flory distribution, as a function of chain-growth probability, a. LTFT low temperature Fischer–Tropsch process, HTFT high temperature Fischer–Tropsch process.

The total weight of carbon in the product spectrum then forms a convergent infinite sum with an analytical solution:

$$\sum_{i=1}^{\infty} C_i = \sum_{i=1}^{\infty} i(1-\alpha)\alpha^{i-1} = \frac{1}{1-\alpha} \tag{8.76}$$

Hence, in a properly normalized distribution of weights we have

$$w_i = i(1-\alpha)^2 \alpha^{i-1} \tag{8.77}$$

where i is the number of carbon atoms, w_i is the weight fraction of chain length i, α is the chain growth propagation probability, and $(1-\alpha)$ the probability that a chain terminates. Figure 8.24 shows the product distributions predicted by Eq. (8.76) for various values of α. The average product length can also be obtained from Eq. (8.76) and equals $1/(1-\alpha)$ for the number-averaged chain length, $\langle L_n \rangle$, and $(1+\alpha)/(1-\alpha)$ for the weight-average $\langle L_w \rangle$. Table 8.7 gives a few examples.

Rearranging Eq. (8.77) into logarithmic form yields

$$\ln(w_i/i) = i \ln \alpha + \ln((1-\alpha)^2/\alpha) \tag{8.78}$$

and plotting selectivities according to (8.78) gives a straight line with $\ln \alpha$ as the slope.

The chain-growth probability depends on the conditions and on the catalyst. Nickel, for example, has a low α and produces predominantly methane, while cobalt and iron under low-temperature conditions may have α-values of 0.9 and higher, implying that the products are largely in the wax region. Decreasing the H_2 : CO ratio, decreasing the temperature, and increasing the pressure all lead to

Table 8.7 Some examples of how average chain lengths, selectivities to higher products C_{5+} and the selectivity towards the undesired methane $(1 - a)^2$ depend on the chain-growth probability, a.

a	$\langle L_n \rangle$	$\langle L_w \rangle$	C_{5+} (wt%)	Methane selectivity $(1 - a)^2$ (wt%)
0.5	2	3	19	25
0.75	4	7	63	6.2
0.8	5	9	74	4
0.9	10	19	92	1
0.95	20	39	98	0.25

longer chains. Furthermore, promoters such as potassium and rare earth oxides tend to increase α.

The strategy in modern GTL and CTL plants is to increase α as much as possible to obtain product slates rich in waxes. These long hydrocarbons are then fed into a hydrocracker, yielding products in the diesel, kerosene, or naphtha range, depending on the severity of the cracking process.

8.3.2.2 Reactor Technology

Three basically different process technologies are currently in operation: low, medium and high temperature, operating at temperature ranges of $\sim 220-240$, 270–280, and around 350 °C, respectively, and at pressures in the range of 25–45 bar. Figure 8.25 summarizes the different reactors that are in use along with typical production capacities in barrels of product per day, as is customary in the XTL or 'Anything-to-Liquids' world.

The low and middle temperature Fischer–Tropsch (LTFT and MTFT, respectively) processes produce long hydrocarbon chains, waxes, which are liquids under reaction conditions. Hence, the reaction occurs in the three-phase domain, with gaseous reactants and light products, liquid waxes, and a solid catalyst. The slurry phase reactors, as used by SASOL (LTFT) and Synfuels China (MTFT), offer excellent isothermal and gradientless operation, while catalysts can be removed and added on line without interruption of the process. The largest reactors of this type are 60 m high and offer production capacities up to 20 000 bbl/day. Fixed-bed multitubular reactors for the LTFT have much smaller capacity, and several are used in parallel, such that units can be stopped for catalyst replacement. Catalysts cannot be used at their full potential, as cooling is less efficient than in the slurry phase operation. Nevertheless, the technology is successfully applied in the largest GTL plant in the world in Qatar. Both cobalt and iron catalysts are in use in LTFT, while MTFT is carried out with iron catalysts.

High-temperature Fischer–Tropsch (HTFT) processes produce shorter products, and are geared more towards gasoline and chemicals (naphtha, olefins). HTFT reactors operate in the gas–solid domain and exclusively use iron catalysts, see Figure 8.26 for a comparison of LTFT and HTFT.

Fischer-Tropsch Reactors

Technology:

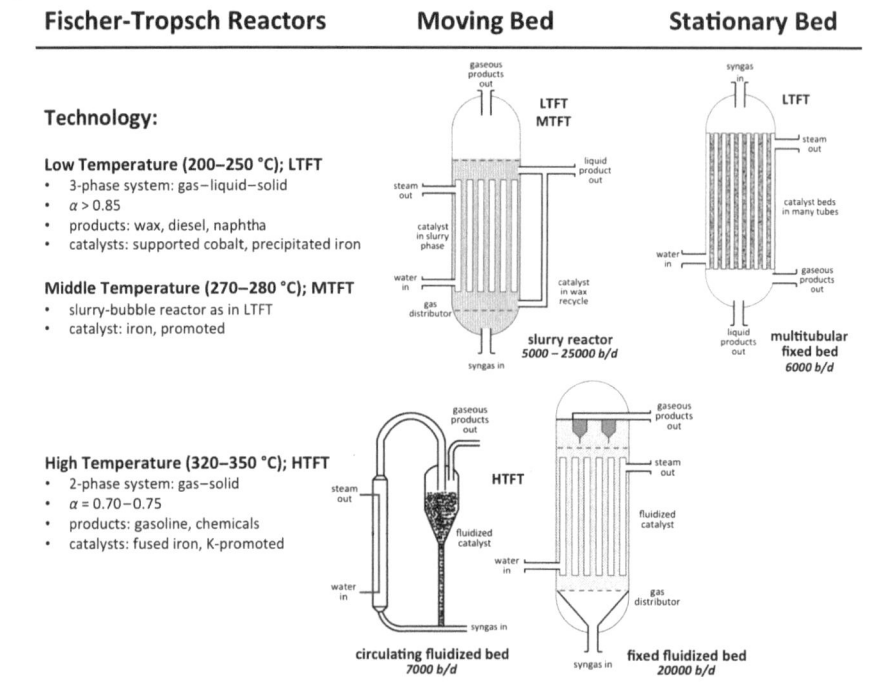

Low Temperature (200–250 °C); LTFT
- 3-phase system: gas–liquid–solid
- $\alpha > 0.85$
- products: wax, diesel, naphtha
- catalysts: supported cobalt, precipitated iron

Middle Temperature (270–280 °C); MTFT
- slurry-bubble reactor as in LTFT
- catalyst: iron, promoted

High Temperature (320–350 °C); HTFT
- 2-phase system: gas–solid
- $\alpha = 0.70$–0.75
- products: gasoline, chemicals
- catalysts: fused iron, K-promoted

Moving Bed: LTFT / MTFT — slurry reactor 5000–25000 b/d; HTFT — circulating fluidized bed 7000 b/d

Stationary Bed: LTFT — multitubular fixed bed 6000 b/d; fixed fluidized bed 20000 b/d

Figure 8.25 Overview of reactors used in present-day Fischer–Tropsch technology, along with typical production capacity in barrels per day (bbl/d; 1 barrel = 159 liter; 1 Mt/year corresponds to approximately 25 000 bbl/d).

Due to the large number of reactions occurring in the Fischer–Tropsch reactor, a good catalyst must satisfy an array of criteria. Of general importance are the distribution of product molecular weights, the degree of branching in the molecules, the content of double bonds, and the content of oxygenated products. Of particular importance are a low selectivity for methane formation and a low propensity to form carbon which leads to catalyst deactivation. Owing to the ease by which the catalyst can be exchanged during operation in the two preferred reactor types, the importance of the stability towards deactivation depends primarily on the price of the catalyst. Thus, the catalyst must be either very inexpensive or very stable towards deactivation.

Industrially used catalysts are based on either iron or cobalt. Ruthenium is an active Fischer–Tropsch catalyst but is too expensive for industrial use. The more noble Ni catalyst produces nearly exclusively methane and is used for the removal of trace of CO in H_2, or for converting gasified biomass into substitute natural gas.

The modern gas-to-liquids plants that were built after the turn of the century in Qatar by Sasol and Shell both use supported cobalt catalysts. Much research on cobalt Fischer–Tropsch catalysts has been carried out, with particular attention for catalyst stability [38, 39] and mechanistic issues [31, 40]. An interesting result was that the Fischer–Tropsch synthesis reaction appears to be considerably

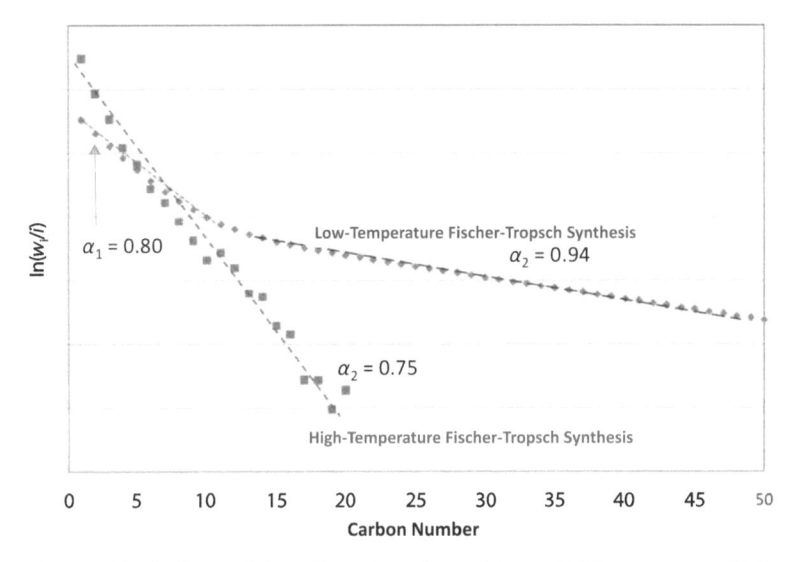

Figure 8.26 Anderson–Schulz–Flory plots of typical low and high temperature Fischer–Tropsch (LTFT and HTFT) experiments (adapted from [27]).

site demanding, as expressed by a strong particle size dependence. Cobalt particles with sizes of about 6 nm and higher exhibit the highest turnover frequencies, while the activity declines for smaller particles, and also the selectivity shifts somewhat to shorter hydrocarbons [41–43]. The reasons for this size dependence are still under debate.

New coal-to-liquid plants have stayed with iron as the catalyst [30, 44, 45], although in principle cobalt could work as well. The complications with iron are that it converts to a mixture of carbides, which are inherently more brittle than iron itself, implying that structural promoters are needed to maintain integrity. As the Fischer–Tropsch synthesis reaction produces water, the catalyst works in a potentially complex mixture of syngas, hydrocarbon products, water, and CO_2. As a result the catalyst can be a dynamic and complicated mixture of metallic, carbidic, and oxidic phases, along with promoters [32, 46]. Discussing the subject goes too far for an introductory textbook like this one.

Chapter 4 illustrates a number of catalyst characterization techniques with measurements on cobalt Fischer–Tropsch catalysts.

8.4 Water–Gas Shift Reaction

In the sections on steam reforming, methanol, and Fischer–Tropsch synthesis, we encountered the water–gas shift reaction several times. Beyond being a side reaction, it is also an interesting process in itself; it is used to modify the composition of the gas coming from steam reforming such that it contains less CO

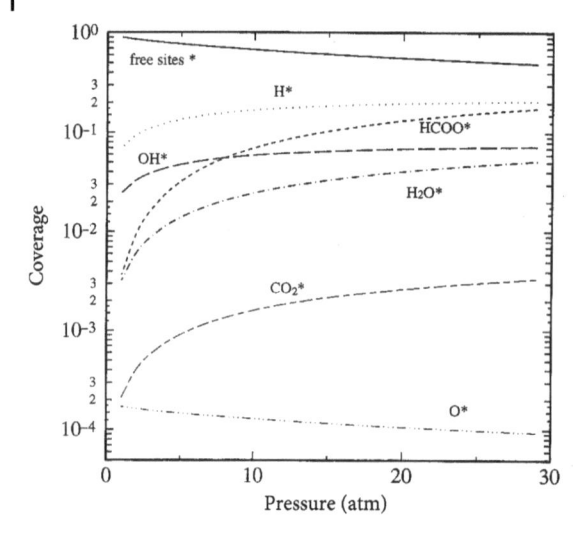

Figure 8.27 Surface coverages of the various intermediates on a copper surface during the water–gas shift reaction at 200 °C in a gas mixture of 33% H_2O, 52% H_2, 13% CO_2, and 1% CO. Note the high coverage of formate at high pressures (reproduced from Ovesen et al. [47] with permission from Elsevier).

and more H_2. This is extremely important when producing H_2 for the ammonia synthesis, where no CO- or oxygen-containing components can be tolerated. The required hydrogen is produced by steam reforming and shifted in two stages. In the first, the synthesis gas is cooled and reacted over a high temperature water–gas shift catalyst operating at 350–400 °C and 20–30 bar. This catalyst consists of Fe_2O_3/Cr_2O_3 and reduces the CO content in the gas to a level of 2–3%, which is limited by equilibrium. To decrease the CO content further, the temperature must be lowered, since the water–gas shift is an exothermic reaction. However, the Fe_2O_3/Cr_2O_3 catalyst is insufficiently active at low temperature. Instead, the much more active $Cu/Zn/Al_2O_3$ is used in the second stage as a low temperature shift catalyst. The lowest feasible temperature at which the gas can be operated without condensation at 20–30 bar is around 200 °C, and the CO content is reduced to below 0.2%. Any further reduction requires other approaches, such as methanation, which we shall discuss in relation to the production of clean, oxygen-free hydrogen for ammonia synthesis or for fuel cell applications. Copper-based catalyst are not used at higher temperatures due to sintering (see also Figure 8.18).

The low temperature water–gas shift reaction is well described by a microkinetic model and follows to a large extent the scheme in Eqs. (8.23)–(8.31) [47]. The analysis revealed that formate may actually be present in nonvanishing amounts at high pressure (Figure 8.27).

8.5
Synthesis of Ammonia

The kinetics of the ammonia synthesis have been discussed as an example of microkinetic modeling in Chapter 7. Here, we present a brief description of the process, concentrating on how process variables are related to the microscopic details and the optimization of the synthesis.

8.5.1
History of Ammonia Synthesis

Since the early nineteenth century, scientists have used the decomposition of ammonia as a convenient test reaction for experiments. However they did not succeed in accomplishing the reverse reaction between N_2 and H_2 to NH_3, in spite of nitrogen fixation being a highly desired goal. The reason for this became clear after the formulation of equilibrium thermodynamics by van't Hoff. It was Fritz Haber who realized that appreciable conversion for an equilibrium reaction

$$N_2 + 3H_2 \rightleftharpoons 2NH_3 \quad \Delta G° = -92.4 \, \text{kJ/mol} \quad (8.79)$$

requires the combination of high pressure and low temperature. As the rate of N_2 decomposition is unmeasurably small, it takes a very active catalyst to enable formation of ammonia. In 1909, Haber and coworkers managed to produce ammonia at a significant rate of 2 kg/day over an osmium-based catalyst at 175 bar.

To convert N_2 and H_2 into ammonia at a reasonable scale, flow reactors are needed that can be operated at high pressures. Until then, high-pressure reactions were mainly carried out in batch processes. Carl Bosch at BASF developed the technology that enabled scaling up to several tons of ammonia per day at 300 bar.

Haber obtained the Nobel Prize in 1918, Bosch in 1931.

The catalyst was reformulated by Alwin Mittasch, who synthesized some 2500 different catalysts and performed more than 6500 tests. They arrived at a triply promoted catalyst consisting of a fused iron catalyst, with Al_2O_3 and CaO as structural promoters and potassium as an electronic promoter. The process was first commercialized by BASF, with the first plant located in Oppau in Germany producing 30 t/day in 1913. The plant initially produced ammonium sulfate fertilizer, but when the First World War broke out it was redesigned to produce nitrates for ammunition. The plant was expanded and in 1915 it produced the equivalent of 230 t ammonium per day.

The development of ammonia synthesis represents a landmark in chemical engineering, as it was the start of large-scale, continuous high-pressure operation in flow reactors, and in catalysis, because the numerous tests of Mittasch provided a systematic overview of the catalytic activity of many substances [48, 49].

Why is the synthesis of ammonia so important? Nitrogen is an essential component of biological systems, for which amino acids are fundamental building blocks. Although nitrogen accounts for 80% of the air, N_2 is among the most stable molecules and is, therefore, not easily activated.

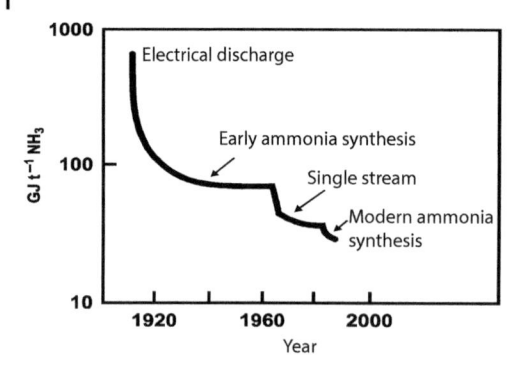

Figure 8.28 The energy consumption of the ammonia synthesis has decreased steadily and is now comparable with the natural process (courtesy of Haldor Topsøe AS).

Nature incorporates nitrogen via three different routes. One-third comes from nitrogen oxidized by, for example, fires and lightening where the thermodynamic conditions (just as in the car engine) favor NO_x formation. Another important resource are bacteria that use the enzyme nitrogenase, which under anaerobe conditions and in the presence of energy, can convert N_2 into ammonia. The active site of this enzyme is a FeMoCo complex. Calculations have suggested that the reaction pathway differs completely from the metal-catalyzed one presented in the previous chapter [10]. In the enzymatic route, hydrogen is added sequentially to the nitrogen molecule, which first dissociates when the fifth H atom is added. It appears that nature has found a smart route that allows this process to proceed at room temperature. The bacteria lives in symbioses with several plants, for example the pea/leguminous plant or alder tree, which by supplying the bacteria with sugar from photosynthesis gets the necessary ammonia to produce amino acids and thereby new enzymes. The enzyme uses at least 16 ATP to produce two molecules of ammonia, and the human-created ammonia process is almost as efficient. Adenosine 5′-triphosphate (ATP) is the energy carrier in living cells. The energy can, for example, be released by the following reaction:

$$ATP + H_2O \rightleftarrows ADP + phosphate \quad \Delta G° = -30.5 \text{ kJ mol}^{-1} \quad (8.80)$$

Although the iron ammonia synthesis catalyst has not changed very much, the energy effectiveness of the process has been improved considerably (Figure 8.28). It now costs only about 400–500 kJ mol^{-1} of ammonia synthesized, which is becoming comparable with the enzymatic process.

Ammonia synthesis is the second largest chemical process, after the production of sulfuric acid (see also Chapter 1. It accounts for about 1% of the total human-related energy consumption. Roughly 80% of the ammonia produced is used for fertilizers (either as liquid ammonia or as more easily handled salts such as ammonium nitrate, ammonium phosphate, etc.) and, as such, ammonia synthesis is indispensable for our society. Other applications of ammonia are nitrogen-containing polymers such as nylons, polyamides, polyurethanes, or explosives (nitroglycerin, trinitrotoluene, TNT).

Figure 8.29 An ammonia plant capable of producing a total of 2 × 1350 t ammonia per day over 2 × 150 t of catalyst. The immense size of the reactors is illustrated by the size of the people indicated by the arrow (courtesy of Haldor Topsøe AS).

8.5.2
Ammonia Synthesis Plant

Preferably, an ammonia plant is constructed at a geographical location where plenty of energy (e.g., as methane) and water are available, and where easy transport of the ammonia by ship is feasible. An ammonia plant that produces roughly 2 × 1350 t ammonia per day over 2 × 150 t of catalyst is shown in Figure 8.29. The facility is visually dominated by the two steam reforming plants, which are easily recognized.

Figure 8.30 shows the scheme for producing ammonia. First, the natural gas is desulfurized and then steam-reformed in the primary reformer into a mixture of unreacted methane (10–13%), CO, CO_2, and H_2 that is then combined with air, which contains the necessary nitrogen for the ammonia process, to react in a secondary reformer. Here, the oxygen reacts with hydrogen and methane in strongly exothermic processes that heat the gas from 800 to about 1000 °C, which results in even stronger conversion of the remaining methane. Next, the gas is cooled by heat exchangers, enabling the heat to be reused for the endothermic primary reformer. The gas, cooled to about 400 °C, contains less than 0.25 wt% methane as it then enters the two-stage water–gas shift, to reduce the CO content in favor of more hydrogen (Section 8.4).

Oxygen-containing molecules cannot be tolerated in the ammonia synthesis, primarily because they form iron oxide that blocks the active surface. First, the CO_2 is removed, through a scrubber, by reaction with a strong base. The remaining CO (and CO_2) is then removed by the methanation reaction, converting the CO into methane and water. Finally, the water is removed by, for example, molecu-

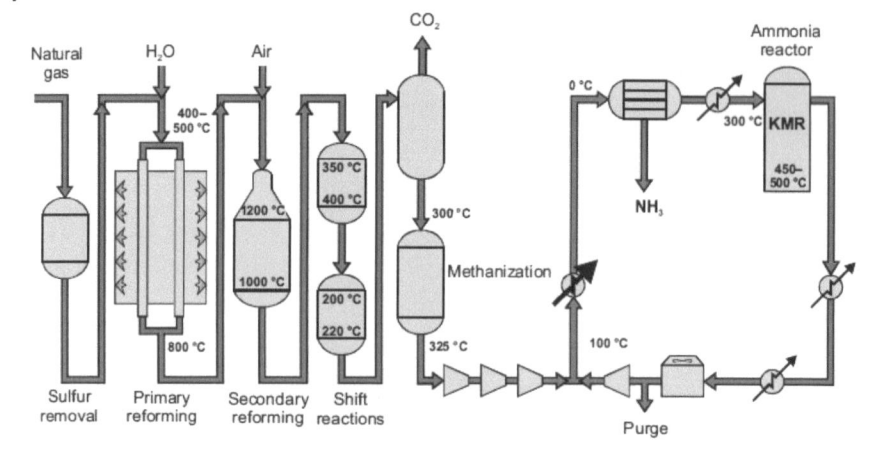

Figure 8.30 Process scheme of ammonia synthesis (courtesy of Haldor Topsøe AS).

Table 8.8 Typical composition of the feed gas when it enters the ammonia synthesis reactor.

Gas	N_2	H_2	CH_4	Ar	CO
%	74.3	24.7	0.08	0.03	1–2 ppm

lar sieves. Methane does not present problems because it interacts weakly with the catalyst surface. The gas mixture (Table 8.8) is compressed to the roughly 200 bar needed for ammonia synthesis and admitted to the reactor.

The gas reacts over the ammonia catalyst in an exothermic process at 450–500 °C, leading to an exit concentration of ammonia of about 15–19%. The ammonia is extracted by condensation and the unreacted gas recycled to the reactor. A fraction is purged to prevent the accumulation of inert components Ar. The ammonia condensation is not complete, meaning that the real inlet gas of the reactor already contains several percent of ammonia.

8.5.3
Operating the Reactor

The ammonia synthesis process is exothermic; hence, heat develops as the reaction proceeds through the reactor. This leads to a lower equilibrium partial pressure of ammonia and, therefore, it is desirable to operate at as low a temperature as possible, while keeping the rate as high as possible. These considerations result in a compromise for an operation temperature between 450 and 500 °C. The ammonia reactor has undergone several changes and has become highly sophisticated in order that as little amount of catalyst as possible is used, so as to minimize the size and, thus, the investment costs of the plant. Figure 8.31 shows a reactor with an adiabatic, two-bed radial flow arrangement and indirect cooling. The reactor is quite large and contains up to several hundred tons of catalyst.

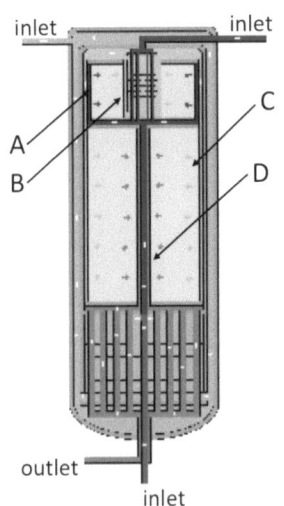

Figure 8.31 Schematic drawing of an adiabatic two-bed radial flow reactor. There are three inlets and one outlet. The major inlet comes in from the top (left) and follows the high-pressure shell (which it cools) to the bottom, where it is heated by the gas leaving the reactor bottom (left). Additional gas is added at this point (bottom right) and it then flows along the center, where even more gas is added. The gas is then let into the first bed (A) where it flows radially inward and reacts adiabatically whereby it is heated and approaches equilibrium (B). It is then cooled in the upper heat exchanger and move on to the second bed (C) where it again reacts adiabatically, leading to a temperature rise, and makes a new approach to equilibrium (D) (courtesy of Haldor Topsøe AS).

The complex construction of the reactor in Figure 8.31 can be rationalized by the concept of optimal operating line or maximal rate line [49]. Consider a plot of ammonia concentration versus temperature (Figure 8.32). The line to the right is the equilibrium line, declining monotonically since the ammonia synthesis is exothermic. If we plot the line of constant rates (for four different rate constants) as a function of ammonia concentration and temperature, we obtain the four curves shown in Figure 8.30. Initially the ammonia concentration goes up with temperature, but eventually it must bend over and follow the equilibrium line. The constant rate is increased by a factor of 10 for each curve, where the uppermost curve is that with the lowest rate. If we now connect the maxima for each curve, we find the optimum operating line, which is the one to follow to obtain the highest yield of ammonia with the smallest amount of catalyst in the reactor. The concept of an optimal operation line is used in many exothermal processes such as methanol synthesis, water–gas shift, and sulfur dioxide oxidation. It typically runs parallel to the equilibrium curve, shifted 30–50 °C towards lower temperatures.

The principle of the adiabatic, two-bed, radial flow reactor shown in Fig. 8.31 is represented schematically in Figure 8.33, which actually shows a three-bed version with indirect cooling stages between the beds. The equilibrium line and the optimal operation line are also shown in the ammonia concentration versus temperature plot. Since ammonia synthesis is exothermic, the temperature rises roughly 14–18 °C/% of ammonia synthesized. On follow a volume of gas through the reactor, on the operating line indicated by the arrow in Figure 8.33, the temperature rises through the first bed, corresponding to the upward line in the lower part of the figure. In the cooling stage, the temperature decreases as indicated by a horizontal shift to the left. On entering the second bed, the temperature rises again until the gas reaches the next cooling stage. In this manner, it is possible to keep the reactor working close to the optimal operation line, by letting the operation line zig-zag around the optimum operating line.

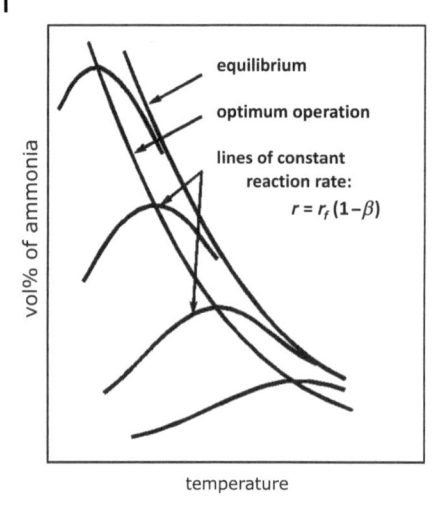

Figure 8.32 The solid and monotonically declining line to the right represents the equilibrium curve. The four curves represent lines with constant rates in the same plot. Since we want to operate at points where of maximum ammonia concentration, the optimal operation line is defined as the line running parallel to the equilibrium curve, passing through all the maxima (reproduced with permission from Dybkjaer [49], copyright 2012 by Springer).

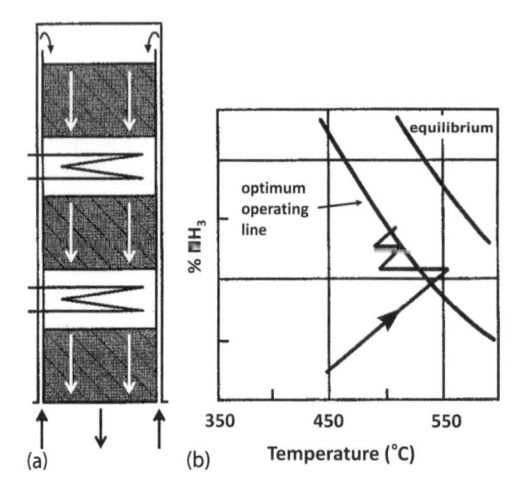

Figure 8.33 (a) Schematic diagram of an adiabatical three-bed, indirectly cooled reactor with two heat exchangers. (b) A diagram showing the equilibrium curve to the upper right, the optimal operating line and the operation line for the reactor are to the left (reproduced from Jacobsen et al. [50] with permission from Elsevier).

This procedure reflects today's technology where, typically, a two- or three-bed reactor is used. Recent investigations lead by Jacobsen have shown how this concept opens up new and interesting possibilities for combining fundamental insight with detailed process design, improving the overall process by optimizing the catalysts in the beds [50].

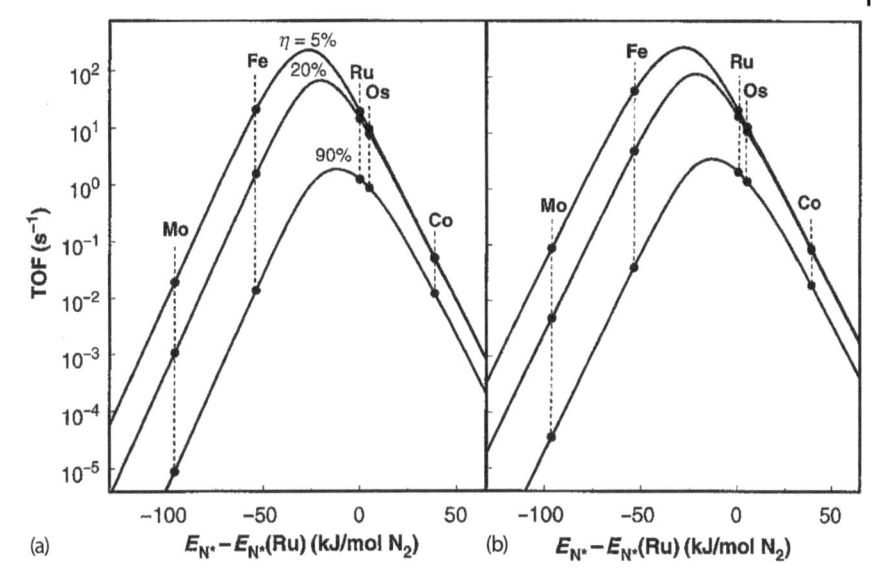

Figure 8.34 Predicted volcano plots for ammonia synthesis, showing the turnover frequency versus the relative bonding strength of N atoms to the surface for ammonia concentrations of 5%, 20%, and 90%: (a) conditions of 420 °C, 80 bar, and a 2 : 1 H_2 : N_2 gas mixture; (b) 450 °C, 200 bar, and a 3 : 1 H_2 : N_2 gas mixture (reproduced from Jacobsen et al. [50] with permission from Elsevier).

8.5.4
Scientific Rationale for Improving Catalysts

In Chapter 6, we discussed the principle of universality in heterogeneous catalysis to show how a linear relation between bonding energy and activation energy explained the volcano plots associated with Sabatier's principle. In Figure 6.42c, it was also shown how the rate of ammonia formation correlates with the adsorption energy of the N atoms and the concentration of gaseous ammonia. More ammonia corresponds to N atoms that are less strongly bound to the surface. Hence, if we can modify the nitrogen bonding energy continuously, for example by alloying, we could optimize the catalyst further.

By utilizing both the microkinetic expression for the ammonia synthesis and the linear Brønsted–Evans–Polanyi relation between the activation energy for dissociation of N_2 and the atomic nitrogen bonding energy, the theoretical turnover frequency (TOF) can be estimated as a function of only one parameter, namely the atomic nitrogen bonding energy ΔE. For each parameter set consisting of gas composition, pressure, and temperature, a volcano plot can be generated by plotting TOF versus ΔE for atomic nitrogen on ruthenium (Figure 8.34).

Figure 8.34 shows how the volcano plots shift towards lower nitrogen bonding energy when the ammonia concentration increases. By repeating this type of calculation as a function of temperature and ammonia concentrations for specific N_2 : H_2 ratios, different volcano plots appear and the nitrogen bonding energies

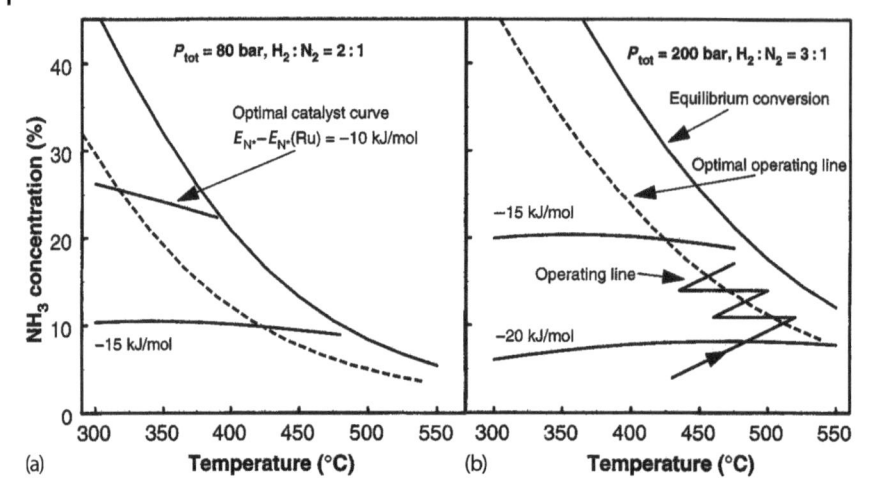

Figure 8.35 Equilibrium curve, optimal operation line, and optimal catalyst curves in an [ammonia] versus temperature plot for two different sets of conditions: (a) 420 °C, 80 bar, and 2 : 1 H$_2$: N$_2$; (b) 450 °C, 200 bar, and 3 : 1 H$_2$: N$_2$. The crossing between the optimal catalyst curve for specific nitrogen bonding energy and the operation curve reveals where the particular catalyst should be located in the bed. The operation curve should be kept as closely as possible to the optimal operation line (reproduced from Jacobsen et al. [50] with permission from Elsevier).

that correspond to maximal TOFs can be found. By plotting the relation between ammonia concentration and the temperature for which the maximum of the volcano plot appears at the *same* nitrogen bonding energy, the *optimal catalyst curve* can be found for that particular bonding energy (Figure 8.35).

The equilibrium curve and the optimal operation line are again plotted in an ammonia concentration versus temperature plot for each of the two sets of conditions in Figure 8.35, but now together with the optimal catalyst curves for a few selected nitrogen bonding energies. The Figure 8.35b also shows the operating line, and it is now possible to estimate which catalyst should be where in the reactor. That is, when the optimal catalyst curve for $E_{N*} - E_{N*}(\text{Ru}) = -20\,\text{kJ}\,\text{mol}^{-1}$ crosses the operating line, then this particular catalyst should be placed at that position in the reactor. Ideally the bonding energy should, thus, vary continuously down through the bed, from roughly $-25\,\text{kJ}\,\text{mol}^{-1}$ at the entrance to $-15\,\text{kJ}\,\text{mol}^{-1}$ at the exit, to obtain the best results.

The important message here is that the overall performance of the reactor may be improved by using an assembly of catalysts that varies though the reactor bed. To what extent such approaches will become viable depends on the cost of varying the catalysts and the savings realized by reducing the size of the high-pressure reactor.

The design sketched above is an elaborate version of the so-called Kellogg Advanced Ammonia Process (KAAP) in which iron-based catalysts are used in the first bed, and ruthenium-based catalysts, which bind nitrogen more weakly, are used in the second, third, and fourth beds [51].

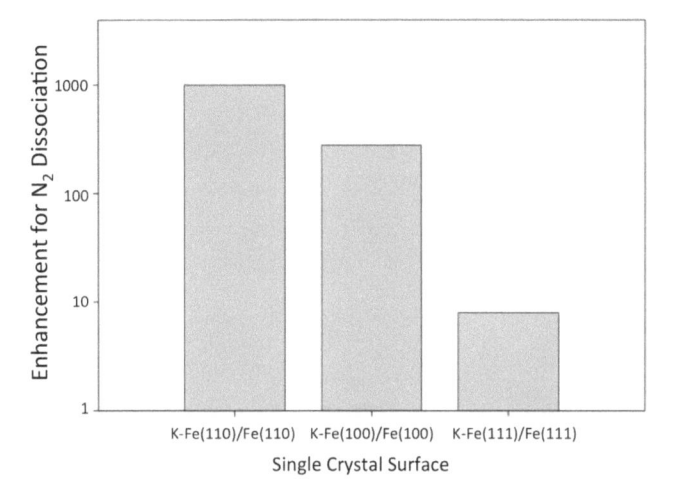

Figure 8.36 Enhancement of the sticking coefficients of nitrogen by promoting the three basal plane of iron with potassium (adapted from Somorjai and Materer [54]).

8.6 Promoters and Inhibitors

In the previous sections, we have dealt mainly with the catalytic activity of pure substances such as metallic iron, ruthenium, copper, platinum, etc. Real catalysts, however, are often much more complex materials that have been optimized by adding remote amounts of other elements that may have a profound impact on the overall reactivity or selectivity of the catalyst. Here, we shall deal with a few prominent examples of such effects.

Promoters are materials that enhance the effect of the catalyst. They can be divided into structural promoters and electronic promoters. Structural promoters are, for example, the low content of Al_2O_3 in the ammonia catalyst, which stabilizes the relatively large metallic iron area (ca. $20\,m^2$) per gram of catalyst. Electronic promoters such as K or Cs (actually different alkalis and earth alkalis may work dependent on the system) enhance the catalyst itself by modifying the surface. Generally, they become strongly polarized when absorbed on the surface, setting up a dipole field on the surface [52]. The dipole will interact with other adsorbates and may, for example, help in dissociating nitrogen if it has an opposite dipole field in the transition state, since this will lower the activation energy (see Mortensen *et al.* [53]). An electronic promoter may also affect the nitrogen coverage if, for example, it interacts with the ammonia, which will tend to have a dipole in the same direction as that induced by the alkalis. This would then lead to a repulsion that reduces the poisoning effect of ammonia. Thus, depending on the signs, the effects may be quite different. The effect of adding potassium is clearly demonstrated in Figure 8.36, where the enhancements in the sticking of nitrogen on basal iron planes are shown.

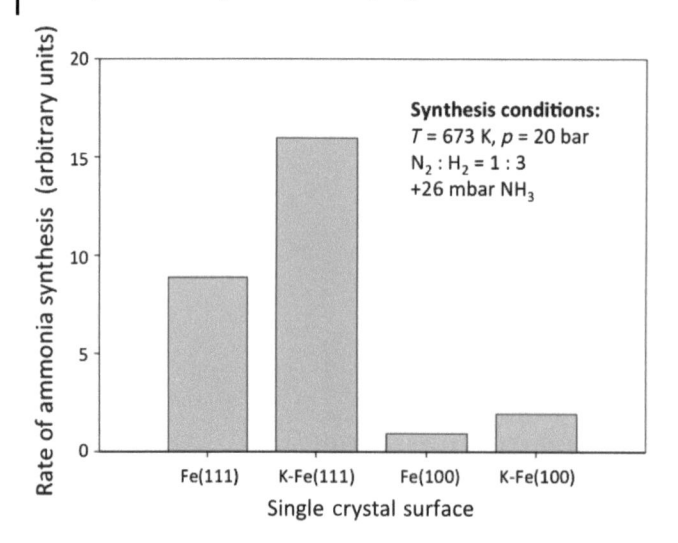

Figure 8.37 Ammonia rate for two types of iron single crystals with and without potassium promotion (adapted from Stongin and Somorjai [55]).

The effect is huge for the close-packed Fe(110) surface, but it should be considered that the sticking here is remote so that the effect of the potassium is basically to equalize the sticking on the three surfaces.

These effects profoundly influence the reactivity, as shown in Fig. 8.37, where the ammonia synthesis rate is plotted for two basal planes of iron and the same iron surfaces modified with 0.1 ML of potassium.

The reactivity of the basal planes varies substantially and is largest for the more open Fe(111) surface, which basically can be considered as being a very rough surface. With the more stable closed-packed surfaces, the reactivity drops considerably and there are many indications here that the reactivity is determined by the same B5 types of sites as was the case for the Ru(100) surface discussed in Chapters 6 and 7 [56]. Thus, the effect of potassium may not only be electronic. It could be speculated that adding potassium to the more close-packed surfaces introduces the much more reactive B5 sites and thereby also acts as a structural promoter, but this is still a field of some controversy.

Alkalis are often mistaken for being promoters in general, but this is not true. For example, alkalis inhibit methane dissociation on Ni since in this case the dipole of methane in the transition state points in the same direction as that of the alkalis absorbed on the surface, leading to a repulsion that increases the barrier for dissociation. This is illustrated schematically in Fig. 8.38 together with DFT calculations of the barriers for dissociation methane on clean and potassium-modified Ni single crystals. Notice that electronic effects tend to have a strong impact as they enter the exponentials through the activation energies while blocking effects are only linear.

For clarity, it is emphasized that the effect occurs because the transition state develops an electric dipole. Neither nitrogen nor methane has a dipole in the gas

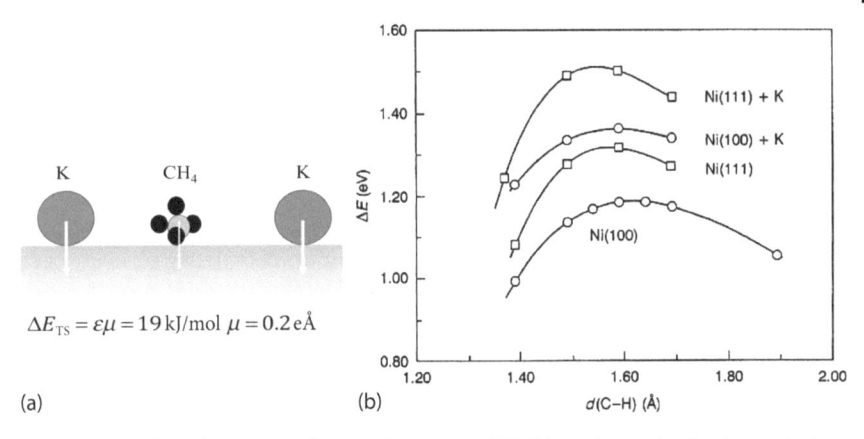

$\Delta E_{TS} = \varepsilon\mu = 19\,\text{kJ/mol} \quad \mu = 0.2\,\text{eÅ}$

(a) (b)

Figure 8.38 Effect of potassium deposited on a surface (a). The potassium sets up a dipole field which may interact with that of adsorbed species and molecules in transition states, resulting in lower or increased activation barriers (b). With methane, the dipoles are in the same direction and will, therefore, lead to repulsion and higher barriers, as shown in the DFT calculations ((b) reproduced from Bengaard et al. [57] with permission from Elsevier).

phase, but when interacting with the metal electrons they develop one. With nitrogen, that dipole is opposite that of the alkali absorbate, while for methane it is in the same direction, leading to promotion and inhibition, respectively.

Another example of potassium as a promoter is in the hydrogenation of CO to give methanol directly, as mentioned earlier [26]. Here the potassium works as a promoter for CO hydrogenation, but with conventional methanol synthesis great efforts are made to avoid the presence of alkalis in the catalyst as they tend to degrade selectivity by promoting the production of higher alcohols, that is, the surface becomes too reactive. Thus, great care has to be exercised to achieve the optimal effect.

The poison may either act electronically or simply by blocking free sites on the surface. Poisons can also be divided into two classes, those adsorbing reversibly and those adsorbing irreversibly. For example, water is a strong poison for ammonia synthesis since it leads to absorbed oxygen on the iron, blocking surface sites just like the nitrogen. It is straightforward to implement the water equilibrium in a microkinetic model for ammonia synthesis and we would simply get an additional term in the denominator of the equation that describes the number of free sites where oxygen would become the MARI. Actual water contents in the range of ppm have a strong impact on the rate of ammonia production. Fortunately, this process is reversible so that when the water is removed from the inlet the oxygen will be reacted off and the catalyst obtains its original activity. The situation is more severe if the gas is contaminated with sulfur or chlorine. These species bind more strongly to the surface and will, therefore, be much more difficult to react off. Sulfur is, in general, a highly undesired element for many catalysts due to its strong absorption energy and, therefore, it has to be removed from the reactant prior to entering the reactor. Often, ZnO guards are used that react with the sulfur-containing compounds to form ZnS.

Finally, there are the absolutely irreversible poisons, that is, those which we will never get off the surface. The adsorption of Au on the Ru steps shown above is a good example, but only of academic interest. A more realistic problem is the lead added to gasoline. If one mistakenly fuels a car equipped with a three-way catalyst with leaded gasoline, then the catalyst will undergo irreversible poisoning as the lead will be deposited on the Pt surface and so completely block it. Lead has (like gold) a low surface energy and will, therefore, always stay on the surface. In general, many precautions have to be taken to avoid poisoning effects and severe cleaning of gases is often necessary before they enter the catalyst. An interesting aspect in realizing that actually very few sites may rule the reactivity of a surface, as was seen for the steps on the ruthenium surface, is that even minute amounts of an additive may block such sites and improve the selectivity of the surface considerably. For example, in considering the hydrogenation reaction, it may be advantageous to block the more reactive step or B5 sites as they only have small effects on the hydrogenation reaction, but have a tremendous impact on, for example, undesired C–C bond scission. This gives us a new set of possibilities for designing new types of catalysts.

8.7
The "Hydrogen Society"

8.7.1
The Need for Sustainable Energy

Our current heavy reliance (> 80%) on fossil fuels will eventually have to come to an end (Table 8.9). This end may not be as soon as indicated in the table, since the predictions are based on known resources. New oil and gas fields are still being discovered and the methods for retrieving oil from known fields are continuously improving. Furthermore, vast reserves, such as tar sand and gas hydrates, await technology to enable their economically and environmentally sound exploitation. The appearance of the fracking technology for retrieving natural gas is a good example. Over the last decade, the annual consumption has increased from 2.4×10^{12} to 3.4×10^{12} m^3 but at the same time the reserves have gone up from 1.4×10^{14} to 2.0×10^{14} m^3. This means that there will be natural gas for another ~ 60 years at the current consumption rate. The same cannot be said for the oil and coal reserves where newly discovered reserves do not match the increased consumption rate over the last decade. Nevertheless, the large coal reserves can also be exploited, for example, through gasification and Fischer–Tropsch synthesis.

An important issue is the extent to which emission of the greenhouse gas CO_2, released by using fossil fuels disturbs our global ecosystem. It is well-established that the average temperature has increased and that this increase runs parallel to

Table 8.9 Known fossil fuel reserves, annual consumption, and years left if consumption continues at the same level (data from www.eia.gov).

	Oil[2013] [L]	Gas[2014] [m^3]	Coal[2011] [t]
World known reserves	~ 1.6×10^{14}	~ 2.0×10^{14}	~ 9.8×10^{11}
Word annual consumption	~ 5.0×10^{12}	~ 3.4×10^{12}	~ 8.3×10^{9}
Years left	~ 32	~ 60	~ 120

the increase of CO_2 in the Earth's atmosphere. However, whether the temperature increase is due to CO_2 alone is a matter of controversy. Alternative explanations have recently appeared, suggesting that solar activity and in particular the solar wind strongly affects cloud formation on Earth and thereby the temperature. Historical climate changes have been found to correlate with solar activity, and strong increases in both temperature and CO_2 in the atmosphere have been deduced from geological investigation. To what extent does CO_2 emitted by fossil fuels add to the natural effect? As long as there is doubt, we should limit the release of CO_2 as much as possible. Schemes have been proposed to continue the use of fossil fuels by separating the CO_2 from exhaust gases and sequestrating it into permanent deposits such as depleted gas wells or other stable geological formations. Such measures would add roughly 30% to the cost of the energy.

What are the opportunities for using forms of energy that do not lead to CO_2 formation? Nuclear power from fission reactors presents problems with the handling and deposition of nuclear waste. Fusion reactors are more appealing, but may need a further several decades of development. However, solar and wind energy offer realistic alternatives.

The Earth receives plenty of energy from the sun. On a clear day, with the sun directly overhead, the incoming energy amounts to $1\,kW\,m^{-2}$. This corresponds to about 6800 times the present energy consumption (see Table 8.10). This solar energy is unevenly distributed over the Earth, but conversion into wind and waves improves the situation.

Replacement of fossil energy by sustainable energy is often dubbed as the Tera Watt Challenge. The unit watt (1 J/s) refers to the rate of energy consumption. At present the world uses approximately 18 TW, and this number is expected to increase to about 30 TW by 2050. How to transform our society such that it can tap into the vast resources deriving from solar energy in an efficient and affordable way is a major challenge for our generation, but one that has to be solved if we want to sustain our lifestyle and expand our economies.

Table 8.10 Solar energy compared to energy consumption [58].

Incoming energy from the Sun on the Earth	3.8×10^{24} J/year	120 000 TW
Human-related energy consumption	5.6×10^{20} J/year	17.7 TW
Electric power consumption	7.1×10^{19} J/year	2.2 TW

N.B. 1 terrawatt = 1 TW = 10^{12} W

8.7.2
Sustainable Energy Sources

Some places on Earth offer energy more or less for free. Wind, flowing water, hot water sources, height differences that enable hydropower installations, all of these represent sustainable energy sources with a relatively minor burden on the environment. Underground thermal sources contribute considerably to energy production in Iceland, where a high level of geothermal activity exists.

Solar energy comes in the form of photons with a wide spread in energy of several eV. Although the Sun's interior, where fusion of hydrogen into helium takes place, is very hot – on the order of 10^8 K – the surface is only about 5800 K. The solar spectrum can, thus, be approximated by the radiation from a black body at that temperature, which through the Planck's law gives us the incoming flux of sunlight as

$$I_E \, d\varepsilon = \frac{2\pi\varepsilon^3}{c^2} \frac{1}{e^{\frac{\varepsilon}{k_B T}} - 1} \, d\varepsilon \tag{8.81}$$

relating the incoming energy flux (J m^{-2} s^{-1}) in the energy range interval $[\varepsilon, \varepsilon + d\varepsilon]$ to the actual energy ε and the temperature T as shown in Figure 8.39.

It is possible to harvest solar energy directly by converting it into electric energy by photovoltaic (PV) devices [59]. Its principle is shown in Figure 8.40. The device consists of two semiconductors that have been doped with either p- or n-type dopants so that their Fermi levels are rather different. The n-type (negatively doped) will have a high lying Fermi level, while that of the p-type (positively doped) material is low. When joined, they form a junction and the Fermi level will equalize in the two semiconductors as charge will flow from n-type to p-type material. This has the interesting consequence that there will be an electrical field in the semiconductor and that conduction and valence band shift, with a region in between where the bands bend. This implies that when an electron in the valence band absorbs a photon it can be excited into the empty conduction band if the energy is sufficiently large; the hole left behind and the electron will then be separated, instead of recombining which would negate the entire process. In the diagram of Figure 8.40, electrons pile up on the right side, while the holes accumulate on the left. One can utilize the energy by letting the electrons flow through an outer circuit from right to left, where they recombine with the holes. However, many energy losses are involved in this process. First of all, all excitation energy

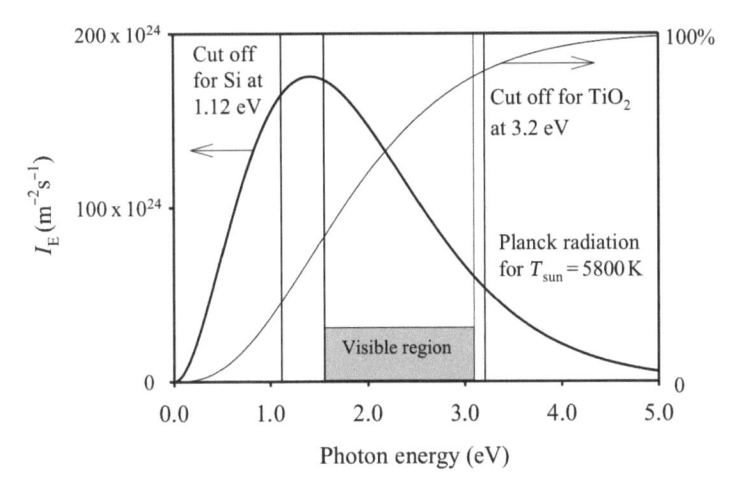

Figure 8.39 Theoretical energy flux for radiation from the Sun as a black body at 5800 K. The spectrum reaching the surface of the Earth is modified through absorption in the infrared region by H_2O and CO_2 and in the ultraviolet by ozone. The integrated flux is shown on the right-hand scale. The region of visible light is indicated as well as the energy required for making electron–hole pairs in silicon and in TiO_2.

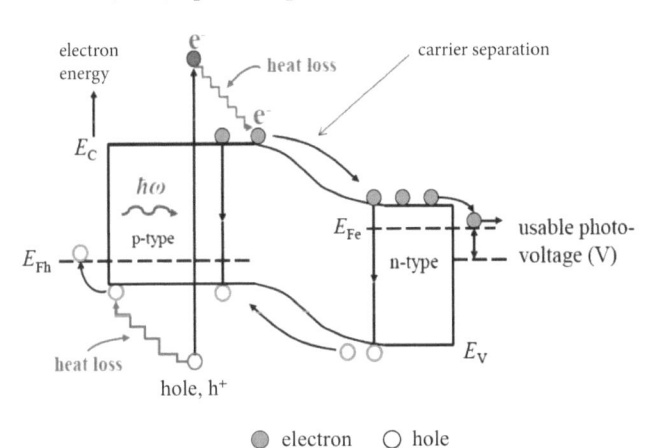

Figure 8.40 The principle of a photovoltaic device is illustrated. Notice it is only the energy difference between the two quasi-Fermi levels of the electrons E_{Fe} and the holes E_{Fh} that becomes available as useable energy and results in an open circuit voltage V. This will be further reduced when a current is extracted. Nevertheless, substantial energy can be harvested this way.

gathered by the electron which is larger than the band gap, is lost. It is, therefore, necessary to have many combinations of band gap in tandem if we want to harvest the sunlight efficiently. Second, the process of band bending itself costs energy. Finally, there will be energy losses due to electron–hole recombinations.

The ultimate theoretical efficiencies of PV devices are as high as 86%, when using an infinite number of photo-absorbers in tandem, and solar concentrators.

This is usually referred to as the Shockley–Queisser limit, named after the authors who reported the concept in 1961 [60]. Today, the best devices, consisting of three junctions in combination with a solar concentrator, approach 45%. They are prohibitively expensive, but the more affordable single-junction devices can achieve 25% and 21% for crystalline and polycrystalline silicon, respectively. Silicon is the optimal material for a single component PV device. This is related to the rather small band gap of 1.1 eV (see Figure 8.39), which allows a large part of the solar spectrum to be used for excitation. On the other hand, a lot of energy is lost, since higher energy photons excite electrons far over the band gap. Nevertheless, the theoretical limit for the efficiency of single-junction silicon is 33%.

Photovoltaic technology is still too expensive to compete with fossil fuels, which is largely because CO_2 emission has no price tag yet. Nevertheless, the cost of PV energy is decreasing steadily and is essentially following Morse's law due to more efficient large-scale manufacturing methods. For instance, the time needed for a device to break even, that is, during which it produces the energy for its own manufacture, has gone down from 4 to about 1.5 years over the last decade.

Wind energy has also received much attention over the last two decades and it has been judged to have a potential of harvesting 70 TW on land only. Large-scale windmills with a capacity of 6 MW or more are claimed to be competitive with conventional power plants relying on fossil fuels. In countries like Denmark, wind energy contributes substantially to electric power generation (42% in 2015), corresponding to 6% of the total energy consumption. However, there are problems associated with the use of wind energy, and there are also limits to where windmills can be placed. At present windmill parks are mainly installed in offshore parks, where winds are stronger than on land, and neighbors take less offense. Of course, investment, maintenance, and service costs are higher. For a windmill, it takes around half a year to generate the energy used to make it. Wind energy comes close to competing with fossil fuels.

The most used sustainable energy resource worldwide is biomass, formed by photosynthesis. Even under optimal conditions, photosynthesis can only harvest about 1% of the solar energy (more realistic numbers are 0.3–0.5%), and store it in, for example, grain, straw, or wood. In terms of molecules, the energy is mostly stored in sugars. This relatively low efficiency sets a limit on the potential use of biomass, in particular since it also competes with the production of food. The advantage of biomass, however, is that enables the storage of energy, which is more difficult for solar and wind electricity.

8.7.3
Energy Storage

Electricity Storage produced from photovoltaics and windmills represents an excellent form of energy, but one that presents challenges for integrating these inherently intermittent energy sources into a network from which we expect extremely high stability, both in terms of voltage and frequency. Whereas large parts of the industry as well as transportation could in principle be electrified, there will be sectors which cannot, and these will continue to make use of liquid or gaseous fuels, think about airplanes and ships.

Efficient ways of storing electric energy include pumping of water uphill in reservoirs and then utilize gravitational energy to generate electricity when it falls down again. Such procedures can be done with efficiencies of 70%.

Alternatively, one may devise routes for storing electrical energy in the form of chemical bonds, in molecules that serve as energy carriers, or as building blocks for chemical production. Hydrocarbons as in gasoline and diesel, or methane as in natural gas possess the highest energy density on a weight basis and would be the preferred energy carriers for storage purposes. Nevertheless, simpler molecules such as hydrogen and methanol are also attractive options. In addition, the latter could be used as a building block for chemicals. These routes will heavily rely on catalysis, which is why we mention them in this book [61–63].

8.7.3.1 Hydrogen by Electrolysis

Using electrons to decompose water into its constituting elements, H_2 and O_2, is obviously an important step in the conversion of electricity into chemicals. This brings us at the subject of electrochemistry and electrocatalysis. While in heterogeneous catalysis we steer reactions in the desired direction through temperature, pressure and composition, in electrocatalysis we have to manipulate the chemical potential of the electrons by applying a voltage on an electrode, and using the right electrocatalyst. Indeed, water can be dissociated into H_2 and O_2 by using a simple 2-V battery at ambient conditions. How much current density can be obtained, depends on the choice of catalyst (provided thermodynamic requirements are fulfilled). For comparison, dissociating water by heterogeneous catalysis would require extremely high temperatures (> 2000 K) to establish thermodynamically favorable conditions.

Electrolysis is performed by mounting two electrodes in an electrolyte and applying an external voltage. As Figure 8.41 shows, there is hardly any overpotential for the hydrogen electrode and the current increases exponentially with either positive or negative potential until the reaction becomes transport limited. This, however, is not the case for the oxygen evolution. The reversible potential is at 1.23 V, but oxygen starts to form only at an overpotential of about 0.3–0.4 V (1.5–1.6 V on the RHE scale). This means that an electrolyzer can achieve an efficiency of roughly $1.2/1.8 \approx 65\%$, depending on the current drawn. Similarly, for a fuel cell where oxygen reacts with hydrogen, one observes a current at 0.8–0.9 V versus RHE, indicating again a loss of 0.3–0.4 V. What is the origin of these overpotentials and how do they relate to catalysis? For this, we need to consider the elementary steps in the reaction.

Formally, the electrolytic splitting of water to produce hydrogen can be written as

$$2H_2O \rightleftharpoons 2H_2 + O_2 \tag{8.82}$$

The free energy required for this reaction is 4.8 eV and since it involves 4 electrons, each of them needs an energy in excess of 1.2 eV. The process can be divided up

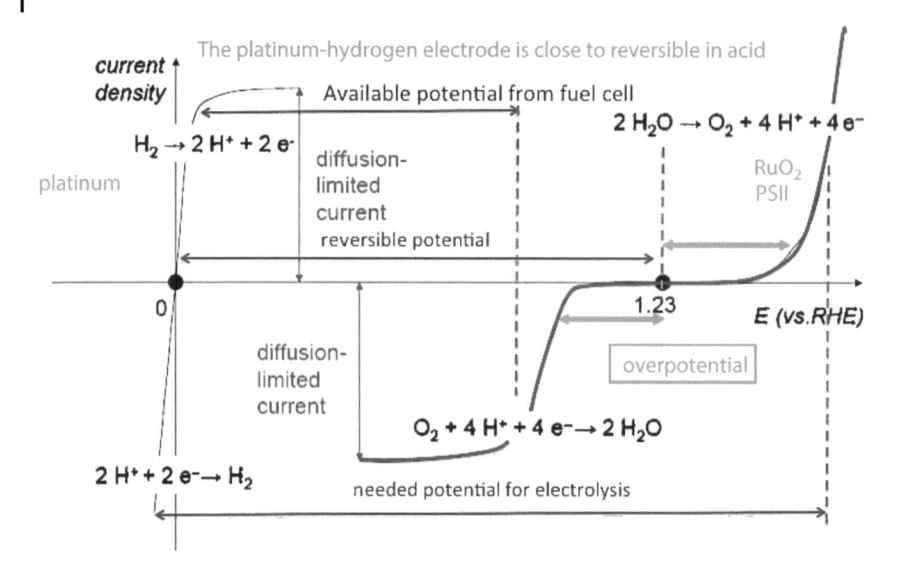

Figure 8.41 Electrocatalysis: current drawn from the anode and the cathode as a function of potential. The thermodynamic potential for dissociating water into hydrogen and oxygen is 1.23 V (i.e., the reversible potential). In practice one needs a potential of about 1.6 V, depending on the current (the lower arrow is the difference between the matching current at the anode (right) and the cathode (left). If operated as a fuel cell, we can obtain the potential indicated by the upper arrow where left now is anode and right is cathode. The x-axis is relative to the hydrogen electrode at standard conditions and pH 0 at 0 V (indicated as RHE). Note that the anode is always where the oxidation takes place, while reduction happens at the cathode (adapted from Koper and Heering [64]).

in two half-cell reactions, that is, the anode reaction for the oxygen evolution:

$$2H_2O \rightleftharpoons 4H^+ + 4e^- + O_2 \tag{8.83}$$

and the cathode reaction for the hydrogen evolution:

$$4H^+ + 4e^- \rightleftharpoons 2H_2 \tag{8.84}$$

The way to think of these reactions is that one controls the chemical activity of the electrons by biasing the electrode. Thus, to promote the anode reaction one needs to efficiently remove the electrons; hence, one "pulls" the reaction in the right direction. A positive bias attracts the electrons, and O_2 can evolve with an adequate electrocatalyst. Similarly, a negative potential on the cathode promotes H_2 formation.

As discussed with Figure 8.41, the bottleneck in water splitting is not the evolution of hydrogen at the cathode, but the oxygen evolution reaction (OER) on the anode Eq. (8.83). It consists of four elementary steps in a catalytic cycle, each involving an electron–proton transfer. This is illustrated in Figure 8.42 [65, 66]. In order for oxygen evolution to proceed efficiently, all such steps must be made exergonic ($\Delta G < 0$). This is illustrated in Figure 8.42a for the ideal catalyst, where

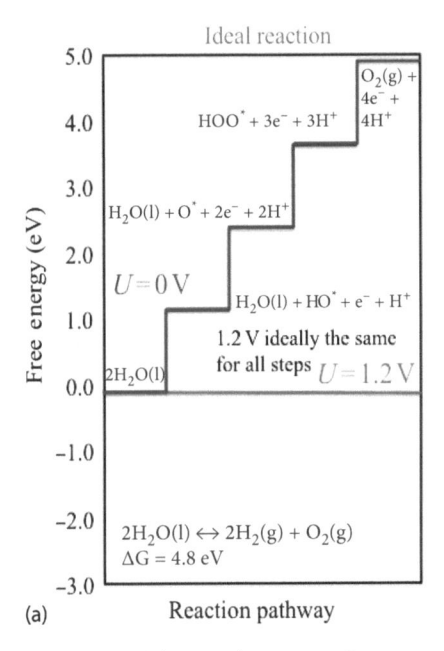

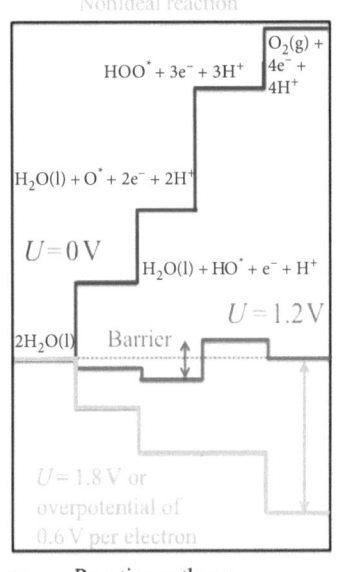

Figure 8.42 Schematic free energy diagram for water splitting. Part (a) shows the ideal case where all steps require the same energy. In reality, some steps take more energy, as is shown in part (b). At $U = 1.8$ V, the free energy change of all the steps in the oxygen evolution become negative, explaining the relatively large overpotential that is needed for this reaction. The diagrams only show the bonding energy of the various intermediates, but there can also be barriers in between the steps, like in heterogeneous catalysis, that would make the reaction even more complex as well as temperature dependent (adapted from Rossmeisl et al. [65]).

the four steps are evenly spaced in energy. The step profile shows the case with no applied potential, $U = 0$ V, where it is an uphill reaction. By supplying a voltage of 1.2 V, we obtain the reversible situation (horizontal line), that is, $\Delta G = 0$, for all steps, and the process will run spontaneously downhill at a slightly higher potential. However, each of the four steps has its own energy requirement.

Figure 8.42b shows the more realistic case. Here, the formation of HOO (or O and OH) is associated with a larger energy step than the other steps. The result is that ΔG of the most difficult step is still positive at the reversible potential $U = 1.2$ V. To let the process occur spontaneously, a higher potential of 1.8 V is needed, implying that 0.6 V, or 33% of the energy, is lost as heat in the process of making oxygen. This effect is further demonstrated in Figure 8.43a, which shows a plot of the theoretical overpotential for this reaction based on density functional theory (DFT) calculations. The theoretical overpotential is shown as a function of two descriptors, the free energy of the intermediates OOH and OH, parameters that have been identified as essential in describing the activity. Calculated values of these parameters for a number of possible catalysts are included in the plot; all the known data are linearly correlated and the best are 0.3–0.4 V off the optimum.

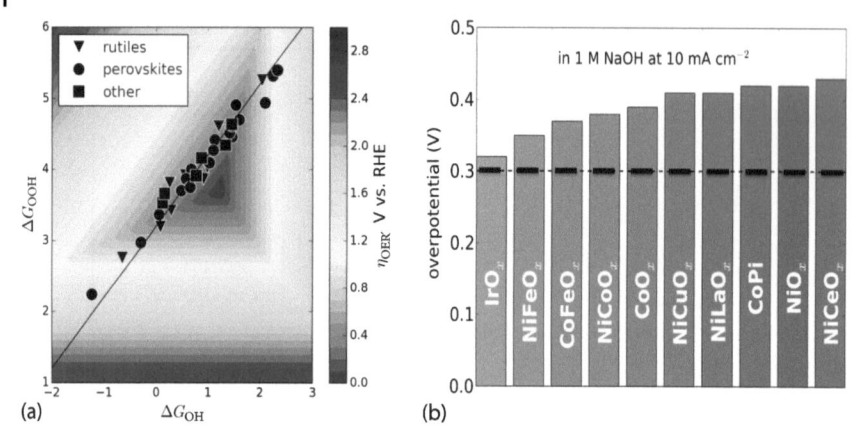

Figure 8.43 (a) Contour map of the theoretical overpotential for the oxygen evolution reaction as a function of two descriptors of catalytic activity, the adsorption energies of OH and OOH intermediates. (a) Data for a number of oxide surfaces are included – all follow a scaling relation well off the optimum in activity (data in (a) adapted from Montoya et al. [67] and (b) reproduced with permission from McCrory et al. [68], copyright 2013, American Chemical Society).

This provides a likely explanation for the experimental observation that it has been impossible, so far, to find catalysts with overpotentials less than 0.3–0.4 V (see the benchmarking data in Figure 8.43b. It seems that we have primarily been optimizing catalysts that belong to a family of materials that follow a "scaling relation" between these two descriptors. Thus, substantial work is now dedicated to find manners of breaking these scaling relations mentioned in Chapter 6.

Most electrolysers are operated with a strong alkaline electrolyte. The reason for this is that under acidic conditions, the only effective catalysts for oxygen evolution are based on IrO_2, which contains the scarce and expensive noble metal iridium. Substantial efforts are directed towards finding alternative, abundant catalysts [67–69].

The challenge of finding an efficient and cheap catalyst in electrochemistry is somewhat different than in heterogeneous catalysts. An important requirement for an electrocatalyst is that it conducts electricity. However, the anode catalyst for oxygen evolution automatically becomes oxidic when it is exposed to a positive potential of 1.5–1.8 V. Most oxides are, however, poor conductors. Furthermore, it is the current density per area of catalyst on the electrode that determines the activity, since current and reactants have to meet here. Table 8.11 presents an overview of the different types of electrolysis [69].

8.7.3.2 Combining Photovoltaics and Electrolysis

The combination of photovoltaics and electrolysis, called photoelectrocatalysis (PEC), appears as a straightforward way to produce hydrogen from sunlight and water. Nevertheless, this design makes the capture of light even more challenging since the PV device must be able to deliver a sufficiently high voltage to split water,

8.7 The "Hydrogen Society" | 373

Table 8.11 Overview of electrolysis (adapted from Sapountzi et al. [69]).

	Low temperature electrolysis			High temperature electrolysis	
	Alkaline (OH⁻) electrolysis	Proton exchange (H⁺) electrolysis		Oxygen ion (O²⁻) electrolysis	
	Liquid	Polymer electrolyte membrane	Solid oxide electrolysis (SOE)		
Operation principles	Conventional	Solid alkaline	H⁺-PEM	H⁺-SOE	O²⁻-SOE	Co-electrolysis
Charge carrier	OH⁻	OH⁻	H⁺	H⁺	O²⁻	O²⁻
Temperature (°C)	20–80	20–200	20–200	500–1000	500–1000	750–900
Electrolyte	Liquid	Solid (polymeric)	Solid (polymeric)	Solid (ceramic)	Solid (ceramic)	Solid (ceramic)
Anodic reaction (OER)	$4OH^- \to 2H_2O + O_2 + 4e^-$	$4OH^- \to 2H_2O + O_2 + 4e^-$	$2H_2O \to 4H^+ + O_2 + 4e^-$	$2H_2O \to 4H^+ + O_2 + 4e^-$	$O^{2-} \to 1/2 O_2 + 2e^-$	$O^{2-} \to 1/2 O_2 + 2e^-$
Anodes	Ni > Co > Fe (oxides) Perovskites: $Ba_{0.5}Sr_{0.5}$-$Co_{0.8}Fe_{0.2}O_{3-\delta}$, $LaCoO_3$	Ni-based	IrC_2, RuO_2, $Ir_x Ru_{1-x}O_2$, Supports: TiO_2, ITO, TiC	Perovskites with protonic-electronic conductivity	LSM-YSZ	LSM-YSZ
Cathodic reaction (HER)	$2H_2O + 4e^- \to 4OH^- + 2H_2$	$2H_2O + 4e^- \to 4OH^- + 2H_2$	$4H^+ + 4e^- \to 2H_2$	$4H^+ + 4e^- \to 2H_2$	$H_2O + 2e^- \to H_2 + O^{2-}$	$H_2O + 2e^- \to H_2 + O^{2-}$ $CO_2 + 2e^- \to CO + O^{2-}$
Cathodes	Ni alloys	Ni, Ni-Fe, $NiFe_2O_4$	Pt/C MoS_2	Ni-cermets	Ni-YSZ Substituted $LaCrO_3$	Ni-YSZ Perovskites
Efficiency (%)	59–70	65–82		Up to 100	Up to 100	–
Applicability	Commercial	Laboratory scale	Near-term commercialization	Laboratory scale	Demonstration	Laboratory scale

Anode $H_2O + 4h^+ \longrightarrow O_2 + 4H^+$

Large band gap
>1.7 eV

"Mo$_3$S$_4$"

Small band gap
~1.1 eV

Cathode $4H^+ + 4e^- \longrightarrow 2H_2$

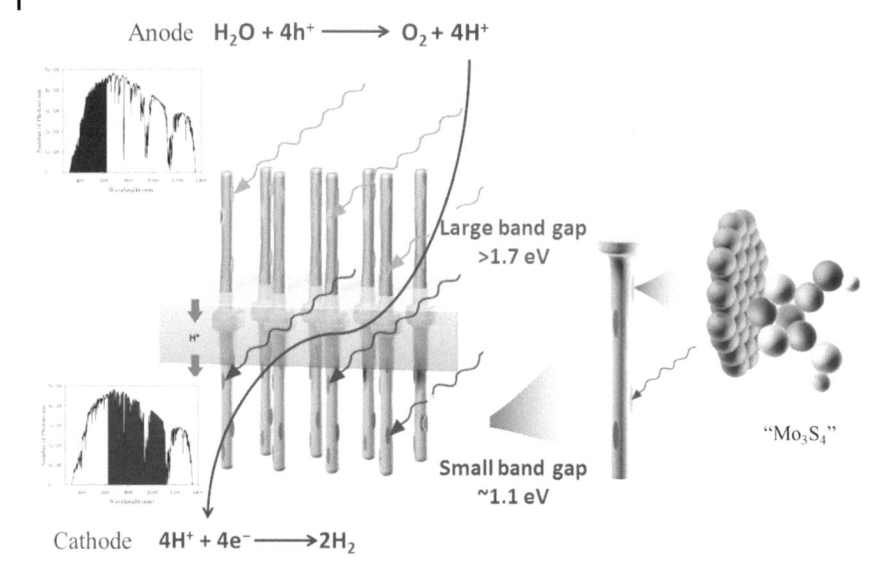

Figure 8.44 Concept of a tandem device combining a low and a high band gap material to respectively harvest red and blue light of the solar spectrum and, thus, achieve higher efficiency than a single component device. The solar spectrum is shown to the left and the light absorbed by the two parts are indicated. Notice how the red light passes through the high band-gap semiconductor. The holes are used to oxidize water and form O_2 on the anode side, while hydrogen evolution is facilitated by the nonprecious molybdenum sulfide catalysts on the cathode side. Oxygen and hydrogen evolution are separated by a nafion membrane that allows protons to pass but not the electrons (reproduced from Hou et al. [70] with permission from the Nature Publishing Group).

which requires at least 1.6 V. Hence, a single device would not be capable of doing this with reasonable efficiency and tandem devices are called for. Figure 8.44 shows a possible approach in which the semiconductors are present in pillars to allow for a long path where light can be absorbed, while the distance for holes and electrons to the catalytic surface is short. In addition the pillar structure yields a large geometric area reducing the current per area and, thus, the overpotential as seen above.

Substantial effort is invested to design and make such devices stable and efficient, while at the same time using materials that are neither prohibitively scarce nor expensive. Devices like the one in Figure 8.44 might someday be rolled out at the scale of km^2 to harvest light directly and produce a storable fuel like hydrogen [70].

8.7.3.3 Producing Fuels with higher Energy Densities

Although hydrogen is a feasible storage medium for chemical energy, it has the disadvantage of being rather voluminous, see Figure 8.45. This is less a problem for stationary applications than for the transport sector. The last two decades have seen intensive research into developing suitable hydrogen storage materials, either chemical compounds or metal hydrides. The present view is that the

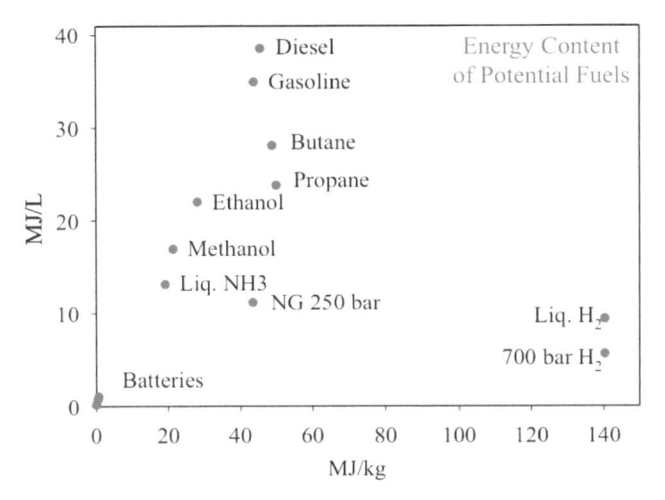

Figure 8.45 Specific energy content per volume plotted against specific energy content per mass. For transportation, the ideal fuel has a high energy content per volume, while for stationary applications this is less relevant. For batteries, the time to recharge in relation to radius of action is important.

smartest way to store hydrogen for fuel-cell driven cars is probably to compress it to 700 bar. Cars with this technology are becoming available on the market and tank stations with compressed hydrogen are starting to appear as well (Denmark presently has seven, with more to come). Developing compressed hydrogen technology for transportation is of course only meaningful if hydrogen is produced from sustainable electricity, for example, excess electricity from windmills.

Figure 8.45 illustrates that traditional fuels like gasoline and diesel have favorable energy densities for automotive applications. A tank full of such fuel easily allows for driving distances of 500–1000 km, which is a challenge for alternatively powered vehicles. From this point view, it thus makes much sense to invest in making cars much more efficient. In addition, we mention that such fuels can be made synthetically via syngas, by producing H_2 from electricity via electrolysis, and having a source of CO_2 available [71, 72]. The scenario in Figure 8.46 provides an illustration. Hydrogen and oxygen are produced from excess electricity, and oxygen released by this process is used for burning biomass which cannot be used directly to highly concentrated CO_2. The latter can be used for hydrogenation into methanol, methane, or higher hydrocarbons, if necessary via syngas. The implication of such scenarios is that the primary electricity generation will have to exceed the total energy consumption considerably, since there will unavoidably be substantial energy losses in the processes for storing energy. Referring back to the Tera Watt Challenge, we need to learn to store energy as indicated above on the TW scale.

Of course, electrolysis does not have to be limited to decomposing water. One can also use the protons in Eq. (8.84) to electrochemically hydrogenate CO_2 or CO to the desired fuels on the cathode side. This could in principle also apply to

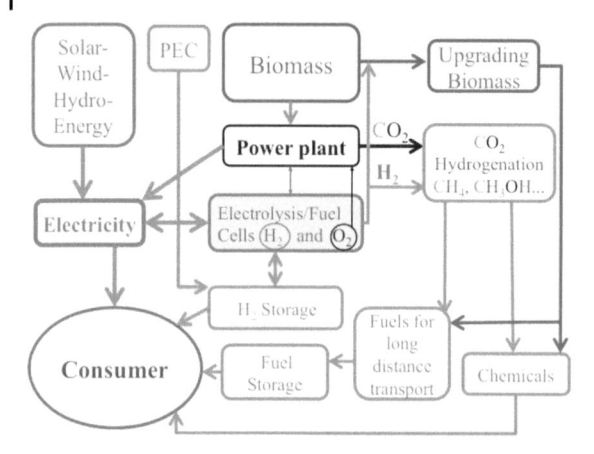

Figure 8.46 Scenario for a future society relying on sustainable energy coming mainly from solar and wind energy in the form of electricity. Note that electrolysis to produce H_2 and O_2 and production of CO_2 from biomass play an essential role in this energy flow scheme.

N_2 hydrogenation, mimicking what nature does at ambient conditions. Such reactions were pioneered by Hori in the 1980s and research has recently been intensified [73]. Hori investigated a broad range of the elements for CO_2 electrochemical hydrogenation and found that materials like Pt made only hydrogen, while others made formate (Hg, Pb) or CO (Au, Pd). Only copper was capable of making the desired hydrocarbons. All these measurements were consistently done at a potential generating a fixed current of 5 mA/cm². This resulted in substantial overpotentials of always more than 1 V, resulting in efficiencies below 40%. Figure 8.47 shows as an example the hydrogenation of CO_2 on polycrystalline copper. The reversible potentials for the various compounds are indicated by vertical lines. It is seen that substantial overpotentials are needed in order to produce any interesting compounds. Thus, there are two great challenges in this field: reducing the overpotentials and improving the selectivity. Recent work has shown that the overpotential for CO production can be reduced to 0.3–0.4 V, new electrocatalysts produce methanol and ethanol from CO at relatively low potentials of −0.3 to −0.4 V versus RHE [74].

In general, it appears easier to find good catalysts for reactions involving fewer electrons than many. Both hydrogen evolution and reduction of CO_2 to CO require two electrons per elementary step:

$$CO_2 + 2H^+ + 2e^- \longrightarrow CO + H_2O \tag{8.85}$$

Whereas production of methanol takes six electrons:

$$CO_2 + 6H^+ + 6e^- \longrightarrow CH_3OH + H_2O \tag{8.86}$$

In this respect, we refer to Figure 8.42, on the oxygen evolution reaction. The reaction proved problematic because three unevenly spaced energy levels were involved. For reaction (8.86), we anticipate five intermediate levels! As mentioned

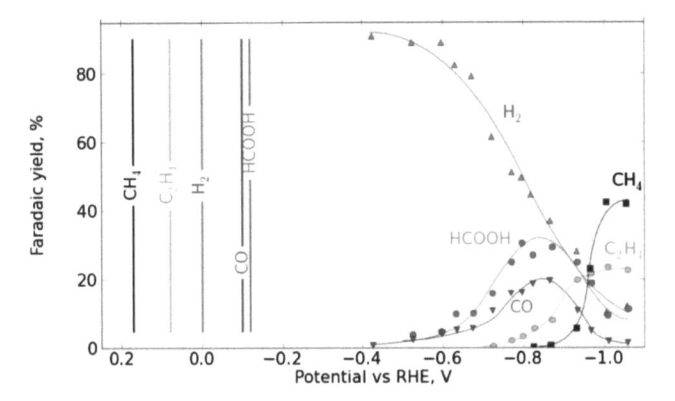

Figure 8.47 Electrochemical hydrogenation of CO_2 on copper (based on data from Hori [73]).

before, materials are now being screened using the approach described in Chapter 6. It becomes increasingly evident that many steps in a mechanism are in fact connected through scaling relations: related intermediates bind with similar characteristics to similar adsorption sites. Therefore, research should aim at breaking through the reasons behind such scaling laws, for example, by exploring the incorporation of more than one site of different geometry in catalytic ensembles.

8.7.4
Hydrogen Fuel Cells

Hydrogen in its role as a fuel can be converted back to water and electricity in a fuel cell, as indicated in Figure 8.41. The principle of the fuel cell was first demonstrated by Sir William Robert Grove in 1839. Today, different schemes exist for utilizing hydrogen in electrochemical cells. We explain the two most important, namely the polymer electrolyte membrane fuel cell (PEMFC) and the solid oxide fuel cell (SOFC).

8.7.4.1 The Proton Exchange Membrane Fuel Cell (PEMFC)

A fuel cell is a layered structure consisting of an anode, a cathode, and a solid electrolyte (Figure 8.48). Hydrogen reacts on the anode, typically Pt nanoparticles deposited on a conducting graphite support, where it is oxidized into protons and electrons on the anode:

$$2H_2 \longrightarrow 4H^+ + 4e^- \qquad (8.87)$$

The electrons are transported through an outer circuit connected to the cathode, where oxygen is reduced to oxygen ions on a similar catalyst system as for the anode:

$$4H^+ + 4e^- + O_2 \rightleftharpoons 2H_2O \qquad (8.88)$$

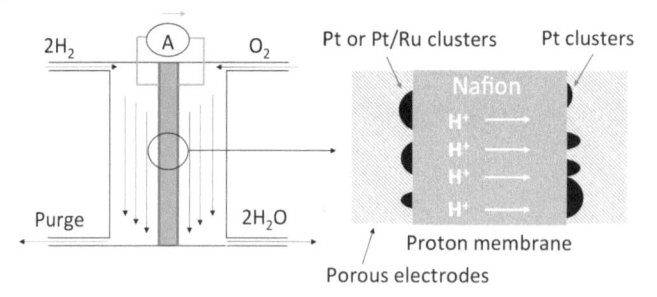

8 Catalysis in Practice: Synthesis Gas and Hydrogen

Figure 8.48 Principle of a polymer electrolyte membrane (PEM) fuel cell. A Nafion membrane sandwiched between electrodes separates hydrogen and oxygen. Hydrogen is oxidized into protons and electrons at the anode on the left. Electrons flow through the outer circuit, while protons diffuse through the membrane, which is only roughly 0.1 mm thick. Oxygen is reduced at the cathode to negative ions, which recombine with the protons to form water. The catalyst particles are deposited on conducting porous electrodes pressed against the membrane to ensure good electric contact.

while the protons are allowed to migrate though the PEM and react on the cathode with the oxygen ions to water leading to the overall reaction

$$2H_2 + O_2 \longrightarrow 2H_2O \tag{8.89}$$

It is seen that this is just the reverse process of the electrolysis in Eqs. (8.82)–(8.84).

Since the electrolyte membrane only allows the conduction of ions, the electrons are forced through an exterior circuit, creating an electromotive force. The voltage generated by such a cell is given by the Nernst equation. For the hydrogen-oxygen reaction we can write

$$\Delta G = \Delta G^\circ + RT \ln \left(\frac{p_{H_2O}}{p_{H_2} \sqrt{p_{O_2}}} \right) \tag{8.90}$$

Since two electrons pass through the circuit per water molecule generated and ΔG is given per mole, the potential is found by dividing ΔG by the charge associated with one mole of water, $-2eN_A$:

$$\varepsilon = \varepsilon^\circ - \frac{RT}{2eN_A} \ln \left(\frac{p_{H_2O}}{p_{H_2} \sqrt{p_{O_2}}} \right) \tag{8.91}$$

The standard cell potential for the hydrogen-oxygen reaction is then determined by the free energy of formation of water (gas) by

$$\varepsilon^\circ = -\frac{\Delta G^\circ}{2eN_A} = -\frac{-228.582 \text{ kJ mol}^{-1}}{2 \cdot 1.60 \times 10^{-19} \text{C} \cdot 6.02 \times 10^{23}} = 1.19 \text{ V} \tag{8.92}$$

Due mainly to kinetic losses, the voltage available for practical use is only 0.7–0.8 V, with current densities up to 0.5 A cm^{-2}. A proton exchange membrane such as Nafion only works under wet conditions, meaning that there must always be

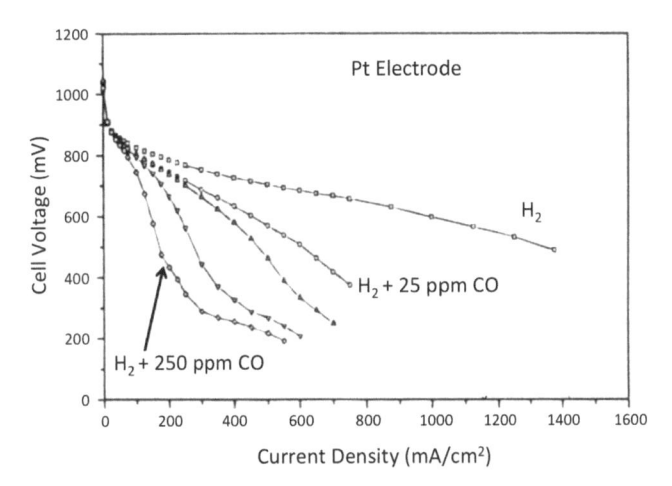

Figure 8.49 PEMFC potential as a function of current density for different CO contents in the hydrogen supply. Note the rapid drop in potential as soon as current is drawn, even for pure hydrogen. This is due to overpotentials in the system, while the monotonic decrease at higher current is attributed to the internal resistance of the PEM (reproduced from Oetjen et al. [76] by permission of the Electrochemical Society).

water present, limiting the application to temperatures below 100 °C. PEMFCs typically operate at around 80 °C, making them very interesting for both small and mobile applications such as cars, portable computers, and mobile phones. They were developed in the 1960s for the Gemini space program. However, the high production costs still prevent them from being a viable alternative to conventional technologies based on the simple combustion of fossil fuels.

A major obstacle is related to the anode material. The active component in the anode is a highly dispersed metal supported on graphite that is pressed against the membrane. Platinum is chosen as the active metal because of its efficiency in dissociating hydrogen, but, unfortunately, platinum is also very sensitive towards trace amounts of impurities (e.g., CO) in the hydrogen gas.

The presence of impurities is an important issue in mobile applications where the hydrogen at least initially will be supplied by the decomposition of hydrocarbons or methanol in on-board reformer systems as long as no appropriate hydrogen storage media are available. In such systems, CO is an unavoidable byproduct, and since CO binds more strongly to Pt than hydrogen, the low operating temperature of the PEMFC requires that the concentration of CO be kept in the low ppm range (see Figure 8.49).

The catalysts at the anode can be made less sensitive to CO poisoning by alloying platinum with other metals such as ruthenium, antimony, or tin [75]. There is a clear demand for better and cheaper catalysts. Another way to circumvent the CO problem is to use proton-exchange membranes that operate at higher temperatures, where CO desorbs. Such membranes have been developed and are now commercially available.

Table 8.12 Energy content of various compounds considered as future energy carriers.

Compound	H$_2$	MgH$_2$	CH$_3$OH	NH$_3$	CO(NH$_2$)$_2$	Gasoline
M_W (g mol^{-1})	2	26.3	32	17	60	–
$\Delta G°$ (kJ mol^{-1})	−228	−228	−684	−326	−654	–
ρ (g mL^{-1})	0.078	1.45	0.79	0.68	1.3	–
$\Delta G°$ (kJ mL^{-1})	8.9	12.6[a]	16.9	13.1	14.1	25 – 30[b]

a) The energy used for releasing the hydrogen from the hydride has not been subtracted.
b) Gasoline is a mixture of a number of hydrocarbons and the energy content has a certain variation.

A third approach is to remove CO catalytically from hydrogen. This calls for reactions other than those discussed in connection for ammonia synthesis, where the last traces of CO are removed by methanation. Such a process can be straightforwardly incorporated in a large industrial plant, but hardly in the small-scale reforming units used to produce hydrogen on-board automobiles. Selective oxidation of CO by oxygen in the presence of hydrogen has appeared feasible if the temperature is sufficiently low to prevent the CO$_2$ formed reacting with H$_2$ in the reverse water–gas shift reaction [77]. A gold-based catalyst successfully accomplishes the reaction at a temperature compatible with fuel cell operation.

It is intriguing that the otherwise inert metal gold becomes catalytically active when the particles are in the 1–3 nm size range. The pioneering work of Haruta revealed that small gold particles oxidize CO even below ambient temperatures [78, 79]. As long as the oxidation of CO is much faster than that of hydrogen, the reaction is sufficiently selective to keep hydrogen losses acceptably low.

Fuel cells are most effective when they operate on pure hydrogen. This could be manufactured from various fuels, of which methanol and gasoline are the most obvious. Reforming or catalytic partial oxidation is required to generate the hydrogen, which must also be purged of CO. Another approach is to use hydrogen-containing fuels that do not contain carbon, such as ammonia.

Table 8.12 lists the energy content of several energy carriers. Safety and ease of handling are important factors in choosing a particular energy carrier. Conventional fuels, such as gasoline, as sources of hydrogen for automobiles have the advantage of a well-established infrastructure. At this stage, however, it seems that the simple solution of compressing hydrogen to 700 bar in a light weight composite container may be a viable solution.

Interestingly, the PEMFC may also operate directly on methanol. Naturally, the problems associated with high coverage of various intermediates will be present, as mentioned above, as well as additional problems such as loss of methanol over the membrane. Nevertheless, it is possible to operate a methanol fuel cell with a voltage around 0.4 V and a reasonable current, to power small mobile devices such as portable computers and cell phones and make them less dependent on electricity.

8.7.4.2 The Solid Oxide Fuel Cell

In the solid oxide fuel cell (SOFC), it is the O^{2-} ion that diffuses through the electrolyte membrane. To give the ions sufficient mobility, high operating temperatures (700–800 °C) are required. This makes the SOFC rather insensitive to the type of fuel used, but introduces a number of materials issues related to the behavior of anode materials such as Ni (which is the preferred), Pd, Pt, and Co in the presence of H_2 and H_2O at elevated temperatures. The electrode metal is dispersed on the electrolyte, which is usually yttria-stabilized zirconia (YSZ). Many of the fundamental questions in SOFC technology are the same as those encountered in catalysis, such as rate of adsorption, dissociation, spill over from the metal to the oxide and vice versa, and diffusion into the electrolyte [80]. For instance, the surface coverage under operating conditions and the influence of the reduction potential on the structure and morphology of the anode material are still largely unknown; such knowledge may be important for the design of new and better SOFCs that operate at lower temperatures.

While the PEM fuel cells appear to be suitable for mobile applications, SOFC technology appears more applicable for stationary applications. The high operating temperature gives it flexibility towards the type of fuel used, which enables, for example, the use of methane. The heat, thus, generated can be used to produce additional electricity. Consequently, the efficiency of the SOFC is $\sim 60\%$, compared with $\sim 45\%$ for PEMFC under optimal conditions.

8.7.4.3 Efficiency of Fuel Cells

Fuel cells convert chemical energy directly into electricity, an inherently efficient process. Hence, the thermodynamically attainable efficiencies are around 100%. The efficiency of a fuel cell is given by

$$\varepsilon_{\max} = \frac{\Delta G}{\Delta H} = 1 - \frac{T \Delta S}{\Delta H} \tag{8.93}$$

whereas that of the Carnot efficiency is

$$\varepsilon_{\max} \equiv \frac{|w_{\max}|}{|q_h|} = 1 - \frac{T_c}{T_h} \tag{8.94}$$

Here T_c is the temperature of the cooling reservoir, that is, the surroundings, while T_h is the temperature of the process, that is, the temperature of combustion. The Carnot efficiency is applicable for conventional heat pump engines. Efficiencies of more than 100% correspond to converting heat from the surroundings into electricity and is only of academic interest, as is the high efficiency listed in Table 8.13.

In practice the situation is less favorable due to losses associated with overpotentials in the cell and the resistance of the membrane. Therefore, the practical efficiency of a fuel cell is around 40–60%. For comparison, the Carnot efficiency of a modern turbine used to generate electricity is on the order of 40–45%. It is important to realize, though, that the efficiency of Carnot engines is in practice

Table 8.13 Theoretical efficiency for converting hydrogen, methane, and methanol into power in a fuel cell.

Temperature (K)	Fuel H_2	CH_4	CH_3OH
300	0.94	1.00	1.02
1000	0.78	1.00	1.08

limited by thermodynamics, while that of fuel cells is largely set by material properties, which may be improved as seen above for electrolysis.

In terms of generating electricity, fuel cells do not offer substantially higher efficiencies than thermal engines, but they do have an advantage over the internal combustion engine of a car. The latter is usually most efficient when the engine operates at rather high revolutions per minute (e.g., 3000 rpm). However, such conditions apply to driving on highways. The engine is much less efficient under urban driving conditions. A car waiting at traffic lights still uses considerable amounts of energy, to keep the engine ready for action. Here, the fuel cell offers a clear advantage as it only supplies energy when it is needed. Favorable overall efficiency and the fact that a fuel cell produces no toxic emissions and very little noise are the main reasons for introducing the fuel cell in the automotive market. Figure 8.50 compares the efficiencies of a car powered by hydrogen and a fuel cell and a conventional car using gasoline.

An interesting point of fuel cells is that they can in principle be used reversibly for both storing electric energy in the form of hydrogen and returning it to the grid when needed (the so-called regenerative fuel cell). There is substantial research in this area since this could be a way to solve the problem of the intermittent sustainable energy supply. It is worth noticing that the solid oxide fuel cell (SOFC) has a thermodynamic advantage at electrolysis, since it operates at high temperature, while the opposite is the case for fuel cell operation. This is related to the fact that producing three moles of gas from two moles of water is more favorable at high temperature for entropy reasons. In fuel cell mode, however, the entropy loss is a disadvantage.

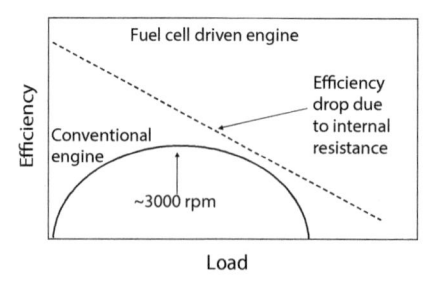

Figure 8.50 Schematic of the efficiencies of a fuel cell driven car and a conventional car with a combustion engine. Note the advantage of the fuel cell car at low load, prevailing under urban driving conditions.

8.7.4.4 New Electrode Materials for Fuel Cell

Just as the oxygen evolution reaction (a four electron reaction) was an issue for electrolysis, the oxygen reduction reaction (ORR) presents challenges in the fuel cell. Substantial efforts have been dedicated to find new and better catalysts than the archetype Pt. It has been realized that the source of the sluggish and substantial overpotential of the ORR reaction is related to the bonding energy of the oxygen-containing species, as was the case for the oxygen evolution reaction (OER) in Figure 8.42. The bond energy between these species and the surface, in particular that of OH, is an excellent descriptor for the ORR reaction and an optimum curve ("volcano curve") explains the activity trends, see Figure 8.51. It is seen that platinum is in fact not optimal for the ORR reaction, as it is too reactive. The best material is one that binds OH and other O intermediates less strongly. All these trend investigations were performed on single crystals or on polycrystalline samples in a half-cell, and not in a real fuel cell, such that the ORR reaction could be studied in isolation. It is now commonly accepted that the activity enhancements as observed in Figure 8.51 are caused by the formation of a Pt overlayer on top of the alloys, which is compressed because the alloys have smaller lattice distances than pure Pt itself. According to the d-band model discussed in Chapter 6, such compressed Pt layers are expected to be less reactive. On top of this effect, the influence of the second metal in the layer underneath may contribute a ligand effect. Some of the alloy components, however, leach out from the surface region, due to the acidic environment; hence not all of the alloys are stable.

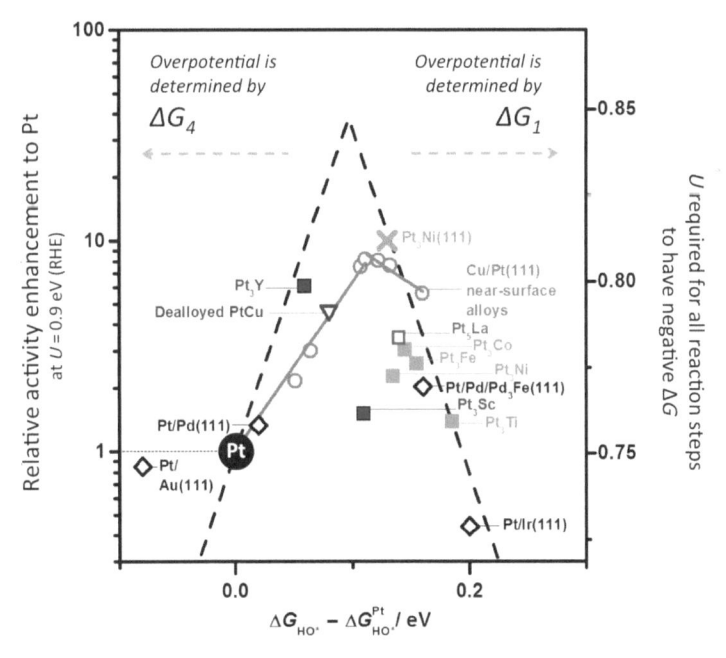

Figure 8.51 The volcano curve for the oxygen reduction reaction (ORR) as a function of OH bonding energy relative to pure Pt (reproduced from Greeley *et al.* [81] with permission from the Nature Publishing Group).

Studying macroscopic materials like those shown in Figure 8.51 is of great interest for testing predictions generated by *in silico* DFT screening. Such procedures enable identification of potential candidates for new electrode materials. However, for practical applications, these materials must be applied as nanoparticles and tested under realistic fuel cell conditions.

Figure 8.52 shows how this has been done with PtY and PtGd alloy nanoparticles, which are substantially more active that Pt nanoparticles, on a per mass of Pt basis (relevant because of the high price of platinum). The scanning transmission electron microscopy elemental maps indicate that the surface of the particles is strongly enriched in Pt, as illustrated in the model of the nanoparticle. EXAFS analysis revealed a contraction of a few percent in the Pt–Pt distance, which explains the higher activity, as described above [82, 83].

The perspective of improving the ORR activity of Pt by an order of magnitude means a reduction in use of Pt needed in fuel cells. At present, loadings of 0.4 and 0.1 mg/cm^2 of pure Pt are required for cathode and anode of a fuel cell, respectively. Since 1 cm^2 of electrode provides about 0.5 W, it is left to the reader to calculate the amount of Pt needed to fuel a 100 kW car. As the annual production of platinum is on the order of 200 t, only a small fraction of today's cars

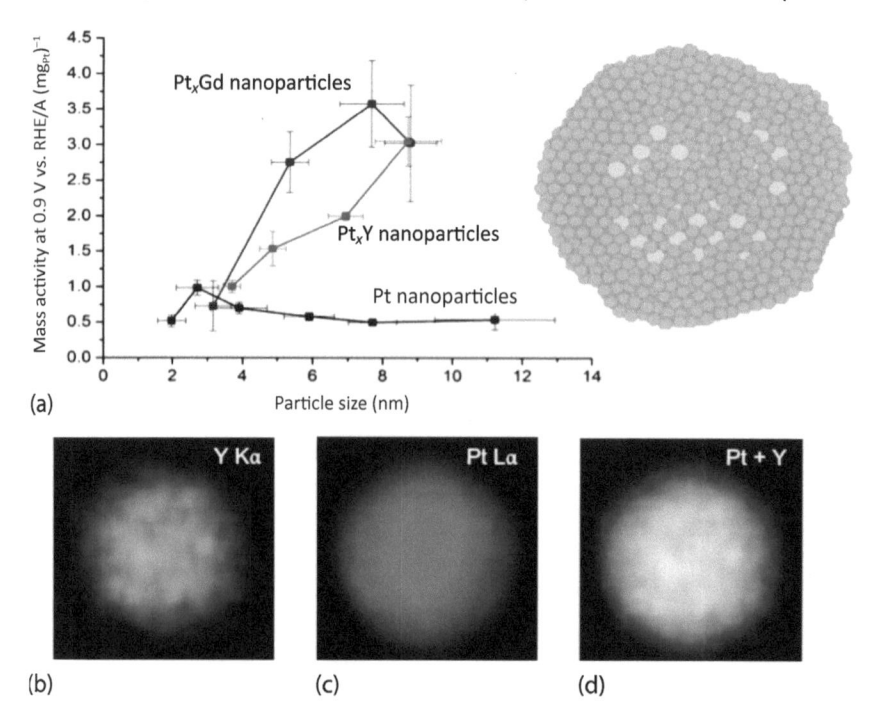

Figure 8.52 (a) The specific mass activity of Pt, Pt$_x$Y, and Pt$_x$Gd nanoparticles in the oxygen reduction reaction as a function of particle size, along with a schematic picture of an alloy particle which is enriched in Pt at the surface. (b–d) The elemental maps confirm the enrichment of Pt at the surface of a Pt$_x$Y nanoparticle (data adapted from [82, 83]).

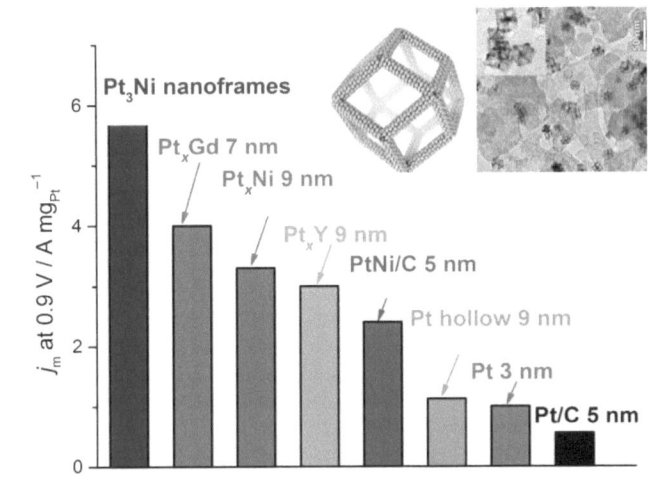

Figure 8.53 ORR performance of a state of the art nanoframe electrocatalysts expressed as current drawn at fixed potential. The nanoframe is shown along with a TEM image (adapted from Chen et al. [84]).

could be provided with Pt-based fuel cells. Hence, finding ways to use Pt more efficiently, like in Figure 8.51 and 8.52 is of the utmost importance. Systems like PtY and PtG alloy nanoparticles are among the most stable and active ORR catalysts known today. The best one reported thus far, however, is a PtNi-based catalyst, shaped in the form of nanoframes [84], see Figure 8.53.

For both types of catalyst, there are still big challenges in scaling up the synthesis of these materials, and to show that they will achieve their promising activity in practice, with sufficient stability to resist corrosion in acidic environments over time. As with all catalysts, particle growth due to Ostwald ripening may occur, which will lead to reduced performance in the fuel cell.

References

1 Rostrup-Nielsen, J.R. and Christiansen, L.J. (2011) *Concepts in Syngas Manufacture*, Imperial College Press, London.
2 Holmblad, P.M., Wambach, J. and Chorkendorff, I. (1995) Molecular-beam study of dissociative sticking of methane on Ni(100). *Journal of Chemical Physics*, **102**, 8255–8263.
3 Goodman, D.W. (1984) Model catalytic studies over metal single-crystals. *Accounts of Chemical Research*, **17**, 194–200.
4 Bengaard, H.S., Nørskov, J.K., Sehested, J., Clausen, B.S., Nielsen, L.P., Molenbroek, A.M., and Rostrup-Nielsen, J.R. (2002) Steam reforming and graphite formation on Ni catalysts, *Journal of Catalysis*, **209**, 365–384.
5 Andersson, M.P., Abild-Pedersen, E., Remediakis, I.N., Bligaard, T., Jones, G., Engbæk, J., Lytken, O., Horch, S., Nielsen, J.H., Sehested, J., Rostrup-Nielsen, J.R., Nørskov, J.K., and Chorkendorff, I. (2008) Structure sensitivity of the methanation reaction: H_2-induced CO dissociation on nickel surfaces. *Journal of Catalysis*, **255**, 6–19.
6 Abild-Pedersen, F., Lytken, O., Engbaek, J., Nielsen, G., Chorkendorff, I., and Nørskov, J.K. (2005) Methane activation on Ni(111): Effects of poisons

and step defects. *Surface Science*, **590**, 127–137.

7 Sinfelt, J.H. (1977) Catalysis by alloys and bimetallic clusters. *Accounts of Chemical Research*, **10**, 15–20.

8 Nielsen, L.P., Besenbacher, F., Stensgaard, I., Laegsgaard, E., Engdahl, C., Stoltze, P., Jacobsen, K.W., and Nørskov, J.K. (1993) Initial growth of Au on Ni(110) – surface alloying of immiscible metals. *Physical Review Letters*, **71**, 754–757.

9 Besenbacher, F., Chorkendorff, I., Clausen, B.S., Hammer, B., Molenbroek, A.M., Nørskov, J.K., and Stensgaard, I. (1998) Design of a surface alloy catalyst for steam reforming. *Science*, **279**, 1913–1915.

10 Rod, T.H. and Nørskov, J.K. (2002) The surface science of enzymes. *Surface Science*, **500**, 678–698.

11 Hickman, D.A. and Schmidt, L.D. (1993) Production of syngas by direct catalytic-oxidation of methane. *Science*, **259**, 343–346.

12 Ertl, G., Knözinger, H., Schüth, F., and Weitkamp, J. (2008) *Handbook of Heterogeneous Catalysis*, Wiley-VCH Verlag GmbH, Weinheim.

13 Bartholomew, C.H. and Farrauto, R.J. (2006) *Fundamentals of Industrial Catalytic Processes*, John Wiley & Sons, Inc., Hoboken.

14 Rasmussen, P.B., Kazuta, M., and Chorkendorff, I. (1994) Synthesis of methanol from a mixture of H_2 and CO_2 on Cu(100). *Surface Science*, **318**, 267–380.

15 Rasmussen, P.B., Holmblad, P.M., Askgaard, T., Ovesen, C.V., Stoltze, P., Nørskov, J.K., and Chorkendorff, I. (1994) Methanol synthesis on Cu(100) from a binary gas-mixture of CO_2 and H_2. *Catalysis Letters*, **26**, 373–381.

16 Nakano, H., Nakamura, I., Fujitani, T., and Nakamura, J. (2001) Structure-dependent kinetics for synthesis and decomposition of formate species over Cu(111) and Cu(110) model catalysts. *Journal of Physical Chemistry B*, **105**, 1355–1365.

17 Yoshihara, J. and Campbell, C.T. (1996) Methanol synthesis and reverse water-gas shift kinetics over Cu(110) model catalysts: Structural sensitivity. *Journal of Catalysis*, **161**, 776–782.

18 Clausen, B.S., Schiotz, J., Grabaek, L., Ovesen, C.V., Jacobsen, K.W., Nørskov, J.K., and Topsøe, H. (1994) Wetting/non-wetting phenomena during catalysis: Evidence from in situ online EXAFS studies of Cu-based catalysts. *Topics in Catalysis*, **1**, 367–376.

19 Ovesen, C.V., Clausen, B.S., Schiotz, J., Stoltze, P., Topsøe, H., and Nørskov, J.K. (1997) Kinetic implications of dynamical changes in catalyst morphology during methanol synthesis over Cu/ZnO catalysts. *Journal of Catalysis*, **168**, 133–142.

20 Topsøe, H., Ovesen, C.V., Clausen, B.S., Topsøe, N.Y., Nielsen, P.E.H., Tornqvist, E., and Nørskov, J.K. (1997) Importance of dynamics in real catalyst systems, in *Dynamics of Surfaces and Reaction Kinetics in Heterogeneous Catalysis* (eds G.F. Froment and K.C. Waught), Elsevier Science BV, Amsterdam, 121–139.

21 Nakamura, J., Nakamura, I., Uchijima, T., Kanai, Y., Watanabe, T., Saito, M., and Fujitani, T. (1996) A surface science investigation of methanol synthesis over a Zn-deposited polycrystalline Cu surface. *Journal of Catalysis*, **160**, 65–75.

22 Behrens, M., Studt, F., Kasatkin, I., Kuehl, S., Haevecker, M., Abild-Pedersen, F., Zander, S., Girgsdies, F., Kurr, P., Kniep, B.-L., Tovar, M., Fischer, R.W., Nørskov, J.K., and Schloegl, R. (2012) The active site of methanol synthesis over $Cu/ZnO/Al_2O_3$ industrial Catalysts. *Science*, **336**, 893–897.

23 Kuld, S., Conradsen, C., Moses, P.G., Chorkendorff, I., and Sehested, J. (2014) Quantification of zinc atoms in a surface alloy on copper in an industrial-type methanol synthesis catalyst. *Angewandte Chemie – International Edition*, **53**, 5941–5945.

24 Berlowitz, P.J. and Goodman, D.W. (1987) The activity of Pd(110) for methanol synthesis. *Journal of Catalysis*, **108**, 364–368.

25 Fajula, F., Anthony, R.G., and Lunsford, J.H. (1982) Methane and methanol synthesis over supported palladium catalysts. *Journal of Catalysis*, **73**, 237–256.

26 Maack, M., Friis-Jensen, H., Sckerl, S., Larsen, J.H., and Chorkendorff, I. (2003) Methanol synthesis on potassium-modified Cu(100) from $CO + H_2$ and $CO + CO_2 + H_2$. *Topics in Catalysis*, **22**, 151–160.

27 Van de Loosdrecht, J., Botes, F.G., Ciobica, I.M., Ferreira, A., Gibson, P., Moodley, D.J., Saib, A.M., Visagie, J.L., Weststrate, C.J., and Niemantsverdriet, J.W. (2013) Fischer–Tropsch Synthesis: Catalysts and Chemistry, in *Comprehensive Inorganic Chemistry: From Elements to Applications* (eds J.A. Reedijk and K. Poeppelmeyer), Elsevier, Amsterdam, 525–557.

28 Dry, M.E. and Hoogendoorn, J.C. (1981) Technology of the Fischer–Tropsch process. *Catalysis Reviews – Science and Engineering*, **23**, 265–278.

29 Dry, M.E. (2002) High quality diesel via the Fischer–Tropsch process – A review. *Journal of Chemical Technology and Biotechnology*, **77**, 43–50.

30 Xu, J., Yang, Y., and Li, Y.-W. (2015) Recent development in converting coal to clean fuels in China. *Fuel*, **152**, 122–130.

31 Khodakov, A.Y., Chu, W., and Fongarland, P. (2007) Advances in the development of novel cobalt Fischer–Tropsch catalysts for synthesis of long-chain hydrocarbons and clean fuels. *Chemical Reviews*, **107**, 1692–1744.

32 De Smit, E. and Weckhuysen, B.M. (2008) The renaissance of iron-based Fischer–Tropsch synthesis: On the multifaceted catalyst deactivation behaviour, *Chemical Society Reviews*, **37**, 2758–2781.

33 Gracia, J.M., Prinsloo, F.F., and Niemantsverdriet, J.W. (2009) Mars-van Krevelen-like mechanism of CO hydrogenation on an iron carbide surface, *Catalysis Letters*, **133**, 257–261.

34 Ozbek, M.O. and Niemantsverdriet, J.W. (2014) Elementary reactions of CO and H_2 on C-terminated $\chi-Fe_5C_2(001)$ surfaces. *Journal of Catalysis*, **317**, 158–166.

35 Grobler, T., Claeys, M., van Steen, E., and Janse van Vuuren, M.J. (2009) GC × GC: A novel technique for investigating selectivity in the Fischer–Tropsch synthesis, *Catalysis Communications*, **10**, 1674–1680.

36 Schulz, H. and Claeys, M. (1999) Reactions of α-olefins of different chain length added during Fischer–Tropsch synthesis on a cobalt catalyst in a slurry reactor. *Applied Catalysis A – General*, **186**, 71–90.

37 Weststrate, C.J., Van Helden, P., and Niemantsverdriet, J.W. (2016) Reflections on the Fischer–Tropsch synthesis: Mechanistic issues from a surface science perspective. *Catalysis Today*, **275**, 100–110.

38 Saib, A.M., Moodley, D.J., Ciobica, I.M., Hauman, M.M., Sigwebela, B.H., Weststrate, C.J., Niemantsverdriet, J.W., and van de Loosdrecht, J. (2010) Fundamental understanding of deactivation and regeneration of cobalt Fischer–Tropsch synthesis catalysts. *Catalysis Today*, **154**, 271–282.

39 Tsakoumis, N.E., Ronning, M., Borg, O., Rytter, E., and Holmen, A. (2010) Deactivation of cobalt based Fischer–Tropsch catalysts: A review. *Catalysis Today*, **154**, 162–182.

40 van Santen, R.A., Ciobica, I.M., van Steen, E., and Ghouri, M.M. (2011) Mechanistic Issues in Fischer–Tropsch Catalysis. *Advances in Catalysis*, **54**, 127–187.

41 Bezemer, G.L., Bitter, J.H., Kuipers, H., Oosterbeek, H., Holewijn, J.E., Xu, X.D., Kapteijn, F., van Dillen, A.J., and de Jong, K.P. (2006) Cobalt particle size effects in the Fischer–Tropsch reaction studied with carbon nanofiber supported catalysts, *Journal of the American Chemical Society*, **128**, 3956–3964.

42 den Breejen, J.P., Radstake, P.B., Bezemer, G.L., Bitter, J.H., Froseth, V., Holmen, A., and de Jong, K.P. (2009) On the origin of the cobalt particle size effects in Fischer–Tropsch catalysis. *Journal of the American Chemical Society*, **131**, 7197–7203.

43 Xiong, H., Motchelaho, M.A.M., Moyo, M., Jewell, L.L., and Coville, N.J. (2011) Correlating the preparation and performance of cobalt catalysts supported on carbon nanotubes and carbon spheres in the Fischer–Tropsch synthesis. *Journal of Catalysis*, **278**, 26–40.

44 Hao, X., Dong, G., Yang, Y., Xu, Y., and Li, Y.W. (2007) Coal to liquid (CTL): Commercialization prospects in China. *Chemical Engineering & Technology*, **30**, 1157–1165.

45 Liu, Z.Y., Shi, S.D. and Li, Y.W. (2010) Coal liquefaction technologies – Development in China and challenges in chemical reaction engineering. *Chemical Engineering Science*, **65**, 12–17.

46 Zhang, C.H., Yang, Y., Teng, B.T., Li, T.Z., Zheng, H.Y., Xiang, H.W., and Li, Y.W. (2006) Study of an iron-manganese Fischer–Tropsch synthesis catalyst promoted with copper. *Journal of Catalysis*, **237**, 405–415.

47 Ovesen, C.V., Clausen, B.S., Hammershoi, B.S., Steffensen, G., Askgaard, T., Chorkendorff, I., Nørskov, J.K., Rasmussen, P.B., Stoltze, P., and Taylor, P. (1996) Microkinetic analysis of the water–gas shift reaction under industrial conditions. *Journal of Catalysis*, **158**, 170–180.

48 Jennings, J.R. (Ed.) (1991) *Catalytic Ammonia Synthesis, Fundamentals and Practice*, Plenum Press, New York.

49 Dybkjaer, I. (2012) *Ammonia Production Processes*, in Ammonia, Catalysis and Manufacture (ed A. Nielsen), Springer, Berlin, 199–327.

50 Jacobsen, C.J.H., Dahl, S., Boisen, A., Clausen, B.S., Topsøe, H., Logadottir, A., and Nørskov, J.K. (2002) Optimal catalyst curves: Connecting density functional theory calculations with industrial reactor design and catalyst selection. *Journal of Catalysis*, **205**, 382–387.

51 Czuppon, T.A., Knez, S.A., Strait, R.B., and Aiche, A. (1997) Commercial experience with KAAP and KRES, in *Ammonia Plant Safety & Related Facilities*, **37**, 34–49.

52 Janssens, T.V.W., Castro, G.R., Wandelt, K., and Niemantsverdriet, J.W. (1994) Surface-potential around potassium promoter atoms on Rh(111) measured with photoemission of adsorbed Xe, Kr, and Ar. *Physical Review B*, **49**, 14599–14609.

53 Mortensen, J.J., Hansen, L.B., Hammer, B., and Nørskov, J.K. (1999) Nitrogen adsorption and dissociation on Fe(111). *Journal of Catalysis*, **182**, 479–488.

54 Somorjai, G.A. and Materer, N. (1994) Surface structures in ammonia synthesis. *Topics in Catalysis*, **1**, 215–231.

55 Strongin, D.R. and Somorjai, G.A. (1988) The effects of potassium on ammonia-synthesis over iron single-crystal surfaces. *Journal of Catalysis*, **109**, 51–60.

56 Egeberg, R.C., Dahl, S., Logadottir, A., Larsen, J.H., Nørskov, J.K., and Chorkendorff, I. (2001) N$_2$ dissociation on Fe(110) and Fe/Ru(0001): What is the role of steps? *Surface Science*, **491**, 183–194.

57 Bengaard, H.S., Alstrup, I., Chorkendorff, I., Ullmann, S., Rostrup-Nielsen, J.R., and Nørskov, J.K. (1999) Chemisorption of methane on Ni(100) and Ni(111) surfaces with preadsorbed potassium. *Journal of Catalysis*, **187**, 238–244.

58 Lewis, N.S. and Nocera, D.G. (2006) Powering the planet: Chemical challenges in solar energy utilization, *Proceedings of the National Academy of Sciences of the United States of America*, **103**, 15729–15735.

59 Bube, R.H. (1998) *Photovoltaic Materials*, Imperial College Press, London.

60 Shockley, W. and Queisser, H.J. (1961) Detailed balance limit of efficiency of p-n junction solar cells. *Journal of Applied Physics*, **32**, 510–519.

61 Olah, G.A. (2013) Towards oil independence through renewable methanol chemistry. *Angewandte Chemie – International Edition*, **52**, 104–107.

62 Schloegl, R. (2011) Chemistry's role in regenerative energy, *Angewandte Chemie – International Edition*, **50**, 6424–6426.

63 Schloegl, R.E. (2012) *Chemical Energy Storage*, De Gruyter, Berlin.

64 Koper, M.T.M. and Heering, H.A. (2010) Comparison of Electrocatalysis and Bioelectrocatalysis of Hydrogen and Oxygen Redox Reactions, in *Fuel Cell Science, Fundamentals, Theory and Biocatalysis* (eds A. Wieckowski and J.K. Nørskov) John Wiley & Sons, Inc., Hoboken.

65 Rossmeisl, J., Logadottir, A., and Nørskov, J.K. (2005) Electrolysis of water on (oxidized) metal surfaces. *Chemical Physics*, **319**, 178–184.

66 Nørskov, J.K., Rossmeisl, J., Logadottir, A., Lindqvist, L., Kitchin, J.R., Bligaard, T., and Jonsson, H. (2004) Origin of the overpotential for oxygen reduction at a fuel-cell cathode. *Journal of Physical Chemistry B*, **108**, 17886–17892.

67 Montoya, J.H., Garcia-Mota, M., Nørskov, J.K., and Vojvodic, A. (2015) Theoretical evaluation of the surface electrochemistry of perovskites with promising photon absorption properties for solar water splitting. *Physical Chemistry Chemical Physics*, **17**, 2634–2640.

68 McCrory, C.C.L., Jung, S., Peters, J.C., and Jaramillo, T.F. (2013) Benchmarking heterogeneous electrocatalysts for the oxygen evolution reaction, *Journal of the American Chemical Society*, **135**, 16977–16987.

69 Sapountzi, F., Gracia, J., Fredriksson, H.O.A., Weststrate, C.J., and Niemantsverdriet, J.W. (2017), Electrocatalysts for the generation of hydrogen, oxygen and synthesis gas, *Progress in Energy and Combustion Science*, **58**, 1–35.

70 Hou, Y., Abrams, B.L., Vesborg, P.C.K., Bjorketun, M.E., Herbst, K., Bech, L., Setti, A.M., Damsgaard, C.D., Pedersen, T., Hansen, O., Rossmeisl, J., Dahl, S., Nørskov, J.K., and Chorkendorff, I. (2011) Bioinspired molecular co-catalysts bonded to a silicon photocathode for solar hydrogen evolution. *Nature Materials*, **10**, 434–438.

71 Graves, C., Ebbesen, S.D., Mogensen, M. and Lackner, K.S. (2011) Sustainable hydrocarbon fuels by recycling CO_2 and H_2O with renewable or nuclear energy. *Renewable & Sustainable Energy Reviews*, **15**, 1–23.

72 Zhan, Z., Kobsiriphat, W., Wilson, J.R., Pillai, M., Kim, I., and Barnett, S.A. (2009) Syngas production by coelectrolysis of CO_2/H_2O: The basis for a renewable energy cycle. *Energy & Fuels*, **23**, 3089–3096.

73 Hori, Y. (2008) Electrochemical CO_2 Reduction on Metal Electrodes, in *Modern Aspects of Electrochemistry* (ed C.G. Vayenas), Springer, Berlin, 89–189.

74 Li, C.W., Ciston, J., and Kanan, M.W. (2014) Electroreduction of carbon monoxide to liquid fuel on oxide-derived nanocrystalline copper. *Nature*, **508**, 504–507.

75 Markovic, N.M. and Ross, P.N. (2002) Surface science studies of model fuel cell electrocatalysts. *Surface Science Reports*, **45**, 121–229.

76 Oetjen, H.F., Schmidt, V.M., Stimming, U., and Trila, F. (1996) Performance data of a proton exchange membrane fuel cell using H_2/CO as fuel gas, *Journal of the Electrochemical Society*, **143**, 3838–3842.

77 Kahlich, M.J., Gasteiger, H.A., and Behm, R.J. (1999) Kinetics of the selective low-temperature oxidation of CO in H_2-rich gas over $Au/\alpha-Fe_2O_3$. *Journal of Catalysis*, **182**, 430–440.

78 Haruta, M. (1997) Size- and support-dependency in the catalysis of gold. *Catalysis Today*, **36**, 153–166.

79 Hutchings, G.J. and Haruta, M. (2005) A golden age of catalysis: A perspective. *Applied Catalysis A – General*, **291**, 2–5.

80 Kek, D., Mogensen, M., and Pejovnik, S. (2001) A study of metal (Ni, Pt, Au)/yttria-stabilized zirconia interface in hydrogen atmosphere at elevated temperature. *Journal of the Electrochemical Society*, **148**, A878–A886.

81 Greeley, J., Stephens, I.E.L., Bondarenko, A.S., Johansson, T.P., Hansen, H.A., Jaramillo, T.F., Rossmeisl, J., Chorkendorff, I., and Nørskov, J.K. (2009) Alloys of platinum and early transition metals as oxygen reduction electrocatalysts. *Nature Chemistry*, **1**, 552–556.

82 Hernandez-Fernandez, P., Masini, F., McCarthy, D.N., Strebel, C.E., Friebel, D., Deiana, D., Malacrida, P., Nierhoff, A., Bodin, A., Wise, A.M., Nielsen, J.H., Hansen, T.W., Nilsson, A., Stephens, I.E.L. and Chorkendorff, I. (2014) Mass-selected nanoparticles of Pt$_x$Y as model catalysts for oxygen electroreduction. *Nature Chemistry*, **6**, 732–738.

83 Velazquez-Palenzuela, A., Masini, F., Pedersen, A.F., Escudero-Escribano, M., Deiana, D., Malacrida, P., Hansen, T.W., Friebel, D., Nilsson, A., Stephens, I.E.L., and Chorkendorff, I. (2015) The enhanced activity of mass-selected Pt$_x$Gd nanoparticles for oxygen electroreduction. *Journal of Catalysis*, **328**, 297–307.

84 Chen, C., Kang, Y., Huo, Z., Zhu, Z., Huang, W., Xin, H.L., Snyder, J.D., Li, D., Herron, J.A., Mavrikakis, M., Chi, M., More, K.L., Li, Y., Markovic, N.M., Somorjai, G.A., Yang, P., and Stamenkovic, V.R. (2014) Highly crystalline multimetallic nanoframes with three-dimensional electrocatalytic surfaces. *Science*, **343**, 1339–1343.

Chapter 9
Oil Refining and Petrochemistry

9.1
Crude Oil

Crude oil is by far the most important resource for modern society. Approximately 660 refineries in the world convert crude oil to transportation fuels (gasoline, diesel, kerosene), lubricants, and to feed stocks for all sorts of chemicals.[1] Catalysts play a key role in these processes [1, 2].

At the end of 2014, the proven oil reserves amounted to roughly 1.7×10^{12} bbl of oil, of which almost half are in the Middle East. The world consumption of oil in 2014 was about 3.7×10^{10} bbl, implying that the proven reserves would be sufficient to allow the use of oil at the present consumption level for another 50 years.[2]

Crude oil is a mixture of many different hydrocarbons, with molecular weights up to about 2000, that vary in both boiling points, from below room temperature to over 600 °C, and in hydrogen-to-carbon atomic ratios (between almost 3 for the smallest to about 0.5 for the largest molecules). Crude oil contains alkanes (paraffins), cycloalkanes (naphthenes), alkenes (olefins), and aromatics, many of these are contaminated by sulfur (0–3%), nitrogen (0–0.5%), oxygen, vanadium, and nickel. The heavier the molecules, the higher the content of these unwanted heteroatoms. The composition of crude oils varies considerably from one well to another. For example, the naphtha content of crudes may vary between 20% for North Sea Oil to virtually zero for certain South American crude oils. Table 9.1 presents the average composition of crude oils and the main characteristics of the various fractions.

A modern refinery is a complicated collection of conversion processes, each tailored to the properties of the feed it has to convert. The scheme shown in Figure 9.1 summarizes the most important operations; some reasons for these processes are given in Table 9.2, along with relevant catalysts. First, the crude oil is distilled to separate it in various fractions, varying from gases, liquids (naphtha,

---

1) http://www.eni-italy.info/ENI_Web.
2) data from BP, www.bp.com/statisticalreview)

Concepts of Modern Catalysis and Kinetics, Third Edition. I. Chorkendorff and J.W. Niemantsverdriet.
© 2017 WILEY-VCH Verlag GmbH & Co. KGaA. Published 2017 by WILEY-VCH Verlag GmbH & Co. KGaA.

9 Oil Refining and Petrochemistry

Table 9.1 Average characteristics of crude oils.

Fraction	Boiling point (°C)	Volume (%)	Density (kg/L)	Sulfur (wt %)
Gas, LPG		1.5	0.5–0.6	0
Light naphtha ($\leq C_5$)	< 80	6	0.66	0
Heavy naphtha (C_5–C_{10})	80–170	15	0.74	0.02
Kerosene	170–220	9	0.79	0.1
Gas oil	220–360	25	0.83–0.87	0.8
Vacuum gas oil (C_{20}–C_{40})	360–530	23	0.92	1.4
Residue ($\geq C_{40}$)	> 530	20	1.02	2.2

kerosene, and gas oil), to the heavy residue (the so-called "bottom of the barrel") that remains after vacuum distillation.

The gaseous fractions, together with the byproducts of cracking processes, are sold as liquefied petroleum gas (LPG) in containers as camping or domestic gas, or in some countries as transportation fuel for cars. All fractions other than LPG are upgraded in subsequent processes. Hydrotreating removes undesired heteroatoms such as sulfur and nitrogen, and the fraction meant to be sold as gasoline undergoes reforming to improve its antiknock properties, as expressed by the octane number.

Upgrading of the heavy residue has become an important issue in refining. This fraction is usually rich in contaminants and has traditionally been used as bitumen for paving roads. Nowadays hydrotreating and hydrocracking, or fluid catalytic cracking for residues of moderate S and N content are applied to convert these residues into smaller fractions that can be used as transport fuel, preferably gaso-

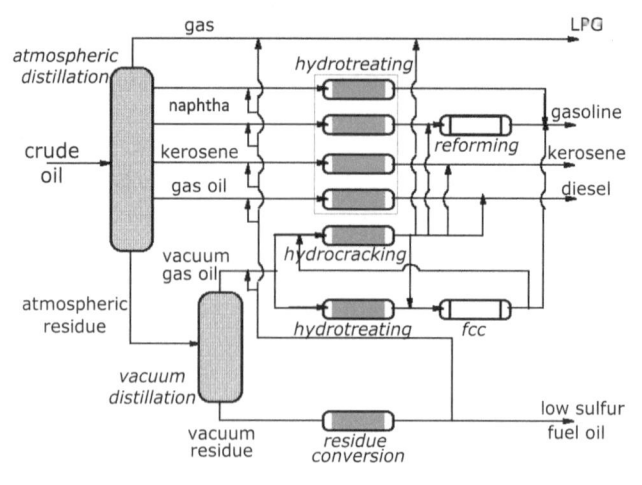

Figure 9.1 Simplified processing scheme of an oil refinery.

Table 9.2 Overview of the major refinery processes.

Process	Purpose	Catalyst
Hydrotreating	Removal of heteroatoms (S, N, O, metals) and hydrogenation of aromatics	
• HDS	Hydrodesulfurization (yields H_2S)	$Co-MoS_2/Al_2O_3$, $Ni-WS_2/Al_2O_3$
• HDN	Hydrodenitrogenation (yields NH_3)	$Ni-MoS_2/Al_2O_3$
• HDO	Hydrodeoxygenation (yields H_2O)	$Co-MoS_2/Al_2O_3$, $Ni-MoS_2/Al_2O_3$
• HDA	Hydrodearomatization (aromatics to naphthenic rings)	$Ni-WS_2/Al_2O_3$
• HDM	Hydrodemetallization (yields metal sulfides)	$Ni-MoS_2/Al_2O_3$ (adsorbent rather than catalyst)
Reforming	Conversion of naphtha into high quality gasoline (higher octane number). Hydrogen is a byproduct that is used in hydrotreating, hydrocracking, and residue conversion	$Pt-Re/Al_2O_3$, $Pt-Ir/Al_2O_3$
Hydrocracking	Conversion of vacuum gas into transportation fuels with hydrogen in a combination of hydrotreating, hydrogenation, and acidic cracking using hydrotreating and dedicated bifunctional cracking catalysts	Ni-Mo, Ni-W, Pt-Pd, acidic supports (alumina, zeolites)
Mild hydrocracking	Low-pressure version of hydrocracking	
Hydrodewaxing	Selective hydrocracking of paraffins from oil fractions or products to prevent wax precipitation (negative influence on cold flow properties)	
Fluidized catalytic cracking (FCC)	Conversion of vacuum gas oil to transportation fuels using acidic catalysts	Zeolite Y
Residue hydroconversion	Conversion of residues using hydrogen to low-sulfur fuel oil, transportation fuel and feedstock for FCC and hydrocracking	

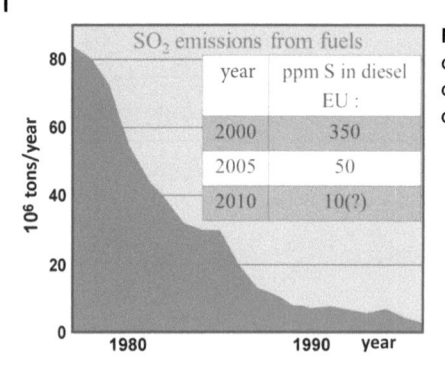

Figure 9.2 SO$_2$ emission from fuels has decreased substantially owing to the application of hydrodesulfurization and DeSOx technology

year	ppm S in diesel EU :
2000	350
2005	50
2010	10(?)

line. In such bulk operations, small improvements have an immediate economic impact.

In addition to the transportation fuels shown as the main products in Figure 9.1, the refinery also produces feedstocks such as ethylene, propylene, and butene for many chemical processes. We discuss some of these along with the major refinery processes in this chapter. Table 9.3 gives an impression of the refining capacity in the world.

9.2
Hydrotreating

As indicated in the scheme of Figure 9.1, hydrotreating, aimed at removing undesirable elements and components from the various oil fractions, is one of the key operations in the refinery. The difference between hydrotreating and hydrocracking is that, although sulfur, nitrogen, and metal removal as well as hydrogenation occur in both, the molecular weight of the hydrocarbon molecules changes only slightly in hydrotreating (e.g., by the removal of S, N, or O atoms, and the addition of a few H atoms), whereas in hydrocracking molecular weights are typically reduced by a factor of two or more. Hydrotreating proceeds under milder conditions (typically 30 bar, 350 °C) than hydrocracking (typically 100 bar, 400 °C). Mild hydrocracking is a variation which allows the use of existing hydrotreaters but nevertheless results in moderate conversion to lighter products.

Over the years, hydrotreating has become increasingly more important in the refinery, and this trend is expected to continue in the future. Three factors are responsible for this:

- The decreasing availability of light crude oils (which contain relatively few contaminants).
- The increasing use of the heavier fractions of the oil (the bottom of the barrel).
- More stringent emission standards for automotive transportation, requiring for example, low-sulfur fuels (eventually lower than 10 ppm).

thiophene

di-methyl di-benzo thiophene

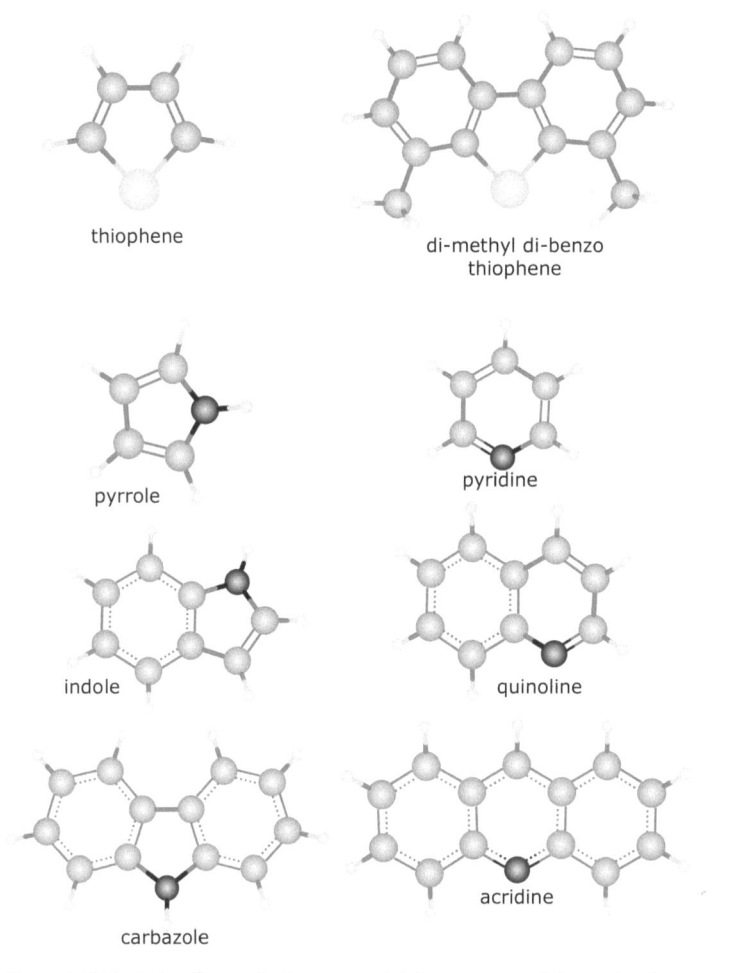

pyrrole

pyridine

indole

quinoline

carbazole

acridine

Figure 9.3 Typical sulfur- and nitrogen-containing compounds in crude oil.

In the 1970s and 1980s, the release of sulfur to the atmosphere in the form of SO_2 was significant, causing forest death and acidification of lakes. Figure 9.2 shows how the implementation of hydrodesulfurization technology and removal of sulfur from power plants (see Chapter 10 changed that situation substantially.

Nowadays, based on the amount of processed material, hydrotreating is the largest process in heterogeneous catalysis. On the basis of catalysts sold per year, hydrotreating ranks third after automotive exhaust and fluid catalytic cracking.

9.2.1
Heteroatoms and Undesired Compounds

Sulfur in crude oil is mainly present in organic compounds such as mercaptans (R–SH), sulfides (R–S–R′), and disulfides (R–S–S–R′), which are all relatively easy

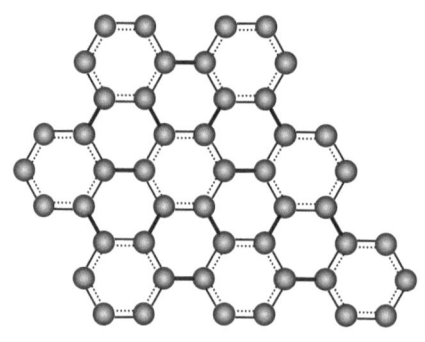

Figure 9.4 Coke is the name for the carbonaceous deposit that builds up on catalysts during the treatment of hydrocarbons. It consists of many aromatic structures, and has a low H : C ratio. Graphite, shown here, is the most extreme form of coke. Coke can be removed by oxidation. Sometimes burning of coke generates heat to drive other processes.

to desulfurize, and thiophene and its derivatives (Figure 9.3). The latter require more severe conditions for desulfurization, particularly the substituted dibenzothiophenes, such as the ones shown in Figure 9.3. Sulfur cannot be tolerated because it produces sulfuric acid upon combustion, and it poisons reforming catalysts in the refinery and automotive exhaust converters (particularly those for diesel-fueled cars). Moreover, sulfur compounds in fuels cause corrosion and have an unpleasant smell.

Nitrogen-containing molecules include the easily decomposed amines (R–NH$_2$) and the more stable aromatic molecules, such as the five-membered ring compound pyrrole and the six-membered ring pyridine, along with their higher derivatives (Figure 9.3). Nitrogen-containing molecules are poisons for acidic catalysts used in cracking and reforming reactions and they contribute to NO$_x$ formation upon combustion.

Oxygen can be present in naphthenic acids, phenols, and furan (analogue structures to thiophene and pyrrole) and higher derivatives. Oxygenated compounds give rise to corrosion and product deterioration. Hydrodeoxygenation is particularly important in the upgrading of biomass.

Unsaturated hydrocarbons, such as olefins, though not present in crude oil, occur in cracked products. They are undesirable because they may oligomerize to larger molecules and coke on acid sites. Olefins are readily hydrogenated under hydrotreating conditions, although H$_2$S and other S compounds inhibit the reaction. Aromatics are present in crude. Polynuclear aromatics are precursors to coke (Figure 9.4), which deactivates reforming and cracking catalysts.

The most important undesired metallic impurities are nickel and vanadium, present in porphyrinic structures that originate from plants and are predominantly found in the heavy residues. In addition, iron may be present due to corrosion in storage tanks. These metals deposit on catalysts and give rise to enhanced carbon deposition (nickel in particular). Vanadium has a deleterious effect on the lattice structure of zeolites used in fluid catalytic cracking. A host of other elements may also be present. Hydrodemetallization is strictly speaking not a catalytic process because the metallic elements remain in the form of sulfides on the catalyst. Decomposition of the porphyrinic structures is a relatively rapid reaction and as a result it occurs mainly in the front end of the catalyst bed, and at the outside of the catalyst particles.

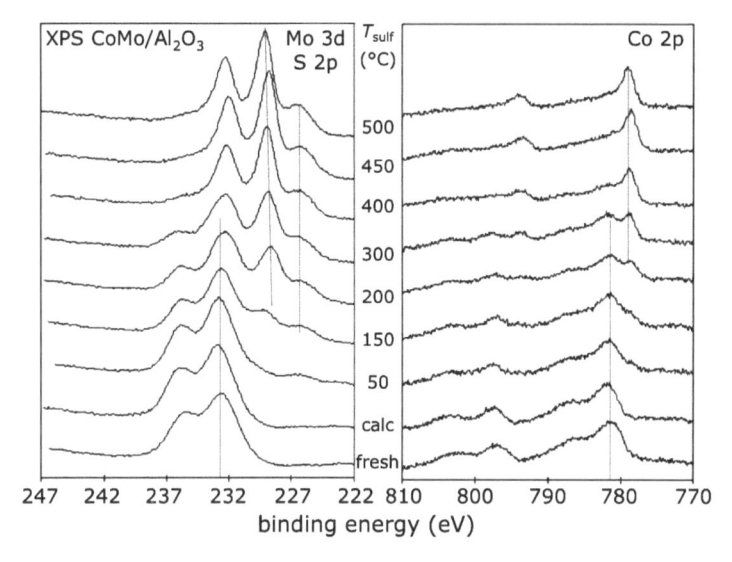

Figure 9.5 XPS Spectra of a CoMo/Al$_2$O$_3$ catalyst show the conversion from the oxidic to the sulfidic phase (reproduced from Muijsers *et al.* [6] with permission from Elsevier).

9.2.2
Hydrotreating Catalysts

Catalysts for hydrotreating usually consist of molybdenum disulfide, MoS$_2$, promoted by cobalt or nickel, on a γ-alumina support [3, 4]. Alternatively, Ni-WS$_2$/γ-Al$_2$O$_3$ may be used, but, as tungsten is significantly more expensive than molybdenum, it is only applied for hydrotreating under severe conditions (e.g., hydrocracking). Noble metal catalysts based on platinum and palladium have been used for the hydrogenation of refractory molecules. As a rule, Ni-promoted catalysts are more appropriate for hydrodenitrogenation and hydrogenation, whereas CoMoS catalysts are preferred for hydrodesulfurization. We focus our description on the CoMoS system, but NiMoS and NiWS catalysts are to a large extent similar.

Owing largely to research in the last 20 years, the sulfided CoMo/Al$_2$O$_3$ system is one of the best-characterized industrial catalysts [5]. A combination of methods, such as Mössbauer spectroscopy, EXAFS, TEM, XPS, and infrared spectroscopy, has led to a picture in which the active site of such a catalyst is known in almost atomic detail.

The catalyst is straightforwardly prepared by impregnating the alumina support with aqueous solutions of ammonium heptamolybdate and cobalt nitrate. The oxidic catalyst thus obtained is converted to the sulfided state by treatment in a mixture of H$_2$S in H$_2$, or in the sulfur-containing feed. X-ray photoelectron spectra of the CoMo/Al$_2$O$_3$ catalyst in different stages of a temperature-programmed sulfidation (Figure 9.5) illustrate the conversion from oxides to sulfides. The sulfur peak at about 225 eV indicates that sulfur uptake starts already at low temperature and is associated with the reduction of Mo^{6+} in MoO$_3$ to Mo^{5+}, in a mixed

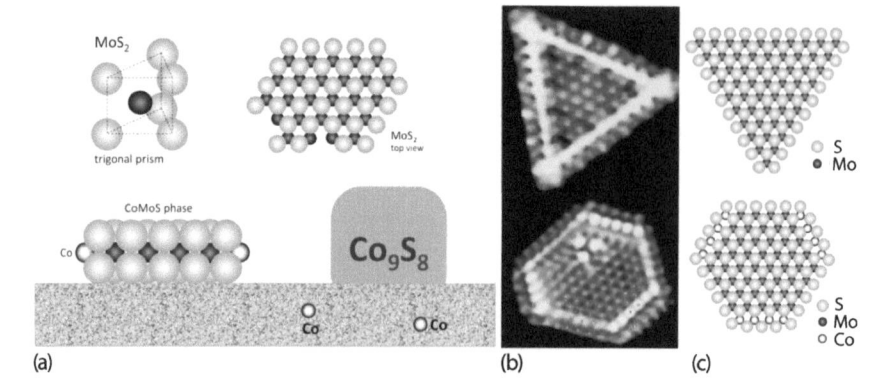

Figure 9.6 (a) Schematic picture of a sulfided CoMoS catalyst, along with a top view of the MoS$_2$ structure, which is built up from trigonal prisms. Cobalt may be present in three states: the active CoMoS in which Co decorates the edges of MoS$_2$; Co$_9$S$_8$, which has little activity; and Co^{2+} ions in the Al$_2$O$_3$ support. (b) Scanning tunneling microscopy images of MoS$_2$ and CoMoS on a gold substrate, along with structure models (c) (adapted from Lauritsen and Besenbacher [8]).

oxysulfidic environment. Conversion to MoS$_2$ takes place at temperatures above 300 °C. Cobalt starts to convert to the sulfidic state around 150 °C, more or less simultaneously with the formation of MoS$_2$. Interaction of the promoter elements with the alumina support plays an important role. On silica and carbon supports, the cobalt sulfidizes already at low temperatures and forms Co$_9$S$_8$, which is not an efficient promoter for the HDS reaction [7].

Molybdenum disulfide has a layered structure. Each layer is a sandwich consisting of Mo^{4+} between two layers of S^{2-} ions (Figure 9.6). The sulfur ions form trigonal prisms and half of the prisms contain a molybdenum ion in the middle. The chemical reactivity of MoS$_2$ is associated with the edges of the sandwich, whereas the basal planes are much less reactive. The edges form the sites where gases adsorb and where the catalytic activity resides.

Extensive spectroscopic research has revealed that the cobalt promoters are also located at the edges, in similar positions to the molybdenum atoms. Its role is most likely to facilitate the formation of sulfur vacancies, which form the active site for HDS. Figure 9.6a summarizes schematically what a working CoMo/Al$_2$O$_3$ hydrodesulfurization catalyst looks like. It contains MoS$_2$ particles with dimensions of a few nanometers, decorated with cobalt to form the catalytically highly active CoMoS phase. It also contains cobalt ions firmly bound to the lattice of the alumina support and it may contain crystallites of the stable bulk sulfide, Co$_9$S$_8$, which has a low activity for the HDS reaction [5].

Also shown in Figure 9.6b are scanning tunneling microscopy images of MoS$_2$ and CoMoS particles on a Au(111) substrate. Such atomic scale images have yielded tremendously important information on these active catalyst structures, and even the adsorption of large S-containing molecules like dibenzothiophene on the edges and corners of the sulfide structures could be visualized [8]. Recent

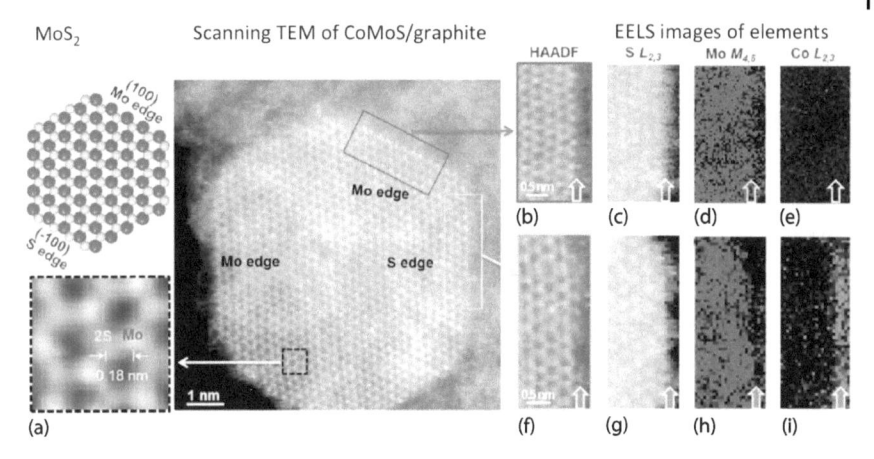

Figure 9.7 (a) Scanning transmission electron microscopy image of CoMoS on a graphite support. Energy loss images allowing for identification of the elements on the Mo edge (b–e) and for the sulfur edge (f–i). Notice how the cobalt is only present at the sulfur edge (reproduced with permission from Zhu et al. [9], copyright 2014, Wiley).

improvements in transmission electron microscopy have made it possible to depict where the various atoms are located in 'close-to-real' systems such as MoS_2 and CoMoS nanoparticles supported on graphite, see Fig. 9.7 [9].

9.2.3
Hydrodesulfurization Reaction Mechanisms

The catalytic site involved in the hydrotreating reactions is generally assumed to be the sulfur vacancy at the edge of the MoS_2 or WS_2 slabs, as illustrated schematically in Figure 9.8. Hence, the catalytic activity not only depends on the ease at which the heteroatom is stripped from the hydrocarbon, but also by the rate at which the heteroatom can be removed from the catalyst to create a vacancy. Indeed, the catalytic activity of a range of transition metal sulfides has been found to correlate with the metal–sulfur bond strength. This results in a so-called volcano plot that is very characteristic for catalysis (Figure 9.9). Among the unpromoted systems, the sulfides of ruthenium, osmium, iridium, and rhodium display the highest activity for desulfurization. However, none of these can be considered for practical use.

In spite of much theoretical work, we still do not have a complete picture of why the Co- and Ni-promoted MoS_2 catalyst is so successful. Interpretations range from the promoter-induced weakening of the metal-to-sulfur bond strength to the presence of unique sulfur species bound between molybdenum and the promoter.

Mechanistically, hydrotreating of ring-type molecules such as thiophene and related molecules involves the hydrogenation of the unsaturated ring followed

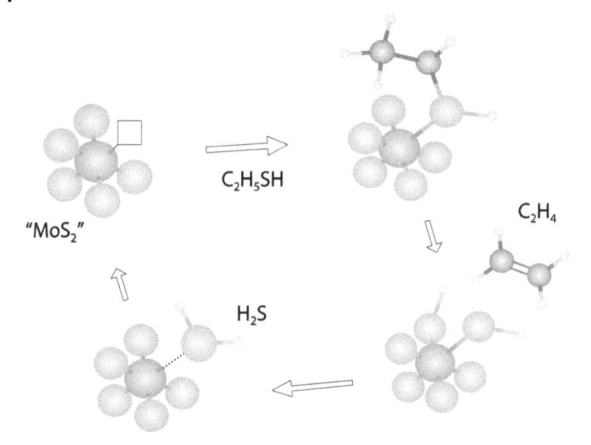

Figure 9.8 Schematic representation of the catalytic cycle for the hydrodesulfurization of a sulfur-containing hydrocarbon (ethane thiol) by a sulfur vacancy on MoS$_2$: the C$_2$H$_5$SH molecule adsorbs with its sulfur atom towards the exposed molybdenum. Next, the C–S bond breaks, ethylene forms by β-H elimination and desorbs subsequently into the gas phase. The catalytic site is regenerated by the removal of sulfur by hydrogen in the form of H$_2$S (adapted from Prins et al. [4]).

by cleavage of the bonds between the heteroatom and its carbon neighbors. Figure 9.10 illustrates this for the hydrodesulfurization of thiophene.

The kinetics of the reaction, as discussed in Chapter 7, are usually described in terms of a less detailed scheme, in which the step from adsorbed thiophene to the

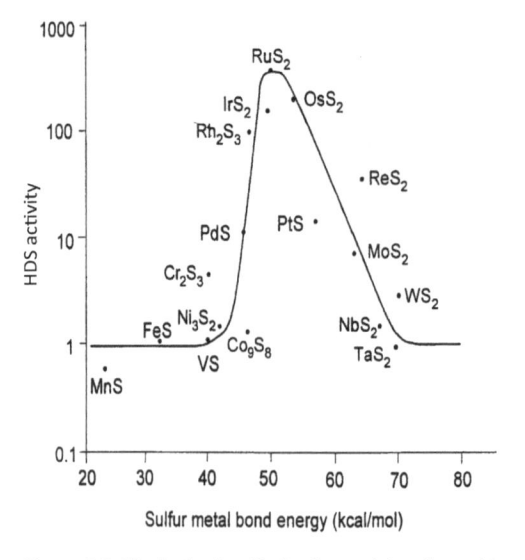

Figure 9.9 The hydrodesulfurization activity of transition metal sulfides obeys Sabatier's principle (see Section 6.5.3.5); the curve is a so-called volcano plot (adapted from Pecoraro and Chianelli [10] and Raybaud et al. [11, 12]).

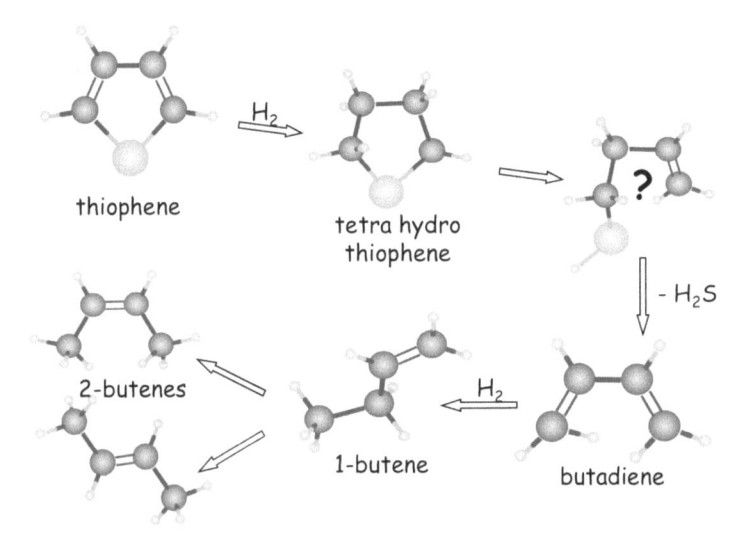

Figure 9.10 Global reaction mechanism for the hydrodesulfurization of thiophene, in which the first step involves a hydrogenation of the unsaturated ring, followed by cleavage of the C–S bond in two steps. Butadiene is assumed to be the first sulfur-free product, although 1-butene is the most abundant product in practice. Hydrogenation then produces the 2-butene isomers in subsequent steps (after Prins et al. [4]).

first S-free hydrocarbon in the cycle, butadiene, is taken in one step. We derived a rate equation of the form

$$r = N_\Delta k \theta_T \theta_H^2 = \frac{N_\Delta k K_T K_{H_2} p_T p_{H_2}}{\left(1 + K_T p_T + K_{H_2S}^{-1} p_{H_2S}\right)\left(1 + K_{H_2}^{1/2} p_{H_2}^{1/2}\right)^2} \qquad (9.1)$$

where p_T, p_{H_2}, and p_{H_2S} are the partial pressures of thiophene, hydrogen, and hydrogen sulfide, respectively, k is the rate constant of the step in which thiophene is desulfurized and hydrogenated to butadiene, K_i are the respective equilibrium constants. N_Δ is the total number of possible sulfur vacancies on the edges of the MoS_2 where thiophene may adsorb, while hydrogen may adsorb anywhere on the edges of the MoS_2.

As the adsorption of hydrogen is rather weak, the corresponding term in the denominator may be omitted. The rate expression shows that the reaction is suppressed by H_2S. Hence, the most active catalysts (which appear at the top of the volcano curve of Figure 9.9)

- Allow for the relatively easy release of sulfur,
- Are not too sensitive to inhibition by H_2S,
- Bind thiophene strongly enough that the molecules is readily hydrogenated and desulfurized.

Table 9.3 Oil refinery throughputs over the past 20 years. Source: 2015 BP Statistical Review of World Energy; www.bp.com

	Thousand barrels per day			
	1994	2004	2014	2014 share (%)
USA	13 400	15 475	15 844	20.6
Canada	1 550	1 957	1 735	2.3
Mexico	1 420	1 284	1 155	1.5
South and Central America	4 525	5 401	4 721	6.1
Europe and Eurasia	23 560	19 898	18 858	24.5
Middle East	4 565	5 795	6 656	8.7
Africa	2 150	2 304	2 255	2.9
Australasia	700	709	607	0.8
China	2 115	5 555	9 986	13.0
India	na	2 559	4 473	5.8
Japan	3 175	4 038	3 289	4.3
Other Asia Pacific	4 635	7 329	7 253	9.4
Total World	61 795	72 305	76 833	100.0

With the increasingly stringent specifications on transportation fuels and emission standards in the world (Table 9.4), hydrotreating will become even more important in the future than it already is. Further improvements in catalyst performance will be needed to successfully desulfurize the refractory molecules from Figure 9.3, such as the di-methyl di-benzo thiophene, and related structures, where access to the thiophenic sulfur is hindered by the substituents on the phenyl rings, and much more difficult than in the situation illustrated in Figure 9.10.

9.3
Gasoline Production

Gasoline as it is available at the fuel station originates from various product streams in the refinery. As Figure 9.11 shows, roughly one third comes from the FCC cracker. This stream is rich in olefins, which is favorable for the octane num-

Table 9.4 Sulfur content (in ppm weight) in gasoline.

	1994	1995	2000	2005	2010
Europe	1000	500	150	50	10
USA	–	–	–	30	10

7%	other
1%	butane
4%	isomerate *(RON = 85)*
6%	hydrocrackate
7%	straight run naphtha *(68)*
11%	MTBE *(116)*
15%	alkylate *(95)*
17%	reformate *(99)*
32%	fcc naphtha *(93–95)*

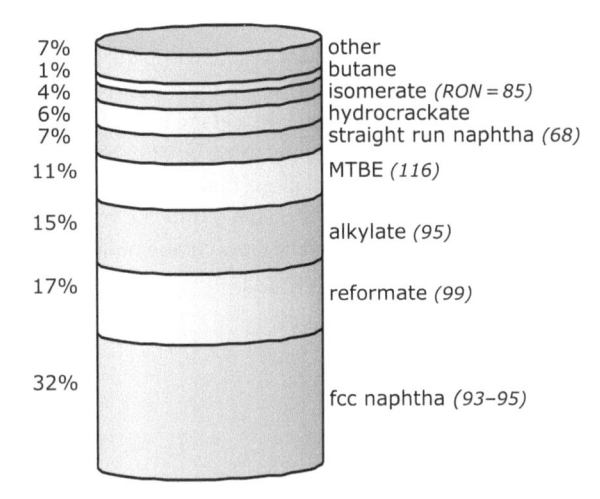

Figure 9.11 Gasoline composition and research octane numbers (RON). The RON scale is set by *n*-heptane (RON = 0) and 2,2,4-trimethylpentane (RON = 100). The RON of a certain oil fraction is the percentage of 2,2,4-trimethylpentane in a mixture with *n*-heptane that has the same antiknock properties.

ber of the fuel. However, FCC naphtha is also the major contributor of sulfur to the gasoline (more than 95%). Further desulfurization of this fraction is possible but leads to hydrogenation of olefins as well, which causes a degradation of the octane number. Hence, developing HDS catalysts with minimal hydrogenation activity, aimed at hydrocarbon fractions that end up in gasoline, is an important goal in refineries.

Other large components in gasoline come from catalytic reforming, alkylation, and the addition of an oxygenated octane booster, methyl tertiary-butyl ether (MTBE). Since the turn of the century, ethanol is increasingly used as octane booster and blended into gasoline at levels of 10–15%. We briefly discuss the most important catalytic processes in the refinery, which together are responsible for roughly 75% of the principal components in gasoline.

Octane and Cetane Numbers

The octane number is a measure of the gasoline's ability to prevent knocking. In gasoline (Otto) engines, ignition is controlled by a spark from the spark plugs, and uncontrolled self-ignition (knocking) should be avoided. The scale is defined by the antiknock properties of *n*-heptane (RON = 0) and 2,2,4-trimethyl-pentane (iso-octane; RON = 100). The research octane number and the motor octane number differ in the conditions under which they are measured (engine speed, temperature, and spark timing). The octane number of a fuel equals the percentage of iso-octane in a mixture with *n*-heptane that has the same antiknock properties as the fuel. Jet fuels with octane numbers above 100 are tested against standards containing octane boosters. The cetane number reflects the ignition properties of diesel fuels. Here, ignition should proceed rapidly after injection into the compressed, hot air. Diesel fuels are tested against a mixture of cetane (hexadecane, fast ignition) and a-methyl naphthalene (slow ignition), the cetane number being the percentage of cetane in the mixture that gives the same ignition behavior as the test fuel. Hence, contrary to gasoline, a good diesel fuel contains many straight-chain alkanes and few aromatics, as in straight-run gas oil (40–50), hydrocracking gas oil (55–60). For comparison, FCC cycle oil has low cetane numbers (0–25) but high octane numbers (95). Typical diesel transport fuels have cetane numbers around 50.

9.3.1
Fluidized Catalytic Cracking

The heavier fractions of the oil refining process (vacuum gas oil and residues) are converted to the more useful naphtha and middle distillates by cracking [1, 13]. At the beginning of the twentieth century, cracking was carried out by mixing heavy fractions with $AlCl_3$ in a batch process. In 1930, Houdry introduced fixed-bed catalytic cracking with acid-treated clays (SiO_2/Al_2O_3) as a catalyst. Synthetic amorphous silica-aluminas formed the next class of catalysts. Nowadays, zeolites are used, which offer high acidity, high thermal stability, and a reduced propensity for coke formation.

Cracking is an endothermic reaction, implying that the temperature must be rather high (500 °C), with the consequence that catalysts deactivate rapidly by carbon deposition. The fluidized catalytic cracking (FCC) process, developed by Standard Oil Company of New Jersey (1940; better known as ESSO, Exxon, and nowadays ExxonMobil), offers a solution for the short lifetime of the catalyst. Although cracking is carried out in a fast moving bed rather than in a fluidized bed reactor, the name fluidized catalytic cracking continues to be used.

The FCC process is used worldwide in more than 400 installations in 2006. Figure 9.12 shows the principle of an FCC unit. The preheated heavy feed (flash distillate and residue) is injected at the bottom of the riser reactor and mixed with the catalyst, which comes from the regeneration section. Table 9.5 gives a typical product distribution for the FCC process. Cracking occurs in the entrained-flow riser reactor, where hydrocarbons and catalyst have a typical residence time of only a few seconds. This, however, is long enough for the catalyst to become entirely covered by coke. While the products leave the reactor at the top, the catalyst flows into the regeneration section, where the coke is burned off in air at 1000 K.

Although cracking occurs on chlorine-treated clays and amorphous silica-aluminas as well, the application of zeolites has resulted in a significant improvement of the gasoline yield. The finite size of the zeolite micropores prohibits the formation of large condensed aromatic molecules. The beneficial shape selectivity improves the carbon efficiency of the process and also the lifetime of the catalyst.

Table 9.5 Approximate product distributions of fluid catalytic cracking for amorphous silica–alumina and zeolite catalysts.

Component	Weight percent Silica–alumina	Zeolite
Fuel gas (H_2, C_1, C_2) and LPG (C_3, C_4)	12–15	15–18
Gasoline	30–35	45–50
Light cycle oil	25	20
Residue	20–30	10–15
Carbon on the catalyst	6–8	5

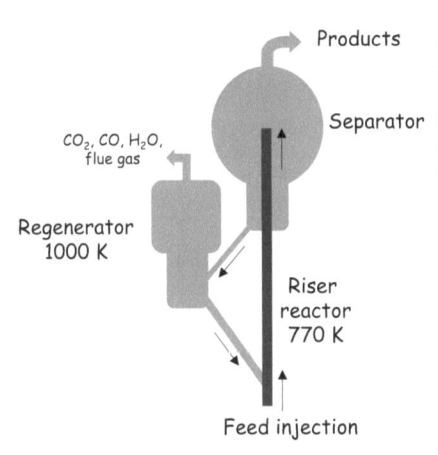

Figure 9.12 Scheme of an FCC unit. Cracking of the heavy hydrocarbon feed occurs in an entrained bed, in which the catalyst spends only a few seconds and becomes largely deactivated by coke deposition. Coke combustion in the regenerator is an exothermic process that generates heat for the endothermic cracking process as well.

The catalyst consists of about 20% of zeolite Y and 80% of matrix material or binder. This matrix not only determines the physical properties of the catalyst, but also acts as a trap for nickel, vanadium, and sodium impurities. Trapping these component is important because nickel enhances coke deposition on the zeolite, sodium deactivates the acid sites, and vanadium destroys the zeolite lattice. In addition, the matrix may also contain cracking activity, decomposing large hydrocarbons that cannot enter the zeolite. As the catalyst spends substantially longer in the regeneration section (10 min) than in the riser reactor (3 s), it must have very high strength and attrition resistance in combination with excellent thermal stability. FCC catalysts represent big business, with an estimated worldwide consumption of about 1000 t/day! The production capacity for FCC catalysts, however, is over 500 000 t/year, implying that the market is very competitive. Figure 9.13 gives an overview of the catalyst production.

FCC units, and in particular the catalyst regenerating section, may give rise to significant pollution. Sulfur in the coke oxidizes to SO_2 and SO_3, while the combustion also generates NO_x compounds. In addition, the flue gas from the regenerator contains particulate matter from the catalyst. The FCC process is also the major source of sulfur in gasoline. Of all sulfur in the feed, approximately 50%

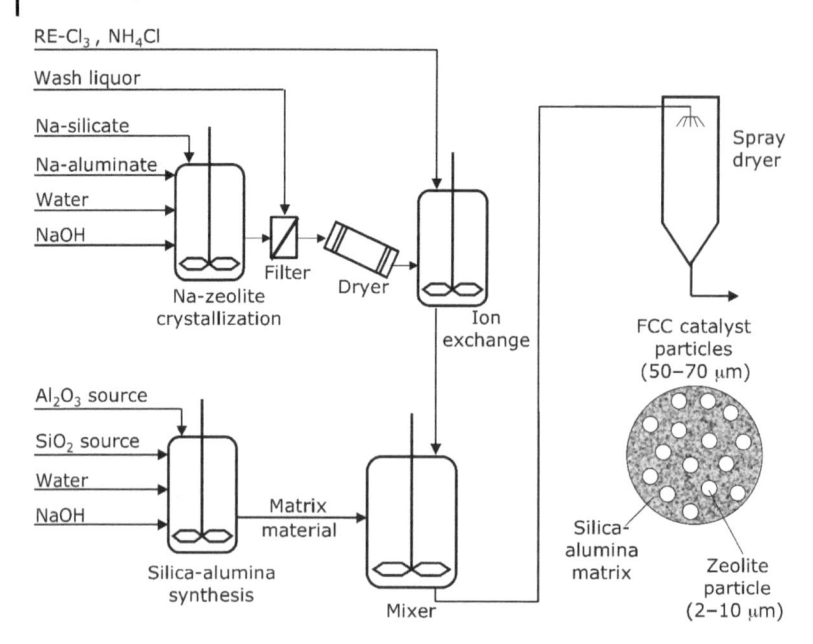

Figure 9.13 Production scheme of an FCC catalyst; RE stands for rare earth elements, which are added to enhance stability (adapted from Moulijn et al. [1]).

ends up as H$_2$S in the light gas – LPG fraction, 43% in the liquid products and 7% in the coke on the spend catalysts.

The reaction mechanism involves carbonium and carbenium ion intermediates. The first and difficult step is the generation of carbonium ions from alkanes:

$$H^+ + C_nH_{2n+2} \longrightarrow [C_nH_{2n+3}]^+ \tag{9.2}$$

The unstable carbonium ion decomposes to a carbenium ion

$$[C_nH_{2n+3}]^+ \longrightarrow [C_nH_{2n+1}]^+ + H_2 \tag{9.3}$$

and, in a cracking step

$$[C_nH_{2n+1}]^+ \longrightarrow [C_{n-x}H_{2n-2x+1}]^+ + C_xH_{2x} \tag{9.4}$$

Once initiated, the reaction propagates as follows:

$$C_mH_{2m+2} + [C_nH_{2n+1}]^+_{ads} \longrightarrow C_nH_{2n+2} + [C_mH_{2m+1}]^+_{ads} \tag{9.5}$$

Decomposition of the adsorbed carbenium ions is the main reaction channel. However, isomerization (aromatization) and oligomerization reactions proceed as well, and are the route to coke formation.

9.3.2
Reforming and Bifunctional Catalysis

Naphtha, neither produced by the initial distillation (straight run naphtha) nor by the cracking of heavy fractions (FCC naphtha) contains sufficient branched hydrocarbons to give it the octane number needed in gasoline. Typical octane numbers (RON) for naphtha are between 20 and 50, whereas the common gasolines have octane numbers between 95 and 98. Reforming converts the mainly straight-chain alkanes (in straight run naphtha) and alkenes (in FCC naphtha) into branched and aromatic hydrocarbons. Unfortunately, these reactions are accompanied by undesirable side processes such as cracking and coke formation. Reforming reactions are catalyzed by platinum on an acidic support. Both are involved in the reaction, making the system a bifunctional catalyst.

As naphtha contains many hydrocarbons in the range of C_5–C_{10}, the reformate product is a complex mixture of a few hundred different molecules. *n*-Hexane is the smallest hydrocarbon that exhibits all classes of reactions in reforming (Figure 9.14). The main reactions are

- Dehydrogenation of alkanes to alkenes
- Aromatization of cyclic products to benzene, toluene, xylenes, etc.
- Hydrogenation of alkenes to alkanes.

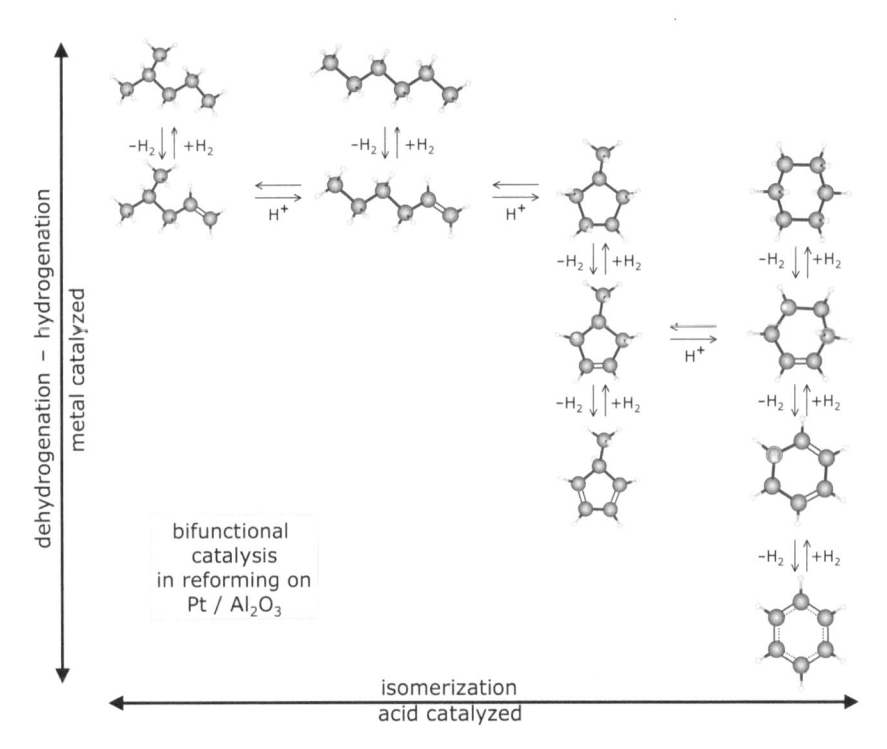

Figure 9.14 Bifunctional catalysis in the reforming of *n*-hexane. Metal-catalyzed dehydrogenation and hydrogenation reactions occur in vertical direction, and acid-catalyzed horizontally. Note also the special role of methyl cyclopentane in the route to cyclic and aromatic products (adapted from Sinfelt [14]).

These reactions are catalyzed by the metal. The other reactions:

- Isomerization of olefins
- Cracking of hydrocarbons

are catalyzed by acidic sites on the alumina support. Most of these reactions (except isomerization and hydrogenation) are strongly endothermic, and net producers of hydrogen. Thermodynamics therefore require a high temperature and a low pressure. However, such conditions also favor coke formation, which severely limits the lifetime of the catalyst. For this reason, hydrogen is recycled at moderate pressures to limit the production of coke. Hence, the actual reforming conditions (500 °C, 5–20 bar, H_2 : hydrocarbon ratio 5 : 10) are a compromise between product quality and yield, on the one hand, and reduced coke formation, on the other. Note that reforming is nevertheless a net producer of hydrogen, which is used in other refinery operations, such as hydrotreating and hydrocracking. Reforming is also an important source of aromatics for the chemical industry. Increasing the temperature and lowering the pressure benefits the overall selectivity towards aromatic molecules.

The catalyst consists of highly dispersed platinum on an alumina support (Figure 9.15). The presence of chlorine on the support enhances the acidity of the latter, which is highly desired for the isomerization properties of the catalyst. The stability of the catalyst against coke formation has been significantly improved by the introduction of a second metal, such as rhenium or iridium. Such catalysts

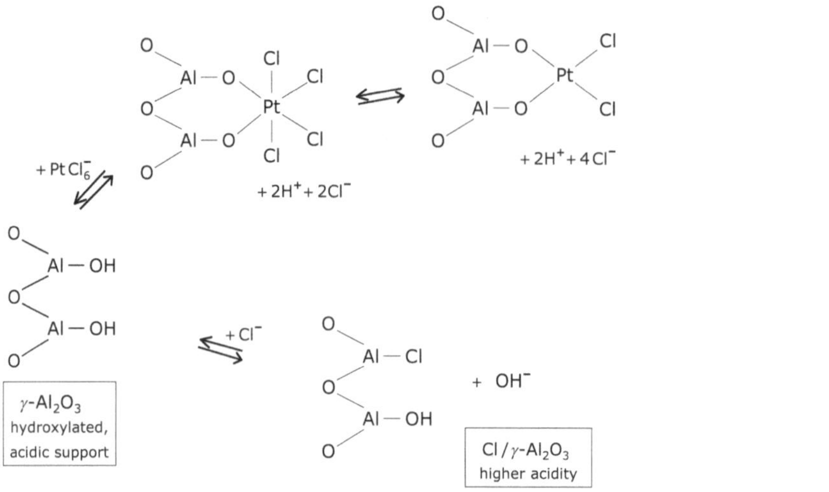

Figure 9.15 Schematic of how a bifunctional Pt/Al_2O_3 catalyst is prepared. The alumina support is impregnated with an aqueous solution of hexachloroplatinic acid (H_2PtCl_6) and HCl. The competitive adsorption between Cl^- and $PtCl_4^-$ helps to disperse the platinum evenly over the support, while the adsorbed chlorine enhances the acidity of the support. Calcination and reduction produce the eventual catalyst.

allow reforming at lower pressures, but they are extremely sensitive to poisoning by S and N compounds, making efficient hydrotreating essential.

To illustrate in a simple example how a bifunctional catalyst operates, we discuss the kinetic scheme of the isomerization of butane. The first step is the dehydrogenation of the (normal) alkane on the metal:

$$\text{n-}C_4H_{10} \longrightarrow C_4H_8 + H_2 \quad \text{on platinum} \tag{9.6}$$

The thus formed alkene adsorbs on an acid site of the alumina support, where it isomerizes to the iso-alkene,

$$\left.\begin{array}{l} C_4H_8 + H^+ \longrightarrow [C_4H_9]^+_{\text{ads}} \\ [C_4H_9]^+_{\text{ads}} \rightleftharpoons [i\text{-}C_4H_9]^+_{\text{ads}} \\ [i\text{-}C_4H_9]^+_{\text{ads}} \longrightarrow i\text{-}C_4H_8 + H^+ \end{array}\right\} \text{ on acid sites} \tag{9.7}$$

Finally, the isomer is hydrogenated to an alkane on the metal (or, alternatively, desorbs as an iso-alkene):

$$i\text{-}C_4H_8 + H_2 \longrightarrow i\text{-}C_4H_{10} \quad \text{on platinum} \tag{9.8}$$

If the catalyst contains sufficient platinum to allow the hydrogenation–dehydrogenation steps to be in equilibrium, the isomerization can be taken as the rate-limiting step, and the rate assumes the form:

$$r = \frac{N_{\text{acid}} k K [n\text{-}C_4H_{10}]/[H_2]}{1 + K[n\text{-}C_4H_{10}]/[H_2]} \tag{9.9}$$

Hence, the rate depends only on the ratio of the partial pressures of hydrogen and butane. Support for the mechanism is provided by the fact that the rate of butene isomerization on a platinum-free catalyst is similar to that of the above reaction. The essence of the bifunctional mechanism is that the metal converts alkanes to alkenes and vice versa, enabling isomerization via the carbenium ion mechanism which allows a lower reaction temperature than reactions involving a carbonium ion formation step from an alkane.

Surface science studies have generated much insight in how hydrocarbons react on the surfaces on platinum single crystals [15]. In addition, reactions of hydrocarbons on acidic sites on alumina or on zeolites have been studied in great detail [16].

We finish this section with an example of a dehydrogenation reaction on a metal, which is instructive as it does not rely on equilibrium adsorption, unlike most other examples given in this book. The dehydrogenation of methylcyclohexane, $C_6H_{11}CH_3$, to toluene, $C_6H_5CH_3$ on platinum is well described by a reaction mechanism consisting of consecutive steps in the forward direction only:

$$C_6H_{11}CH_3 + * \xrightarrow{k_1} C_6H_{11}CH_{3,\text{ads}} \tag{9.10}$$

$$C_6H_{11}CH_{3,\text{ads}} \xrightarrow{k_2} C_6H_5CH_{3,\text{ads}} + 3H_2\uparrow \tag{9.11}$$

$$C_6H_5CH_{3,\text{ads}} \xrightarrow{k_3} C_6H_5CH_3 + * \tag{9.12}$$

The second step is of course not elementary, but for the purpose of our discussion this is not relevant. Because toluene is a more unsaturated hydrocarbon than either methyl cyclohexane or the intermediate, we expect that toluene will be in the majority on the surface.

The rate of reaction is

$$r = V\frac{d[C_6H_5CH_3]}{dt} = Nk_3\theta_T \tag{9.13}$$

where θ_T is the toluene coverage. The balance of occupied and free sites becomes

$$\theta_T + \theta_M + \theta_* = 1 \tag{9.14}$$

but because adsorbed toluene dominates, we may assume

$$\theta_T + \theta_* \approx 1 \tag{9.15}$$

Application of the steady state assumption to θ_T and θ_M yields

$$\frac{d\theta_T}{dt} = k_2\theta_M - k_3\theta_T = 0 \tag{9.16}$$

$$\frac{d\theta_M}{dt} = k_1[C_6H_{11}CH_3]\theta_* - k_2\theta_M = 0 \tag{9.17}$$

$$\rightarrow k_3\theta_T = k_1[C_6H_{11}CH_3](1-\theta_T) \rightarrow \theta_T = \frac{k_1[C_6H_{11}CH_3]}{k_3 + k_1[C_6H_{11}CH_3]} \tag{9.18}$$

where θ_M is the methyl cyclohexane coverage. Substitution in the rate of reaction leads to the rate expression:

$$r = V\frac{d[C_6H_5CH_3]}{dt} = \frac{Nk_1[C_6H_{11}CH_3]}{1 + \frac{k_1}{k_3}[C_6H_{11}CH_3]} \tag{9.19}$$

A high coverage of toluene implies that the right term in the denominator is substantially larger than unity. In this case, the order of reaction in methyl cyclohexane becomes zero. In practice, we could say that the reaction is limited by the desorption of toluene from the surface.

9.3.3
Alkylation

Alkylates account for about 15% of the gasoline (Figure 9.11). Alkylation is the reaction between iso-butane [CH_3–CH–$(CH_3)_2$] and small olefins such as propene, butane, and pentene to branched alkanes, which improves the octane number of the gasoline [17]. The process, which is exothermic and therefore preferably carried out at low temperature, is catalyzed by acids in the liquid phase. Reactant iso-butane is furnished by the hydrocracking and FCC processes, or made separately by isomerization of butane. Alkylation is accompanied by undesired

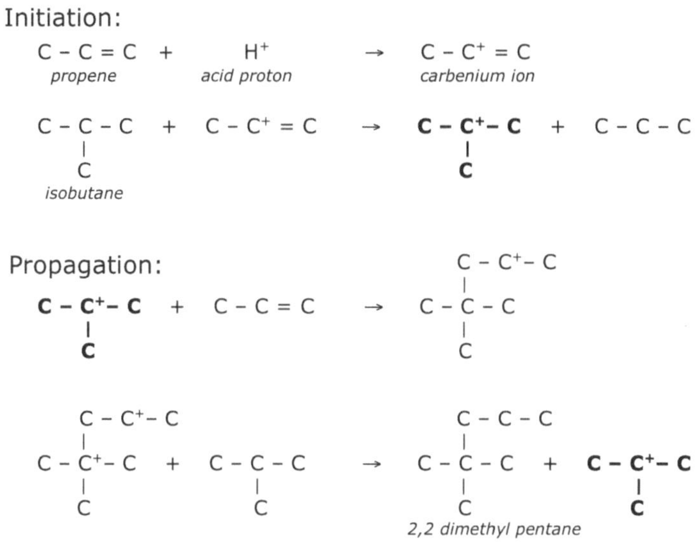

Initiation:

$$C - C = C + H^+ \rightarrow C - C^+ = C$$
propene   acid proton   carbenium ion

$$\begin{array}{c} C - C - C \\ | \\ C \end{array} + C - C^+ = C \rightarrow \begin{array}{c} \mathbf{C - C^+ - C} \\ | \\ \mathbf{C} \end{array} + C - C - C$$
isobutane

Propagation:

$$\begin{array}{c} \mathbf{C - C^+ - C} \\ | \\ \mathbf{C} \end{array} + C - C = C \rightarrow \begin{array}{c} C - C^+ - C \\ | \\ C - C - C \\ | \\ C \end{array}$$

$$\begin{array}{c} C - C^+ - C \\ | \\ C - C^+ - C \\ | \\ C \end{array} + \begin{array}{c} C - C - C \\ | \\ C \end{array} \rightarrow \begin{array}{c} C - C - C \\ | \\ C - C - C \\ | \\ C \end{array} + \begin{array}{c} \mathbf{C - C^+ - C} \\ | \\ \mathbf{C} \end{array}$$

2,2 dimethyl pentane

Figure 9.16 Alkylation mechanism of isobutane and propene is a chain reaction with the isobutene carbenium ion as the chain carrier (indicated in bold) (adapted from Moulijn et al. [1]).

oligomerization reactions of the olefins; therefore, the feed contains a large excess of isobutane. Alkylate mixtures possess octane numbers of about 87 (RON) and, therefore, contribute greatly to the antiknock properties of the gasoline.

Typical conditions in a process using H_2SO_4 as the catalyst are 5–10 °C, 2–5 bar, i-butane/olefin ratios of 10, and a residence time in the reactor of 20–30 min. The catalyst consumption rate is high (about 100 kg of acid per ton of alkylate product). Processes based on HF are more favorable in several respects: these run at 25–40 °C (hence cryogenic cooling is not necessary), 10–20 bar, with shorter residence times, and acid consumption rates below 1 kg of HF per ton of alkylate.

The reactions proceed via carbenium ions in a chain mechanism, initiated by the reaction between an olefin and an acid to C–C$^+$–C, which then reacts with iso-butane to C–C$^+$(C)–C. This carbenium ion is the central species in propagation steps to alkylated products such as 2,2-dimethylpentane and related products (Figure 9.16).

The major disadvantage of the alkylation process is that acid is consumed in considerable quantities (up to 100 kg of acid per ton of product). Hence, solid acids have been explored extensively as alternatives. In particular, solid super acids such as sulfated zirconia (SO_4^{2-}/ZrO_2), show excellent activities for alkylation, but only for a short time because the catalyst suffers from coke deposition due to oligomerization of alkenes. These catalysts are also extremely sensitive to water.

Another compound which greatly enhances the octane rating of the gasoline is MTBE (octane rating 116), methyl tertiary-butyl ether, which is produced by the alkylation of isobutene with methanol over an acidic polystyrene resin catalyst named Amberlyst. Although MTBE was among the fastest growing chem-

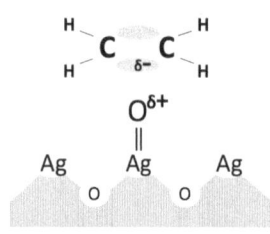

Figure 9.17 Ethylene hardly adsorbs on clean silver, but it does interact with pre-adsorbed oxygen atoms. At low coverage, the O atoms preferably interact with the C–H bond of ethylene, leading to its decomposition into fragments that oxidize to CO_2 and H_2O, but at higher coverages the oxygen atoms become electrophilic and interact with the π-system of ethylene to form the epoxide [18].

icals at the end of the twentieth century and it was widely applied in gasoline as an antiknock agent to increase the octane rating, it has been phased out because it was held responsible for polluting groundwater near fuel stations. Other boosters are ETBE (ethyl t-butyl ether; RON = 118), TAME (t-amyl methyl ether; RON = 111), and ethanol (RON = 113). Nowadays, ethanol is increasingly used for increasing the octane rating in the USA and Europe.

9.4
Petrochemistry: Reactions of Small Olefins

Small olefins, notably ethylene (ethene), propene, and butene, form the building blocks of the petrochemical industry. These molecules originate among others from the FCC process, but they are also manufactured by steam cracking of naphtha. A wealth of reactions is based on olefins. As examples, we discuss here the epoxidation of ethylene and the partial oxidation of propylene, as well as the polymerization of ethylene and propylene.

9.4.1
Ethylene Epoxidation

Ethylene oxide is an important intermediate for ethylene glycol (antifreeze) and for plastics, plasticizers, and many other products [18]. In Chapter 1, we already explained that the replacement of the traditional manufacturing process – which generated 1.5 mol of byproducts per 1 mol of epoxide – by a catalytic route based on silver catalysts was a major success story with respect to clean chemistry (Figure 9.17).

The reaction mechanism is fairly straightforward. It starts with the dissociative adsorption of oxygen on silver:

$$O_2 + 2* \longrightarrow 2O* \tag{9.20}$$

In fact, this step is crucial for the next one, as ethylene hardly adsorbs on clean silver:

$$C_2H_4 + * \longrightarrow C_2H_4* \tag{9.21}$$

The surface reaction between the two adsorbed species to ethylene oxide, which desorbs instantaneously, completes the cycle:

$$C_2H_4* + O* \longrightarrow C_2H_4O + 2* \qquad (9.22)$$

This mechanism may suggest that all reactants end up in the epoxide, but unfortunately this is not correct. Total oxidation of both the reactants and the product are competing processes, as expressed in the following overall scheme:

$$CH_2=CH_2 \longrightarrow H_2\overset{O}{\overset{/\backslash}{C}}=CH_2$$
$$\searrow \qquad \swarrow$$
$$CO_2 + H_2O \qquad (9.23)$$

To avoid the consecutive reaction of the desired product to CO_2, the catalyst has a low surface area and minimal porosity, to ensure a short residence time in the catalyst bed.

The reaction is exothermic and so to avoid serious temperature excursions the reactor consists of a bundle of narrow tubes, each a few centimeters in diameter, surrounded by a heat transfer medium. The catalyst consists of relatively large silver particles on an inert α-Al_2O_3 support. The surface area is below $1\,m^2\,g^{-1}$. Promoters such as potassium and chlorine help to boost the selectivity from typically 60% for the unpromoted catalysts to around 90%, at ethylene conversion levels on the order of 50%.

Trace amounts of chlorine suppress the total oxidation. This effect was discovered inadvertently when the selectivity of the process in an industrial plant rose spontaneously from one day to the other. Analysis of the catalyst revealed traces of chlorine, originating from a newly commissioned neighboring chlorine plant. Consequently, small amounts of a chlorine-containing compound, such as ethylene dichloride, are nowadays added to the feed.

Because of its industrial importance and its relative simplicity of its reaction mechanism and the catalyst system, much fundamental work has been done on this reaction [18].

9.4.2
Partial Oxidation and Ammoxidation of Propylene

Partial oxidation of propylene results in acrolein, $H_2C=CHCHO$, an important intermediate for acrylic acid, $H_2C=CHCOOH$, or in the presence of NH_3, in acrylonitrile, $H_2C=CHCN$, the monomer for acrylic fibers. Mixed metal oxides are used as the catalysts [19].

Catalytic oxidations on the surface of oxidic materials usually proceed according to the Mars–Van Krevelen mechanism [20, 21], as illustrated in Figure 9.18 for the case of CO oxidation. Instead of a surface reaction between CO and an adsorbed O atom, CO_2 is formed by reaction between adsorbed CO and an O atom

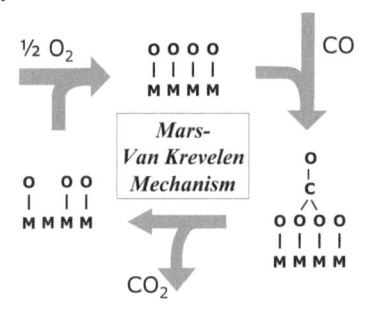

Figure 9.18 Mars–Van Krevelen mechanism for the oxidation of CO on a metal oxide surface. A characteristic feature is that lattice oxygen is used to oxidize the CO, leaving a defect that is replenished in a separate step by oxygen from the gas phase.

from the metal oxide lattice. The vacancy formed is filled in a separate reaction step, involving O_2 activation, often on defect sites.

The partial oxidation of propylene occurs via a similar mechanism, although the surface structure of the bismuth-molybdenum oxide is much more complicated than in Figure 9.18. As Figure 9.19 shows, crystallographically different oxygen atoms play different roles. Bridging O atoms between Bi and Mo are believed to be responsible for C–H activation and H abstraction from the methyl group, after which the propylene adsorbs in the form of an allyl group ($H_2C=CH-CH_2$). This is most likely the rate-determining step of the mechanism. Terminal O atoms bound to Mo are considered to be those that insert in the hydrocarbon. Sites located on bismuth activate and dissociate the O_2 which fills the vacancies left in the coordination of molybdenum after acrolein desorption.

Acrolein, in turn, can be oxidized further to acrylic acid. The catalyst for this step is a mixed vanadium-molybdenum oxide.

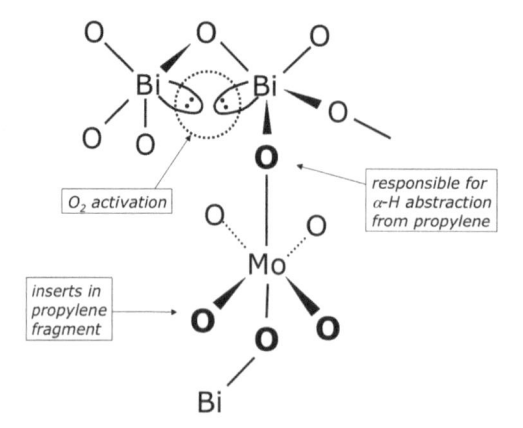

Figure 9.19 Schematic picture of the different oxygen sites involved in the partial catalytic oxidation of propylene to acrolein on Bi_2MoO_6, along with their conceived role in the reaction mechanism (adapted from Glaeser et al. [22]).

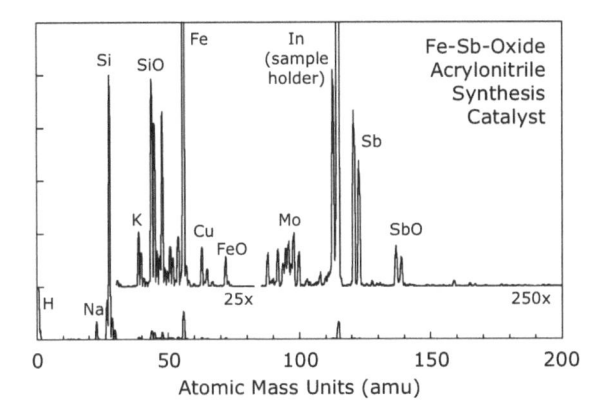

Figure 9.20 Secondary-ion mass spectrum of a promoted Fe-Sb oxide catalyst used for the selective oxidation of propylene and ammonia to acrylonitrile, showing the presence of Si, Cu, and Mo along with traces of alkali in the catalyst. Note the visibility of Mo, Sb, and SbO isotope patterns which assist in the identification of peaks in a mass spectrum (reproduced from [23] with permission from Wiley-VCH).

Another industrially important reaction of propylene, related to the one above, is its partial oxidation in the presence with ammonia, resulting in acrylonitrile, $H_2C=CHCN$. This ammoxidation reaction is also catalyzed by mixed metal oxide catalysts, such as bismuth-molybdate or iron antimonate, to which a large number of promoters is added (Figure 9.20). Being strongly exothermic, ammoxidation is carried out in a fluidized-bed reactor to enable sufficient heat transfer and temperature control (400–500 °C).

Partial oxidations over complex mixed metal oxides are far from ideal for single-crystal like studies of catalyst structure and reaction mechanisms, although several detailed (and by no means unreasonable) catalytic cycles have been postulated. Successful catalysts are believed to have surfaces that react selectively with adsorbed organic reactants at positions where oxygen of only limited reactivity is present. This results in the desired partially oxidized products and a reduced catalytic site, exposing oxygen deficiencies. Such sites are reoxidized by oxygen from the bulk that is supplied by gas-phase O_2 activated at remote sites.

9.4.3
Polymerization Catalysis

Most small olefins produced in the chemical industry are used to make polymers, with a global production on the order of 100 million tons per year. Polymers are macromolecules with molecular weights of typically 10^4 to 10^6 and consist of linear or branched chains, or networks built up from small monomers such as ethylene, propylene, vinyl chloride, styrene, etc. The vast majority of polymers are made in catalytic processes. Here, we concentrate on ethylene polymerization over chromium catalysts as an example [24].

The Phillips Process, ethylene polymerization with chromium catalysts, accounts for roughly one third of all high-density polyethylene produced worldwide. The process was invented in 1951 and has been improved continuously such that a "family" of different chromium-based catalysts is now available to produce about 50 different types of polyethylene, which vary with respect to molecular weight, chain branching and incorporated comonomers such as butene, hexane, or octene. This versatility accounts for the commercial success of the Phillips ethylene polymerization process. The properties of the desired polymer product can be tailored by varying parameters like calcination temperature, polymerization temperature, and pressure, by adding titanium or fluorine as a promoter to the support, by varying the pore size of the silica support, or even changing the support to alumina, which leads to higher molecular weight.

The chromium catalyst for polymerization differs from catalysts in other chemical reactions in the sense that it is eventually consumed. During the reaction, the polymer molecules fill up the pores and exert considerable pressure upon the support. Consequently the catalyst breaks up and remains in a finely dispersed form in the end product.

Due to its small chromium content, the Phillips catalyst presents a real challenge for characterization. Typical loadings are only about 0.5 wt%, or on the order of one chromium atom per nm^2. The catalyst is made by impregnation of the silica support with a solution of chromic acid in water. After drying, chromium is present as a hydrated Cr(VI) oxide. Calcination is an essential step in the preparation, and it needs to be carried out at a temperature of at least 550 °C to yield a catalyst with acceptable polymerization activity. The high temperature ensures that the Cr^{6+} ion is bound to two silicon ions via oxygen bridges (Figure 9.21). Furthermore, it is present in an isolated state, making the chromium system a true single-site catalyst. The isolated Cr^{6+} species, however, is still fully coordinated by oxide ions and is, therefore, inactive for reactions of ethylene. Reduction of the chromium species by ethylene at 150–160 °C produces a Cr^{2+} or Cr^{3+} species that is coordinatively unsaturated and serves as the catalytic site.

The reaction mechanism consists of three regimes:

- Initiation: activation of the first ethylene molecule and formation of an alkyl species on the chromium.
- Chain growth by repetitive insertion of ethylene molecules into the Cr-alkyl bond. The chain growth process is illustrated schematically in Figure 9.22.
- Termination by formation of a vinyl end group, upon which the polymer detaches from the Cr ion, and a new ethyl group forms to start a new polymer chain.

Initiation is slow, while chain growth occurs very rapidly. Activators, such as triethylaluminum, can be added to assist in the reduction of the Cr^{6+} site and to provide the initial ethyl group, such that the slow initiation step is avoided.

Figure 9.23 shows scanning force microscopic images of the polyethylene product formed on a planar model version of the Phillips catalyst. To appreciate their meaning, one should be aware that the polymer forms at a reaction temperature

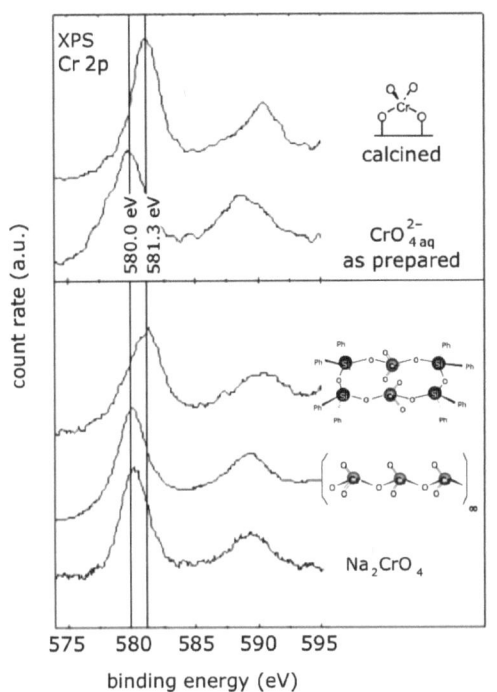

Figure 9.21 XPS spectra of a chromium polymerization catalyst along with chromium(VI) reference compounds for comparison of the state of chromium in the catalyst. Impregnated chromate in the freshly prepared catalyst shows the same binding energy as alkali chromates/dicromates or bulk CrO_3. Upon calcination, the binding energy of chromium in the catalyst increases significantly, by 1.3 eV. This unusually high binding energy is typical for chromate(VI) forming ester bonds to silica as in the cluster compound $[CrO_2(OSi(C_6H_5)_2OSi(C_6H_5)_2O)]_2$ (reproduced with permission from Thüne et al. [25], copyright 1997, American Chemical Society).

of 160 °C, that is, above the melting point. Upon cooling the polyethylene crystallizes in the form of lamellar, sheet-like structures, which attempt to order in a parallel way. This process is impeded by structural imperfections and crosslinks between the different lamellae. After a relatively short reaction time, a closed layer of polyethylene lamellae of 80 nm thick has formed. In this layer, shorter paraffins coexist with polymers. In addition to stacked, plate-like parts characteristic of paraffins and edge-on grown polyethylene lamellae, some crystal bundles of polyethylene grown in sheaf-like fashion are also present. After longer reaction times, the layer becomes considerably thicker, and the morphology changes drastically. Figure 9.23 shows ordering of the polyethylene into domains of roughly 50 μm in diameter. The film is on average 250 nm thick. In Figure 9.23b, polyethylene lamellae grow sheaf-like from a nucleation center. In fact, the lamellae try to order parallel, but succeed only partially, because the lamellae are imperfect,

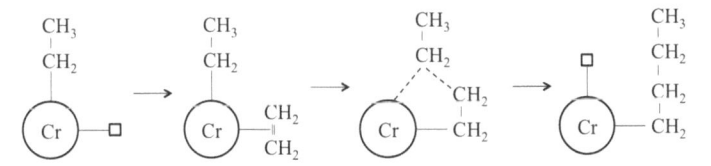

Figure 9.22 The Cossee–Arlman mechanism of chain growth in ethylene polymerization involves the insertion of ethylene in the growing alkyl chain; the square denotes a vacancy on the chromium ion [26].

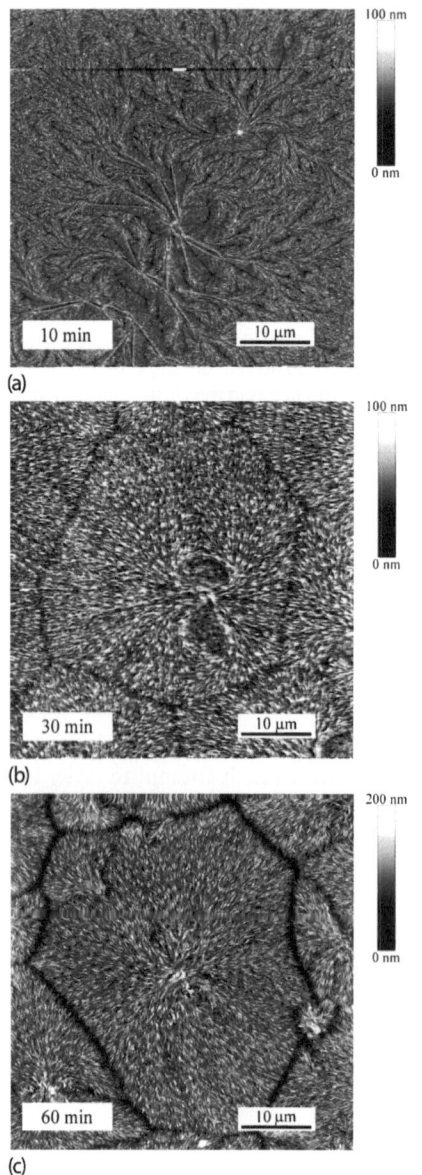

Figure 9.23 Scanning force microscopy images of polyethylene films formed on a planar chromium polymerization catalyst. The small white stripes are lamellar crystals. These form the well-known spherulite superstructure upon crystallization from the melt. Depending on the layer thickness, spherulite growth stops at different stages of development (reproduced from Thüne et al. [27] with permission of Elsevier).

due to, for example, loops of polymer molecules and loose ends that stick out. Figure 9.23c shows the final stage of spherulite formation, in a layer of 420 nm thickness. The nucleation center can still be recognized, but has been covered by a dome-like structure, as is usually observed for crystalline polyethylene in thick layers [27].

References

1. Moulijn, J.A., Makkee, M., and Van Diepen, A.E. (2013) *Chemical Process Technology*, John Wiley & Sons, Inc., Hoboken.
2. Bartholomew, C.H. and Farrauto, R.J. (2006) *Fundamentals of Industrial Catalytic Processes*, John Wiley & Sons, Inc., Hoboken.
3. Bensch, W. (2013) Hydrotreating: Removal of Sulfur from Crude Oil Fractions with Sulfide Catalysts, in *Comprehensive Inorganic Chemistry II (Second Edition): From Elements to Applications* (eds J.A. Reedijk and K. Poeppelmeier), Elsevier, Amsterdam, pp. 287–321.
4. Prins, R., de Beer, V.H.J., and Somorjai, G.A. (1989) Structure and function of the catalyst and the promoter in Co-Mo hydrodesulfurization catalysts. *Catalysis Reviews – Science and Engineering*, 31 1–41.
5. Topsøe, H., Clausen, B.S., and Massoth, F.E. (1996) *Hydrotreating Catalysis*, Springer, Berlin.
6. Muijsers, J.C., Weber, T., van Hardeveld, R.M., Zandbergen, H.W., and Niemantsverdriet, J.W. (1995) Sulfidation study of molybdenum oxide using $MoO_3/SiO_2/Si(100)$ model catalysts and Mo_3^{IV}-sulfur cluster compounds. *Journal of Catalysis*, 157 698–705.
7. Coulier, L., de Beer, V.H.J., van Veen, J.A.R., and Niemantsverdriet, J.W. (2000) On the formation of cobalt-molybdenum sulfides in silica-supported hydrotreating model catalysts. *Topics in Catalysis*, 13 99–108.
8. Lauritsen, J.V. and Besenbacher, F. (2015) Atom-resolved scanning tunneling microscopy investigations of molecular adsorption on MoS_2 and CoMoS hydrodesulfurization catalysts. *Journal of Catalysis*, 328 49–58.
9. Zhu, Y.Y., Ramasse, Q.M., Brorson, M., Moses, P.G., Hansen, L.P., Kisielowski, C.F., and Helveg, S. (2014) Visualizing the stoichiometry of industrial-style Co-Mo-S catalysts with single-atom sensitivity. *Angewandte Chemie – International Edition*, 53 10723–10727.
10. Pecoraro, T.A. and Chianelli, R.R. (1981) Hydrodesulfurization catalysis by transition-metal sulfides. *Journal of Catalysis*, 67 430–445.
11. Raybaud, P., Kresse, G., Hafner, J. and Toulhoat, H. (1997) Ab initio density functional studies of transition-metal sulphides: I. Crystal structure and cohesive properties. *Journal of Physics – Condensed Matter*, 9 11085–11106.
12. Raybaud, P., Hafner, J., Kresse, G., and Toulhoat, H. (1997) Ab initio density functional studies of transition-metal sulphides: II. Electronic structure. *Journal of Physics – Condensed Matter*, 9 11107–11140.
13. Von Ballmoos, R., Harris, D.H., and Magee, J.S. (1997) Catalytic Cracking, in *Handbook of Heterogeneous Catalysis* (eds. G. Ertl, H. Knözinger, and J. Weitkamp), Wiley-VCH Verlag GmbH, Weinheim, Vol 4, 1955–1986.
14. Sinfelt, J.H. (1964) Bifunctional Catalysis. *Advances in Chemical Engineering*, 5, 37–74.
15. Somorjai, G.A. and Li, Y. (2010) *Introduction to Surface Chemistry and Catalysis*, John Wiley & Sons, Inc., Hoboken.
16. van Bekkum, H., Flanigan, E.M., and Jansen, J.C. (1991) Introduction to Zeolite Science and Practice, *Studies in Surface Science and Catalysis*, Vol. 58, Elsevier, Amsterdam.
17. Corma, A. and Martinez, A. (1993) Chemistry, catalysts, and processes for isoparaffin-olefin alkylation – Actual situation and future-trends. *Catalysis Reviews – Science and Engineering*, 35 483–570.
18. Van Santen, R.A. and Kuipers, H. (1987) The mechanism of ethylene epoxidation. *Advances in Catalysis*, 35 265–321.
19. Gates, B.C. (1992) *Catalytic Chemistry*, John Wiley & Sons, Inc., New York.
20. Doornkamp, C. and Ponec, V. (2000) The universal character of the Mars and Van Krevelen mechanism. *Journal of Molecular Catalysis A: Chemical*, 162 19–32.

21 Vannice, M.A. (2007) An analysis of the Mars–van Krevelen rate expression. *Catalysis Today*, **123** 18–22.
22 Glaeser, L.C., Brazdil, J.F., Hazle, M.A., Mehicic, M., and Grasselli, R.K. (1985) Identification of active oxide ions in a bismuth molybdate selective oxidation catalyst. *Journal of the Chemical Society, Faraday Transactions I*, **81** 2903–2912.
23 Niemantsverdriet, J.W. (2007) *Spectroscopy in Catalysis, an Introduction*, Wiley-VCH Verlag GmbH, Weinheim.
24 McDaniel, M.P. (2010) A review of the Phillips supported chromium catalyst and its commercial use for ethylene polymerization, in *Advances in Catalysis* (eds B.C. Gates, H. Knözinger and F.C. Jentoft), **53**, pp. 123–606.
25 Thune, P.C., Verhagen, C.P.J., van den Boer, M.J.G., and Niemantsverdriet, J.W. (1997) Working surface science model for the Phillips ethylene polymerization catalyst: Preparation and testing. *Journal of Physical Chemistry B*, **101** 8559–8563.
26 Arlman, E.J. and Cossee, P. (1964) Ziegler-Natta catalysis III. Stereospecific polymerization of propene with the catalyst system $TiCl_3-AlEt_3$. *Journal of Catalysis*, **3** 99–104.
27 Thune, P.C., Loos, J., Lemstra, P.J., and Niemantsverdriet, J.W. (1999) Polyethylene formation on a planar surface science model of a chromium oxide polymerization catalyst. *Journal of Catalysis*, **183** 1–5.

Chapter 10
Environmental Catalysis

10.1
Introduction

Today's society asks for technology that has a minimum impact on the environment. Ideally, chemical processes should be "clean" in that harmful byproducts or waste are avoided. Moreover, the products (e.g., fuels) should also not generate environmental problems when they are used. The hydrogen fuel cell (Chapter 8) and the hydrodesulfurization process (Chapter 9) are good examples of such technologies where catalysts play an essential role. However, harmful emissions cannot easily be avoided, for example, in power generation and automotive traffic, and here catalytic clean-up technology helps to abate environmental pollution. This is the subject of this chapter.

Traffic and industry are the most important sources of air pollution in the Western world. They are responsible for the emission of carbon monoxide, nitrogen oxides (NO_x), sulfur oxides (SO_x), and all sorts of organic compounds. Catalysis has become indispensable in converting these environmentally harmful molecules into more benign species such as N_2, H_2O, and CO_2, although the latter of course contributes to the greenhouse effect. Environmental catalysts are now used in a wide range of applications, for example, in large-scale processes such as cleaning flue gases from power plants, in the transport sector, to catalysis in charcoal broilers in restaurants to decompose aromatic molecules formed during grilling of steaks and hamburgers, in toilets to reduce unpleasant odors, to remove volatile organic compounds (VOCs) in industry, and to decompose ammonia in waste water streams. Here, we concentrate on some of the major processes, namely automotive exhaust catalysis and NO_x-removal systems for cleaning flue gases from power plants. We briefly describe the processes, the catalysts, and discuss the reaction mechanisms and kinetics.

Concepts of Modern Catalysis and Kinetics, Third Edition. I. Chorkendorff and J.W. Niemantsverdriet.
© 2017 WILEY-VCH Verlag GmbH & Co. KGaA. Published 2017 by WILEY-VCH Verlag GmbH & Co. KGaA.

Table 10.1 Typical concentrations of exhaust gas constituents [1].

Constituent	Concentration
Hydrocarbons[a]	750 ppm
NO_x	1050 ppm
CO	0.68 vol%
H_2	0.23 vol%
O_2	0.51 vol%

a) Based on C_3

10.2
Air Pollution by Automotive Exhaust

Transportation is a major source of air pollution, particularly in urban environments. The total number of cars, trucks, buses, and motorcycles in the world in 1990 was estimated at about 650 million. All these cars produce exhaust, and an estimated 500 000 people per year, worldwide, die as a result of transportation-related air pollution. Transportation consumes more than half of the world's oil production as gasoline, diesel, and kerosene. It also generates roughly a quarter of the total carbon dioxide emission (on the order of 200 g km^{-1} for an average middle class passenger car).

Typical concentrations of car exhaust gas constituents present in addition to CO_2 and H_2O are given in Table 10.1. The origin of these molecules is evident: CO, the most immediately toxic exhaust component, is due to incomplete oxidation of hydrocarbons. Hydrogen is always present at approximately 1/3 the concentration of CO, and originates from the cracking of hydrocarbons. Nitric oxide (almost exclusively NO) is formed during combustion of fuel at high temperatures. NO becomes NO_2 in the atmosphere and is largely responsible for the brownish color of smog that is sometimes visible above cities on sunny days.

Unburnt gasoline and cracked hydrocarbons such as ethylene and propylene are also substantial constituents of exhaust. Gasoline contains additives such as benzene, toluene, and branched hydrocarbons to achieve the necessary octane numbers. The direct emission of these volatile compounds, for example, at gas stations, is a significant source of air pollution. Leaded fuels, containing antiknock additions such as tetraethyl lead, have been abandoned because lead poisons both for human beings and the three-way catalyst, especially for the removal of NO by rhodium.

In the USA, the Clean Air Act of 1970 established airquality standards for six major pollutants: particulate matter, sulfur oxides, carbon monoxide, nitrogen oxides, hydrocarbons, and photochemical oxidants. It also set standards for automobile emissions – the major source of carbon monoxide, hydrocarbons, and nitrogen oxides. An overview of the major standards is given in Table 10.2. The levels of, for example, the European Union (1996) are easily achieved with the present

Table 10.2 Emission standards (g km^{-1}) for automotive exhaust from gasoline-fuelled cars.

	CO	Hydrocarbons	NO$_x$	Hydrocarbon + NO$_x$
USA	2.11	0.25	0.62	–
USA, 1994	2.11	0.16	0.25	–
EU, 1996	2.2	–	–	0.5
Japan	2.1	0.25	0.25	–
Low emission vehicle	2.11	0.05	0.12	–
Ultra-low emission vehicle	1.06	0.02	0.12	–
Zero-emission vehicle	0	0	0	–

catalysts. The more challenging standards, up to those for the ultralow emission vehicle, are within reach, but zero-emission will probably only be attainable for a hydrogen-powered vehicle.

10.2.1
The Three-Way Catalyst

The three-way catalyst, consisting of Pt and Rh particles supported on a ceramic monolith, represents a remarkably successful piece of catalytic technology. It enables the removal of the three pollutants CO, NO, and hydrocarbons by the overall reactions shown in Table 10.3.

In addition, NO is reduced by H$_2$ and by hydrocarbons. To enable the three reactions to proceed simultaneously – notice that the two first are oxidation reactions, while the last is reduction – the composition of the exhaust gas needs to be properly adjusted to an air-to-fuel ratio of 14.7 (Figure 10.1). At higher oxygen content, the CO oxidation reaction consumes too much CO and hence NO conversion fails. If, however, the oxygen content is too low, all NO is converted, but hydrocarbons and CO are not completely oxidized. An oxygen sensor (λ probe) is mounted in front of the catalyst to ensure the proper balance of fuel and air via a microprocessor-controlled injection system.

Catalytic treatment of motor vehicle exhaust has been applied in all passenger cars in the USA since the 1975 models. The first cars with electronic feedback systems and three-way catalysts were 1979 Volvos, sold in California. Today all new gasoline cars sold in the Western world are equipped with catalytic converters.

Table 10.3 Reactions in the three-way catalyst.

Reaction	Most efficient catalysts
CO + O$_2$ $\longrightarrow$ CO$_2$	Pt, Pd
C$_x$H$_y$ + O$_2$ $\longrightarrow$ CO$_2$	Pt, Pd
NO + CO $\longrightarrow$ N$_2$ + CO$_2$	Rh, Pd

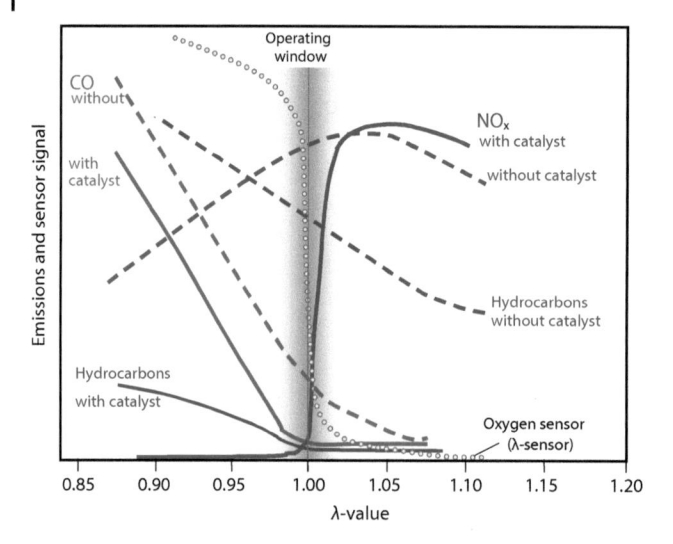

Figure 10.1 Emissions of CO, NO$_x$, and hydrocarbons along with the signal from the oxygen sensor as a function of the air/fuel composition; $\lambda = 1$ corresponds to the air-to-fuel ratio of 14.7. Note that the three pollutants can only be converted simultaneously in a very narrow operating window of air-to-fuel ratios.

It certainly improves the air quality, but also adds to the fuel consumption as the restricted air/fuel ratio limits the efficiency of the engine.

The principle of the probe is shown in Figure 10.2. It is a simple oxygen sensor made in a similar manner as the solid oxide fuel cell discussed in Chapter 8. An oxide that allows oxygen ions to be transported is resistively heated in order to ensure sufficiently high mobility and a short response time (~ 1 s).

The oxygen content in the exhaust is measured against a suitable reference, in this case atmospheric air. The response is given by the Nernst equation:

$$\varepsilon = \frac{RT}{4F} \ln\left(\frac{P_{O_2 \text{ reference}}}{P_{O_2 \text{ exhaust}}}\right) \tag{10.1}$$

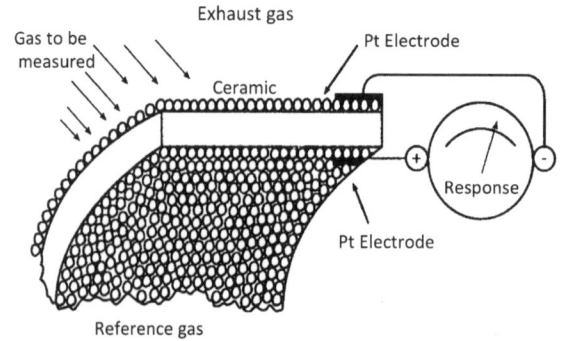

Figure 10.2 Principle of the λ-probe oxygen sensor used to regulate the injection system to obtain the correct air-to-fuel ratio in the exhaust gas.

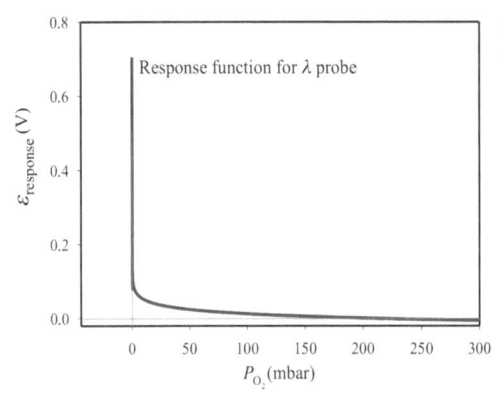

Figure 10.3 Response of the λ-probe oxygen sensor.

and is shown in Figure 10.3. Note the strong response at low oxygen content, ensuring a large variation in signal around the working point of the stoichiometric ratio. The equation suggests that the voltage goes towards infinity when the oxygen vanishes on one side. This naturally does not happen since then the oxides will then start to be reduced and become conducting for electrons. Typically the readout varies between 0 and 1 V when going from equal to zero oxygen partial pressure on the inside.

The λ-probe relies on diffusion of atomic oxygen through a solid electrolyte and, therefore, it will have a certain response time. Reducing the thickness of the oxide membrane and increasing the temperature both shorten the response time, but a certain delay cannot be avoided. For example, if the driver suddenly steps on the gas pedal, the exhaust becomes reducing. Consequently, sulfur deposited in the catalyst becomes hydrogenated to H_2S, causing the characteristic "rotten eggs" smell (this smell sometimes arises during the startup of a cold engine). New types of sensors with faster response are, therefore, being explored to avoid these effects. Ideally these should be placed right after each cylinder and, therefore, they should be capable of withstanding high temperatures.

10.2.1.1 The Catalytic Converter

Figure 10.4 shows some typical automotive converters. A critical consideration in the design is that they must not obstruct the flow of the exhaust, otherwise the engine would stall. Hence, the reactor must have a rather open structure. This is achieved by applying the catalytically active particles on a multichannel array, called monolith. This type of catalyst is very suitable for operation under conditions where the rate is limited by transport outside the catalyst (see also Figure 5.37).

Figure 10.5 shows schematically how the catalyst is built up. The major supporting structure is the monolith, covered by a 30–50 μm layer of porous 'washcoat'. The latter is the actual support and consists largely of γ-Al_2O_3 (70–85%) and other oxides such as cerium oxide (10–30%), lanthanum oxide, or alkaline earth oxides (BaO). Some formulations use NiO as sulfur getter in the outer layers of the washcoat. Denser oxides such as α-Al_2O_3 or ZrO_2 are sometimes used as a diffusion

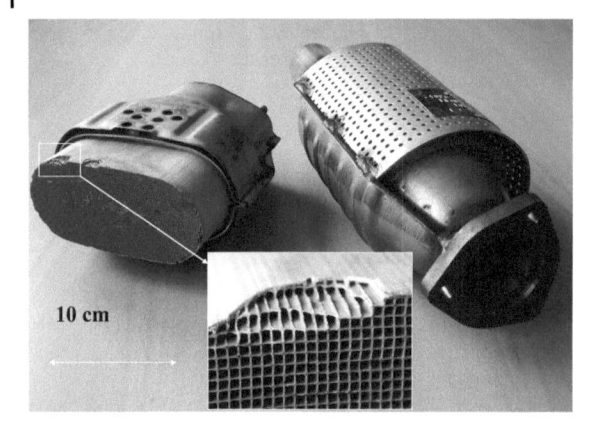

Figure 10.4 Typical automotive exhaust converters. The one to the left has been cut open to reveal the monolith. The insert shows a blow up of the upper part of the monolith where a part has been chipped off.

barrier, to prevent incorporation of rhodium in the support at high temperatures. Only 1–2% of the washcoat's weight corresponds to noble metals (Pt, Pd, Rh). Some manufactures use all three, but most converters contain Rh together with Pt. Recently, all-palladium converters have been also introduced.

Numerous permutations in composition exist, but usually the precise composition, particularly that of the washcoat, is a commercial secret. Detailed accounts of the three-way catalyst have been given by Heck and Farrauto [2] and by Gandhi [2, 3]. Here, we briefly describe the functions of the catalyst ingredients.

- Alumina, present in the gamma modification, is the most suitable high surface area support for noble metals. The γ-Al_2O_3 in washcoats typically has a surface area of 150–175 $m^2\,g^{-1}$. However, at high temperatures γ-alumina transforms into the alpha phase, and stabilization to prevent this is essential. Another concern is the diffusion of rhodium into alumina, which calls for the application of diffusion barriers.
- Ceria is a partially reducible oxide (its reduction is promoted by the noble metals). When the air/fuel oscillation swings to the lean side, it takes up its maximum capacity of oxygen; this oxygen is available for CO_2 formation when the

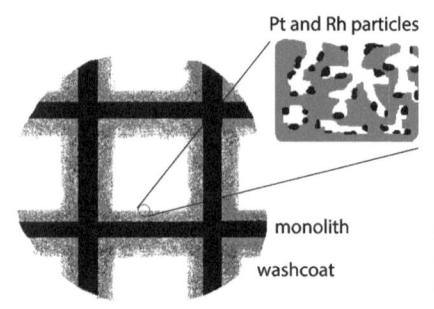

Figure 10.5 Monolith, washcoat, and noble metal particles in an automotive exhaust catalyst.

composition swings to the rich side. Ceria, thus, counteracts the effect of oscillating feed gas compositions. It further stabilizes the high surface area of the γ-Al_2O_3 by inhibiting its phase change to the α-phase and it impedes the agglomeration of the noble metals and the loss of noble metal surface area by acting as a diffusion barrier. Oxygen vacancies, conceivably also in regions where CeO_x is in contact with noble metal, are active sites for CO oxidation by O_2, or by H_2O through the water–gas shift reaction [4, 5].

- Lanthanum oxide is valence invariant and does not exhibit any oxygen storage capacity, but it effectively stabilizes γ-Al_2O_3. It spreads over the alumina surface and provides a barrier against dissolution of rhodium in the support.
- Platinum serves as the catalyst for oxidation of CO and hydrocarbons. It is relatively insensitive to contamination by lead or sulfur. At high temperatures, it is not known to dissolve in the washcoat, but sintering into larger particles may lead to a substantial loss of platinum surface area with dramatic consequences for the overall oxidation activity.
- Rhodium is the crucial ingredient of the three-way catalyst. The metal, a byproduct of Pt extraction, is mined in South Africa ($\pm$ 2/3) and Russia ($\pm$1/3). However, the average mining ratio equals Pt:Rh = 17 : 1 is much lower than that in the catalyst. In 1991, about 87% of the world rhodium production went into catalysts (Pt: 37%). Although expensive, rhodium is difficult to replace owing to its unique properties with respect to NO surface chemistry. Loss of rhodium activity is due to particle growth under reducing conditions (> 900 K), and to diffusion into the alumina support under oxidizing conditions (> 900 K).
- Palladium can be present in addition to Rh and Pt but may also replace them. Palladium is as good for oxidation as platinum (even for the oxidation of saturated hydrocarbons) but it is somewhat less active for NO reduction. Hence, noble metal loadings of a Pd-only catalyst are 5–10 times higher than for a Pt-Rh catalyst. Palladium is less resistant to residual lead in gasoline than Pt; however, in the US, gasoline is essentially Pb-free. Palladium catalysts also require higher ceria loadings to help prevent high-temperature deactivation.

The ideal operating temperatures for the three-way catalyst lie between 350 and 650 °C. After a cold start, it takes at least a minute to reach this temperature, implying that most CO and hydrocarbons emission takes place directly after the start. Temperatures above 800 °C should be avoided in order to prevent sintering of the noble metals and dissolution of rhodium in the support.

10.2.1.2 Demonstration Experiments

Platinum's catalytic oxidation is easily shown by warming alcohol (a few milliliters) in an Erlenmeyer flaks, by holding the flask in your hands or with a lighter, and taking a 1mm Pt wire (rolled to a 0.1 mm flat foil at the end) and heating it in a flame until it glows, then place it in the flask. The exothermic oxidation of the alcohol vapor will make it glow. Under the right conditions, the reaction will continue for hours. You will probably notice that the reaction is not entirely selective towards CO_2 as you can normally smell compounds like acetic acid and acetaldehyde.

This simple experiment, illustrating that alcohol can be oxidized without the development of a flame, was already performed by Davy around 1825 and formed the basis for the first practical application of a catalyst, namely Davy's mine lamp.

A slightly more elaborate experiment can be performed if a used automotive catalyst is available. A piece of monolith (e.g., 5×2.5 cm) fitted in a glass tube serves as the converter. The exhaust is simulated by bubbling air through alcohol (e.g., by using an aquarium pump); the alcohol content is conveniently varied by heating. The catalyst does not work at room temperature, so it must be heated with a heat gun. Once ignited it oxidizes the alcohol, demonstrating the principle. Dependent on the alcohol concentration, the catalyst may glow in the same manner as the platinum wire mentioned above.

10.2.1.3 Catalyst Deactivation

In the USA, three-way catalysts have to maintain high activity and meet the emission standards of Table 10.2 after 50 000 miles or 5 years. Because catalysts do deactivate with use, fresh catalysts are designed such that they perform well below the emission standards. The extent to which a three-way catalyst deactivates depends on many factors. The wide range of vehicle operating conditions due to differences in style of driving is an important one.

The major causes of degradation are thermal damage, poisoning by contaminants, and mechanical damage of the monolith. High temperatures lead to sintering of the noble metal particles and may also induce reactions between the metals and the support. High temperatures can, for example, be caused by fast driving or by repeated misfiring of the engine, resulting in the (exothermic) oxidation of large amounts of unburned fuel over the catalyst. Excessively high temperatures can damage the support, by promoting the transition to α-alumina and loss of surface area. Shock and high temperatures may produce channels for the exhaust to pass through the system without contacting the catalyst.

Lead and phosphorous are poisons for the catalyst. Lead may still be present at very low levels in unleaded gasoline, but in general it does not present a problem. Misfuelling, that is, using leaded fuel, however, is a serious reason for deactivation, affecting the NO reduction activity of the three-way catalyst irreversibly. The oxidation activity is temporarily lower after misfuelling but recovers, usually to values within the emission standards. This is easily understood as lead has a low melting point and thus a low cohesive energy. Thus, it will have a low surface energy and a strong tendency to decorate the surface, blocking the active sites. This effect may be lessened under conditions where other adsorbates are present like CO and oxygen, but it may have permanent effects on, for example, the Rh steps sites, which, as with N_2 dissociation on Ru, may be instrumental in dissociating the NO resulting in the irreversible behavior.

Phosphorous is present in engine oils ($\pm$ 1 g L^{-1}). It binds strongly to the alumina support and may eventually also block the noble metal surfaces. Sulfur, although a potential poison for all metals, interacts relatively weakly with platinum and rhodium, but is a source of concern with palladium catalysts. Sulfur oxidized to SO_2 blocks the noble metal area at low temperatures (below 300 °C). A small fraction that is oxidized further to SO_3 may react with alumina to $Al_2(SO_4)_3$, causing loss of surface area of the washcoat. In addition, alkali and halide residues from catalyst precursors have been found to negatively affect the stability of the washcoat and the metal particles, which is why it is preferable to have catalyst precursors that are free of these elements. Other contaminants that may be present in gasoline, such as organosilicon compounds or additives based on manganese, also negatively affect the performance of a three-way catalyst. The sensitivity of the oxygen sensor to contaminants is another notable consideration.

10.2.2
Catalytic Reactions in the Three-Way Catalyst: Mechanism and Kinetics

CO oxidation and the reaction between CO + NO have been extensively studied. Much less is known about hydrocarbon oxidation, and the role of hydrocarbons in reducing NO is only beginning to be explored. Surface science studies with reactions on well-defined single-crystal surfaces have contributed significantly to our understanding, for a review see Nieuwenhuys [6].

10.2.2.1 The CO Oxidation Reaction
The oxidation of CO is treated in detail in Chapter 2, Eqs. (2.184)–(2.187). Assuming that the CO and O recombination into CO_2 is rate limiting gave the following rate expression:

$$V\frac{dp_{CO_2}}{dt} = Nk_3^+ \theta_{CO}\theta_O = \frac{Nk_3^+ K_{CO}\sqrt{K_{O_2}}p_{CO}\sqrt{p_{O_2}}}{\left(1+\sqrt{K_{O_2}p_{O_2}}+K_{CO}p_{CO}\right)^2} \quad (10.2)$$

where V is the reacting volume, N the number of catalytically active sites, k_3^+ the forward rate constant of the surface reaction between CO and an O atom, θ the surface coverage of the indicated molecule, K the equilibrium constant of adsorption, and p the partial pressure of the indicated gas. Here, the coverage of CO_2 has been neglected and we assume that we are far from equilibrium so that the back reaction can be neglected.

The energetics of the CO oxidation reaction is illustrated in Figure 10.6. The activation energy of the homogeneous gas phase reaction between CO and O_2 is largely determined by the 500 kJ mol^{-1} needed to break the O–O bond of O_2, and, hence, dissociation of O_2 represents the rate-determining step for this reaction. The catalyst, however, easily dissociates the O_2 molecule. Here, the rate-determining step is the surface reaction between CO∗ and O∗, which has an activation energy of about 100 kJ mol^{-1}. CO oxidation, thus, illustrates nicely that the essential action of the catalyst lies in the dissociation of a bond. Once this has been accomplished, the subsequent reactions follow, provided the intermediates are not held too strongly by the catalyst, as expressed in Sabatier's Principle.

The temperature dependence of the reaction rate is of interest, revealing a very general phenomenon in catalysis (see also Chapter 2, Figure 2.12). At low temper-

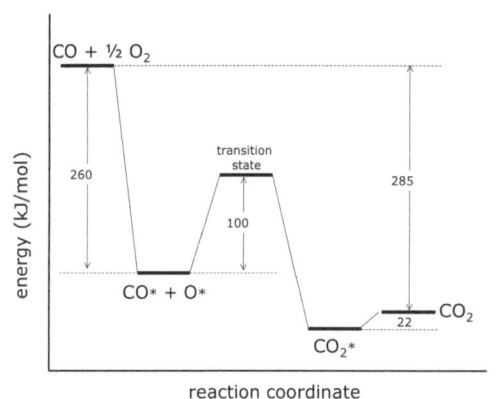

Figure 10.6 Approximate energy diagram of the CO oxidation on palladium. Notice the largest energy barrier is the CO + O recombination (adapted from Engel and Ertl [7]).

atures, the surface is predominantly covered by CO, and the denominator of the rate expression is dominated by the term $K_{CO}[CO]$, giving rise to a negative order in CO. At temperatures above the desorption temperature of CO, oxygen tends to build up on the surface giving rise to a rate expression in which the CO term in the denominator becomes insignificant and the order in CO becomes positive. The term $(K_2[O_2])^{1/2}$ is not negligible, though.

Figure 10.7 shows the temperature dependence of the CO oxidation rate on a rhodium surface, as reported by Bowker *et al.* [8]. It expresses that the rate of reaction maximizes when both reactants, adsorbed CO and O, are present in comparable quantities at a temperature where the activation barrier of the reaction can be overcome. At low temperatures, the reaction is negatively affected by the lack of oxygen on the surface, while at higher temperatures the adsorption/desorption equilibrium of CO shifts towards the gas phase side, resulting in low coverages of CO. As discussed in Chapter 2, this type of non-Arrhenius-like behavior with temperature is generally the case for catalytic reactions.

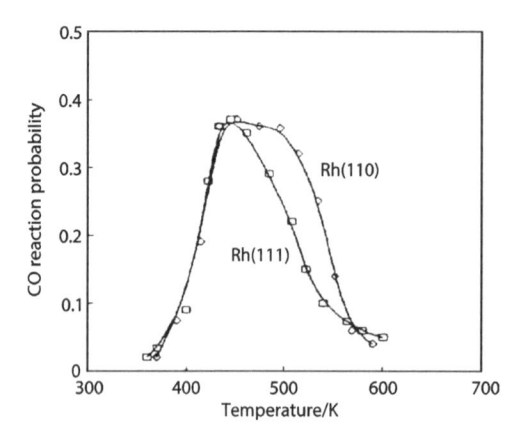

Figure 10.7 CO_2 formation rate from CO and O_2 over Rh(111) and Rh(110) surfaces. Note the similarity to the simple model used to describe the rate in Figure 2.12 (reproduced with permission from Bowker *et al.* [8], copyright 1993 by Springer).

10.2.2.2 Is CO Oxidation a Structure-Insensitive Reaction?

CO oxidation is often quoted as a structure-insensitive reaction, implying that the turnover frequency on a certain metal is the same for every type of site, or for every crystallographic surface plane. Figure 10.7 shows that the rates on Rh(111) and Rh(110) are indeed similar on the low-temperature side of the maximum, but that they differ at higher temperatures. This is because on the low temperature side the surface is mainly covered by CO. Hence, the rate at which the reaction produces CO_2 becomes determined by the probability that CO desorbs to release sites for the oxygen. As the heats of adsorption of CO on the two surfaces are very similar, the resulting rates for CO oxidation are very similar for the two surfaces. However, at temperatures where the CO adsorption–desorption equilibrium lies more towards the gas phase, the surface reaction between O∗ and CO∗ determines the rate, and here the two rhodium surfaces show a difference (Figure 10.7). The apparent structure insensitivity of the CO oxidation appears to be a coincidence that is not necessarily caused by equality of sites or ensembles thereof on the different surfaces.

Temperature-programmed reactions between small amounts of adsorbed species are an excellent way to study the intrinsic reactivity of catalytic surfaces. Such experiments on rhodium (100) and (111) surfaces covered by small amounts of CO∗ and O∗ showed a profound difference in CO_2 formation rate [9]. Hence, care should be taken to interpret apparent structure sensitivity found under normal operating conditions of high pressure and coverage in terms of the intrinsic reactivity of sites. From the theory of chemisorption and reaction discussed in Chapter 6, it is hard to imagine how the concept of structure insensitive can be maintained on the level of individual sites on surfaces, as atoms in different geometries always possess different bonding characteristics.

10.2.2.3 The CO + NO Reaction

The undoubtedly structure-sensitive reaction NO + CO has a rate that varies with rhodium surface structure. A temperature-programmed analysis (Figure 10.8) gives a good impression of the individual reaction steps: CO and NO adsorbed in relatively similar amounts on Rh(111) and Rh(100) give rise to the evolution of CO, CO_2, and N_2, whereas desorption of NO is not observed at these coverages. Hence, the TPRS experiment of Figure 10.8 suggests the following elementary steps:

$$NO + * \rightleftharpoons NO* \tag{10.3}$$

$$CO + * \rightleftharpoons CO* \tag{10.4}$$

$$NO* + * \longrightarrow N* + O* \tag{10.5}$$

$$CO* + O* \longrightarrow CO_2 + 2* \tag{10.6}$$

$$N* + N* \longrightarrow N_2 + 2* \tag{10.7}$$

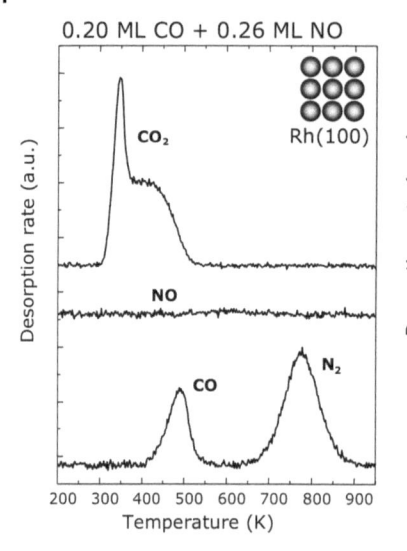

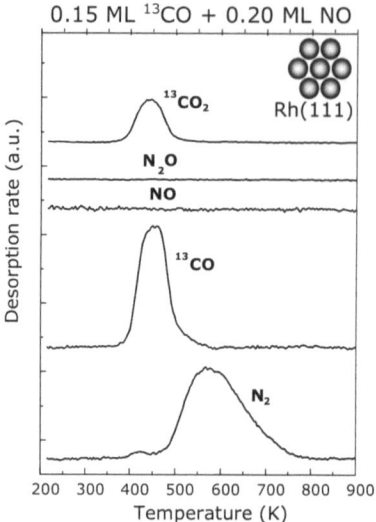

Figure 10.8 Temperature-programmed reaction of NO and CO on two surfaces of rhodium. The initially molecularly adsorbed NO dissociates entirely at relatively low temperatures, but NO does not desorb. Note the difference in selectivity and reactivity between the surfaces: on Rh(100) most of the CO oxidizes to CO_2 and the reaction already starts at 300 K. On Rh(111) most of the CO desorbs unreacted and CO_2 formation does not begin until about 400 K. N_2 formation, however, proceeds faster on Rh(111) than on Rh(100) (reproduced with permission from Hopstaken and Niemantsverdriet [10], copyright 2000, by the American Vacuum Society).

Breaking of the N–O bond by the rhodium surface is the most essential step in the catalytic reduction of NO (see also Chapter 7. Although rhodium is sufficiently reactive to achieve this (even without promoters), dissociation can nevertheless be severely impeded if the surface coverage is too high (as Figure 7.12 shows). In the low coverage regime, however, such effects play no role.

Table 10.4 lists the rate parameters for the elementary steps of the CO + NO reaction in the limit of zero coverage. Parameters such as those listed in Table 10.4 form the highly desirable input for modeling overall reaction mechanisms. In addition, elementary rate parameters can be compared to calculations on the basis of the theories outlined in Chapters 3 and 6. In this way, the kinetic parameters of elementary reaction steps provide, through spectroscopy and computational chemistry, a link between the intramolecular properties of adsorbed reactants and their reactivity: statistical thermodynamics furnishes the theoretical framework to describe how equilibrium constants and reaction rate constants depend on the partition functions of vibration and rotation. Thus, performing spectroscopy of adsorbed reactants and intermediates provide the input for computing equilibrium constants, while calculations on transition states of reaction pathways, starting from structurally, electronically and vibrationally well-characterized ground states, enable the prediction of kinetic parameters.

Table 10.4 Kinetic parameters of the elementary steps involved in the NO + CO reaction [11, 12].

Kinetic parameters CO + NO	Rh(100) E_{act} (kJ/mol)	v (s^{-1})	Rh(111) E_{act} (kJ/mol)	v (s^{-1})
Dissociation of NO$_{ads}$	37 ± 3	$10^{11 \pm 1}$	65 ± 6	$10^{11 \pm 1}$
Desorption of NO$_{ads}$	106 ± 10	$10^{13.5 \pm 1}$	113 ± 10	$10^{13.5 \pm 1}$
Desorption of CO$_{ads}$	139 ± 3	$10^{14 \pm 0.3}$	155 ± 5	$10^{15 \pm 1}$
Reaction of CO$_{ads}$ + O$_{ads}$ ⟶ CO$_2$	103 ± 5	$10^{12.7 \pm 0.7}$	67 ± 3	$10^{7.3 \pm 0.2}$
Reaction of N$_{ads}$ + N$_{ads}$ ⟶ N$_2$	215 ± 10	$10^{15.1 \pm 0.5}$	118 ± 10	$10^{10 \pm 1}$

10.2.2.4 The CO + NO Reaction at Higher Pressures

The NO + CO reaction is only partially described by the reactions (10.3) to (10.7), as there should also be steps to account for the formation of N$_2$O, particularly at lower reaction temperatures. Figure 10.9 shows rates of CO$_2$, N$_2$O, and N$_2$ formation on the (111) surface of rhodium in the form of Arrhenius plots. Comparison with similar measurements on the more open Rh(110) surface confirms again that

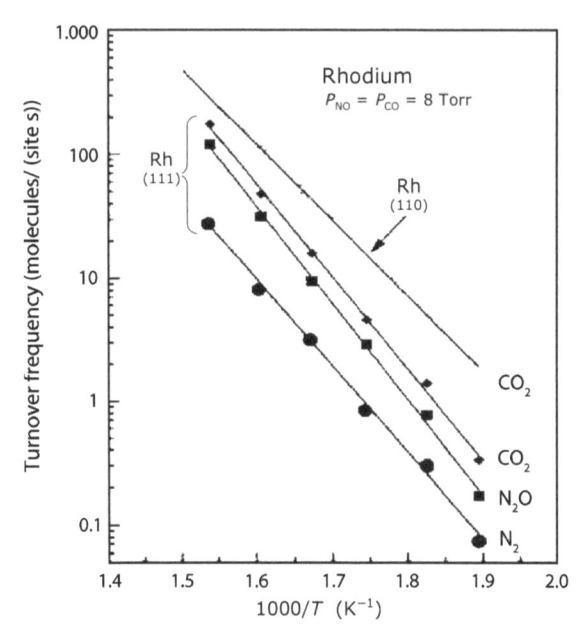

Figure 10.9 Arrhenius plots of the NO + CO reaction over rhodium(111) and (110) (reproduced from Peden et al. [13] with permission of Elsevier).

the reaction is strongly structure sensitive. As N$_2$O is undesirable, it is important to know under what conditions its formation is minimized. First, the selectivity to N$_2$O, expressed as the ratio given in Eq. (10.8), decreases drastically at higher temperatures where the catalyst operates. Second, real three-way catalysts contain rhodium particles in the presence of CeO$_x$ promoters, and these appear to suppress N$_2$O formation [14]. Finally, N$_2$O undergoes further reaction with CO to give N$_2$ and CO$_2$, which is also catalyzed by rhodium.

$$S(N_2O) = \frac{[N_2O]}{[N_2O] + [N_2]} 100\% \tag{10.8}$$

The mechanism of the NO + CO reaction at realistic pressures is, thus, very complicated. In addition to all the reaction steps considered above, one also has to take into account that intermediates on the surface may organize into islands or periodically ordered structures. Monte Carlo techniques are needed in order to account for these effects. Consequently, we are still far from a complete kinetic description of the CO + NO reaction. For an interesting review of the mechanism and kinetics of this reaction we refer to Zhdanov and Kasemo [15].

10.2.2.5 Reactions Involving Hydrocarbons

Hydrocarbons in the exhaust react with oxygen and with NO. Although these reactions have had much less attention than oxidation of CO and reduction of NO by CO, the reactions of hydrocarbons are important in the overall reaction mechanism of the three-way converter, particularly because the converter is by no means a homogeneously mixed reactor. Hence, zones exist where, for example, hydrocarbon fragments and nitrogen atoms are co-adsorbed on the noble metal surface of the catalyst, which might in principle lead to undesirable byproducts such as HCN. Fortunately, the presence of oxygen appears to prevent such reactions in practice. A kinetic description of reactions involving hydrocarbons is difficult to give due to the complicated decomposition pathways of the hydrocarbons on noble metal surfaces.

10.2.2.6 The NO$_x$ Storage-Reduction Catalyst for Lean-Burning Engines

One of the most straightforward methods to reduce carbon dioxide emissions is to enhance the fuel efficiency of engines. The three-way catalyst, although very successful at cleaning up automotive exhaust, dictates that engines operate at air-to-fuel (A/F) ratios around 14.7 : 1. Unfortunately, this is not the optimum ratio with respect to fuel efficiency, which is substantially higher under lean-burn conditions at A/F ratios of about 20 : 1, where the exhaust becomes rich in oxygen and NO$_x$ reduction is extremely difficult (Figure 10.1).

A dream reaction that would resolve the problem immediately is the direct decomposition of NO into N$_2$ and O$_2$. This process is strongly exothermic, that is, $\Delta H^0_{NO} = -91$ kJ/mol, and the equilibrium constant strongly favors the decomposition of NO into molecular oxygen and nitrogen at low temperatures. If this reaction were feasible, cooling the gas down to 500 °C would be sufficient to lower the equilibrium pressure of NO to approximately 1 ppm in a typical exhaust gas

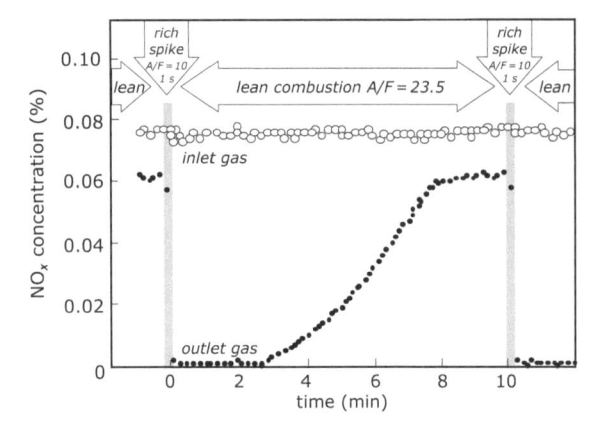

Figure 10.10 Principle of operation of NO$_x$ storage catalyst. During lean combustion, NO is oxidized to NO$_2$ and stored by BaO as barium nitrates. Once the getter is saturated, a short rich excursion of the air–fuel (A/F) mixture reduces the nitrates and the cycle starts anew. Note that in an operating system, the cycle time from lean to rich conditions will be much shorter than indicated in this figure.

containing, for example, 80% N$_2$ and 2% O$_2$. There are literature reports that Cu-zeolites can decompose NO directly to molecular oxygen and nitrogen, but unfortunately the zeolite is not stable under humid conditions [16]. Thus, a suitable catalyst for this reaction has not yet been found. Metals are unlikely candidates as a surface that has sufficient activity to break the NO bond also has a high affinity for O atoms. For example, in a TPD experiment of NO on rhodium, the NO dissociates readily, the N$_2$ desorbs at reasonable temperatures, between 500 and 800 K (depending on the surface structure), but the O$_2$ does not desorb until well above 1000 K, which falls outside the normal operating window of the three-way catalyst.

The NO$_x$ storage-reduction (NSR) catalyst, developed by Toyota and other companies, offers a solution based on a two-step process, in which the engine switches periodically between a long lean-burn stage and a very short fuel-rich stage [17]. The NSR catalyst combines the oxidation activity of platinum with a NO$_x$ storage compound based on barium oxide. Figure 10.10 illustrates the principle of operation.

In the lean-burn stage, all exhaust components are oxidized by the platinum particles in the catalyst. In particular, NO is oxidized to NO$_2$. The latter reacts with BaO getter to form Ba(NO$_3$)$_2$. In the rich mode, which only lasts seconds, the exhaust stream is deficient in oxygen, and reducing components such as CO, H$_2$, and hydrocarbons are present that reduce the barium nitrate and react to N$_2$, CO$_2$, and H$_2$O.

Barium oxide is not a catalyst; all reactions involving this component are entirely stoichiometric. Nevertheless, as Figure 10.10 illustrates, even when the barium storage function is saturated, the NO$_x$ content in the outlet gas from the catalyst is lower than in the inlet, owing to the capability of platinum to reduce NO_x by hydrocarbons in oxygen-rich exhausts.

Sulfur sensitivity is the major difficulty limiting the general application of the NSR catalyst because SO_2 reacts with BaO in a similar way as NO_2 does. However, barium sulfate ($BaSO_4$) is more stable than barium nitrate and hence SO_2 irreversibly deactivates the NSR catalyst. Interestingly, high H_2 levels in the exhaust improve the catalyst's resistance to sulfur. To increase the hydrogen content of the exhaust, NSR catalysts contain rhodium in combination with ZrO_2, which offers sulfur-resistant steam-reforming activity, enabling H_2O and hydrocarbons to form H_2 and CO_2.

At present, the NSR concept is only applicable in markets where low-sulfur fuels (30 ppm S or less) are available, as in Japan and Sweden. The first NSR catalyst was applied by Toyota in 1994 and through the end of 2000 about 300 000 cars in Japan had been equipped with it. As sulfur specifications of fuels will be tightened in the future, the NSR technology will find wider application, as it allows gasoline-fueled engines to operate under conditions of increased fuel efficiency, implying that CO_2 emissions will be lower.

10.2.3
Concluding Remarks on Automotive Catalysts

The three-way catalyst and the NO_x storage-reduction catalyst represent remarkably successful catalytic technology. The catalysts are unique in that they have to operate under a wide range of conditions, depending on type of use, personal driving style, local climate, etc. This in contrast to the usual situation in industry, where conditions are optimized and kept constant.

The three-way catalyst is 'over designed' to meet specifications after several years of usage. If these specifications become stricter, the TWC (Three-way catalyst) system can be further improved to meet these requirements. For example, the highly demanding Californian standards for ultralow emission vehicles (Table 10.2) can easily be met by a three-way catalyst that has a provision for preheating before start-up. Another elegant method to reduce hydrocarbon emissions during the cold start phase is to pass the exhaust stream through a zeolite before it enters the catalytic converter. At low temperatures hydrocarbons adsorb on the interior surface of the zeolite. As the exhaust stream becomes warmer when the engine heats up, the hydrocarbons gradually desorb from the zeolite and are oxidized in the catalyst. Close coupling of the catalyst to the engine also helps to heat it up faster after a cold start. We refer the reader to the book of Heck and Farrauto for an overview of new developments [2].

Table 10.5 Composition of flue gas from coal-fired power plants [18].

NO (ppm)	NO$_2$ (ppm)	SO$_2$ (ppm)	SO$_3$ (ppm)	H$_2$O (%)	O$_2$ (%)	CO$_2$ (%)	Dust (low) (mg/m^3)	Dust (high) (g/m^3)
400–700	2–5	500–2000	2–20	6–8	4–5	10–12	5–20	10–20

10.3
Air Pollution by Large Stationary Sources

10.3.1
Selective Catalytic Reduction: The SCR Process

Large stationary installations for generating power constitute another source of potentially harmful atmospheric emissions. Primary technology may prevent the emission of sulfur if oil is used as the fuel, but the NO$_x$ problem persists due to the combustion at high temperatures. Coal, which is often used in power plants, has sulfur as a major contaminant. Power plants are operated under lean conditions, implying that the flue gas contains an excess of oxygen. As discussed above, technology exists to clean the exhaust from lean-burning engines in cars by applying oscillations with a brief "rich" mode in between lean modes. Such technology is not feasible though in huge plants that produce 500–1000 MW electricity. Hence, a catalytic process that runs, in principle, under steady state conditions is called for, although it needs to be able to respond to changes in flue gas composition.

Table 10.5 shows typical flue gas compositions from coal-fired boilers. From the environmental point of view, SO$_2$, which will be oxidized to SO$_3$ in the air and contribute to acid rain, is the most serious concern. Acid rain was a major problem until the 1960s and 1970s. Nowadays, catalytic technology for oxidizing SO$_2$ in flue gas to SO$_3$ (this process involves V$_2$O$_5$ in a melt) and then hydrating it to H$_2$SO$_4$, has largely eliminated the SO$_2$ emissions and turned coal-fired power plants into manufacturers of sulfuric acid.

NO, however, can only be removed by adding a reductant, ammonia, and using a catalyst. The process is called selective catalytic reduction, or SCR [19–21]. The catalyst consists of vanadia and titania and works in the temperature interval 600–700 K according to the overall reaction

$$4NH_3 + 4NO + O_2 \longrightarrow 4N_2 + 6H_2O \tag{10.9}$$

The ammonia is either injected as pure ammonia under pressure or in an aqueous solution at atmospheric pressure. Instead of ammonia, urea can be used. The challenge of the process is to efficiently remove as much NO$_x$ as possible at full conversion of the reductant, as emission of NH$_3$ from the SCR reactor would of course be highly undesirable.

The combined approach of removing both the sulfur and the NO$_x$ from the flue gas is called SNOX (Haldor Topsøe A/S) or DESONOX (Degussa). An example

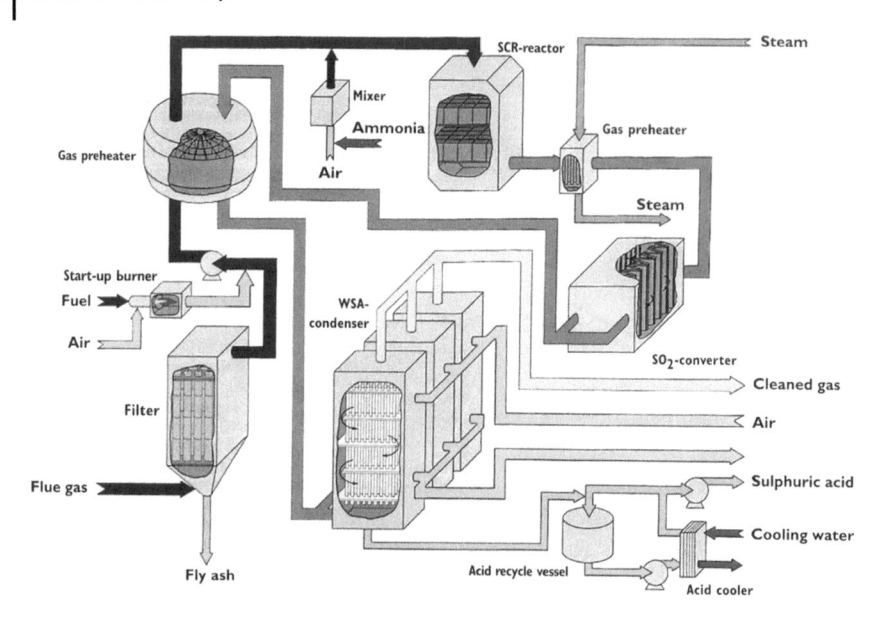

Figure 10.11 Schematic diagram of the SNOX process used to remove both SO$_2$ and NO$_x$ from the flue gas (courtesy of Topsøe A/S).

of the setup for this process is shown in Figure 10.11, where 99% of the NO$_x$ is converted in the SCR reactor and the SO$_2$ is converted into sulfuric acid.

10.3.1.1 The Catalyst for the SCR Process

As with the automotive exhaust converter, the SCR catalyst is designed to handle large flows of gas (e.g., 300 N m^3 s^{-1} for a 300 MW power plant) without causing significant pressure drop. Figure 10.12 shows a reactor arrangement with about 250 m^3 of catalyst in monolithic form, sufficient for a 300 MW power plant.

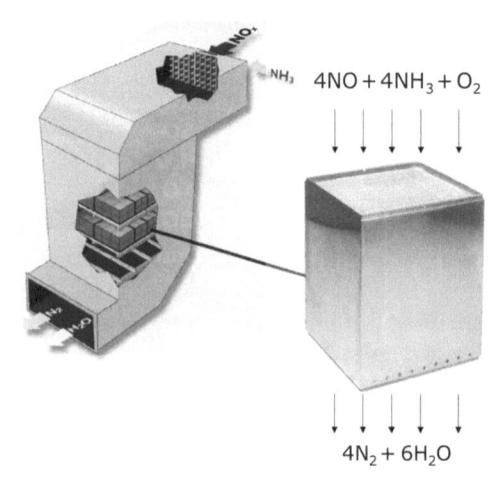

$$4NO + 4NH_3 + O_2$$
$$4N_2 + 6H_2O$$

Figure 10.12 An SCR (selective catalytic reduction) reactor is loaded with monolith assemblies, one of which is shown to the right. Each unit is about 40 × 40 × 50 cm^3. More than 3000 units are required for a 300 MW power plant (courtesy Topsøe A/S).

Table 10.6 Composition of various fuels used in power plants. The Orimulsion is a very heavy fuel, which is retrieved by pumping steam down in the oil-field, thereby giving a mixture of heavy oil and water, which is only fluid at elevated temperature.

	Unit	Orimulsion™	Coal	Heavy fuel oil
Calorific value	MJ/kg	27.5	25	40
Water content	%	30	10	0
Ash	%	0.08	13	0.05
Sulfur	%	2.7	1	2.5
Carbon	%	60	64	86
Hydrogen	%	7.3	4	11
Nitrogen	%	0.5	1	0.4
Vanadium	ppm	300	1–30	50
Nickel	ppm	65	0–10	15
Magnesium	ppm	20	No data	No data

A range of different monolith-catalyst combinations exists to cope with the various sorts of fuel that can be used in a power plant, such as oil, coal, or biomass. Dust, which is a particular problem, is filtered out of the flue gas by electrostatic precipitators either before (low dust operation) or after the SCR reactor (high dust operation).

The catalytically active material on the monolith also comes in many variations. Formulations based on iron, chromium, and vanadium as the active components supported on TiO_2, Al_2O_3, SiO_2, and zeolites have been reported; see the review by Bosch and Janssen [21].

The commonly used catalyst today is a vanadia on a titania support, which is resistant to the high SO_2 content. Usually the titania is in the anatase form since it is easier to produce with large surface areas than the rutile form. Several poisons for the catalyst exist, for example, arsenic and potassium. The latter is a major problem with biomass fuel. In particular, straw, a byproduct from grain productions, seems to be an attractive biomass but contains potassium, which is very mobile at reaction temperatures and tends to condense at places where the gas cools, in particular in the SCR catalyst, where it may block the channels of the monoliths, thereby reducing its effective life time to several months only. How to deal with such poisons is a challenging problem for the future use of biomass in power plants. Schemes for regenerating and protecting the SCR catalyst are under development.

Table 10.6 shows some of the major components of fuels that are used in power plants. The coal and heavy fuel are the conventional fuels for power plants, while the Orimulsion is a relative new product from Venezuela, which is attractive owing to a higher hydrogen content that leads to reduced emission of CO_2.

The vanadium content of some of these fuels presents an interesting problem. When the vanadium leaves the burner, it may condense on the surface of the heat

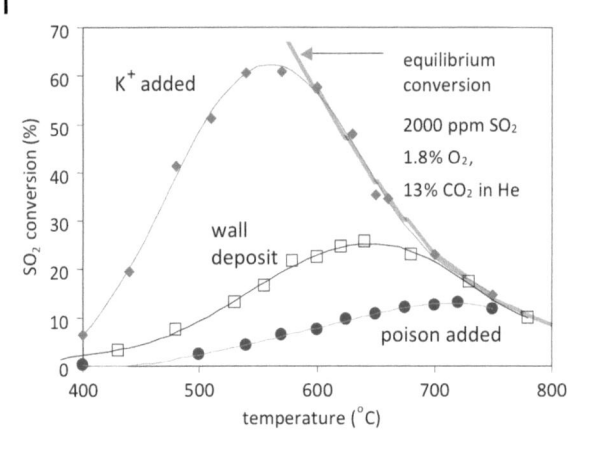

Figure 10.13 Vanadia wall deposits in a power plant firing Orimulsion fuel catalyze the premature oxidation of SO_2 in heat exchangers. Note that potassium enhances the undesired conversion, while a selective poison diminishes the effect to some extent. Nevertheless, in all cases the conversion goes to equilibrium. As SO_2 oxidation is exothermic, the equilibrium concentration decreases rapidly with increasing temperature (adapted from [22]).

exchanger in the power plant. As vanadia is a good catalyst for oxidizing SO_2, this reaction may occur prior to the SCR reactor. This is clearly seen in Figure 10.13 which shows the SO_2 conversion by wall deposits in a power plant that has used the vanadium-containing Orimulsion as a fuel. The presence of potassium actually increases this premature oxidation of SO_2. The problem arises when ammonia is added, since SO_3 and NH_3 react to form ammonium sulfate, which condenses and gives rise to deposits that block the monoliths. Note that ammonium sulfate formation also becomes a problem when ammonia slips through the SCR reactor and reacts downstream with SO_3.

10.3.1.2 SCR Reaction Kinetics

Understanding the kinetics of the SCR process helps greatly in developing new and better catalysts. For efficient operation, one of the important issues is to maximize NO_x conversion while avoiding significant slip of the injected ammonia. Figure 10.14 shows the NO_x conversion as a function of the NH_3/NO_x ratio for two different space velocities, along with ammonia slip. Keeping the NH_3 slip below 10 ppm dictates the space velocity, or rather the amount of catalyst that is needed.

For practical purposes, the reaction kinetics are described by a power rate law with reaction orders between 0.5 and 1.0 for ammonia and −0.1 to 1.0 for NO_x. Although such rate laws represent a useful parameterization for industrial use, a microkinetic model provides a much better basis for detailed fundamental insight in the reaction.

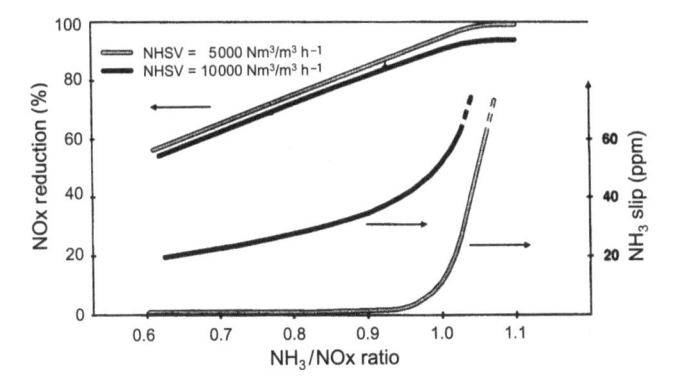

Figure 10.14 NO$_x$ reduction and ammonia slip during SCR as a function of the NH$_3$/NO$_x$ ratio for two different space velocities. The maximum admissible level of ammonia is 10 ppm, which dictates the space velocity, or rather the amount of catalyst needed (data from Dumesic et al. [23] and Topsøe et al. [24]).

Dumesic et al. [23] proposed a model involving six steps based on the general Mars van Krevelen mechanism for oxidations:

1) $$NH_3 + V^{5+}-OH \rightleftharpoons V-ONH_4 \qquad (10.10)$$
2) $$V-ONH_4* + V=O \rightleftharpoons V-ONH_3-V^{4+}-OH \qquad (10.11)$$
3) $$NO + V-NH_3-V^{4+}-OH \longrightarrow N_2 + H_2O + V^{5+}-OH \qquad (10.12)$$
4) $$2V^{4+}-OH \rightleftharpoons H_2O + V^{3+} + V=O \qquad (10.13)$$
5) $$O_2 + 2V^{3+} \longrightarrow 2V=O \qquad (10.14)$$
6) $$H_2O + V^{5+}-OH \rightleftharpoons V^{5+}-OH_3O \qquad (10.15)$$

Step 1 represents adsorption of ammonia and step 2 its activation. The irreversible step 3 is obviously not elementary in nature, but unfortunately much information on the level of elementary steps is not available. Step 4 describes water formation and step 5 is the reoxidation of the V^{3+} site. Step 6 describes the blocking of sites by adsorption of water. The model, thus, relies on partially oxidized sites and vacancies on an oxide, similarly to the hydrodesulfurization reaction described in Chapter 9. The reactions can be summarized in the cyclic scheme of Figure 10.15.

The SCR catalyst is considerable more complex than, for example, the metal catalysts we discussed earlier. Also, it is very difficult to perform surface science studies on these oxide surfaces. The nature of the active sites in the SCR catalyst has been probed by temperature-programmed desorption of NO and NH$_3$ and by *in situ* infrared studies. This led to a set of kinetic parameters (Table 10.7) that can describe NO conversion and NH$_3$ slip (Figure 10.16). The model gives a good fit to the experimental data over a wide range, is based on the physical reality of the SCR catalyst and its interactions with the reacting gases, and is, therefore, preferable over a simple power rate law in which the catalysis happens in a "black

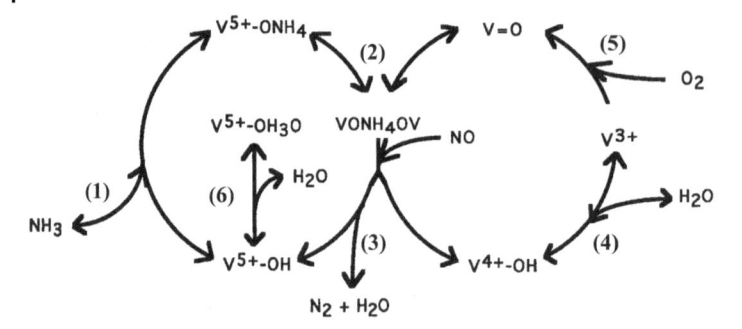

Figure 10.15 Catalytic cycle for the SCR process over acidic V^{5+}–OH sites and redox V=O sites (reproduced from Dumesic et al. [23] with permission from Elsevier).

box." Nevertheless, several questions remain unanswered, such as what are the elementary steps and what do the active sites look like on the atomic scale.

The SCR process would be even more attractive if ammonia could be avoided. Numerous investigations have been performed using more easily handed hydrocarbons, but no process has yet been found that can compete with the ammonia (or urea) in the SCR process.

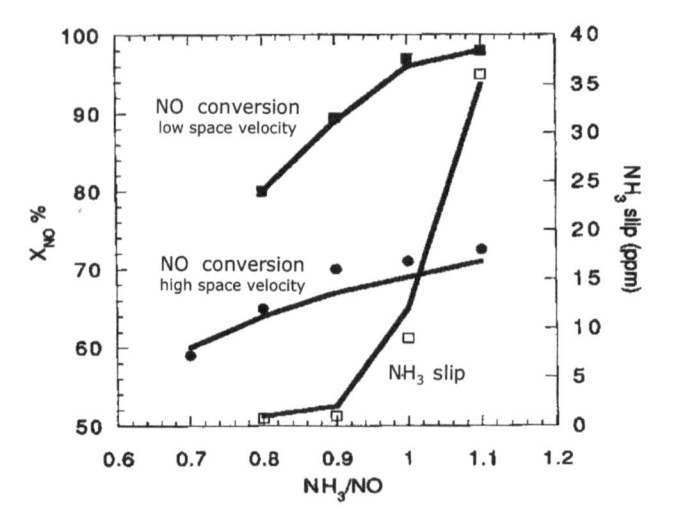

Figure 10.16 NO conversion and ammonia slip as a function of the NH_3/NO ratio in the presence of O_2 and H_2O over a V_2O_3/TiO_2 catalyst at 623 K. The lines represent the model based on reactions (10.10)–(10.15) and the parameters in Table 10.7 (reproduced from Dumesic et al. [23] with permission from Elsevier).

Table 10.7 Kinetic parameters for the reactions (10.10)–(10.15) as used in Figure 10.16. Prefactors are in s^{-1} for surface reactions and $s^{-1}\,bar^{-1}$ for steps involving gaseous species.

Step	k_i^+ (s^{-1} or $s^{-1}\,bar^{-1}$)	E_i^+ (kJ mol^{-1})	k_i^- (s^{-1} or $s^{-1}\,bar^{-1}$)	E_i^- (kJ mol^{-1})
1	8×10^6	0	1×10^{13}	84
2	1×10^{11}	91	1×10^{11}	133
3	1.3	23.1	–	–
4	1×10^{11}	68.5	1.9×10^4	0
5	8×10^2	0	–	–
6	8×10^7	0	1×10^{13}	69.3

10.3.2
The SCR Process for Mobile Units

The three-way catalyst discussed earlier in this chapter is suitable for gasoline powered cars, but does not work for diesels, as these always work under lean (oxygen-rich) conditions. SCR technology has no problem with oxygen-rich exhausts because the reducing agent needed for the NO_x is added to the exhaust. In principle, SCR technology is used on mobile NO_x producers, such as ships, and in particular, ferries, and on trucks. For safety reasons ammonia is less acceptable, but urea, which is readily converted into ammonia and CO_2, presents a safe alternative. Figure 10.17 illustrates the principle of operation.

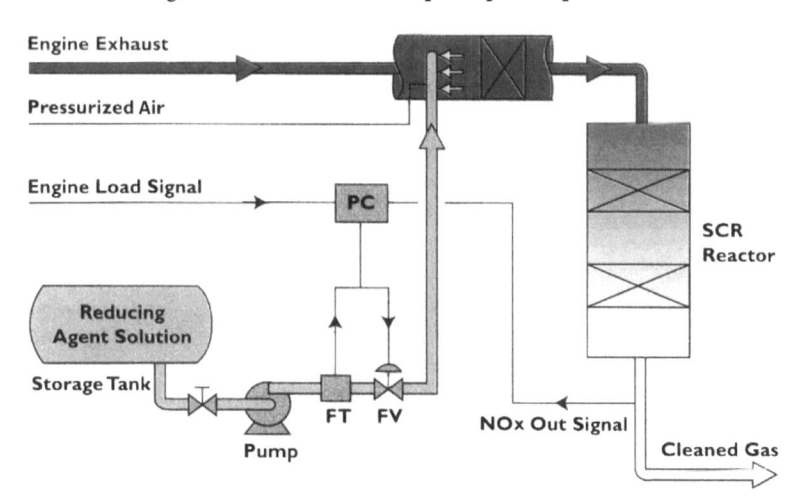

Figure 10.17 Principle of selective catalytic reduction using, for example, urea or a solution of ammonia urea as the reduction agent for application of the SCR reaction on mobile diesel units such as ferries or trucks (with courtesy of Topsøe A/S).

The next step is to scale down this technology further such that it can be implemented more widely on heavy duty trucks. This is an area that is currently undergoing strong development, in striving to comply with the increasing demands for cleaner exhausts and more stringent protection of the environment.

References

1 Taylor, K.C. (1993) Nitric-oxide catalysis in automotive exhaust systems. *Catalysis Reviews – Science and Engineering*, **35**, 457–481.
2 Heck, R.M. and Farrauto, R.J. (2009) *Catalytic Air Pollution Control – Commercial Technology*, John Wiley & Sons, Inc., Hoboken.
3 Gandhi, H.S., Graham, G.W., and McCabe, R.W. (2003) Automotive exhaust catalysis. *Journal of Catalysis*, **216**, 433–442.
4 Mullins, D.R. (2015) The surface chemistry of cerium oxide. *Surface Science Reports*, **70**, 42–85.
5 Trovarelli, A. (1996) Catalytic properties of ceria and CeO_2-containing materials. *Catalysis Reviews – Science and Engineering*, **38**, 439–520.
6 Nieuwenhuys, B.E. (1999) The surface science approach toward understanding automotive exhaust conversion catalysis at the atomic level, in *Advances in Catalysis* (eds W.O. Haag, B.C. Gates, and H. Knözinger), Vol. 44, pp. 259–328.
7 Engel, T. and Ertl, G. (1978) Molecular-beam investigation of catalytic-oxidation of CO on Pd(111). *Journal of Chemical Physics*, **69**, 1267–1281.
8 Bowker, M., Guo, Q.M., Li, Y.X., and Joyner, R.W. (1993) Structure sensitivity in CO oxidation over rhodium. *Catalysis Letters*, **18**, 119–123.
9 Hopstaken, M.J.P. and Niemantsverdriet, J.W. (2000) Structure sensitivity in the CO oxidation on rhodium: Effect of adsorbate coverages on oxidation kinetics on Rh(100) and Rh(111). *Journal of Chemical Physics*, **113**, 5457–5465.
10 Hopstaken, M.J.P. and Niemantsverdriet, J.W. (2000) Reaction between NO and CO on rhodium(100): How lateral interactions lead to auto-accelerating kinetics. *Journal of Vacuum Science & Technology A – Vacuum Surfaces and Films*, **18**, 1503–1508.
11 Borg, H.J., Reijerse, J., van Santen, R.A., and Niemantsverdriet, J.W. (1994) The dissociation kinetics of NO on Rh(111) as studied by temperature-programmed static secondary-ion mass-spectrometry and desorption. *Journal of Chemical Physics*, **101**, 10052–10063.
12 Hopstaken, M.J.P. and Niemantsverdriet, J.W. (2000) Lateral interactions in the dissociation kinetics of NO on Rh(100). *Journal of Physical Chemistry B*, **104**, 3058–3066.
13 Peden, C.H.F., Belton, D.N., and Schmieg, S.J. (1995) Structure sensitive selectivity of the NO–CO reaction over Rh(110) and Rh(111). *Journal of Catalysis*, **155**, 204–218.
14 Oh, S.H. (1990) Effects of cerium addition on the CO-NO reaction-kinetics over alumina-supported rhodium catalysts. *Journal of Catalysis*, **124**, 477–487.
15 Zhdanov, V.P. and Kasemo, B. (1997) Mechanism and kinetics of the NO–CO reaction on Rh. *Surface Science Reports*, **29**, 35–90.
16 Centi, G. and Perathoner, S. (1995) Nature of active species in copper-based catalysts and their chemistry of transformation of nitrogen-oxides. *Applied Catalysis A – General*, **132**, 179–259.
17 Matsumoto, S., Ikeda, Y., Suzuki, H., Ogai, M., and Miyoshi, N. (2000) NO_x storage reduction catalyst for automotive exhaust with improved tolerance against sulfur poisoning. *Applied Catalysis B – Environmental*, **25**, 115–124.
18 Topsøe, N.Y. (1997) Catalysis for NO_x abatement: Selective catalyic reduction of NO_x by ammonia: Fundamental and industrial aspects. *CaTTech*, **1**, 125–134.
19 Epling, W.S., Campbell, L.E., Yezerets, A., Currier, N.W., and Parks, J.E.

(2004) Overview of the fundamental reactions and degradation mechanisms of NO_x storage/reduction catalysts. *Catalysis Reviews – Science and Engineering*, **46**, 163–245.

20 Roy, S. and Baiker, A. (2009) NO_x Storage-Reduction Catalysis: From mechanism and materials properties to storage-reduction performance. *Chemical Reviews*, **109**, 4054–4091.

21 Bosch, H. and Janssen, F.J.J.G. (1988) Catalytic reduction of nitric oxides – a review of the fundamentals and technology. *Calatlysis Today*, **2**, 369–532.

22 Rasmussen, S.B., Hagen, S.U., Masters, S.G., Hagen, A., Stahl, K., Eriksen, K.M., Simonsen, P., Jensen, J.N., Berg, M., Fehrmann, R., and Chorkendorff, I. (2003) Catalytic and chemical properties of boiler deposits from Orimulsion fuel, *PowerPlant Chemistry*, **5**, 360–369.

23 Dumesic, J.A., Topsøe, N.Y., Topsøe, H., Chen, Y., and Slabiak, T. (1996) Kinetics of selective catalytic reduction of nitric oxide by ammonia over vanadia/titania. *Journal of Catalysis*, **163**, 409–417.

24 Topsøe, N.Y. (1994) Mechanism of the selective catalytic reduction of nitric-oxide by ammonia elucidated by in-situ online Fourier-transform infrared-spectroscopy. *Science*, **265**, 1217–1219.

Appendix

Some Useful Fundamental Constants

Avogadro number	N_A	6.022×10^{23} mol^{-1}
Planck's constant	h	6.626×10^{-34} J s
$h/(2\pi)$	$\hbar$	1.055×10^{-34} J s
Velocity of light	c	2.998×10^{8} m s^{-1}
Elementary charge	e	1.602×10^{-19} C
Atomic mass unit	amu	1.661×10^{-27} kg
Electron mass	m_e	9.110×10^{-31} kg
Proton mass	m_p	1.673×10^{-27} kg
Neutron mass	m_n	1.675×10^{-27} kg
Faraday's number	F	9.649×10^{4} C mol^{-1}
Boltzmann's constant	k_B	1.381×10^{-23} J K^{-1}
Gas constant	R	8.314 J mol^{-1} K^{-1}
Vacuum permeability	μ_0	1.257×10^{-6} V s^2 C^{-1} m^{-1}
Vacuum permittivity	ε_0	8.854×10^{-12} C V^{-1} m^{-1}
Gravitational constant	G	6.674×10^{-11} m^3 kg^{-1} s^{-2}

Concepts of Modern Catalysis and Kinetics, Third Edition. I. Chorkendorff and J.W. Niemantsverdriet.
© 2017 WILEY-VCH Verlag GmbH & Co. KGaA. Published 2017 by WILEY-VCH Verlag GmbH & Co. KGaA.

Energy Conversion Factors

	kJ	kJ mol^{-1}	eV	cm^{-1}	K	a. u.	erg
kJ	1	6.022×10^{23}	6.242×10^{21}	8.064×10^{3}	7.241×10^{25}	2.294×10^{20}	1.000×10^{10}
kJ mol^{-1}	1.661×10^{-24}	1	1.037×10^{-2}	1.339×10^{-20}	1.203×10^{2}	3.795×10^{-4}	1.661×10^{-14}
eV	1.602×10^{-22}	9.647×10^{1}	1	8.065×10^{3}	1.160×10^{4}	3.675×10^{4}	1.602×10^{-12}
cm^{-1}	1.240×10^{-4}	7.467×10^{19}	1.240×10^{-4}	1	1.439×10^{0}	4.556×10^{-6}	1.986×10^{-16}
K	1.381×10^{-26}	8.314×10^{-3}	8.617×10^{-5}	6.950×10^{-1}	1	3.116×10^{-6}	1.380×10^{-16}
a. u.	4.359×10^{-21}	2.635×10^{3}	2.721×10^{1}	2.195×10^{5}	3.158×10^{5}	1	4.360×10^{-11}
erg	1.000×10^{-10}	6.022×10^{13}	6.242×10^{11}	5.035×10^{15}	7.244×10^{15}	2.294×10^{10}	1

Other Useful Relations

$$1 \text{ bar} = 760 \text{ mmHg} = 105\,000 \text{ Pa} = 105\,000 \text{ kg m}^{2}\text{ s}^{-2}$$

$$F \frac{p}{\sqrt{2\pi m k_B T}} = 2.632 \times 10^{29} \frac{p_{[\text{bar}]}}{\sqrt{m_{[\text{amu}]} T_{[K]}}} \text{ m}^{-2}\text{s}^{-1}$$

1 mol at 300 K and 1 bar $\sim$ 25 L

1 meV $\sim$ 8 cm^{-1}

$k_B T$ at 300 K $\sim$ 25 meV = 1/40 eV

Questions and Exercises

Questions

Chapter 1

1.1 How important is the chemical industry for the economy of your country? Try to find its contribution to your country's gross national product.

1.2 Can you indicate how important catalysis is for the chemical industry in your country?

1.3 Try to list the top ten catalytic processes in your country.

1.4 Explain why catalysis is often described as an enabling discipline for chemistry.

1.5 Consult the website of, for example, the American Chemical Society to find the most recent versions of Tables 1.3 and 1.4 and comment on the major differences.

1.6 Give a definition of catalysis.

1.7 What is the essence of catalysis on a molecular level, that is, in terms of bonds in and between reacting molecules and the surface? Where is the essential influence of the catalyst?

1.8 Several textbooks introduce the concept of catalysis with a potential energy diagram in which an energy barrier separates the products and the reactants, and then that a catalyst lowers this barrier. Do you approve of this representation? Explain your answer.

1.9 A catalyst affects the kinetics of a chemical reaction, but not the thermodynamics. Can you explain why this is so? What is the consequence for an equilibrium reaction to which a catalyst is added?

Concepts of Modern Catalysis and Kinetics, Third Edition. I. Chorkendorff and J.W. Niemantsverdriet.
© 2017 WILEY-VCH Verlag GmbH & Co. KGaA. Published 2017 by WILEY-VCH Verlag GmbH & Co. KGaA.

1.10 Explain the differences and similarities between homogeneous and heterogeneous catalysis.

1.11 To which of the two categories in Section 1.3 does biocatalysis belong – or should it be considered a separate category?

1.12 Why are enzymatic catalysts often much more efficient than other catalysts?

1.13 Give some examples of industrially applied biocatalytic processes.

1.14 Explain the concepts of atom efficiency and environmental friendliness.

1.15 What is an E factor? Which types of processes usually have the highest E factors?

1.16 Explain what catalysis means on the different length scales indicated in Figure 1.8.

1.17 Referring to Figure 1.2, sketch a few situations that involve an unsuccessful combination of catalyst and reacting molecules.

1.18 List the most important scientific journals in heterogeneous catalysis.

1.19 Browse through a recent issue of *Journal of Catalysis, ACS Catalysis, Applied Catalysis A, Catalysis Today,* and *Catalysis Letters.* Can you give a short characteristic of each journal in terms of the type of articles they mostly publish?

1.20 Try to list the most important journals where work in homogeneous catalysis is reported. Why do the heterogeneous and homogeneous catalysis communities publish in different types of journals?

1.21 Explain the difference between letters, full papers, reviews, and conference proceedings.

1.22 What is the impact factor of a journal? Is it important?

1.23 What is the citation half-life and what does it tell you about the status of a journal?

1.24 Citation classics are papers that are exceptionally often cited, for example, 100 times or more. Try to identify one or more citation classics in the area of catalysis you are most interested in. How often do you think that the average paper in catalysis will be cited?

1.25 Use an information service such as Web of Science, Scopus, SciFinder, or Google Scholar to make a top 10 of countries that are most productive in publishing articles in catalysis (e.g., for the past five years). Reflect on the outcome and consider if it is according to your expectation.

1.26 Now do the same for the most cited articles in catalysis over the past five years. Discuss the differences with the result from the previous question.

1.27 How would you best explain catalysis in simple terms to friends who know little to nothing about chemistry or chemical engineering?

1.28 Suppose you would want to lobby for catalysis with your favorite politician, for example, to convince him or her that academic research and education in catalysis is a wise long-term investment in the economy of your country. How would you explain to this person what catalysis is, and why it is important?

Chapter 2

2.1 Explain in your own words why kinetics is an important discipline for catalysis.

2.2 How old are catalysis and kinetics as scientific disciplines?

2.3 Can a catalyst change the composition of a gas mixture in equilibrium? Explicitly include the role of kinetics in your consideration.

2.4 Nickel metal successfully catalyzes the hydrogenation of double bonds in unsaturated hydrocarbons such as propylene and butane, but iron does not. Can these metals catalyze the dehydrogenation of alkanes such as propane and butane to their respective alkenes and alkynes?

2.5 How do reaction dynamics and reaction kinetics differ?

2.6 What is the difference between a rate constant and an equilibrium constant?

2.7 What is the meaning of chemical potential and how does it depend on the pressure?

2.8 What are the optimal conditions for an exothermic reaction and why can they not always be fulfilled?

2.9 What can you say about the optimal conditions for conversion for a process where the number of product molecules is less than the number of reactant molecules, while the reaction is (a) endothermic and (b) exothermic?

2.10 Explain the principle of chemical equilibrium and how it relates to enthalpy and entropy. What does entropy reflect and what it is a measure of?

2.11 Give a definition of an elementary step and point out how it differs from an overall reaction.

2.12 What is a power rate law? What do you think is the historical origin of the power rate law?

2.13 How appropriate is the power rate law for describing the kinetics of a catalytic reaction?

2.14 How do we define the reaction order of a given reaction in a certain reactant or product?

2.15 What is the fugacity of a gas and when is it used?

2.16 Explain the term apparent activation energy and discuss how well an apparent activation energy satisfies the Arrhenius equation.

2.17 What is the difference between a differential and an integral rate equation?

2.18 For what sort of industrial processes can the kinetics be described by the steady state approximation?

2.19 Are situations conceivable in which the steady state approximation can be applied to the kinetics of a batch reaction?

2.20 Explain the relation between the reaction order of a certain component in a catalytic reaction and the surface coverage of this component.

2.21 What is the mean-field approximation in the kinetics of catalytic reactions, and when does it break down? In such cases, is the rate larger or smaller than expected on basis of the mean-field approximation?

2.22 Derive the Langmuir adsorption isotherm for molecular adsorption of CO on a metal with equivalent adsorption sites. Do the same for dissociative adsorption of H_2, for dissociative adsorption of CO, and, finally, for the case when CO and H_2 adsorb together on the same surface (CO molecularly and H_2 dissociatively).

2.23 What is the essential difference between Langmuir–Hinshelwood and Eley–Rideal mechanisms? Which of the two is the more likely mechanism? Can you give an example for each type where you believe it would be the most likely mechanism? Justify your choices, please.

2.24 Can a catalytic reaction be an elementary reaction?

2.25 In solving the kinetics of a catalytic reaction, what is the difference between the complete solution, the steady state approximation, and the quasi-equilibrium approximation? What is the MARI (most abundant reaction intermediate species) approximation?

2.26 Discuss the meaning of the concept "order of reaction" for a catalytic reaction (hint: compare a catalytic reaction to a homogeneous gas phase reaction and discuss the principle of reaction order for each case).

2.27 Can a reaction order of a catalytic reaction depend on pressure and temperature?

2.28 Derive Equation (2.47) for the activation energy, by starting either from the Arrhenius law, or from the concept of an Arrhenius plot.

2.29 Explain the concept of apparent activation energy for a catalytic reaction.

2.30 Derive the apparent activation energy for a system consisting of a pre-equilibrium followed by a rate determining step (i.e., for which the rate expression has the form $r = kKp_1p_2$).

2.31 Does an apparent activation energy depend on pressure and temperature?

2.32 Why can reaction orders in catalytic reactions change sign when the reaction conditions change?

2.33 Can the activation energy of a catalytic reaction have a negative value? If so, under what conditions?

2.34 Why does the quasi-equilibrium approximation fail in the low-pressure limit of a reactant?

2.35 Under what conditions may reactions start to oscillate? Give some examples of oscillating reactions.

2.36 Compare the Michalis–Menten expression for the rate of an enzyme-catalyzed reaction with the Langmuir–Hinshelwood expression for the same reaction on a metal surface. Are the two expressions equivalent?

2.37 Suppose two mechanisms have been proposed for a certain catalytic reaction. Discuss to what extent it is possible to prove that the mechanisms are right or wrong on the basis of a kinetic analysis.

2.38 In question 2.37, to what extent could the use of isotopically labeled components help?

Chapter 3

3.1 What is the aim of reaction rate theory? Do you think that the theory is of practical use for scientists working in catalysis? Justify your answer.

3.2 The formal definition of a partition function implies that it is a summation over an infinitely high number of terms. Explain why the partition function is, nevertheless, a useful quantity.

3.3 What is the minimum value a partition function can assume? Why? And the maximum?

3.4 Discuss the basic assumption underlying the Boltzmann distribution of energies for an ensemble of molecules.

3.5 How many degrees of freedom does a molecule consisting of N atoms possess?

3.6 Write down the complete partition function for a two-atomic heteronuclear molecule such as CO in the gas phase.

3.7 Which requirements must be fulfilled to write the partition function as a product of the different degrees of freedom?

3.8 Write down the partition function for an ensemble consisting of N molecules of CO within a gas-phase volume V.

3.9 Under what conditions are the partition functions for translation, rotation, and vibration of an adsorbed molecule (a) close to unity, (b) moderate, and (c) large?

3.10 Why can ammonia not be synthesized from N_2 and H_2 by heating the gas to temperatures where N radicals are present?

3.11 Explain the concepts of collision theory. For what type of reactions do you expect collision theory to be valid?

3.12 How many collisions occur, roughly, in a liter of gas at atmospheric pressure, and what fraction of these collisions will normally give rise to reaction (assuming commonly applied reaction temperatures and barrier energies)?

3.13 Why does the collision theory of reaction rates conflict with equilibrium thermodynamics?

3.14 Give an example of a monomolecular elementary reaction (a unimolecular reaction). How does a molecule in a unimolecular reaction acquire sufficient energy to overcome the barrier energy for reaction?

3.15 Why can unimolecular reactions exhibit second-order kinetics at low pressures?

3.16 Give a concise description of transition state theory. How can the necessary parameters to make a quantitative prediction of reaction rate be obtained?

3.17 Sketch plausible transition states for (a) the dissociation of a molecule in the gas phase, (b) the reaction cyclopropane to give propene, (c) the isomerization of CH_3CN to CH_3NC, (d) the desorption of an atom from a surface, and (e) the dissociation of an adsorbed molecule such as CO on a metal surface.

3.18 What is a tight or a loose transition state? How can one infer the nature of a transition state from the value of the pre-exponential factor?

3.19 Explain the concept of a sticking coefficient within the transition state theory. Why is the sticking coefficient always smaller than unity for a direct adsorption process?

3.20 On the basis of entropy changes, why is the direct adsorption of a molecule on a surface site less probable than indirect adsorption through a precursor state?

3.21 Suggest a transition state for the desorption of a molecule when the pre-exponential factor is 10^{16} s^{-1}. Do the same for a desorption when the prefactor is 10^{13} s^{-1}.

3.22 Discuss the validity and usefulness of the Arrhenius equation in terms of your knowledge of transition state theory.

3.23 What are the fundamental differences between collision theory and transition state theory?

Chapter 4

4.1 Sketch and describe the morphology (or some different possibilities if you wish) of a typical heterogeneous catalyst on a high-surface area support.

4.2 What are the aims of catalyst characterization in the context of (a) industrial catalysis and (b) fundamental research?

4.3 For each (or a selection) of the following techniques, give a concise description of the principles, the sort of information it yields about a supported catalyst, and a brief assessment of strengths and weaknesses. The use of clear diagrams and schemes is highly recommended.

a) X-ray diffraction
b) X-ray photoelectron spectroscopy
c) Infrared spectroscopy
d) Temperature-programmed reduction and oxidation
e) Temperature-programmed desorption
f) Transmission electron microscopy
g) Extended X-ray absorption fine structure
h) Raman spectroscopy
i) Mössbauer spectroscopy
j) Ion scattering spectroscopy or low-energy ion scattering
k) Secondary ion mass spectrometry.

4.4 Why are electron and ion spectroscopies generally surface-sensitive techniques when applied in the low-energy regime.

4.5 When a high-energy electron beam in an electron microscope hits a sample, a wealth of information becomes available through a number of scattering, diffraction, and decay processes (Fig. 4.15). Indicate how these may be used to obtain additional information about the sample.

4.6 Which of the techniques listed in question 4.3 requires a vacuum as a measurement environment, and why? Which can be used under (simulated) reaction conditions?

4.7 Why are typical surface science techniques such as low-energy electron diffraction, scanning tunneling, and atomic force microscopy generally unsuitable for studying supported catalysts?

4.8 Explain the principles of low-energy electron diffraction and compare the technique to X-ray diffraction.

4.9 What is the significance of a "reciprocal lattice?"

4.10 Explain the principles of the scanning probe microscopies STM and AFM, and discuss the type of information these techniques provide. What are the major differences between the two?

4.11 Discuss strategies for devising model systems of catalysts that allow surface science methods to be applied in catalysis research.

4.12 Propose a strategy (or strategies) for bridging the gap between the world of adsorption and reaction on well-defined single-crystal surfaces and the world of supported catalysts in high-pressure reactors.

4.13 How many of the characterization techniques listed in Figure 4.3 fit into the general scheme of Figure 4.2?

4.14 Which of the techniques in Figure 4.3 is suitable for *in situ* characterization of catalysts?

4.15 If you were to start a research group on a heterogeneous catalysis project to be specified by you, and you were allowed to choose four techniques from the list of Figure 4.3, which would you choose? Explain your answer.

Chapter 5

5.1 What are the major requirements of a solid catalyst that is to be applied in an industrial process?

5.2 Draw the simple (111), (100), and (110) surfaces of the face-centered cubic metals, the (110) and (100) surfaces of body-centered cubic metals, and the (001) surface of a hexagonally-closed packed surface. Try to list them in order of increasing surface free energy, supposing that the metal–metal bonds in the interior are the same for the metals.

5.3 What is the area of the bcc (100) unit cell in terms of the lattice constant a? What is the area of the fcc (111) unit cell?

5.4 Why is the fcc (100) surface more reactive than the fcc (111) surface?

5.5 Draw the following structures:

a) An fcc (110) surface with a (2 × 1) adsorbate overlayer.
b) An fcc (111) surface with a (2 × 2) adsorbate overlayer.
c) The same as (b), but with 3 adsorbate molecules per (2 × 2) super cell.
d) An fcc (111) surface with a ($\sqrt{3} \times \sqrt{3}$)$R30°$ adsorbate overlayer.

5.6 Discuss the most important differences between the surfaces of metals and those of oxides.

5.7 Explain the concepts of Lewis and Brønsted acidity. What are the consequences for adsorption on surfaces with Lewis acidity?

5.8 Discuss the role of the surface free-energy in phenomena such as alloy segregation, surface reconstruction, faceting, and sintering of small particles.

5.9 What can you say about the optimal conditions for conversion for a process where the number of product molecules is less than the number of reactant molecules, while the reaction is (a) endothermic and (b) exothermic? And which of the requirements for a successful catalyst would you expect to be dominant in cases (a) and (b)?

5.10 How does a gas environment (air, vacuum) affect the surface composition of an alloy?

5.11 When we deposit a small amount of iron atoms on a nickel crystal, and allow the system to equilibrate, what will happen with the iron atoms? What happens with Ag on Ru, Ag on Ni, and Co on Cu?

5.12 What determines the shape of a metal particle in vacuum? What determines the shape of a metal particle on a substrate?

5.13 Compare the morphology of a small metal particle with a diameter of a few nanometers with that of a larger particle of – say – 15 nm. What do you expect for the types of sites present at the surface and their relative abundances?

5.14 Give the essential assumptions made in deriving the BET isotherm. How does the BET isotherm differ from the Langmuir isotherm?

5.15 What information can be obtained from a BET adsorption isotherm?

5.16 Refer to catalogues or websites of catalyst support producers and compile an overview of commercially available support materials and the range of surface areas per gram in which these are available.

5.17 Propose a simple recipe for preparing the reduced form of a 2 wt% Pd on SiO_2 catalyst, using a support with a surface area of 200 $m^2\ g^{-1}$ and a pore volume of 0.5 mL g^{-1}. For what type of reaction(s) can this catalyst be used?

5.18 Discuss whether the catalyst obtained under question 5.17 would be suitable for a high temperature oxidation process, and how it would have to be modified to be optimal for such an application?

5.19 How can the active metal area of a supported catalyst be measured and distinguished from the total surface area?

5.20 Propose a way to apply titanium oxide in very highly dispersed form onto an alumina support.

5.21 What is a zeolite? How does it differ from a catalyst support?

5.22 Explain why alumina-containing zeolites possess chemically active surfaces but why an all-silica zeolite does not.

5.23 Discuss the pros and cons of using a zeolite in comparison with a supported catalyst.

5.24 Suppose you wish to use a bifunctional catalyst consisting of a very finely dispersed metal in a zeolite. How can you verify if the metal is inside or (perhaps partially) outside the zeolite framework?

5.25 Why are catalyst powders usually pressed into bodies of particular shapes?

5.26 Explain the concepts behind the Thiele diffusion modulus for a spherical particle. Why is this important for the application of a catalyst?

5.27 Describe qualitatively the consequences of transport limitations on the apparent activation energy of a catalytic process by using an Arrhenius plot. What is the best temperature to run this reaction in an industrial application?

5.28 Can you give examples of situations where the overall rate of a catalytic process is limited by transport of the reactants outside the catalyst particles?

5.29 What is the difference between the intrinsic and extrinsic rate of a catalyst?

5.30 How would you design the architecture of a catalyst to be used under conditions where there are severe limitations on the transport of gases into the catalyst particle?

5.31 What is the most important property to consider when designing a catalyst for an industrial process: (a) the rate per catalytically active site; (b) the rate per unit weight of catalyst; or (c) the rate per unit volume of catalyst? Explain your answer.

5.32 Suppose you prepared an iron oxide catalyst supported on an alumina support. Your aim is to use the catalyst in the metallic form, but you want to keep the iron particles as small as possible, with a degree of reduction of at least 50%. Hence, you need to know the particle size of the iron oxide in the unreduced catalyst, as well as the size of the iron particles and their degree of reduction in the metallic state. Refer to Chapters 4 and 5 to devise a strategy to obtain this information. (Unfortunately for you, it appears that electron microscopy and X-ray diffraction do not provide useful data on the unreduced catalyst.)

Chapter 6

6.1 What is the essential difference between physisorption and chemisorption?

6.2 Why do atoms always feel a strong repulsion when they approach a surface or another atom too closely?

6.3 Explain the origin of the Van der Waals interaction.

6.4 Describe qualitatively what happens when two atom approach each other, (a) when the outer atomic orbitals are partly occupied; (b) when the outer orbitals are entirely filled.

6.5 Why is the splitting between bonding and antibonding molecular orbitals not symmetrical around the atomic levels?

6.6 Explain the relevance of atomic orbital overlap and of molecular orbital filling for the strength of the bond between two atoms.

6.7 Why do s electrons form a broad band in a metal, while d electrons give rise to relatively narrow bands? What do you expect for bands formed from f electrons?

6.8 Why are the d-bands of metals on the right of the periodic table narrower than those on the left? Why do d-bands broaden on going down in the periodic table through the transition metals?

6.9 Explain the terms Fermi level, vacuum level, and work function. What are the corresponding properties in terms of molecular orbitals?

6.10 Draw the density of states for a free-electron gas as a function of energy.

6.11 How can one modify the work function of a surface?

6.12 Why is the d-band of a metal narrower at the surface than in the interior?

6.13 Draw a simple version of the density of states for the electron bands of a metal (a good conductor), a semiconductor, and a perfect insulator.

6.14 Why are insulators in general unreactive? Can insulators become reactive?

6.15 Explain why transition metals with approximately half-filled d-bands have the highest melting points. Why does the melting point increase on going down through a column in the periodic table?

6.16 What happens with the outer orbitals of an atom when it approaches a metal surface? Discuss the role of the atom's ionization potential and electron affinity in relation to the work function of the metal for the strength of the eventual chemisorption bond.

6.17 Why can a molecule such as CO easily be dissociated on a metal surface but not in the gas phase?

6.18 Draw a schematic molecular orbital diagram for the adsorption of a diatomic molecule on a d-metal.

6.19 Why does the work function change when an atom or molecule adsorbs on a surface?

6.20 Describe the trend in adsorption energy for atoms such as N and O when going from left to right through the transition metals in the periodic table. Do the same for going vertically through the transition metals.

6.21 Why does CO bind relatively strongly to metals such as Ni, Pd, and Pt, but not on Cu, Ag, and Au?

6.22 Why does CO dissociate readily on iron and not at all on platinum even though the heats of adsorption of CO on these metals are similar?

6.23 Why is gold relatively inert among the metals, in contrast to platinum? Why is mercury more reactive than gold?

6.24 How does the strain or compression of metal atoms in a surface influence the adsorption energy and the reactivity?

6.25 How does the chemical reactivity of a metal atom depend on its coordination number?

6.26 Why do many catalytic reactions exhibit "volcano" behavior as a function of d-band filling of the metal catalyst?

6.27 What will happen to the reactivity of Ag if we deposit 1 ML of Ag on Ni(111) or Au(111)? Hint: refer to Figure 6.33.

6.28 Place the following fcc metal surfaces in order of decreasing reactivity: (111), (110), (100), (001), (557). Do the same for these surfaces of bcc metals.

6.29 Explain the Brønsted–Evans–Polanyi relation in a simple potential energy scheme for an elementary reaction step.

6.30 Are there fields other than catalysis where trends in surface reactivity may be of value?

6.31 What is a descriptor?

6.32 Explain the principle of using scaling relations for predicting the activity of various metals.

6.33 Explain when it is necessary to break the scaling relations and give some examples of how it could be done.

Chapter 7

7.1 What is a sticking coefficient? How can it be measured?

7.2 How can one determine the activation energy for an activated adsorption process?

7.3 What is the essential difference between first- and second-order adsorption processes?

7.4 What does thermalization of a gas mean? Why do experiments on sticking in which gases are not thermalized lead to erroneous results?

7.5 List some adsorption systems with extremely low and very high sticking coefficients. Can you rationalize these values with transition state theory (Chapter 3)?

7.6 Describe the experimental set-up for temperature-programmed desorption from a single crystal surface.

7.7 Derive the rate expression for temperature-programmed desorption (i.e., the rate versus temperature at a constant heating rate).

7.8 Give examples of desorption systems following first-, second-, and zero-order kinetics. Can you give a physical interpretation for the latter?

7.9 Refer to Chapter 3 and summarize the values of pre-exponential factors you may expect for the desorption of gases.

7.10 Give a brief overview of methods that can be used to derive the activation energy of desorption from TPD experiments.

7.11 Discuss in a qualitative sense how a first-order TPD spectrum (e.g., of CO) is affected by the following:

a) The presence of two adsorption sites with distinct heats of adsorption for the adsorbent.
b) Attractive interactions between the adsorbed molecules.
c) Repulsive interactions between the adsorbed molecules.
d) CO-adsorption of a small amount of a promoter such as potassium, which stabilizes the CO.

7.12 Can TPD be used to identify surface species and/or surface reaction pathways?

7.13 Suppose we successfully measured the sticking coefficient and the activation energy for adsorption of a certain molecule, as well as the rate of desorption. Is it then possible to estimate the equilibrium constant for adsorption/desorption?

7.14 Explain how the kinetic parameters of an elementary step can be derived from temperature-programmed experiments with surfaces on which the reacting species have been pre-adsorbed.

7.15 Give at least two reasons why it is important to know the kinetic parameters of elementary surface reactions in catalytic mechanisms.

7.16 Explain the principles of microkinetic modeling and its relevance for research in catalysis.

7.17 Why is dissociative adsorption often a rate-limiting step in many catalytic processes?

7.18 Are the heat of adsorption of a molecule, the activation energy for its dissociation, and the heats of adsorption for the dissociation products correlated?

7.19 Compare the pros and cons of kinetic analysis by fitting a Langmuir–Hinshelwood model to measured data and by microkinetic analysis.

Chapter 8

8.1 Discuss the importance of the steam-reforming process for the production of hydrogen and synthesis gas. Is this process endothermic or exothermic? What is the rate-limiting step for the steam reforming of methane?

8.2 Nickel catalysts used in steam reforming are more resistant to deactivation by carbon deposition if the surface contains sulfur, or gold. Explain why these elements act as promotor. Would you prefer sulfur or gold as a promotor? Explain your answer.

8.3 Very often, light gases such as methane that are liberated during oil production are flared. Why is this? Is steam reforming of such gases an option?

8.4 Give a number of sources for the production of synthesis gas, and also a number of applications of synthesis gas.

8.5 Describe the potential of the Fischer–Tropsch process as a source of transportation fuels.

8.6 Why are iron, cobalt, and ruthenium successful catalysts for the Fischer–Tropsch synthesis, while nickel, palladium, manganese, and molybdenum are not?

8.7 The active form of iron in the Fischer–Tropsch synthesis is that of an iron carbide, while cobalt and ruthenium work in the metallic state. Discuss the possible reasons for this difference.

8.8 Discuss why iron catalysts always need a potassium promoter, while cobalt and ruthenium do not.

8.9 Describe the series of processes used in gas-to-liquids technology. Why is the Fischer–Tropsch process applied in its low-temperature mode? Do the same for the coal-to-liquids process (which generally needs one or two additional steps in the process).

8.10 Compare the high- and low-temperature Fischer–Tropsch processes (HTFT and LTFT). What are the major differences?

8.11 Give a short description of the methanol synthesis, and answer the following:
a) What is the rate-limiting step in the mechanism?
b) What are the dominant surface species?
c) Can methanol be synthesized from CO and why would that be interesting?
d) Why does methanol synthesis depend on the oxidation/reduction potential of the reactants?

8.12 Summarize the key points of the water–gas shift reaction. What are the dominant surface species?

8.13 Why are the methanol and the water–gas shift process always coupled – or are there examples where they are not?

8.14 The ammonia synthesis process consists of a series of catalytic reactions that aim to make a mixture of N_2 and H_2 without components that would deactivate the catalyst. Ammonia is formed only in the last reactor.

a) What is (are) the source(s) of hydrogen in the ammonia synthesis gas?
b) List the catalytic steps that are needed to produce clean hydrogen.
c) What are the major poisons that need to be removed?
d) What is the rate-determining step of the ammonia synthesis reaction?
e) Describe the function of promoters in the ammonia synthesis catalyst.
f) Explain why it is important to develop an ammonia synthesis catalyst with high activity at low temperatures.
g) Why does the ammonia reactor contain more than one catalyst bed and why is it cooled?
h) Can ammonia be made under ambient conditions of pressure and temperature?

8.15 Explain the concept of the optimal operation line for a catalytic process.

8.16 What is the difference between electronic and structural promotors?

8.17 Explain how electronic promotors assist the dissociation of molecules such as N_2.

8.18 Why is sulfur a poison for the ammonia synthesis?

8.19 Discuss the relevance of developing new energy carriers in the future. Which sustainable sources of energy can be considered, and what is their potential for practical use?

8.20 Discuss how big of a fraction of our energy consumption do we in principle need to store if we would have to rely on sustainable energy alone?

8.21 Explain why in electrochemistry one can get a reaction going that is practically impossible under normal thermochemical conditions?

8.22 What is electrolysis and why is it important in energy technology?

8.23 Explain why improving catalysts for the oxygen evolution reaction in electrolytic processes is important even if hydrogen is the desired product.

8.24 Explain the phenomenon of overpotential and what are the possible reasons for it.

8.25 Discuss the possible role of synthesis gas and derived downstream processes such as Fischer–Tropsch, methanol, and ammonia synthesis is strategies for storage of electricity obtained from sustainable sources (wind, sun).

8.26 Can liquid H_2 be used as an aviation fuel? Has it ever been used?

8.27 Explain why photocatalysts based on TiO_2 have limited efficiency for splitting of water.

8.28 Explain why silicon is such a good semiconductor for photovoltaic applications.

8.29 Why are several semiconductors in tandem more efficient than one? Why is this nevertheless not implemented at large scale?

8.30 Discuss whether a photo-electrocatalysis (PEC) system works better than the simpler combination of a photovoltaic device and an electrolyzer?

8.31 How does a fuel cell work? What are SOFC and PEM fuel cells?

8.32 Compare the efficiency of a power plant with that of a fuel cell. Are there fundamental differences in the efficiencies that these systems can obtain?

8.33 Explain the role of catalysis in fuel cell technology.

8.34 Why is the PEM fuel cell so sensitive to CO, while the SOFC cell is not?

8.35 If fuel cell technology were introduced on a large scale for automotive transportation, would you prefer a fuel distribution system in which gasoline fuel remains the major energy carrier but is reformed on-board to hydrogen, or one in which hydrogen is provided at fuel stations? Explain your choice.

8.36 Are there any fundamental limitations in the implementation of PEM fuel cells in automobiles?

8.37 Why do Pt nanoparticles with a compressed Pt overlayer become a better catalyst than normal Pt nanoparticles?

8.38 Describe the importance of catalysis in future strategies for sustainable energy solutions.

Chapter 9

9.1 Describe briefly how crude oil is processed in a refinery. What are the major catalytic processes?

9.2 Which are the major polluting constituents of crude oil?

9.3 Give a brief description of hydrotreating.

9.4 Explain why hydrodesulfurization catalysts are used in the sulfidic form. Would it be possible to use metal catalysts for this process?

9.5 Explain why hydrodesulfurization catalysts are used in the sulfidic form. Would it be possible to use metal catalysts for this process?

9.6 Can you think of reasons why substituted dibenzothiophenes are more difficult to desulfurize than thiophene, or simple thiols (see Figure 9.3)?

9.7 Depending on the choice of catalyst, hydrodesulfurization can be accompanied by hydrogenation to various extents. In which of the product streams in the refinery would you choose for hydrogenative HDS and in which would you not?

9.8 Hydrodesulfurization removes sulfur from the crude oil components in the form of H_2S. How is this byproduct used further?

9.9 What are the major constituents of gasoline as to their origin in the refinery? Which product streams enhance gasoline quality most?

9.10 Give a brief description of the FCC process. What is the life-time of the catalyst?

9.11 The mechanism of cracking on zeolites proceeds differently than on a metal surface. What are the differences and why do they arise?

9.12 Explain the concept of bifunctional catalysis in reforming.

9.13 What is alkylation? Why is it important? Why is it an environmentally unfriendly process? What can be done to improve the process?

9.14 Describe a few partial oxidation processes that are found downstream from the refinery.

9.15 Explain the Mars–van Krevelen mechanism. In what sense does it differ from a metal-catalyzed reaction?

9.16 Can the hydrodesulfurization reaction also be considered as a Mars–van Krevelen reaction?

9.17 Describe the mechanism of catalytic ethylene polymerization.

9.18 Discuss to what extent the Phillips ethylene polymerization catalyst satisfies all of the criteria that define a catalyst. Compare this polymerization catalyst with the FCC catalyst and Pt-based reforming catalysts.

Chapter 10

10.1 Explain what primary and secondary measures are in the context of environmental pollution control.

10.2 Why is the automotive exhaust catalyst called three-way catalyst?

10.3 Which metals are used in the automotive catalyst and what reactions do they catalyze?

10.4 Why does an automotive exhaust catalyst have a control system to regulate the composition of the air-fuel mixture? How is this control performed?

10.5 Under what driving conditions does a car emit the most polluting exhaust?

10.6 Why is lead a more severe poison for (parts of) the three-way catalyst than sulfur?

10.7 Can diesel engines be equipped with a three-way catalyst?

10.8 Describe how NO_x can be removed from the exhaust when a car operates under lean-burn conditions (i.e., oxygen rich). Why is it attractive to drive cars under such conditions?

10.9 What is the typical composition of flue gas of power plants?

10.10 Describe the SCR process for the removal of NO_x from stationary power plants. Which reactants are usually used for the SCR process?

10.11 How can sulfur oxides emissions from power plants be reduced?

10.12 Is SCR technology suitable for application in mobile sources of NO_x such as ships, trucks, and passenger cars?

Exercises

Exercises for Chapter 2

Exercise 2.1 Reactivity and Steady State
Which of the following statements are right? Correct the wrong statements.

a) For a reaction at steady state, product and reactant are in equilibrium.
b) The entropy production of a reaction at steady state is at minimum.
c) If a reaction is at equilibrium, the forward and reverse rates are zero.
d) A reaction system at equilibrium has maximum entropy.
e) A reaction system at equilibrium has a minimum free energy.
f) A reaction system at equilibrium has a maximum entropy production.
g) For every reaction, the overall order in the reactants is equal to the sum of the stoichiometric coefficients.
h) A successful catalyst accelerates the forward reaction of reactants to products but inhibits the reverse reaction.

Exercise 2.2 Affinity and Extent of Reaction
For the following reaction at room temperature (300 K):

$$A + B \rightleftharpoons 2C + D$$

the rate constant of the forward reaction is $0.1 \text{ mol L}^{-3} \text{ s}^{-1}$. At time $t = 0$, the reaction mixture consists of A and B in the proportion of $1 : 2$. At $t = t_1$, the concentration of A decreased to 20% of the total mixture and at equilibrium to 10%. The affinity of a reaction is defined as

$$A \equiv -\sum_i v_i \mu_i$$

which can easily be rewritten as

$$A(t) = -\sum_i v_i \mu_i(t) + \sum_i v_i \mu_i^{eq}$$

since the latter term equals zero at equilibrium. This can then be converted into

$$-A(t) = \sum_i v_i RT \ln\left[\frac{c_i(t)}{c_i^{eq}}\right] \quad \text{or} \quad -\frac{A(t)}{RT} = \ln\left\{\prod_i \left[\frac{c_i(t)}{c_i^{eq}}\right]^{v_i}\right\}$$

meaning that the rate can in general be written as

$$r(t) = k^+ \left(1 - e^{\frac{-A}{RT}}\right) [A][B]^2$$

Calculate:

a) The rate constant of the reverse reaction.
b) The affinity of the reaction at the times $t = 0$, t_1, and ∞ (use $RT = 2.5$ kJ mol^{-1}, 300 K).
c) The extent of reaction and the deviation from equilibrium at these three points in time.

Exercise 2.3 N$_2$O$_5$ Decomposition

N$_2$O$_5$ decomposes according to the following reaction:

$$2N_2O_5 \longrightarrow 4NO_2 + O_2$$

The rate of reaction is found to satisfy the expression

$$r = \frac{d[O_2]}{dt} = k[N_2O_5]$$

Show that the following set of elementary steps leads to the observed rate equation.

$$N_2O_5 \underset{}{\overset{K}{\rightleftharpoons}} NO_2 + NO_3$$

$$NO_2 + NO_3 \overset{k_1}{\longrightarrow} NO_2 + O_2 + NO_2$$

$$NO + NO_3 \overset{k_2}{\longrightarrow} 2NO_2$$

Exercise 2.4 Steady State Assumption

a) Formulate the steady state approach for the following system of coupled reactions:

$$R \longrightarrow I_1 \longrightarrow \ldots \longrightarrow I_i \longrightarrow \ldots \longrightarrow I_n \longrightarrow P$$

and discuss briefly in which reaction situations the use of the steady state approach is appropriate.
b) Discuss briefly to what extent a steady state reaction is in equilibrium. Include considerations of entropy and entropy production in your answer.
c) The dehydrogenation of methylcyclohexane, $C_6H_{11}CH_3$, to toluene, $C_6H_5CH_3$, plays a role in gasoline reforming. The reaction is successfully catalyzed by platinum and proceeds according to the following mechanism:

$$C_6H_{11}CH_3 + * \overset{k_1}{\longrightarrow} C_6H_{11}CH_3*$$

$$C_6H_{11}CH_3* \overset{k_2}{\longrightarrow} C_6H_5CH_3* + 3H_2$$

$$C_6H_5CH_3* \overset{k_3}{\longrightarrow} C_6H_5CH_3 + *$$

Derive an expression for the rate of toluene formation, assuming that the reactions occur in the forward direction only and that the surface coverage of toluene is much larger than that of methylcyclohexane, while desorption of toluene determines the overall rate.

d) Give the range of orders of the reaction in methylcyclohexane.

Exercise 2.5 Steady State Assumption

The decomposition of acetaldehyde:

$$CH_3CHO \xrightarrow{500\,°C} CH_4 + CO$$

proceeds via methyl radicals, $CH_3^{\bullet}$

1. $CH_3CHO \rightarrow CH_3^{\bullet} + CHO^{\bullet}$
2. $CH_3^{\bullet} + CH_3CHO \rightarrow CH_4 + CH_3CO^{\bullet}$
3. $CH_3CO^{\bullet} \rightarrow CO + CH_3^{\bullet}$
4. $2CH_3^{\bullet} \rightarrow C_2H_6$

Derive the rate expression for the formation of CH_4 by using the steady state assumption.

Exercise 2.6 Steady State Assumption in the Kinetics of Chain Reactions

The chain reaction $H_2 + Br_2 \longrightarrow 2HBr$ proceeds through the steps:

1. $Br_2 \longrightarrow 2Br^{\bullet}$
2. $Br^{\bullet} + H_2 \longrightarrow HBr + H^{\bullet}$
3. $H^{\bullet} + Br_2 \longrightarrow HBr + Br^{\bullet}$
4. $2Br^{\bullet} \longrightarrow Br_2$

Show that the reaction rate assumes the following form:

$$\frac{d[HBr]}{dt} = \frac{k[H_2][Br_2]^{1/2}}{1 + k'[HBr]/[Br_2]}$$

in which the rate constants k and k' depend on the rate constants of the elementary steps. (Hint: assume equilibrium between molecular and atomic bromine.)

Exercise 2.7 Heterogeneous Catalysis

a) What are the most important steps in a heterogeneously catalyzed reaction?
b) Compare the changes in potential energy during a catalytic reaction with those of a gas-phase reaction by using a simple energy diagram.
c) Briefly discuss Sabatier's principle.
d) What is autocatalysis? Give a (symbolic) example of an autocatalytic reaction.
e) Explain why the reaction between adsorbed CO and NO on a rhodium surface that is entirely covered by these gases exhibits explosive behavior.

Exercise 2.8 Potential Energy Profiles

The reaction

$$AB + C \longrightarrow A + BC \quad -50\,kJ\,mol^{-1}$$

Adsorbing atom or molecule	ΔH_{ads} (kJ mol^{-1})	Reaction step	E_{act} (kJ mol^{-1})
AB	50	AB + C ⟶ A + BC	500
C	75	AB* ⟶ A* + B*	75
A	200	BC* ⟶ B* + C*	100
B	125		
BC	50		

has a relatively high activation energy of 500 kJ mol^{-1} in the gas phase. By using a suitable catalyst the activation energy drops significantly. The catalytic reaction proceeds via dissociation of AB after molecular adsorption on the surface.

a) Draw the potential energy profiles of the catalyzed and gas-phase reactions in an energy diagram, in which energy values are clearly indicated, by using the data of the accompanying table.
b) Which species is the most stable surface intermediate?
c) Discuss which step is most likely to be the rate-limiting step.

Exercise 2.9 Langmuir Adsorption Isotherms

Derive Langmuir's adsorption isotherm for the following cases:

a) Molecular adsorption of CO.
b) Dissociative adsorption of CO.
c) Competitive adsorption of molecularly adsorbed CO and dissociatively adsorbed H_2, without further reaction.
d) Give the expression for the fraction of unoccupied sites for (c).
e) Suppose the catalytic formation of methanol from CO and H_2 occurs through a mechanism in which the reaction between adsorbed CO and the first H atom determines the rate (the reverse reaction may be ignored), while all subsequent reaction steps are fast, except for the desorption of methanol, which may be considered at equilibrium with the gas phase.
   - Devise the mechanism.
   - Derive an expression for the rate of reaction.
   - Discuss the range of values that the orders in hydrogen, carbon monoxide, and methanol may assume if hydrogen adsorbs more weakly than the other gases.

Exercise 2.10 Problems in PEM Fuel Cells

PEM fuel cell technology relies heavily on dissociating H_2 on a Pt catalyst so that the hydrogen atoms can migrate through the membrane as protons and recombine with oxygen on the other side to form water. The energy released by this process corresponds to roughly 1.2 eV per proton and the process is thought to become viable for portable power generation and in mobile units such as cars. In general the process works quite well when using pure H_2 but in practice hydrogen is generated from processes where also CO is present.

In the following, we shall examine the importance of CO in this context. The elementary reaction can be written as

1. $H_2 + 2* \rightleftharpoons 2H*$
2. $CO + * \longrightarrow CO*$

1. Determine the coverages of atomic hydrogen and CO for the following three gas mixtures: $p_{tot} = 1$ bar and $p_{CO} = 1, 10, 100$ ppm. The operational temperature of the fuel cell is 80 °C and the following information on the adsorption on Pt is available:

$$k_1^+ = \frac{2S_0^{H_2}}{N_0\sqrt{2\pi m_{H_2} k_B T}}, \quad k_1^- = \nu \exp\left(\frac{-E_H^d}{RT}\right)$$

$$k_2^+ = \frac{S_0^{CO}}{N_0\sqrt{2\pi m_{CO} k_B T}}, \quad k_2^- = \nu \exp\left(\frac{-E_{CO}^d}{RT}\right)$$

where the parameters are

$S_0^{H_2} = S_0^{CO} = 1.00$, $k_B = 1.38 \times 10^{-23}$ J K^{-1}, $m_{H_2} = 2 \times 1.6 \times 10^{-27}$ kg, $m_{CO} = 28 \times 1.6 \times 10^{-27}$ kg, $N_0 = 1.5 \times 10^{19}$ m^{-2}, $\nu = 1 \times 10^{13}$ s^{-1}, $R = 8.31$ J K^{-1} mol^{-1}, $E_{CO}^d = 100$ kJ mol^{-1}, $E_H^d = 80$ kJ mol^{-1}

2. In reality CO adsorption is set too low here. It should rather be 120–130 kJ mol^{-1}. Suggest how to make the cell less sensitive to the CO blocking. Discuss which parameters are important.

Exercise 2.11 Methanol Synthesis

Recent research on the catalytic synthesis of methanol from CO_2 and H_2 over a copper catalyst has shown that the rate of reaction is first order in CO_2 and 3/2 in H_2.

$$CO_2 + 3H_2 \longrightarrow CH_3OH + H_2O$$

The mechanism is thought to involve dissociation of hydrogen, which reacts with molecularly adsorbed CO_2 to form formate adsorbed on the surface. The adsorbed formate is then further hydrogenated into adsorbed di-oxo-methylene, methoxy, and finally methanol, which then desorbs. The reaction is carried out under conditions where the surface is predominately empty and the oxygen generated by the process is quickly removed as water. Only the forward rate is considered and the process is assumed to go through the following elementary steps:

1. $H_2 + 2* \rightleftharpoons 2H*$
2. $CO_2 + * \rightleftharpoons CO_2*$
3. $CO_2* + H* \rightleftharpoons HCOO*$

4. $HCOO* + H* \rightleftharpoons H_2COO* + *$
5. $H_2COO* + H* \rightleftharpoons H_3CO* + O*$
6. $H_3CO* + H* \rightleftharpoons H_3COH* + *$
7. $H_3COH* \rightleftharpoons H_3COH + *$

1. Give the rest of the mechanism that removes the adsorbed oxygen.
2. Determine which step is most likely to be rate limiting if a simple model can explain the observed reaction orders.
3. Derive the rate expression, assuming that the rate-limiting step proceeds in the forward direction only and that all other steps are in equilibrium.
4. At which mol-fraction of H_2 does the rate have a maximum?

Exercise 2.12 CO Oxidation

Consider the oxidation on a Pt surface of CO by O_2 to give CO_2, which happens in an automotive catalyst: $CO + \frac{1}{2}O_2 \rightleftharpoons CO_2$

a) Give the four elementary steps involved in this process. It can be assumed here that the adsorption of oxygen is a direct process.
b) Write an expression for the rate assuming that the recombination of adsorbed carbon and oxygen is the rate-limiting step. The partial pressure of CO_2 cannot be neglected in this example.
c) Show that the rate can be expressed as a deviation from equilibrium as

$$r = r^+ - r^- = k^+ K_{CO} p_{CO} \sqrt{K_{O_2} p_{O_2}} \left(1 - \frac{p_{CO_2}}{\sqrt{p_{O_2}} p_{CO} K_G}\right) \theta_*^2$$

where K_G is the equilibrium constant for the overall reaction.
d) Determine an expression for the reaction order in CO, O_2, and CO_2, as well as the apparent activation energy for this process.

Exercise 2.13 Steam Reforming Reaction

Here we shall have a closer look at the steam reforming process, which is used in large-scale industrial production of syn-gas and hydrogen.

$$CH_4 + H_2O \rightleftharpoons CO + 3H_2$$

which is performed at rather high temperature (1000–1200 K) and at moderate pressures. The catalyst for this process is Ni (Ru) on Al_2O_3 or Al_2MgO_4 supports, and the reaction orders were found to be $n_{CH_4} > 0$; $n_{H_2O} < 0$; and $n_{H_2} > 0$.

1. Why is this process performed under moderate pressure and at relatively high temperatures?
2. Propose a mechanism where the rate-limiting step is the recombination of adsorbed carbon C* and adsorbed oxygen O* and write up an equation for the rate. In the following, we assume that only one adsorbate dominates the surface. The so-called MARI for the most abundant reaction intermediate. Here, we assume that it is oxygen O*. Is that reasonable?

3. Do the reaction orders from the above model match the observed orders?
4. We could also have assumed that methane dissociation is rate limiting. Write up the rate again, assuming that oxygen is the MARI.
5. Does the rate equation fit the observed reaction rates?

Exercise 2.14 HDS Reaction

Hydrodesulfurization (HDS) is a very important large-scale process used in refineries to remove sulfur from oil products. It is actually one of the largest catalytic processes. As a model system for this process, we shall consider the HDS of thiophene, which is a model sulfur-containing molecule in oil products. The catalyst is a CoMo-sulfide supported on alumina. The overall reaction looks like

$$C_4H_4S + 2H_2 \rightleftharpoons C_4H_6 + H_2S$$

where the products are butadiene and dihydrogen sulfide. One of the key intermediate is a situation where the thiophene ring is opened and the intermediate is adsorbed to the surface as

$$H_2C=CH-CH=CH-S-*$$

The measured rate follows an equation like

$$r = \frac{kp_T}{1+K_1 p_T} \frac{(K_2 p_{H_2})^\alpha}{1+(K_2 p_{H_2})^\alpha}$$

where p_T is the partial pressure of thiophene and p_{H_2} is the partial pressure of hydrogen.

1. Why does the reverse rate appear to be missing in the observed kinetics?
2. Propose a mechanism with the above key intermediate and find an expression for the rate taking the reverse rate into consideration. Hint: assume that hydrogen and thiophene adsorb on different sites.
3. Show that your rate reduces to the one above when the process is carried out in excess hydrogen.
4. What are the MARIs of this reaction?

The rate has a peculiar dependence on hydrogen. Why does this dependence suggest that we have S* and H* adsorbed on different sites?

Exercise 2.15 Hydrogenation of Ethane

Here we assume that the hydrogenation of C_2H_6 to CH_4 goes through the mechanism:

1. $C_2H_6 + 2* \underset{}{\overset{Q}{\rightleftharpoons}} C_2H_5* + H*$
2. $C_2H_5* + H* \underset{}{\overset{RLS}{\rightleftharpoons}} 2CH_3*$

3. $CH_3* + H* \overset{Q}{\rightleftharpoons} CH_4 + 2*$

4. $H_2 + 2* \overset{Q}{\rightleftharpoons} 2H*$

where step 2 is rate limiting and steps 1, 3, and 4 are in quasi-equilibrium.

a) If a small amount of D_2 is added to the reactants, $C_2H_{6-n}D_n$ can be observed. Explain the origin of this compound.

b) Write an equation for the rate of step 2.

c) In the quasi-equilibrium limit the following expressions are obtained:

1) $\theta_{C_2H_5} = \dfrac{K_1 P_{C_2H_6} \theta_*^2}{\theta_H}$

2) $\theta_{CH_3} = \dfrac{P_{CH_4} \theta_*^2}{K_3 \theta_H}$

3) $\theta_H = \sqrt{K_4 P_{H_2}} \cdot \theta_*$

Mention some critical phenomena, which cannot be studied under the above assumption.

d) Write an expression for the rate and show that, in the limit where the coverages of $C_2H_5^*$ and CH_3^* are small, it can be written as

$$r = k_2^+ K_1 P_{C_2H_6}\left(1 - \dfrac{P_{CH_4}^2}{K_G P_{C_2H_6} P_{H_2}}\right)\theta_*^2$$

where

$$\theta_* \approx \dfrac{1}{1 + \sqrt{K_4 P_{H_2}}}$$

Exercises for Chapter 3

Exercise 3.1 Average Molecular Velocities

a) Calculate the average velocity v of N_2 molecules at room temperature (298 K).
b) At which temperature is the average velocity of N_2 equal to that of He at 298 K?
c) Calculate the average translational energy of 1 mol N_2 at 100, 298, and 1000 K.

Exercise 3.2 Collisions in the Gas Phase

Given a mixture of $N_2 + 3H_2$ at 1 bar and 25 °C:

a) Calculate the number of molecules N_2 and H_2 in 1 m³ (1 g mol of ideal gas at 25 °C has a volume of 24.7894 L).
b) The collision diameter of H_2 is 0.271 nm and that of N_2 is 0.373 nm. How many collisions per second are there between the H_2 molecules?
c) How many collisions per second are there between N_2 molecules?

d) How many collisions per second between N_2 and H_2 molecules?
e) What is the total number of collisions in 1 m³ of this mixture?

Exercise 3.3 Collision Theory of Reaction Rates
Consider the bimolecular collision:

$$NO(g) + Cl_2(g) \longrightarrow NOCl(g) + Cl(g)$$

which has a reaction diameter $d = 3.5$ Å. Determine the pre-exponential factor for this reaction as a function of temperature.

Exercise 3.4 General Aspects of Partition Functions
Write the partition function for 1 mole of helium and find expressions for

a) The energy E.
b) The pressure p.
c) The chemical potential expressed by the standard chemical potential μ^0 and the pressure, and determine an expression for μ^0.
d) The entropy S.

Exercise 3.5 Partition Functions
a) Calculate the translational partition function of a nitrogen molecule at 298 K in a 24.79 L container (equal to the standard molar volume of an ideal gas under these conditions).
b) How does q_{trans} change if T increases, p decreases, or V increases? What is the value of q_{trans} for an N atom and an N_3 molecule under the same conditions as for N_2 in part (a)?
c) The stretch frequency of N_2 is 2330 cm^{-1}; calculate the vibrational partition function of N_2 with respect to the vibrational ground state.
d) Calculate the rotational partition function of N_2 at 298 K. The moment of inertia is 1.407×10 kg m² and the symmetry number of N_2 is 2. How does q_{rot} change with increasing temperature, decreasing pressure, or increasing volume?
e) Calculate the total partition function of N_2.

Exercise 3.6 Rotational Partition Functions
The rotational microwave spectrum of a diatomic molecule has absorption lines (expressed as reciprocal wavenumbers cm^{-1}) at 20, 40, 60, 80, and 100 cm^{-1}. Calculate the rotational partition function at 100 K from its fundamental definition, using $kT/h = 69.5$ cm^{-1} at 100 K.

Exercise 3.7 Vibrational Partition Functions
Calculate the vibrational partition function with respect to the vibrational ground state (i.e., the lowest occupied state) and the fraction of molecules in the ground state at 300, 600, and 1500 K for the following molecules, using $kT/h = 208.5$ cm^{-1} at 300 K:

Molecule	v (cm^{-1})
I_2	213
Cl_2	557
O_2	1556
HCl	2886
H_2	4160

Exercise 3.8 Partition Function, Average Energy, and Equilibrium
Molecule A occurs in two energy states separated by ΔE.

a) Derive an expression for the partition function of A and calculate its limiting values at low and high temperature (i.e., 0 and ∞ K).
b) Calculate the average energy of the molecule and give the limiting values at low and high temperature.
c) Suppose that A is in equilibrium with an isomer B, which possesses the following energy levels with respect to the ground state of isomer A: $\Delta E/2$, $3\Delta E/4$, and ΔE. Derive an expression for the equilibrium constant $K = [B]/[A]$ and calculate the limiting values of K at low and high temperature. Discuss the meaning of the obtained values. Hint: use the following equations:

$$q = \sum_i e^{-\varepsilon_i/kT} ; \quad \bar{\varepsilon} = kT^2 \frac{\partial}{\partial T} \ln(q) ; \quad K = \prod_j q_j^{v_j}$$

Exercise 3.9 Equilibrium Constants from Partition Functions
Two isomers A and B are in equilibrium and possess the following spectroscopically determined energy levels:

$$\varepsilon_i^A = i\Delta E ; \quad \varepsilon_i^B = \left(\frac{i}{2}+1\right)\Delta E ; \quad i = 0, 1, 2.$$

Calculate the equilibrium constant for the reaction $A \rightleftarrows B$ at

a) Low temperature: $T = 0.1\,\Delta E/k$
b) High temperature: $T = 2\,\Delta E/k$
c) Very high temperature: $T = 10\,\Delta E/k$
d) What are the limiting values of K for this particular reaction?

Exercise 3.10 Equilibrium Constants for Reactions on Steps
In the following, we consider nitrogen atoms adsorbed on a ruthenium surface that is not completely flat but has an atomic step for each one hundred terrace atoms in a specific direction. The nitrogen atoms bond stronger to the steps than to the terrace sites by 20 kJ mol^{-1}. The vibrational contributions of the adsorbed atoms can be assumed to be equal for the two types of sites. (Is that a good assumption?)
Determine how the coverage of the step sites varies with terrace coverage.

Hint: write the partition functions for the atoms occupying step and terrace sites and equal their chemical potentials.

Exercise 3.11 Transition State Theory

a) Give the general equation for the reaction of a molecule R via the transition state $R^{\#}$ to a product P according to transition state theory; indicate carefully in which direction the reaction steps are allowed to proceed and provide rate and equilibrium constants where appropriate. Draw an energy diagram which clearly shows the energy levels of R, $R^{\#}$, and P, as well as the barrier energy ΔE.
b) Give the general expression for the reaction rate according to transition state theory in terms of partition functions (no explicit expressions) and ΔE.
c) What is the essential difference between the transition state theory and the collision theory of reaction rates?
d) Suppose an adsorbing molecule has a sticking coefficient for (nondissociative) adsorption of the order of 10^{-3} and that the process is not activated, that is., $\Delta E = 0$. What type of a transition state can be envisaged for the adsorption process? What type of transition state corresponds to adsorption with a sticking coefficient of unity?
e) Why, in general, is the rate of dissociative adsorption considerably smaller than the rate of associative adsorption?

Exercise 3.12 Equilibrium Constants for Adsorption

Here, we consider a simple Langmuir isotherm where an atom A is in equilibrium with adsorbed A* on the surface.

$$A + * \underset{k^-}{\overset{k^+}{\rightleftharpoons}} A*$$

a) Write an expression for the coverage of A in terms of the equilibrium constant K_A and the pressure p_A.
b) We now investigate what K_A really is made of. Write the partition functions for the atoms in the gas phase and in the adsorbed phase. The adsorbed atoms are assumed to be adsorbed on localized sites with an adsorption energy ΔE (relative to the vibrational ground state) and each with a vibration mode orthogonal to the surface and two frustrated horizontal vibrational modes. Assume chemical equilibrium and derive an expression for θ_A. In the following, it can be assumed that $h\nu_\perp / k_B T \ll 1$ and the partition function of the two frustrated modes are close to unity. (Is that a reasonable assumption?)
c) Using the expression derived under (a) and that k^+ and k^- can be expressed as

$$k_A^+ = \frac{S_0^A}{N_0 \sqrt{2\pi m_A k_B T}}, \quad k_A^- = \nu \exp\left(\frac{-E_{CO}^d}{RT}\right)$$

derive an expression for S_0^A and estimate its value assuming that each site takes up $9\,\text{Å}^2$, $T = 300$ K, and $m = 16$ amu (CH_4).

Exercise 3.13 Dissociation in the Gas Phase
Consider the monomolecular dissociation reaction

$$AB \underset{k^-}{\overset{k^+}{\rightleftharpoons}} A + B$$

a) Give the general form of the reaction rate expression for dissociation according to transition state theory, in terms of partition functions and barrier energy (explicit expressions for the partition functions are not requested).

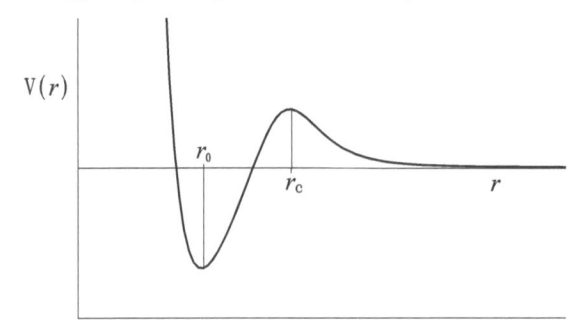

b) Define the concepts of "loose" and "tight" transition states and indicate what they imply for the locations of r_c with respect to r_0 in the potential curve for the dissociation of A–B, where r is the distance between A and B.

c) The rate of dissociation of this monomolecular reaction shows, at low pressures, a second-order dependence of the rate in [AB], whereas at higher pressures the reaction is first order in AB. Explain why.

Exercise 3.14 Transition States and Pre-exponential Factors
Sketch the transition state and give an order of magnitude for the pre-exponential factor for the following reactions:

a) Dissociation of ethane into two methyl radicals: $C_2H_6 \longrightarrow 2CH_3^\bullet$
b) Dissociation of fluoroform into an CF radical and HF: $CF_3H \longrightarrow CF^\bullet + HF$
c) *cis–trans* Isomerization of deuterated ethylene: CHD = CHD
d) Isomerization of cyclopropane to propylene.
e) Isomerization of methyl isocyanide to methyl cyanide (acetonitrile).

Exercise 3.15 Dissociation of Molecular Oxygen
a) Derive an expression for the rate of O_2 dissociation in the gas phase according to transition state theory, by using the notation given in the figure and the explicit expressions for the partition functions of translation, rotation, and vibration.
b) Rewrite your expression in the form of the Arrhenius expression and give the activation energy and pre-exponential factor.

$$q_{\text{trans}} = l(2\pi mkT)^{1/2}/h$$

per degree of freedom

$$q_{vib} = 1$$

for both O_2 and $O_2^{\#}$

$$q_{rot} = 8\pi^2 \mu r^2 kT/h^2$$

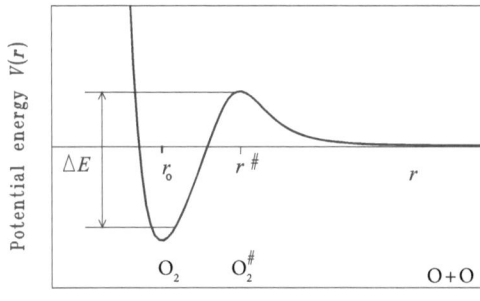

[Figure: Potential energy $V(r)$ vs Distance r, showing a potential well for O_2, a transition state $O_2^{\#}$ at $r^{\#}$, an activation energy ΔE, and dissociation to $O+O$.]

Exercise 3.16 Desorption of Molecular Hydrogen

Hydrogen adsorbs dissociatively on almost all metals. At 500 K the H atoms diffuse freely over the surface. Desorption occurs associatively by recombination of two H atoms, while desorption of atomic hydrogen can be ignored.

a) Calculate the partition function of adsorbed hydrogen atoms at 500 K, assuming that an adsorption site occupies an area of 10^{-15} cm^2.
b) Derive an expression for the rate of desorption and for the rate constant of desorption, k_{des}, in the transition state theory, in terms of partition functions (explicit expressions for the partition functions are not required).
c) Calculate the pre-exponential factor of H_2 desorption, assuming that the transition state is rigid with a total partition function $q'^{\#} = 3$ at 500 K; $q_{trans} = 1(2\pi mkT)^{1/2}/h$ per degree of freedom.

Exercise 3.17 Thermal Desorption of Silver from Ruthenium

The thermal desorption of Ag from a Ru(001) substrate occurs at around 1000 K. For silver coverages between 0.15 and 1 ML the desorption is successfully described by (Niemantsverdriet, J.W., Dolle, P., Markert, K., and Wandelt, K. (1987) *J. Vac. Sci. Technol. A*, **5**, 875)

$$r = 5 \times 10^{13} \theta_{Ag}^n \exp\{-290\,000/RT\}; \quad n \approx 0 \quad \text{and} \quad E_{act} = 290 \text{ kJ mol}^{-1}.$$

a) What type of a transition state may be envisaged for this desorption process?
b) What is the physical meaning of $n \approx 0$ in the rate expression?

Exercise 3.18 Cyclopropane Isomerization

The gas-phase isomerization reaction of cyclopropane to propylene satisfies the following rate expression:

$$r = 3 \times 10^{15} [\text{cyclopropane}] e^{-274\,000/RT} \, (E_{act} \text{ in kJ mol}^{-1})$$

a) Draw an energy diagram for the reaction.
b) Give a qualitative description of the transition state.
c) Which degrees of freedom affect the pre-exponential factor most?
d) Give the transition state expression for the rate of this reaction, as well as the activation energy and pre-exponential factor.
e) Why does the rate constant of this reaction decrease at low pressures?

$$q_{trans} = V\frac{(2\pi mkT)^{3/2}}{h^3} \; ; \quad q_{vib} = \prod_i \frac{1}{1-\exp\{-hv_i/kT\}} \; ;$$

$$q_{rot} = \frac{\sqrt{\pi}}{\sigma}\left(\frac{8\pi^2 kT}{h^2}\right)^{3/2}\sqrt{I_A I_B I_C}$$

where I_x is the moment of inertia for each of the three rotations.

Exercise 3.19 Molecular Desorption
A simplified expression for the desorption of a molecule from an immobile adsorption state is

$$k_{des} = \frac{kT}{h}l^2\left(\frac{2\pi m_{cm}kT}{h^2}\right)\frac{8\pi^2\mu r^2 kT}{h^2}e^{-\frac{E_{des}}{kT}}$$

Write the expression as an Arrhenius equation

$$k_{Arr} = v_{eff}e^{-\frac{E_{act}}{kT}}$$

and give the pre-exponential factor and the activation energy.

Exercise 3.20 Medium Effects
The rate of the gas-phase reaction $CH_3NC \longrightarrow CH_3CN$ exhibits a first-order dependence on CH_3NC pressure at 1 bar, but is second order below 1 mbar.

a) Give the rates for both cases in terms of a power rate law.
b) Explain the second-order dependence observed at low pressure by writing out the reaction steps in the transition state theory, and by using the steady state assumption.

Exercise 3.21 Collision Theory

E_A (kJ mol^{-1})	k a.u.
25	0
35	0
40	1
45	10
50	12
60	9
75	6

The energy dependence of the rate of the homogeneous gas phase reaction A + B ⟶ P has been investigated at 500 K by varying the kinetic energy of A in a molecular beam experiment. The results are given in the accompanying table. Discuss these results in terms of collision theory. How large is the energy barrier? What is the activation energy at 500 K?

$$k_\sigma = \pi d^2 \sqrt{\frac{8kT}{\pi \mu}} e^{-E_b/kT}$$

Exercises for Chapter 5

Exercise 5.1 Ammonia Synthesis Catalyst
An ammonia catalyst has been prepared by reducing magnetite (Fe_3O_4) in which 3 wt% of Al_2O_3 is dissolved. Reduction takes place under such conditions that the magnetite does not change size. During reduction only the iron will be reduced to pure Fe while nothing happens to the Al_2O_3.

1. Explain this phenomenon.
2. Estimate the pore volume of a 1 g of catalyst (or the relative pore volume in %) after the reduction, utilizing that $M_{WFe} = 55.85$ g mol^{-1}, $\rho_{Fe_3O_4} = 5.18$ g cm^{-3}, $\rho_{Fe} = 7.86$ g cm^{-3}, and $\rho_{Al_2O_3} = 3.97$ g cm^{-3}.
3. The surface energies of iron follow the order $\gamma_{Fe(100)} < \gamma_{Fe(110)} < \gamma_{Fe(111)}$. Discuss this order and the consequences it ought to have for the reduced iron surface. In the following we assume that the reduced iron consists solely of the Fe(100) surface. (Is this a reasonable assumption?)
4. The reduced magnetite with alumina was found to have a N_2 BET surface area of 29 m^2 per gram of catalyst. When adsorbing N_2 dissociatively, it was found that 2.2 mL [standard conditions, that is, 273 K and 1 bar (= 100 000 Pa)] of N_2 could be adsorbed per gram of catalyst. Assuming that the atomic nitrogen forms a c(2 × 2) overlayer on the Fe(100) surface determine the iron area per gram of catalyst. The lattice distance of iron is 0.286 nm.
5. What is the difference between this area and the BET area?

Exercise 5.2 BET Method
At the Department of Chemical Engineering, new catalytically active materials have been produced by burning volatile metal compounds in a flame. This produces an aerosol of very small particles that can be collected on a filter. Especially if the particles are cooled very fast, it is possible to obtain a large area per gram of material. In the following, Al_2O_3 is produced by this method.
The area of the synthesized material is measured using the BET method of adsorbing molecular N_2, and the following data were obtained at liquid N_2 temperature. The equilibrium pressure under those conditions is $P_0 = 773.81$ Torr and the catalyst sample weighs 24.9 mg (STP stands for standard temperature and pressure).

P (Torr)	38.65	67.61	96.67	125.75	154.62
V_a (mL STP)	0.4912	0.5407	0.5815	0.6184	0.6565

1. Determine the area of this sample if we assume that each adsorbed N_2 molecule takes up $0.164\,nm^2$ on the surface.
2. Determine the ratio between desorption from the 1st and 2nd layer. Can you rationalize the finding?
3. We now impregnate the surface with Cu. Discuss how this will influence the surface area.
4. We now put 1 g of the impregnated catalyst into a plug flow reactor where it is activated by reduction in a steam of hydrogen. It is assumed that all the copper is reduced to small particles, which expose the most stable surface structure. We then subject it to a mild oxidation so that only the surface is oxidized. The saturation coverage of oxygen under these conditions is 0.5 with respect to Cu surface atoms. After this, the reactor is purged with He so that all traces of excess oxygen are removed from the surface. The catalyst is then exposed to a stream of hydrogen, which reacts with the surface oxygen on the copper to form water, which can then be measured. It is found that 6.7 mL (at standard gas phase conditions) of water is produced. Determine the active Cu surface area per gram of catalyst. Cu is a fcc metal with a lattice distance $a_{Cu} = 3.61$ Å.
5. We now run an experiment where the catalyst sits in a flow of oxygen and hydrogen at 400 K. The flow is set to be $10\,mL\,s^{-1}$ (STP) and the water content is found to be 1% by volume. Determine the turnover frequency (TOF) per Cu site.
6. The overall activation energy was then determined by measuring the rate as a function of temperature. How do you extract the overall activation energy?
7. In the temperature interval 400–600 K, the activation energy is determined to be $50\,kJ\,mol^{-1}$, while in the regime 600–800 K it is only $27\,kJ\,mol^{-1}$. The researcher is very happy that he/she can reduce the activation energy. Can you rationalize these findings and should the researcher be happy?

Exercise 5.3 Effectiveness Factor

In this exercise, we shall estimate the influence of transport limitations when testing an ammonia catalyst such as that described in Exercise 5.1 by estimating the effectiveness factor ε. We are aware that the radius of the catalyst particles is essential so the fused and reduced catalyst is crushed into small particles. A fraction with a narrow distribution of $R = 0.2$ mm is used for the experiment. We shall assume that the particles are ideally spherical. The effective diffusion constant is not easily accessible but we assume that it is approximately a factor of 100 lower than the free diffusion, which is in the proximity of $0.4\,cm^2\,s^{-1}$. A test is then made with a stoichiometric mixture of N_2/H_2 at 4 bar under the assumption that the process is far from equilibrium and first order in nitrogen. The reaction is planned to run at 600 K, and from fundamental studies on a single crystal the TOF is roughly 0.05

per iron atom in the surface. From Exercise 5.1, we utilize that 1 g of reduced catalyst has a volume of 0.2 cm^3 g^{-1}, that the pore volume constitutes 0.1 cm^3 g^{-1}, and that the total surface area, which we will assume is the pore area, is 29 m^2 g^{-1}, and that of this is the 18 m^2 g^{-1} is the pure iron Fe(100) surface. Note that there is some dispute as to which are the active sites on iron (a dispute that we disregard here).

1. Estimate the pore area per volume catalyst (S).
2. Estimate the average pore radius R_p.
3. Estimate the rate constant for the process.
4. Estimate the Thiele diffusion modulus (Φ_s).
5. Estimate the effectiveness factor ε and discuss the result.
6. The plug flow reactor, which has the dimension of $r = 2$ mm is now loaded with 1.0 g of catalyst. Will this set-up fulfill the requirements formulated above?

Estimate the space velocity needed to ensure that the conversion does not exceed 5 % conversion at the exit of the plug flow reactor.

Exercises for Chapter 7

Exercise 7.1 Sticking of N$_2$ on Fe(100); Continuation of Exercise 5.1

1. TPD of the nitrogen-saturated Fe(100) surface shows a symmetric feature with a peak maximum at 740 K if we use a heating ramp of 2 K s^{-1}. Estimate the activation energy for desorption assuming second-order desorption.
2. The sticking coefficient of N$_2$ on this surface was measured to be $S_0(T) = S_0^0[\exp(-\Delta E_{act}/k_B T)]$, where $S_0(T) = 2.5 \times 10^{-5}$ at 500 K and $\Delta E_{act} = 0.03$ eV. Draw a potential energy diagram for N$_2$ adsorption on Fe(100) and estimate the enthalpy of adsorption of N on the Fe(100) surface.
3. Explain in simple terms why the sticking is so low, despite the low activation energy for dissociation.
4. Determine the equilibrium nitrogen coverage on the Fe(100) surface when it is exposed to 1 bar of N$_2$ at 800 K.
5. Hydrogen desorbs from the Fe(100) surface in a TPD experiment at around 320 K. Discuss qualitatively which species would be the MARI (atomic H or atomic N) under the ammonia synthesis conditions (700 K, $P_{tot} = 200$ bar, stoichiometric mixture).

Exercise 7.2 Sticking of Methane on Ni(100)

The attached figure displays the uptake of carbon on a Ni(100) surface when exposed to methane at different temperatures. The methane dissociates and dehydrogenates on the surface, resulting in the carbon overlayer. The reaction is assumed to be

1. $CH_4 + 2* \longrightarrow CH_3* + H*$
2. $CH_3* + * \longrightarrow CH_2* + H*$
3. etc.

where the first step is rate limiting.

1. Determine the number of sites available on the Ni(100) surface given that the density of Ni is $\rho_{Ni} = 8.90\,\text{g cm}^{-3}$ and $M_W = 58.71\,\text{g cm}^{-3}$ and it is assumed that each Ni atom constitutes a site. Ni is a fcc metal.
2. Determine the absolute values of the sticking coefficients at the various temperatures.
3. Determine the apparent activation energy for the sticking. In the experiment, the dosage of $0.2\,\text{bar s}^{-1}$ took 152 s. Determine which pressure of methane (in torr) was used in these experiments.

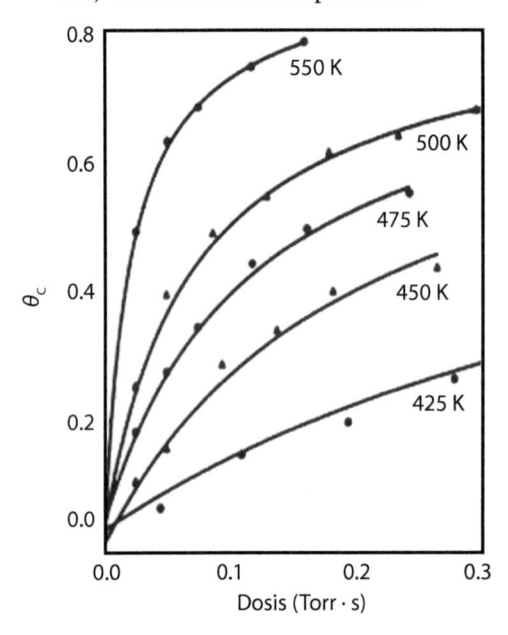

Exercise 7.3 Hydrogen Adsorption/Desorption and Equilibrium

The sticking coefficient of H_2 on a metal has been determined through an adsorption experiment. The metal surface is assumed to have $N_0 = 1.5 \times 10^{19}$ sites m^{-2} and each adsorption site is assumed to be occupied by one hydrogen atom when the surface is saturated. The experiment was performed by exposing the surface to a known pressure of hydrogen over a well-defined period of time (dosis) and then sequentially determining how much was adsorbed by, for example, TPD. All adsorption experiments where performed at such low temperatures that desorption could be neglected.

a) Based on the reaction $H_2 + 2* \rightleftharpoons 2H*$, write an expression for the uptake rate of hydrogen atoms on the surface in terms of hydrogen coverage θ_H.
b) The hydrogen coverage as a function of hydrogen dosage in units of bar s^{-1} is given in the accompanying figure. Determine the initial sticking coefficient at

the different temperatures.

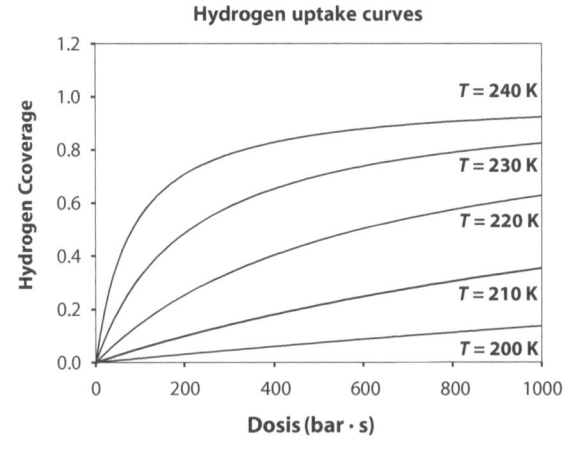

Hydrogen uptake curves

[Graph showing Hydrogen Coverage vs Dosis (bar·s) with curves for T = 240 K, T = 230 K, T = 220 K, T = 210 K, T = 200 K]

c) Determine the activation energy for the hydrogen sticking on this metal.

d) In the following, the surface has been completely saturated with hydrogen atoms and we shall now estimate the desorption energy of the hydrogen molecules by recording a TPD spectra as shown above. It can be assumed that the desorption curve is completely symmetrical and that the desorption rate has a maximum at 408 K. Under the assumption that the prefactor is $\nu = 1 \times 10^{13}$ s^{-1} and $\beta = 2$ K s^{-1}, estimate the desorption energy.

TPD of H$_2$

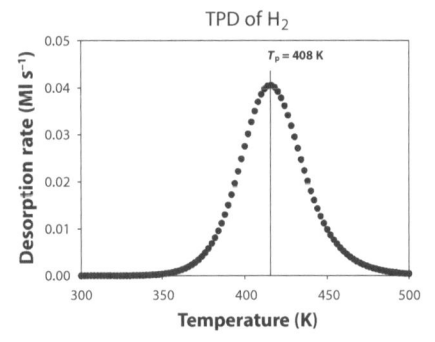

[Graph showing Desorption rate (Ml s^{-1}) vs Temperature (K) with peak at T_p = 408 K]

e) A catalyst is now prepared by impregnating an Al$_2$O$_3$ support with the metal. The metal is reduced and we wish to determine the area of the support and the active metal area. Explain how you could estimate these areas under the assumption that H$_2$ does not react with the support and that N$_2$ does not dissociate on the metal.

Questions and Exercises

General Exercises

Exercise G.1 Catalytic NO Reduction

A catalyst used for cleaning exhaust gases from automobiles consist, among other things, of Rh particles on an Al_2O_3 support material. The Rh particles expose primarily Rh(111) and secondarily Rh(100) surface structures. Rh is a FCC metal with a lattice distance of $a = 0.381$ nm.

a) Why do those two surfaces dominate? Which of the two surfaces do you think is the most reactive? In the following, we shall assume that the Rh(100) surface is dominant. The BET surface area of the catalyst is 180 m² g⁻¹.
b) Give a short description of the interaction on which the BET method is founded and describe which area is being measured by this method.

NO is now chemisorbed on the Rh particles at a temperature where it does not adsorb on the Al_2O_3. The saturation coverage of NO on Rh(100) corresponds to one NO molecule per two rhodium surface atoms, with NO sitting in a $c(2 \times 2)$ surface structure. After having saturated the catalyst with NO, a temperature-programmed desorption experiment (TPD) is performed with a heating rate of 2 K min⁻¹. NO is seen to desorb with a maximal rate at 460 K. The total NO gas that desorbs amounts to 18.5 mL per gram catalyst ($P = 1$ bar and $T = 300$ K). It can be assumed that NO does not dissociate on the Rh(100) surface.

c) Estimate the activation energy for desorption of NO from Rh(100).
d) Determine the area of the Rh(100) surface per gram catalyst.

It is now assumed that the adsorption of NO on Rh(100) is associative and not activated. The sticking coefficients $S(T)$ can be set to unity for all temperatures. Furthermore, we shall assume that the activation energy for the desorption of NO from Rh(100) is 140 kJ mol⁻¹ and that two Rh atoms constitute an adsorption site.

e) Determine the NO coverage on the catalyst if it is exposed to a partial pressure of $P_{NO} = 1$ mbar at $T = 900$ K.

In reality, does NO dissociate into adsorbed oxygen and nitrogen at those temperatures? The catalyst is usually used under conditions where nitrogen desorbs and oxygen is removed by reaction with, for example, CO. The dissociation of NO is an activated process.

f) Explain briefly why Rh is the best metal among the following from the second series of transition metals: Mo, Ru, Rh, Pd, Ag.

Exercise G.2 Oxidation of Hydrogen on Pt

The synthesis of water from H_2 and O_2 on a platinum surface is assumed to proceed via the following elementary steps:

1. $H_2 + 2* \rightleftharpoons 2H*$
2. $O_2 + 2* \rightleftharpoons 2O*$
3. $O* + H* \rightleftharpoons HO* + *$

4. $HO* + H* \rightleftharpoons H_2O* + *$
5. $H_2O* \rightleftharpoons H_2O + *$

a) Can you suggest other elementary steps that could be worth considering in describing the water synthesis?
b) In the following, we shall only consider the above five elementary steps and assume that 1, 2, 4, and 5 all are in pseudo-equilibrium. Write an expression for the rate-limiting step (3) containing both the forward and reverse rate. The expression should only contain the coverages of species entering in step (3).
c) Show that the rate can also be expressed as

$$r = k_3^+ \sqrt{K_1 K_2 P_{H_2} P_{O_2}} \left(1 - \frac{P_{H_2O}}{K_G P_{H_2} \sqrt{P_{O_2}}}\right) \theta_*^2$$

and find an expression for K_G in terms of the equilibrium constants.
d) We shall now assume that oxygen is the MARI (most abundant reaction intermediate species). Give an expression for θ_* under this assumption.
e) Assuming that oxygen is the MARI, find the reaction order for H_2, O_2, and H_2O in the limit far from equilibrium.
f) We now go to the other extreme and assume that the oxygen coverage is very small under the reaction conditions. Determine for which gas mixture of H_2 and O_2 that we will have a maximal rate in the limit far from equilibrium.

Impregnating Al_2O_3 with Pt produces a catalyst. The activity of one gram of catalyst is now measured in a plug flow reactor under conditions where oxygen is the MARI and the activity is high. It is seen that the rate increases with temperature, but a simple linear relation is not found between $\ln(r)$ and $1/T$ in an Arrhenius plot.

g) Give at least one *qualitative* explanation for a deviation from linearity between $\ln(r)$ and $1/T$.

Exercise G.3 N_2O_5 Decomposition
a) Nitrogen pentoxide decomposes following the reaction:

$$N_2O_5 \longrightarrow NO_2 + NO_2 + 1/2 O_2$$

This reaction is first order and has the rate constants k, measured rates between 273 and 338 K, listed in the table.

T(K)	273	298	308	318	328	338
k (s^{-1})	7.87×10^{-7}	3.46×10^{-5}	13.5×10^{-5}	49.8×10^{-5}	150×10^{-5}	487×10^{-5}

Estimate the pre-exponential factor A and the activation energy E_a for the reaction.

b) The following mechanism has been suggested for the decomposition of N_2O_5:
   1) $N_2O_5 \rightleftharpoons NO_2 + NO_3$
   2) $NO_2 + NO_3 \longrightarrow NO + NO_2 + O_2$
   3) $NO + NO_3 \longrightarrow NO_2 + NO_2$

   Note that it is assumed that reaction (1) is reversible, while reactions (2) and (3) are irreversible. Show, using steady state approximations for the intermediates, that the overall reaction rate can be written as

   $$-\frac{d[N_2O_5]}{dt} = k[N_2O_5]$$

   and find k expressed as functions of k_1, k_{-1}, k_2, and k_3.
c) Estimate from transition state theory the temperature dependence of the pre-exponential factor in the Arrhenius expression for the reaction $NO + NO_3 \longrightarrow NO_2 + NO_2$. In other words: for $A \propto T^n$. You can assume that both NO_3 and the activated complex are nonlinear. Furthermore, you can assume that $h\nu \ll k_B T$, and that the electronic degeneracies are all unity.
d) N_2O_5 decomposition is a reaction that proceeds with similar rate constants in solution for different solvents and in the gas phase. What is characteristic for this type of reaction?

Exercise G.4 Automotive Exhaust Catalysis
In the following, we shall examine a model system for the car catalyst where CO and NO react to yield the more environmentally friendly products CO_2 and N_2. The reaction shall be split up into the following elementary steps, which all are assumed to be in quasi-equilibrium, except step 2, which is assumed to be the rate-limiting step:

1. $NO + * \rightleftharpoons NO*$
2. $NO* + * \rightleftharpoons O* + N*$
3. $CO + * \rightleftharpoons CO*$
4. $CO* + O* \rightleftharpoons CO_2* + *$
5. $CO_2* \rightleftharpoons CO_2 + *$
6. $N* + N* \rightleftharpoons N_2* + *$
7. $N_2* \rightleftharpoons N_2 + *$

a) Write the overall net reaction and express the equilibrium constant K_G in terms of the partial pressures of the participating gases.
b) Can you give an example of at least one other elementary step that could be of relevance for this process?
c) Show that, with the above assumptions, the rate can be written as

$$r = k_2^+ K_1 p_{NO} \theta_*^2 \left(1 - \frac{1}{K_G}\frac{\sqrt{p_{N_2}} p_{CO_2}}{p_{NO} p_{CO}}\right)$$

d) Express K_G in terms of K_1–K_7.

In the following, we shall assume that adsorbed oxygen is the MARI (most abundant reaction intermediate species), that is, $\theta_* \cong 1 - \theta_O$:

e) Find an expression for θ_* in terms of equilibrium constants and partial pressures alone.
f) Assuming that oxygen is the MARI, determine the reaction order for all the participating gases, that is, find n_{NO}, n_{CO}, n_{N_2}, and n_{CO_2}.

The apparent activation energy is now measured under conditions where the surface is estimated to be clean, that is, $\theta_* \cong 1$.

g) Derive a theoretical expression for the apparent activation energy, under these conditions, in terms of the activation energy for the rate-limiting step and the change in enthalpy for the steps in quasi-equilibrium.

Exercise G.5 Methanation Reaction

To create a more reactive Ni surface, a pseudomorfic overlayer of Ni(fcc) has been grown on a Ru(0001) surface. Ru is a hcp metal and the Ru(0001) surface has the same structure as a Ni(111) surface, except that the interatomic distances are larger. The unit cell of Ni(111) on Ru is, therefore, 6.36×10^{-20} m^2 instead of the usual 5.37×10^{-20} m^2 found on a normal Ni(111) crystal.

a) Do we expect the reactivity to go up or down for the Ni(111) overlayer deposited on the Ru? Explain why.
b) We now perform a CO TPD experiment, with a heating ramp of 2 K s^{-1}, with the clean Ni(111) surface and the Ni(111) overlayer on Ru(0001). In the obtained TPD spectra, the CO TPD maximum shift from $T_m = 500$ K to $T_m = 550$ K for the clean Ni(111) and Ni(111) overlayer on Ru(0001), respectively. Determine, assuming a first-order desorption rate and a pre-exponential factor of 10^{13} s^{-1}, what this shift corresponds to in terms of bonding energy.

A real catalyst is now prepared and it is assumed that, as described above, all the metal particles consist of Ru covered by Ni, which exposes solely the Ni(111) surface. We now want to estimate the area of the metal by adsorbing CO at 300 K. It is found that the surface is saturated with CO at this temperature when 3.4 mL CO (measured at 1 bar and 300 K) has been adsorbed on 1 g of the catalyst.

c) Calculate the total metal area per gram of catalyst when it is assumed that the saturation coverage of CO on the Ni(111) on Ru surface is 0.5 with respect to the Ni atoms.

We now want to estimate the CO coverage when the catalyst is located in a plug flow reactor with a partial pressure of $P_{CO} = 0.01$ bar at $T = 1000$ K. The desorption energy is estimated to be 147 kJ mol^{-1} and the pre-exponential factor is set to the usual 10^{13} s^{-1}, while the sticking coefficient is estimated to be 0.2 and independent of temperature. For simplicity, we assume that each Ni atom can adsorb a CO molecule.

d) Estimate the CO coverage under those conditions.

A mixture of CO and H$_2$ is now passing over the 1 g of catalyst under the above conditions ($T = 1000$ K, $P_{tot} = 1.00$ bar) and the dominant product CH$_4$ is measured downstream from the reactor. The flow rate is 100 mL min^{-1} and the concentration of methane is 10%, both measured at 1 bar and 300 K. We can assume ideal gas behavior.

e) Determine the turnover frequency (TOF) for methane formation per Ni atom if we assume that each gram of catalyst contains 15 m^2 of active Ni(111) on Ru.

f) In the above calculation, we calculated the TOF per Ni atom present on the ideal flat Ni(111) surface. Methane formation, however, involves the dissociation of CO. Can you think of other more realistic sites where this reaction could take place and what would that mean for the TOF for those sites?

Exercise G.6 Pyrolysis of Ethane

Pyrolysis of ethane is an important industrial process. The following data were obtained at 1150 K and 60 bar:

t (s)	0.06	0.15	0.36	0.62	0.79	0.98
[C$_2$H$_6$] (mol m^{-3})	381	318	254	191	159	127

a) Determine the reaction order for the pyrolysis of ethane.

b) The following simple mechanism has been proposed for the pyrolysis of C$_2$H$_6$ at high pressure:

(R1) $C_2H_6 \longrightarrow CH_3 + CH_3$
(R2) $C_2H_6 + CH_3 \longrightarrow CH_4 + C_2H_5$
(R3) $C_2H_5 \longrightarrow C_2H_4 + H$
(R4) $C_2H_6 + H \longrightarrow C_2H_5 + H_2$
(R5) $C_2H_5 + C_2H_5 \longrightarrow C_4H_{10}$
(R6) $C_2H_5 + C_2H_5 \longrightarrow C_2H_6 + C_2H_4$

A steady state analysis of this mechanism shows that the reaction is 1/2 order in ethane at low degrees of conversion.

If the pressure for the process is lowered, the reaction (R3) will shift from a first-order reaction (high-pressure limit) to a second-order reaction (low-pressure limit). If (R3) is now considered a second-order reaction and assuming that the other pressure dependent reactions do not shift regime, determine expressions for d[C$_2$H$_6$]/dt, d[CH$_3$]/dt, d[C$_2$H$_5$]/dt, and d[H]/dt.

c) Find, by using steady state analyses for the radicals, the reaction order for conversion of C$_2$H$_6$ at low degrees of conversion and low pressure (R3 second order). It can be assumed that only reaction (R4) contributes significantly to the consumption of ethane and that formation by reaction (R6) can be neglected.

d) The reaction
   (R7) $C_2H_5 + H \longrightarrow C_2H_6$
   has been proposed as an important chain-terminating reaction in the pyrolysis of ethane, but no measurements of the reaction rate are available. Estimate, based on collision theory, the rate constant for the reaction at 1000 K, assuming that the activation energy for the reaction is zero. The molecular radii for C_2H_5 and H can be assumed to be 0.4 and 0.2 nm, respectively.

Exercise G.7 Catalytic Synthesis of Hydrogen Peroxide

In the following we shall study a model system for the synthesis of hydrogen peroxide (H_2O_2) over a heterogeneous catalyst containing a hypothetical metal M. It is proposed to split the reaction into the following elementary steps, which are all assumed to be in quasi-equilibrium except for step 3, which is assumed to be the rate-limiting step:

1. $H_2 + 2* \rightleftharpoons 2H*$
2. $O_2 + * \rightleftharpoons O_2*$
3. $O_2* + H* \rightleftharpoons HO_2* + *$
4. $HO_2* + H* \rightleftharpoons H_2O_2* + *$
5. $H_2O_2* \rightleftharpoons H_2O_2 + *$

There are no good catalysts for this reaction and the H_2O_2 is, therefore, made by chemical synthesis. A major problem is that oxygen, in particular atomic oxygen, bonds too strongly to the potentially catalytic metal surfaces.

a) Which part of the transition metals series would be good candidates for the metal M?
b) Write a plausible mechanism in which step 2 involve dissociative adsorption of oxygen and where water is not an intermediate.
c) The enthalpy of formation for hydrogen peroxide is $\Delta^\circ_{H_2O_2} = -136$ kJ mol^{-1}. What are the optimal conditions for synthesizing H_2O_2 in terms of temperature and pressure considering that step 3 is activated?
d) We shall now return to the above reaction scheme. Show that when step 3 is the RLS and that the rate can be written as

$$r = k_3^+ \sqrt{K_1 p_{H_2}} K_2 p_{O_2} \theta_*^2 \left(1 - \frac{1}{K_G} \frac{p_{H_2O_2}}{p_{H_2} p_{O_2}}\right)$$

and find an expression for K_G in terms of K_1–K_5.
e) Find an expression for θ_* in terms of equilibrium constants and partial pressures alone.

In the following, we shall assume that adsorbed molecular oxygen is the MARI (most abundant reaction intermediate species), that is, $\theta_* \cong 1 - \theta_{O_2}$:

f) Assuming that oxygen is the MARI, determine the reaction order for all the participating gases, that is, find n_{H_2}, n_{O_2}, and $n_{H_2O_2}$.

The catalyst is now operated in the zero conversion limit and at such high temperatures that the surface can be considered to be free of reaction intermediates, that is, $\theta_* \cong 1$.

g) Determine for which ratio of H_2 and O_2 the rate will be maximal under such conditions.

Exercise G.8 Effectiveness of Pt catalyst

In the following, we shall consider a Pt catalyst supported on Al_2O_3 that is used for combustion. The catalyst is crushed and a fraction consisting of ideal spherical particles with a radius of 0.1 mm is used for testing. The catalyst has a density of $2\,g\,cm^{-3}$ and a BET surface area of $180\,m^2\,g^{-1}$. One gram of catalyst was placed in a plug flow reactor and the Pt area was first determined by oxidizing the Pt surface, and only the Pt surface, using N_2O and by then subsequently reacting off the oxygen, monitoring how much water was produced from the 1 g of catalyst. It can be assumed that the Pt only exposes a Pt(100) surface and that the saturation coverage of oxygen by dosing N_2O is a $(\sqrt{2} \times \sqrt{2})R45°O-Pt(100)$ structure. Pt has a lattice distance of $a = 3.92$ Å. The reduction resulted in 21.4 mL H_2O vapor at 298 K and 1 bar.

a) What is the area of the Pt per gram of catalyst?
b) Which surface would you expect to be most stable under vacuum conditions and would the surface only consist of this type of site?

It is now assumed that each active site consists of four Pt atoms and the reactivity of 1 g of catalyst is tested under conditions where the rate is first order in oxygen concentration. The flow over the reactor is set to 100 mL min^{-1} with 21% oxygen, the temperature 500 K, the pressure to 1 bar, and the TOF (turnover frequency per site) per Pt site under the chosen conditions is known from surface science experiments to be $0.001\,s^{-1}$. The amount of oxygen converted is considered negligible.

c) Determine the rate and the concentration of the product in the gas flow under these conditions.

The production rate can also be written as $r = VSkC_0$, where V is the volume of the catalyst, S the Pt area per catalyst volume, k the rate constant, and C_0 the concentration of the reactant.

d) Determine k at 500 K.

By varying the temperature (keeping C_0 constant) the following data points were found for temperature and rate.

T(K)	420	460	540	580	660	780	900	980
Rate	1.7×10^{-9}	3.5×10^{-8}	3.6×10^{-6}	2.3×10^{-5}	3.5×10^{-4}	2.8×10^{-3}	1.0×10^{-2}	2.0×10^{-2}

e) Make an Arrhenius plot and give a plausible explanation for the curve and determine the apparent activation energy for the process.
f) Determine the efficiency of the catalyst at 900 K, assuming that the effective diffusion constant is independent of temperature and can be set to 0.003 cm^2 s^{-1}. Describe how you can make the catalyst more cost efficient if it is supposed to work at high temperatures.

Exercise G.9 Uncertainties in Determining Rates and Activation Energies

It can be difficult to estimate theoretically the bond lengths and vibrational frequencies for the activated complex and the energy barrier for its formation. It is of interest to assess how the uncertainty in these parameters affect the rate constant predicted from transition state theory (TST). For the exchange reaction

(R1) $H + H_2 \rightleftharpoons H_3^{\#} \longrightarrow H_2 + H$

the rate constant at 1000 K estimated from TST is 7.8×10^{-13} cm^{-3} s^{-1} while the experimental value is 2.1×10^{-12} cm^{-3} s^{-1}. The TST estimate is based on a linear symmetric configuration of the activated complex with H–H distances of 0.93 Å and vibrational frequencies of 2193 cm^{-1} (bending modes) and 978 cm^{-1} (stretching mode).

1. The uncertainty in theoretical estimates of the activation energy for a reaction is seldom less than 8 kJ mol^{-1}. Evaluate whether the difference between k_{exp} and k_{TST} can be attributed to this uncertainty.
2. Evaluate whether the difference between k_{exp} and k_{TST} can be attributed to a 10% uncertainty in the H–H bond length of the activated complex.
3. Evaluate whether the difference between k_{exp} and k_{TST} can be attributed to an uncertainty of 100 cm^{-1} in the vibrational frequencies of the activated complex.
4. The pre-exponential factor for the H + H$_2$ reaction has been determined to be approximately 2.3×10^{14} mol^{-1} cm^3 s^{-1}. Taking the molecular radii for H$_2$ and H to be 0.27 and 0.20 nm, respectively, calculate the value of the probability factor P necessary for agreement between the observed rate constant and that calculated from collision theory at 300 K.

Questions and Exercises

Index

a

ab inito method 280
acetaldehyde 6, 427
acetic acid 427
acetylene 346
acid rain 437
acrolein 413, 414
acrylonitrile 415
actinide 243
activation barrier 80
activation energy 3, 36, 43, 79, 105, 111, 120, 124, 219, 275, 284, 290, 294, 302, 310, 359, 361, 429
adenosine 5′-triphosphate (ATP) 354
adhesion energy 189
adiabatic two-bed radial flow reactor 357
adsorbate–metal complex 251
adsorbate-induced surface reconstruction 181
adsorption 126, 203, 283
– adsorbate sites 179
– associative 115
– direct process 113, 115, 119
– dissociation 267
– indirect process 113
– isotherm for physisorption 196
– of a molecule on a transition metal 254
– of atoms 113
– of molecules 113, 118
– on a free-electron metal 252
– precursor-mediated 118
– reaction between adsorbates 121
adsorption–desorption condition 195
air pollution 422, 437
air-to-fuel ratio 423, 434
alcohol 428
alkali 362
– absorbate 363
alkaline earth oxide 425

alkaline electrolyte 372
alkylation 410
alkylidyne 346
– chain-growth mechanism 347
alloying 329
all-palladium converter 426
AlPO family 208
alumina 191, 199, 204, 426
– transitional 200
aluminophosphate (AlPO) 207
aluminum oxide 199
Amberlyst 411
amino acids 73
ammonia synthesis 23, 31, 34, 174, 206, 276
– production 355
– reactor 356
– synthesis 307, 353
– – kinetic model 308
– – plant 355
ammonium sulfate fertilizer 353
ammoxidation of propylene 413
Anderson–Schulz–Flory distribution 348
apparent activation energy 37, 43, 51, 63, 67, 310, 312
aromatization 406
Arrhenius energy 124
Arrhenius equation 2, 36, 65, 79, 105, 109, 112, 212, 219, 284
Arrhenius plot 286, 293, 302, 433
association reaction 301
atom efficiency 10
atomic adsorbate 253
atomic adsorption 262
atomic force microscope(AFM) 164
atomic nitrogen 311
Auger electron 136
Auger electron spectroscopy 284
autocatalytic reaction 69
automobile emission 422

Concepts of Modern Catalysis and Kinetics, Third Edition. I. Chorkendorff and J.W. Niemantsverdriet.
© 2017 WILEY-VCH Verlag GmbH & Co. KGaA. Published 2017 by WILEY-VCH Verlag GmbH & Co. KGaA.

498 | Index

automotive converter 425
automotive exhaust converter 1, 319, 438
autothermal reforming (ATR) 320, 322
average energy 81, 82
average vibrational energy 90

b

back donation 254
backscattering 145
bacteria 354
band narrowing 241
barrier energy 270
base chemicals 11
batch process 404
batch reaction 67
batch reactor 42, 47, 306
– coupled reaction 45
Belousov–Zhabotinsky reaction 70, 71
binding energy 137, 139
biocatalysis 5
biomass 368, 396, 439
Bloch wave 241
Blochs theorem 241
Boltzmann
– constant 36, 84, 109
– distribution 80, 83, 86, 89, 237
bonding energy 245, 258, 260, 359
Bose–Einstein distribution 89
boson 237
bottom of the barrel 392
Bragg relation 131
Bravais lattice 180
bridging-the-gap strategy 295
bright field image 146
Broglie wavelength 89
broken-bond calculation 185
Brønsted acid 201, 208
Brønsted base 183, 201
Brønsted–Evans–Polanyi relation 275, 276, 295, 301, 359
Brunauer–Emmet–Teller (BET) isotherm 192
butadiene 305, 401

c

calcination 416
capillary pore condensation 191
car exhaust gas constituents 422
carbenium ion mechanism 409
carbon 201, 324
– deposition 325
– filament growth 327
– graphitic 327

carbon black 191
carbon dioxide emission 422, 434
carbonium ion 406
Carnot efficiency 381
catalysis 1, 319
– as a nultidisciplinary science 16
– bio-enzymatic 50
– definition 2
– green chemistry 9
– homogeneous 5, 50
– in journals 18
– publications 20
– time scales 17
catalyst 1, 17
– activity measurements 213
– characterization 129
– coprecipitated 203
– deactivation 428
– selectivity 174
– shaping of support 201
– space velocity 174
– support 197
– supported 203
– testing 210
– – ten commandments 211
– three-way 423
– unsupported 206
catalysts
– for hydrotreating 397
– testing
– – consequences of transport limitations 222
catalytic converter 425, 436
catalytic partial oxidation (CPO) 320
catalytic reaction 48, 55
– thermodynamic data 30
cathodoluminescence 145
ceria 426
cerium oxide 425
chain reaction 45
Chan–Aris–Weinberg procedure 293
chemical bonding 231
chemical industry 9, 11
chemisorption 195, 211, 228, 246, 254, 255, 266, 431
– stress and strain 263
chemisorption bond 159
chlorine 255, 408, 413
chromium polymerization catalyst 417
Claus process 200
clock reconstruction 182
CO oxidation 65
CO oxidation reaction 429

coal 439
coal-fired boiler 437
coal-to-liquid plant 351
cohesive energy 245
coke 18, 396, 404
collision theory 79, 80, 100, 105, 115
- equilibrium constants 106
- reaction probability 104
CoMoS hydrotreating catalyst 148
compensation effect 293, 294
competitive inhibition 77
compressed hydrogen technology 375
computational chemistry 24, 225, 278, 323, 432
conductor 243
configurational entropy 84
continuously stirred tank reactor (CSTR) 211, 212
conversion process 391
coprecipitation 203
core electron 279
correlation energy 279
Cossee–Arlman mechanism 417
Coulomb interaction 149
crack gas 332
cracking process 392, 404
cristoballite 199
crude oil 391
cubic rock salt structure 182
cycloalkane 391
cyclopropane 106

d
Davy's mine lamp 428
dehydrogenation
- of methylcyclohexane 409
dehydrogenation of ethyl fragments 271
delta function 251
density functional theory (DFT) 266, 274, 278
density functional theory-generalized gradient approximation (DFT-GGA) 280
density of states (DOS) 235, 237
desorption 123, 124, 127, 192, 195, 289
- energy 293
- first-order 291
- of molecules 123
- pre-exponential factor 124
- second-order 291
- zero-order 291
desulfurization 396, 399
dibenzothiophene 396, 398
diesel 375, 443

diffraction 132
diffractogram 132
dimethylpentane 411
dioxomethylene 335
direct adsorption 115, 119
dissociation energy 98
dissociative adsorption 54, 267
Doppler effect 149

e
E factor 11
egg-shell catalyst 204
electrocatalysis 369
electrochemistry 369, 372
electrolysis 369, 372, 375
electrolyzer 369
electron 135, 144, 162
- backscattered 146
- band 234
- energy loss spectroscopy 145
- kinetic energy 227
- microscopy 144
- spin resonance (ESR) 169
electron microscope 71
electron–hole recombination 367
electron–proton transfer 370
electronic partition function 92
electronic promoter 361
elementary surface reaction 283, 303
Eley–Rideal mechanism 56
endergonic reaction 4
endothermic reaction 4, 31
energy barrier 109
energy diagram 2
energy-dispersive X-ray analysis (EDX) 147
energy storage 368
enthalpy 3, 31
entropy 69, 81, 96, 110, 312
- production 68
environmental quotient (EQ) 11
enzyme 5, 17, 33, 73
- alcohol dehydrogenase 6
- catalase 6
- nitrogenase 354
enzyme-catalyzed reaction 73
epichlorohydrin process 9
equilibrium 28, 31, 43, 67, 127, 283
- constant 93, 96, 122, 125, 306, 342
- thermodynamics 106
Erlenmeyer flask 427
ethanol 278, 376
ethylene 271, 288, 422
- dehydrogenation 276

– glycol (antifreeze) 412
– hydrogenation 269, 276
ethylene epoxidation 412
ethylene glycol 9, 77
excess electricity 375
exchange energy 279
exhaust gas constituents 422
exothermic reaction 4, 31, 43, 175, 315, 352, 413
extended X-ray absorption fine structure (EXAFS) 139
extrinsic precursor 115

f

face-centered cubic (fcc) 149
Faujasite (FAU) 208
Fermi distribution 236, 237
Fermi level 235, 238, 242, 253, 255, 257, 366
Fermi–Dirac distribution 89
fermion 237
fertilizer 23, 354
first-order kinetics 286
Fischer–Tropsch synthesis 148, 151, 332, 343, 364
– high-temperature 349
– low and middle temperature 349
– reaction and mechanism 345
fixed-bed multitubular reactor 349
flame hydrolysis 198
flood gun 138
flow reactor 41, 50
flue gas 320
flue gas composition 437
fluidized catalytic cracking 404
formaldehyde 332
fossil fuel 368
fossil fuel reserve 365
Fourier transform method 133
Fourier transformation 142
fracking technology 364
free energy 4, 67, 106, 369
free-electron gas 235, 238
free-electron metal 238, 243
fuel 12
– efficiency 434
fuel cell
– efficiency 381
– new electrode material 383
fugacity coefficient 34
fumed silica 198
furan 396

g

gamma radiation 148
gas
– fugacity 33
– molecular adsorption 288
– nonideal 33
– stoichiometric mixture 35
gas hydrate 364
gas–surface interaction 225
gas-hourly space velocity (GSVH) 214
gasoline 364, 375, 382, 402, 410, 422
– production 402
gas-to-liquids (GTL) plant 344, 350
Gibbs free energy 28, 29, 93, 109
gibbsite 200
graphite 326
green chemistry 9
greenhouse effect 421
Green's function 247, 248
ground state vibration energy 98

h

Hagg carbide 151
Hamiltonian 229, 240, 246
harmonic oscillator 158
heat pump engine 381
heavy fuel 439
Heisenberg uncertainty principle 250
heteronuclear diatomic molecule 232
hexachloroplatinic acid 204
hexane 407
highest occupied molecular orbital (HOMO) 235
homogeneous catalysis 5, 50
homonuclear diatomic molecule 230, 240
hybridization energy 257, 258, 262, 263
hydrocarbon 27, 185, 369, 376, 423, 434
– oxidation 429
– zeolite-catalyzed reaction 207
hydrocracker 349
hydrocracking 394
hydrodemetallization 396
hydrodenitrogenation (HDN) 157
hydrodeoxygenation 396
hydrodesulfurization (HDS) 157, 395, 397, 421
– of thiophene 400
– reaction mechanisms 399
hydrogen 142, 157, 271, 319, 401
– atom 226
– by electrolysis 369
– fuel cell 421
– on-board automobile 380

- society 364
- sulfide 304, 401
hydrogen and nitrogen 32
hydrogen atom 97
hydrogen fuel cells 377
hydrogenation 272
- of butadiene 304
- of ethylene 269
hydrogen-oxygen reaction 378
hydrotreating 392, 394
- catalyst 397
hydroxyl group 203
hydroxylamine 48
hyperfine interaction 149

i
ideal gas 94, 96
ideal gas mixture 29
image dipole 226
impact factor 19
impregnation 203
indirect atomic adsorption 113
inert gas 287
infrared spectroscopy 158, 170, 297, 397
insulator 243
integral rate equation 39
interface energy 189
internal rotation 112
invertase 77
ion exchange 203
ion scattering spectroscopy (ISS) 153
ion spectroscopy 151
iron catalyst 33, 156
iron catalyst for ammonia 148
irreversible step approximation 61
iso-alkene 409
iso-butane 410
isomerization 38, 406
- of butane 409
isotherm 53

j
jellium 254, 257
jellium model 238

k
Kellogg Advanced Ammonia Process (KAAP) 360
Kelvin equation 191
kerosene 392
kinetic energy 135, 140
kinetics 23, 283
- first-order 38
- measurements 43

Knudsen diffusion 215
Kohn–Sham equation 279
Kronecker delta 241
Kronig–Kramer transformation 249

l
Lagrange multiplier 84
lambda probe 424, 425
Langmuir
- adsorption isotherm 53
- isotherm 59, 66, 116, 127, 196, 283
- kinetics 328
- unit 286
Langmuir–Hinshelwood
- approach 301
- equation 75
- kinetics 23, 56
- model 304, 307
- rate equation 305
lanthanide 243
lanthanum oxide 425, 427
laser spectroscopy 25
Laue X-ray diffraction 178
laws of Fick 214
LCAO approximation 240
lead 428
leading-edge analysis 293
Lennard-Jones potential 158, 228
Lewis acidity 182, 201
Lindemann theory 80, 106, 107
Lindemann–Christiansen hypothesis 106
linear combination of atomic orbitals (LCAO) 229
Lineweaver–Burk plot 75
liquefied petroleum gas (LPG) 392
lithium 255
loose transition state 110
Lorentzian function 247
Lorentzian shape 250, 251
low electron energy diffraction (LEED) 161
- spot profile analysis (SPALEED) 163
low-energy ion scattering (LEIS) 151, 153

m
macropore 190
Madelung sum 137
magnetic force microscopy 167
magnetic hyperfine splitting 150
Mars–van Krevelen mechanism 413, 441
mass spectrometer 155
mass spectrometry 302
Maxwell–Boltzmann distribution 80, 86, 101, 103

mean-field approximation 52, 113, 295
mercaptan 395
mesoporous material 196
metal
– surface crystallography 176
methanation 323, 352, 355, 380
methane 32, 175, 284, 356
– adsorption 323
– catalytic partial oxidation 331
– direct oxidation 331
– direct use 330
– dissociation on nickel 274
– dissociative adsorption 285
– partial oxidation 320
methanol 5, 77, 147, 331, 376, 380
– synthesis 332, 341
method of Lagrange undetermined multipliers 84
methyl cyclohexane 410
methyl tertiary-butyl ether (MTBE) 403, 411
Michaelis constant 50, 74
Michaelis–Menten expression 75
microcalorimetry 289
microkinetic modeling 306, 315, 333, 336, 363
minimum energy configuration 267
mobile precursor 286
Mössbauer spectroscopy 148, 150, 170, 397
mole fraction 32, 341
molecular beam experiment 25
molecular chemisorption 261
molecule 17, 23, 38, 88, 103
– average vibrational energy 90
– bonding 229
– in equilibrium 93
molybdenum disulfide 397, 398
molybdenum oxide 182
mono-atomic solid 141
monolith 425, 428, 439
monooxygenase 331
Monte Carlo simulation 53, 113, 295, 297, 299, 300
mordenite (MOR) 208
Morse potential 158
Most Abundant Reaction Intermediate (MARI) approximation 61
motor vehicle exhaust 423
multicomponent catalyst 143
multilayer 274

n

Nafion membrane 378
naphtha 321, 391, 406, 412
naphthene 391
naphthenic acid 396
natural gas 321
– steam reforming 319
nearest neighbor (NN) interaction 296
Nernst equation 378, 424
Newns–Anderson model 246, 252, 256, 268
next-nearest neighbor (NNN) interaction 296
nickel 268, 274, 396, 405
– crystal 323
– steam-reforming catalyst 326
nitrogen 23, 32, 100, 353
– atoms 434
– desorption 313
– dissociation 312
nitrogen-containing molecule 396
nitrogen-containing polymer 354
NO dissociation 273
NO_x storage-reduction (NSR) catalyst 435
non-Arrhenius-like behavior 430
nonideal gas 33
nuclear magnetic resonance (NMR) 169
nuclear partition function 92
nuclear power 365
nuclear waste 365

o

octane number 407, 422
oil refinery 319
oil refining 391
olefin 346, 396, 402, 403, 410
oligomerization 406
one-particle Green's function 248
optimal catalyst curve 360
organic chemistry 10
oscillating reaction 69, 70
overpotential 371
overtone 158
oxogas 332
oxygen 100, 142, 157
– atom 257
– evolution reaction (OER) 370, 383
– reduction reaction (ORR) 383
– sensor 423
oxygen ion 377
oxygen-rich exhaust 443
ozone 5, 44

p

palladium 269, 343, 427
paraffin 391, 417
partition function 80, 81, 95, 114, 120, 124
– electronic 92

- infinite number of equidistant energy levels 83
- nuclear 92
- of atoms and molecules 83
- of rotation 98, 111
- of translation 85, 88, 111
- of vibration 89, 97, 311
- two-level system 81

Pauli principle 227, 236, 243
Pauli repulsion 228, 232, 237
petrochemistry 391, 412
petroleum 12, 15
phenol 396
Phillips process 416
phosphorous 428
photoelectron 130, 135, 139
photoemission 137
photon 134
photon energy 140
photosynthesis 1, 368
photovoltaic (PV) devices 366, 367
photovoltaics 372
physisorbed precursor state 118
physisorption 196, 225, 265
piezoelectric tube 165
Planck's constant 86, 88, 140, 158
Planck's law 366
platinum 7, 261, 268, 379, 408, 427
platinum catalyst 48
platinum tetraamine ion 204
plug flow reactor (PFR) 211, 215, 301, 314, 339
polar surface 182
polyethylene 416, 417
polymer 11, 415
polymer electrolyte membrane fuel cell 377
polymerization catalysis 415
pore diffusion 220
pore system of a support 190
potassium 362, 413
potential energy diagram 122
power plant 437
power rate law 26, 51, 304, 440
pre-exponential factor 79
prefactor 36, 37, 112
- of desorption 125
promoter 48, 361
propylene 412
- ammoxidation 413
- partial oxidation 413
proton exchange membrane fuel cell (PEMFC) 377

pseudo-first order equation 40
pseudomorfic overlayer 263, 264, 274

q
quadrupole doublet 150, 151
quantum mechanics 24
quasi-equilibrium approximation 59, 61, 65

r
Raman spectroscopy 169
rate equation 44
rate-determining step 60
rate-limiting step (RLS) 62
reaction dynamics 24
reaction enthalpy 36
reaction equation 25, 26
reaction order 63, 67, 310
reaction rate theory 79
reactor technology 349
recoilless fraction 150
refinery process 393
reforming reaction 407
regenerative fuel cell 382
repulsion 227, 300
Reynolds number 69
rhenium 408
rhodium 155, 167, 169, 257, 269, 321, 422, 427, 431, 433
Richardson–Dushman formula 239
rotational partition function 90, 98
ruthenium 274, 287, 321, 324, 350, 364, 379
ruthenium surface 266
Rutherford backscattering 151, 154

s
Sabatier's principle 4, 272, 274, 276, 359, 429
saturation coverage 284, 292
scaling relation 276, 278, 372
scanning
- electron microscopy (SEM) 145
- force microscopy (SFM) 164, 418
- tunneling microscopy (STM) 164, 238, 329, 398

Scherrer formula 133
Schrödinger equation 235
Schrödinger-like one-electron equation 279
secondary ion mass spectrometry (SIMS) 151, 297, 303
second-order kinetics 286
selective catalytic reduction 437
semiconductor 243
shale gas 319
Shockley–Queisser limit 368

silica 197, 204
– xerogel 198
silicon 256, 368
slurry phase reactor 349
SNOX 437
solar activity 365
solar energy 366
solid oxide fuel cell (SOFC) 381, 382
solid state theory 233
solid surface 233
SPARG (Sulfur PAssivated ReforminG) process 328
spectroscopy 24, 432
spin-orbit splitting 136
steady state approximation 23, 45, 50, 58, 107
steady state assumption 42
steam reforming 321, 326, 352
sticking coefficient 105, 116, 118, 120, 283, 284, 311
Stirling's approximation 84, 94
stoichiometric gas mixture 337
stoichiometric reaction 9
structural promoter 197
sulfated zirconia 411
sulfide 184, 395
sulfur 157, 321, 345, 363, 405, 428
– layer 184
– passivation 328
sulfur-free diesel fuel 344
surface
– area measurements 191
– collision rate 103
– crystallography 176
– free energy 185, 189
– inhomogeneities 287
– reaction
– – lateral interactions 295
– reactivity 225, 256
– – tuning 273
– science technique 160
sustainable energy 364–366
synchrotron 144
synchrotron radiation 133
syngas 319, 375
– production 320
synthesis gas 27
– reactions 332
synthesis gas and hydrogen 319

t
tar sand 364
Taylor expansion 232

Taylor series 232
temperature-programmed
– analysis 431
– desorption (TPD) 123, 130, 289
– – compensation effect 293
– reaction (TPR) 133, 155
– reaction spectroscopy (TPRS) 290, 301
– sulfidation (TPS) 157
Tera Watt Challenge 375
thermal desorption spectroscopy (TDS) 290
thermal equilibrium 284
thermodynamic equilibrium 27
thermodynamics 25, 30, 106, 222, 408
Thiele diffusion modulus 214, 217, 221
– for a pore 220
thiol 321
thiophene 304, 306, 321, 399
– catalytic hydrodesulfurization 304
three-way catalyst 423, 436
toluene 410
total energy 84
total partition function of the system 87
transition metal 243, 251
– atomic adsorption 253
transition state theory 79, 80, 107, 108, 111, 113, 118, 292, 311
– of surface reactions 113
translational energy 99
translational partition function 88
transmission
– electron microscopy (TEM) 145, 147
– infrared spectroscopy 160
transportation fuel 391
transportation limitation 222
triethylaluminum 416
tungsten 239
turnover frequency (TOF) 174, 212, 213, 223, 359

u
ultrahigh vacuum (UHV) 178
ultrahigh vacuum chamber 290
urea 443
urea hydrolysis 205
urease 74

v
Van der Waals constant 227
Van der Waals interaction 226, 228, 265
Van't Hoff equation 30
vibrational partition function 89, 97
vibrational spectroscopy 261
volcano curve 383
volcano plot 273, 276, 359, 399

w

washcoat 426
water glass 198
water splitting 370
water–gas 332
water–gas shift reaction 277, 321, 333, 341, 345, 351, 380
wind energy 365, 368
Wood notation 180
work function 239
Wulff construction 148, 187, 189, 336

x

X-ray
– absorption near-edge spectroscopy (XANES) 133, 139, 143
– diffraction (XRD) 131, 156
– diffractogram 201
– photoelectron spectroscopy (XPS) 130, 134
– spectroscopy 136

y

yttria-stabilized zirconia (YSZ) 381

z

Zeeman effect 150
zeolite 206, 404
zero-order reaction 40